T0303880

Advances in Physical Ergonomics and Safety

Advances in Human Factors and Ergonomics Series

Series Editors

Gavriel Salvendy
Professor Emeritus
School of Industrial Engineering
Purdue University

Chair Professor & Head
Dept. of Industrial Engineering
Tsinghua Univ., P.R. China

Waldemar Karwowski
Professor & Chair
Industrial Engineering and
Management Systems
University of Central Florida
Orlando, Florida, U.S.A.

3rd International Conference on Applied Human Factors and Ergonomics (AHFE) 2010

Advances in Applied Digital Human Modeling
Vincent G. Duffy

Advances in Cognitive Ergonomics
David Kaber and Guy Boy

Advances in Cross-Cultural Decision Making
Dylan D. Schmorrow and Denise M. Nicholson

Advances in Ergonomics Modeling and Usability Evaluation
Halimahtun Khalid, Alan Hedge, and Tareq Z. Ahram

Advances in Human Factors and Ergonomics in Healthcare
Vincent G. Duffy

Advances in Human Factors, Ergonomics, and Safety in Manufacturing and Service Industries
Waldemar Karwowski and Gavriel Salvendy

Advances in Occupational, Social, and Organizational Ergonomics
Peter Vink and Jussi Kantola

Advances in Understanding Human Performance: Neuroergonomics, Human Factors Design, and Special Populations
Tadeusz Marek, Waldemar Karwowski, and Valerie Rice

4th International Conference on Applied Human Factors and Ergonomics (AHFE) 2012

Advances in Affective and Pleasurable Design
Yong Gu Ji

Advances in Applied Human Modeling and Simulation
Vincent G. Duffy

Advances in Cognitive Engineering and Neuroergonomics
Kay M. Stanney and Kelly S. Hale

Advances in Design for Cross-Cultural Activities Part I
Dylan D. Schmorrow and Denise M. Nicholson

Advances in Design for Cross-Cultural Activities Part II
Denise M. Nicholson and Dylan D. Schmorrow

Advances in Ergonomics in Manufacturing
Stefan Trzcielinski and Waldemar Karwowski

Advances in Human Aspects of Aviation
Steven J. Landry

Advances in Human Aspects of Healthcare
Vincent G. Duffy

Advances in Human Aspects of Road and Rail Transportation
Neville A. Stanton

Advances in Human Factors and Ergonomics, 2012-14 Volume Set: Proceedings of the 4th AHFE Conference 21-25 July 2012
Gavriel Salvendy and Waldemar Karwowski

Advances in the Human Side of Service Engineering
James C. Spohrer and Louis E. Freund

Advances in Physical Ergonomics and Safety
Tareq Z. Ahram and Waldemar Karwowski

Advances in Social and Organizational Factors
Peter Vink

Advances in Usability Evaluation Part I
Marcelo M. Soares and Francisco Rebelo

Advances in Usability Evaluation Part II
Francisco Rebelo and Marcelo M. Soares

Advances in Physical Ergonomics and Safety

Edited By

Tareq Z. Ahram

and

Waldemar Karwowski

CRC Press
Taylor & Francis Group
Boca Raton London New York

CRC Press is an imprint of the
Taylor & Francis Group, an **informa** business

CRC Press
Taylor & Francis Group
6000 Broken Sound Parkway NW, Suite 300
Boca Raton, FL 33487-2742

Printed in the United States of America on acid-free paper
Version Date: 20120529

International Standard Book Number: 978-1-4398-7038-9 (Hardback)

Visit the Taylor & Francis Web site at
http://www.taylorandfrancis.com

and the CRC Press Web site at
http://www.crcpress.com

Table of Contents

Section I: Safety and Health

Preface

The discipline of human factors and ergonomics (HF/E) is concerned with the design of products, process, services, and work systems to assure their productive, safe and satisfying use by people. Physical ergonomics involves the design of working environments to fit human physical abilities. By understanding the constraints and capabilities of the human body and mind, we can design products, services and environments that are effective, reliable, safe and comfortable for everyday use. A thorough understanding of the physical characteristics of a wide range of people is essential in the development of consumer products and systems. Human performance data serve as valuable information to designers and help ensure that the final products will fit the targeted population of end users. Mastering physical ergonomics and safety engineering concepts is fundamental to the creation of products and systems that people are able to use, avoidance of stresses, and minimization of the risk for accidents.

This book focuses on the advances in the physical HF/E and safety, which are a critical aspect in the design of any human-centered technological system. The ideas and practical solutions described in the book are the outcome of dedicated research by academics and practitioners aiming to advance theory and practice in this dynamic and all-encompassing discipline.

A total of sixty-six chapters presented in this book are organized into four sections:

Section I. Safety and Health (12 chapters)
Section II. Work Related Musculoskeletal Disorders (17 chapters)
Section III. Ergonomics and System Design (18 chapters)
Section IV. Ergonomic Evaluation Methods & Task Analysis (19 chapters)

Each section contains chapters that have been reviewed by members of the Editorial Board. Our sincere thanks and appreciation to the Board members as listed below:

We hope that this book, which is the international state-of-the-art in physical domain of human factors and safety, will be a valuable source of theoretical and applied knowledge enabling human-centered design of variety of products, services and systems for global markets.

March 2012

Tareq Ahram and Waldemar Karwowski
University of Central Florida
Orlando, Florida, USA

Editors

Section I

Safety and Health

CHAPTER 1

Decision Support in Safety Management

Toni Waefler, Simon Binz, Kathrin Gaertner, Katrin Fischer

Institute Humans in Complex Systems
School of Applied Psychology
University of Applied Sciences Northwestern Switzerland
Olten, Switzerland
toni.waefler@fhnw.ch

ABSTRACT

Decision-making in safety management requires a substantial information base. Part of this base is the assessment of an organization's safety state and of the effect of safety promoting measures. However, respective assessment processes are focusing mainly reactive, countable safety indicators. Against this background the S-MIS project developed a safety assessment process including proactive safety indicators referring to human and organizational factors that are not quantifiable directly. The process guides a group of safety experts from industries through seven steps in which their tacit knowledge is made explicit, and in which they build consensus regarding indicator selection, operationalization, safety assessment, data interpretation, and decision-making.

Keywords: safety management, safety assessment, safety indicators, decision-making, modeling

1 INTRODUCTION

The aim of safety management systems is to provide integrated organizational mechanisms designed to control safety risks as well as ongoing and future safety performance (Cooper, 1998). One type of instruments required to fulfill this purpose focuses on assessing both, the safety level of an organization as well as the effectiveness of safety-related measures an organization takes in order to improve its safety. In practice, the information basis regarding these two aspects is often very vague. These difficulties are due to a multiplicity of factors contributing to safety, a lack of knowledge regarding their complex interrelations, and a lack of methods for objective, reliable, and valid measurement of these factors. However, in order to proactively mitigate risks or to provide resilient processes able to cope with safety threats, a more reliable information basis would be very helpful.

Against this background the project 'Safety Management Information System' (S-MIS) aims at developing an information system that supports decisions in safety management on the basis of clearly defined indicators and a structured process of applying the indicators. S-MIS process development is based upon literature regarding safety indicators and safety assessment as well as upon case studies aiming at describing respective processes in practice. In a following pilot project together with an industrial partner operating in a high-risk industry the S-MIS process was tested and fine-tuned.

2 INDICATORS FOR SAFETY ASSESSMENT

In order to assess an organizations safety level, generally two approaches are applied (Choudhry et. al., 2006). On the one side safety is assessed on the basis of past safety-related occurrences such as incidents and accidents. Such an assessment focuses on safety outcomes and is based on retrospective data, respective indicators are referred to as reactive indicators, downstream indicators or lagging indicators (Mohamed, 2002; Hinze, 2005). Representatives of this kind of indicators are for example number of accidents or incidents or follow-up costs of accidents. Indicators like these focus on safety outcomes. They implicate that safety can be measured on the basis of (near) system failures. However, the suitability of reactive indicators for making prognosis is very restricted, as the absence of occurrences in the past does not guarantee for an absence of occurrences in the future. Furthermore, such indicators do not allow for preventive risk mitigation, as they respond to occurrences that already happened.

Against this background safety indicators are searched for, which allow for an assessment of organizational conditions and factors, thereby uncovering hazards before they manifest in occurrences. Hence, newer literature advocates a shift from reactive indicators to proactive indicators (or upstream indicators, leading indicators) referring to aspects such as safety climate (Flin et al., 2000; Mohamed, 2002), safety culture (Cooper, 2000; Choudhry and Fang, 2005), hazard identification or observable safe behavior (Strickoff, 2000; Cooper and Phillips, 2004). This proactive approach focuses on aspects that refer to a safe operating in the future, and not to a system failure in the past. According to Grabowski et al. (2007) proactive indicators are suitable to identify hazards, to assess risks, and consequently to mitigate risks. Other authors such as Hollnagel et al. (2006) go further by advocating proactive indicators suitable to detect resilience. Resilience is the ability of a system to keep control under changing conditions. Indicators focusing on resilience therefore not only help to detect pathogens before they manifest in occurrences. They rather refer to an organization's ability to actively sustain safety.

The distinction of proactive and reactive indicators is not always clear-cut (Hopkins, 2009). Subject to different perspectives a proactive indicator can be considered a reactive indicator. The proportion of unsafe human acting for example is sometimes considered a reactive indicator of the human's working context and a precursor of safety related occurrences. However, manifest unsafe acting can have an impact on human motivation to act safely. From that point of view the proportion of unsafe human acting is a proactive indicator referring to a construct like safety motivation.

Hopkins (2009) points to another problem when assessing safety on the basis of indicators. There is a danger that the application of indicators misleads to managing the indicator rather than managing safety. Backlog in equipment maintenance for example is a frequently used indicator for assessing safety. However, the objective to reduce the backlog may lead to a quicker maintenance with reduced quality. Doing so, the indicator shows better results, but the safety is still vulnerable.

3 APPLICATION OF SAFETY INDICATORS IN PRACTICE

In order to better understand safety assessment in practice we conducted case studies with two industrial partners operating in high-risk industries (a description of the methods applied as well as detailed results have been published before in Waefler et al. 2008). In general, results of the two case studies show that safety assessment processes are only partly standardized. Even when processes are well systematized with clearly assigned responsibilities, variations occur in everyday practices. Furthermore, interfaces between distinct safety management processes are very often informal and flexible; i.e. actual coordination at the interfaces depends on subjective comprehension and commitment.

According to the S-MIS objectives, more detailed analysis concentrated on processes concerning: (i) selection of relevant safety indicators, (ii) methods for safety assessment, (iii) evaluation, aggregation and interpretation of safety data, and (iv) judgment and decision-making.

We found that applied indicators almost exclusively concentrate on hard and countable facts like numbers of classified incidents or missed requirements. Accessorily the safety controlling consults qualitative data from further sources (e.g. safety meetings, walkabouts, spontaneous conversations) in order to question and double-check interpretations of measured values. However, these qualitative assessments are taken in an ad-hoc manner without consideration of validated, social science based methods for data collection and analysis. As a consequence the set of applied indicators, almost exclusively comprises reactive indicators capturing outcome measures. There is no systematic attention paid to the underlying factors that enable employees to safe(r) operations and decisions. Thus more proactive indicators are needed that allow for valid measures of safety enablers.

Besides the evaluation of the set of considered indicators, also an in-depth process analysis has been conducted. This analysis mainly refers to decision-making, and hence to process steps like selecting, filtering, evaluating and aggregating data in order to develop and judge options for required actions. Within the scope of the case study evidence has been found that data interpretation, aggregation, judgment and decision-making are driven by distributed subjective assumptions of the safety experts involved. These assumptions include subjective views on the safety related importance of measured indicators, on reasonable ways to interpret information as well as on cause-effect relations and interactions of different indicators. Decision-makers seem to apply tacit knowledge gained during years of safety management practice and remarkable insider knowledge of the plant's functioning. In decision-making this knowledge very often comes into play as affectively mediated 'gut feeling', i.e. underlying knowledge and assumptions can only partly be explained. However, such implicit assumptions and knowledge may also lead to biased judgments and decisions.

4 CONSEQUENCES FOR THE S-MIS PROCESS

In general, knowledge regarding interactions of the factors contributing to safety is far too scarce to allow for designing a management information system that provides safety managers with decisions. Therefore, S-MIS is a process rather than an IT, designed in a way minimizing cognitive biases in decision-making. The IT-part of S-MIS of course supports data collection and analysis as well as the development of a database that allows for meta-analysis and eventually for benchmarking.

However, as the case studies show, the processes of indicator selection, safety assessment, data interpretation and decision-making are to a large part driven by tacit knowledge. Consequently these processes are vulnerable to cognitive biases. Therefore the main aim of the S-MIS process is a continuous and critical reflection of the subjective assumptions that are currently driving information search, information assessment and decision-making. Such reflection is considered to provide to an organization's heedfulness.

In its core S-MIS supports a modeling process. However, not the one true safety model to be programmed into an automated decision engine is envisioned, but rather a continuous process that allows for both, (i) making explicit individual mental models regarding causes and effects, and (ii) the joint elaboration of a shared safety model. This model is guiding indicator identification, safety assessment and decision-making. The shared model itself is subject of a continuous critical reflection and hence of continuous re-construction – if required. Since this continuous modeling is a social and collective process, it also provokes a systematic exchange of the (tacit) knowledge required when taking safety related decisions. It hence supports knowledge management making the organization less dependent of experiences distributed throughout the organization and hidden in individual minds.

5 THE S-MIS PROCESS

On the basis of the literature as well as on the case studies it was decided, that the S-MIS process needs to support stakeholders:

- to identify and measure those indicators appropriate to assess system safety, considering the organization's specific characteristics,
- to aggregate the collected data into an information basis,
- to assess and to interpret the information, and
- to take well-founded safety related decisions.

The S-MIS process is participatory mainly due to the following two reasons: (i) the literature does not provide sufficient knowledge regarding relevant safety indicators and their interrelations, and (ii) safety practitioners show an extensive tacit knowledge regarding factors having an impact on safety. The process incorporates seven steps. A core group of safety practitioners is guided through these seven steps by facilitators. Steps 1 - 4 aim at defining and operationalizing safety indicators on the basis of the safety practitioners' expertise. In step 5 the safety state of the organization is assessed by the means of the safety indicators. Finally, steps 6 - 7 serve to interpret the results of the assessment (cf. step 5) thereby considering interrelations as well as different weights of the indicators, and to take safety-related decisions. The seven steps are described below in more detail. Results from a pilot study are presented for illustrative purpose.

5.1 Step 1: Safety perspectives

In the first step the core team's members elaborate a joint understanding of safety. To do so, the facilitators introduce different sociotechnical models of occurrence emergence (e.g. from Reason, 1990; Hollnagel, 2004; Hollnagel et al., 2006; Dekker, 2011). The members of the core group are then instructed to identify examples on the basis of their organizational experience, which correspond to the introduced models. Through this, the core group members get sensitive to the sociotechnical nature of occurrence emergence.

5.2 Step 2: Basic indicator model

The aim of step 2 is to identify main, high-level indicators on the basis of the core group members' implicit and explicit knowledge. To do so, the knowledge elicitation method MITOCAR is applied (Pimay-Dummer, 2006). The members of the core group individually have to complete sentences such as: "Humans act safely under the condition ... if ... ". By the means of a qualitative content analysis the core group member's main mental concepts of safety enablers are identified. Subsequently the core team members cluster these concepts. As a result the basic indicator model is elaborated. It has a hierarchic structure with "safe acting" on the top and approximately 10 to 15 indicator clusters on the next hierarchic level. Examples of indicator clusters (subsequently referred to as basic indicators) identified in the pilot project are: "general environment of the organization", "culture", "the workforce's engagement and commitment to perform", "working conditions", "human competences", and "leadership and management".

5.3 Step 3: Model differentiation

In this step the clusters of indicators elaborated in step 2 are further differentiated. The result is a multi-level, hierarchical model of indicators with "safe acting" on the top and "basic indicators", "sub-indicators", and "concepts" as underlying levels. Indicators can represent contrasting kinds of organizational conditions and factors, categorized as safety-promotive as well as safety-obstructive. This is to make sure, that the indicator model does not focus on system deficiencies only, but also on resilience. Note that steps 2 and 3 are carried out in an iterative rather than in a sequential manner. Examples of sub-indicators identified in the pilot project are: "motivation", "methods to improve motivation", and "attitude" (all referring to the basic indicator "the workforce's engagement and commitment to perform"). Examples of concepts are: "cost pressure", and "availability of resources" (corresponding to sub-indicator "economic factors").

5.4 Step 4: Operationalization and measuring methods

To assess the organizations safety on the basis of the differentiated indicator model from step 3, the indicators need to be operationalized. To do so, all indicators on the lowest level of the model (concepts) are translated into a number of statements, the value of which can be estimated on a scale, reaching from "totally true" to "totally untrue" accompanied by a numerical scale reaching from 1 to 6. The core group members individually phrase the statements. Subsequently the facilitators check all statements considering methodological principles (Jonkisz and Mossbruger, 2007). Finally, in a series of workshops the core group agrees on those statements being integrated into a questionnaire, which is the basis of the next step. Examples of statements generated in the pilot project are: "Unconventional thinkers are welcome in working teams", or "technical systems are designed user-friendly".

5.5 Step 5: Safety assessment

The application of the indicators to assess an organization's safety involves two sub-steps. Following the Delphi method (e.g. Schmidt, 1997) a focus group is formed consisting of people being representative for the organizational unit to be assessed. In a first sub-step all members of this focus group complete the questionnaire, which consist of the statements elaborated in step 4. In a second sub-step the members of the focus group join together to discuss the results of the survey in a workshop. The main aim of the discussions is to identify qualitative reasoning for the indicators' values. The focus group is also allowed to change the questionnaire-based value of an indicator if it seems reasonable. However, such a change needs to be made in consensus and must be well-founded by qualitative reasoning. Examples of such reasoning in the pilot project, referring to the statement "the meaning of organizational change isn't understood by employees" are: "sometimes, the sense of changes is not explained by detail" or "not all changes have a positive impact on the workforce".

5.6 Step 6: Indicator weights and interrelations

In parallel to step 5 the core group further elaborates the model by weighting the indicators and by determining the indicators' interrelations. Thus it is possible to simulate with the model and therewith to visualize the collective behavior of all the assumptions that have been incorporated into the model (cf. step 7).

The weighting includes comparing the impact indicators of one hierarchical level have with reference to their hierarchically super-ordinated indicator. For each indicator the core group needs to elaborate a consensus concerning the relative weight. To determine the indicators' relations all indicators (same model level and same superior indicator) are confronted with each other. For each pair of indicators the core group elaborates a consensus regarding the interrelation's vectored magnitude and lag.

5.7 Step 7: Interpretation and decision-making

The final step in the S-MIS process aims at interpreting the results of the safety assessment (step 5) and to decide on consequences. As the model not only incorporates the hierarchical structure of the indicators, but also indicator weights and interrelations it allows for sensitivity analysis and impact analysis. This is helpful to identify those safety-related measures that are most suitable to reach the objectives without creating unwanted side or long-term effects.

6 CONCLUSIONS

The S-MIS project aims at providing industry with an IT-supported process suitable for a more reliable safety assessment including proactive indicators and taking into account the sociotechnical nature of organizations. There is a number of challenges S-MIS faces, the main of which are the following: (i) regarding safety enablers and their complex interrelations science does not provide sufficient insights, (ii) on the other hand safety experts in industrial practice seems to have an extensive tacit knowledge in this regard, (iii), this tacit knowledge is distributed among different safety experts and hence a process of consensus building is required, (iv) a large number of relevant indicators - especially proactive indicators - refer to human and organizational factors and hence are not quantifiable directly but require operationalization, and (v) the sheer amount of interrelated indicators makes it very difficult for decision-makers to keep an overview and to take side effects as well as long-term effects into consideration when taking decisions.

In the pilot project the seven steps of the S-MIS process proofed to be suitable to tackle these challenges and hence to provide decision-makers in safety management with a substantially better information base, which is enriched with a detailed reasoning for a concrete safety assessment. However, the pilot project also showed that accomplishing the S-MIS process as whole is quite laborious and time-consuming. The process therefore still needs to be optimized.

REFERENCES

Choudhry, R.M. and Fang, D.P., 2005. The nature of safety culture: a survey of the state-of-the-art and improving a positive safety culture. In: *Proceedings of the 1st International Conference on Construction Engineering and Management.* Seoul, Korea 16–19 October 2005.

Choudhry, R.M., Fang, D. and Mohamed, S., 2006. The nature of safety culture. A survey of the state-of-the-art. *Safety Science,* 45 (10), pp.993-1012.

Cooper, D., 1998. *Improving Safety Culture. A practical Guide.* Chichester: Wiley.

Cooper, M.D., 2000. Towards a model of safety culture. *Safety Science*, 36, pp.111–136.

Cooper, M.D. and Phillips, R.A., 2004. Exploratory analysis of the safety climate and safety behavior relationship. *Journal of Safety Research*, 35, pp.497-512.

Dekker, S., 2011. *Drift into Failure.* Surrey: Ashgate.

Flin, R., 2007. Measuring safety culture in healthcare: A case for accurate diagnosis. *Safety Science*, 45, pp.653–667.

Hinze, J.W., 2005. A paradigm shift: leading to safety. In: Proceedings of the CIB W 99, *4th Triennial International Conference: Rethinking and Revitalizing Construction Safety, Health, Environment and Quality*. Port Elizabeth, South Africa 17–20 May 2005.

Hollnagel, E., 2004. *Barriers and Accident Prevention.* Surrey: Ashgate.

Hollnagel, E., Woods, D.D. and Leveson, N., 2006. *Resilience Engineering. Concepts and Precepts.* Aldershot (Hampshire): Ashgate.

Hopkins, A., 2009. Thinking about process safety indicator. *Safety Science,* 47, pp.460–465.

Jonkisz, E. and Moosbrugger, H., 2007. Planung und Entwicklung von psychologischen Tests und Fragebogen. In: Moosbrugger, H. and Kelava, A. ed. 2007. *Testtheorie und Fragebogenkonstruktion.* Heidelberg: Springer Medizin Verlag, pp. 27-72.

Mohamed, S., 2002. Safety climate in construction site environments. *Journal of Construction Engineering and Management,* 128 (5), pp.375–384.

Pirnay-Dummer, P., 2006. *Expertise und Modellbildung: MITOCAR.* Ph. D. Universität Freiburg.

Reason, J., 1990. *Human Error.* Cambridge: Cambridge University Press.

Schmidt, R.C., 1997. Managing Delphi surveys using nonparametric statistical techniques. *Decision Sciences,* 28, pp.763-774.

Strickoff, R.S., 2000. Safety performance measurement: identifying prospective indicators with high validity. *Professional Safety,* 45.

Waefler, T., Ritz, F., Gaertner, K. and Fischer, K., 2008. Decision-Making in Safety Management. . In: *Proceedings of AHFEI.* Las Vegas, USA 14-16 July 2008.

CHAPTER 2

Ergonomic Evaluation of Scaffolding Task Interventions for Power Plant Maintenance

Shruti Gangakhedkar, David Kaber, Matt Diering, Prithima Reddy

[1]Environment, Health and Safety, The Boeing Company; [2,4]North Carolina State University; [3]3M US

Everett, WA; Raleigh, NC; Greenville, SC, USA

shruti.b.gangakhedkar@boeing.com, dbkaber@ncsu.edu, mrdiering@mmm.com, prithimareddy@gmail.com

ABSTRACT

An investigation of a local power utility's injury database identified maintenance operators to have high incidence of ergonomics-related injuries. Job analyses revealed high risks for scaffolding tasks, particularly walk-board tie-down to frames and frame tube coupling. The present study involved: (1) recommendations of interventions to reduce worker exposure to risk factors; and (2) experiments to quantify intervention effectiveness in terms of operator posture and performance. Standard operating procedure at the utility involved use of #9 gauge wire to walk-boards to frames. Ergonomic analysis revealed extreme wrist positions and high rotational forces. Replacement of wire tie-downs with plastic zip-ties was proposed. In scaffold frame assembly, metal tubes were clamped together using metal couplers with a nut and bolt mechanism requiring ratcheting. Ergonomics analysis revealed very high torques at the wrist. A coupler utilizing a single lever clamping mechanism was designed to eliminate the repetitive ratcheting and excessive torque. Experiments were conducted with nine male workers from the power utility to test the proposed interventions using electro-goniometers for capturing wrist angles as well as forearm pronation in the frame coupling. Time-to-task completion (TTC) was also recorded. Results revealed the plastic zip ties to reduce wrist deviations and decrease TTC compared with metal wire ties. In frame tube coupling, the new lever coupler decreased the maximum wrist angle response; however, there was no significant reduction in forearm pronation and the coupler was found to slightly increase TTC compared with ratcheting couplers. The latter finding was partly attributed to operator lack of familiarity with the new coupler. Based on the experiments, the zip ties were recommended for use by the utility and the lever couplers were recommended for further examination as an alternative to standard scaffolding couplers. The ergonomic interventions designed and tested here may be applicable to other domains, including construction.

Keywords: Job screening, ergonomics-related risk factors, scaffolding, working posture, electrogoniometry, ergonomic interventions

1 PRELIMINARY JOB ANALYSIS OF MAINTENANCE OPERATIONS IN A POWER UTILITY

Power utilities are complex systems requiring extensive maintenance operations with plant shut down occurring during several months each year. A local utility initiated a partnership with North Carolina State University's Ergonomics Lab and the Ergonomics Center of North Carolina (ECNC) with the objective of reducing maintenance-related musculoskeletal injuries. An examination of the utility's injury and illness logs spanning a 5.5 year period revealed maintenance operators to have high-incidence rates, in-line with speculation by corporate environment safety and health personnel. Incidents with ergonomic root causes were subsequently identified and maintenance jobs were found to account for a substantial portion of ergonomics-related injuries (50% of tendinitis cases and 24% of strains/sprains).

On this basis, a two-step approach to identify high-risk operations was conducted. An initial screening of all maintenance operations for ergonomics risk factors was conducted with a tool developed by the ECNC. The tool combined elements of the Rapid Upper Limb Assessment (RULA) method, an observational study method focusing on upper-limb risk related to arm posture and forces, and the NIOSH lifting equation used to determine acceptable lifting limits (McAtamneya, & Corlett, 1993, Waters, et al. 1994). The screening assigned a rating for hazards of extreme posture, force and repetition for 10 different body areas by breaking down each job into its components tasks. Maintenance jobs with "high" overall job risk scores were scaffolding, shipping-receiving and cable pulling.

Next, a detailed risk assessment, assigning a rank of "Passed", "Cautioned" or "Failed" to each task element in the job, using computational aids (e.g.: RULA for hand-intensive jobs) was performed. Results identified scaffolding operations to include the most tasks posing ergonomics risk. Two tasks that received "Failed" rankings were: (1) tightening and loosening of scaffold joints – which required an estimated 300 in-lbs wrist torque exceeding a recommendation of 190.75 in-lbs (Mital & Channaveeraiah, 1988); and (2) tying-down of walk-boards with #9 gauge wire – which involved "very bad" hand and wrist postures (Moore & Garg, 1995).

Previous research (Elders & Burdorf, 2001) has investigated the relation between physical, psychological and individual risk factors in scaffolding operations to the incidence of low-back pain. High correlations were found between reporting of manual handling, awkward back posture, strenuous arm positions and the level of perceived exertion. This is similar to the results of our detailed risk assessment revealing posture, force and repetition levels for certain scaffolding operations to pose excessive demands on operators.

Van der Beek et al. (2005) conducted a study evaluating revised and practitioner approaches to applying the NIOSH lifting equation, including the Arbouw method, to assess the impact of manual material handling tasks in scaffolding. They found the lifting equation to be a good predictor of injury risk in scaffold operations. The Arbouw and practitioners methods were quick but difficult to apply. This research supported the use of ergonomic methods for assessing scaffolding operations. Another study by Vink et al. (1997) applied a participatory ergonomics approach to improving work methods and reducing injuries among scaffolders. The study concluded that scaffolders had greater complaints of injuries to the shoulders, wrists, elbows and upper-back as compared to other laborers.

Although the literature review confirmed the findings of the above detailed ergonomics analysis, we did not find specific studies focusing on walk-board tie-down and frame tube coupling tasks. Therefore, the objectives of this study were to: (1) conceptualize maintenance task interventions in order to reduce the incidence of injuries in the two scaffolding jobs identified through the detailed analysis; and (2) conduct experiments to measure the impact of proposed interventions on physiological and performance response measures.

2 METHOD

Nine male scaffolders familiar with scaffold assembly/disassembly operations were recruited for the study from the utility's employee population. Average on-the-job experience was 11.5 years (SD = 7.3). Anthropometric measures of height, weight, grip strength, back strength and upper-arm strength were recorded.

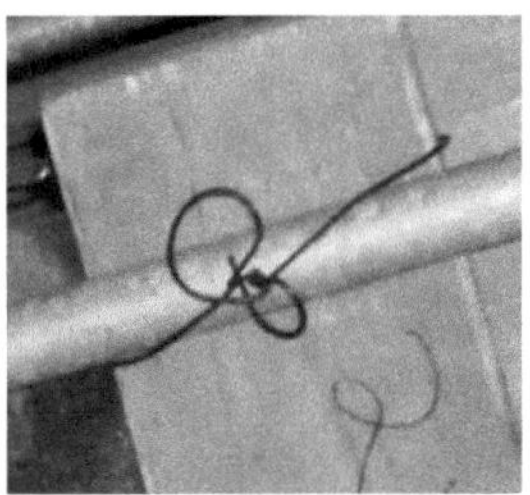

Figure 1, Left: Use of #9 gauge wire to tie-down a wood walk-board; Right: Use of right-angle couplers to secure two tubes in a scaffold frame assembly.

Two tasks were examined in this study, including: (1) Walk-board tie-down, which involved securing a wood or metal plank to a scaffold frame using the method of looping and tightening with a #9 gauge wire (see Figure 1, left side); and (2) Tube coupling, which involved tightening and loosening of scaffold frame couplers holding two tubes at right angles to each other (see Figure 1, right side). Tie-down of walk-boards to frames is critical as any vertical or lateral travel may place operators at risk or compromise the structural integrity of the scaffold. Tube coupling is an integral component of structure assembly as it serves to connect vertical and horizontal members to provide a stable frame.

Two experiments, one for each scaffolding task, were conducted with participants using current equipment and the proposed interventions in different frame configurations. The first experiment focused on the impact of walk-board tie-down methods on operator posture and performance. Following a detailed investigation of alternatives for board tie-down, heavy duty industrial plastic zip-ties were selected. These ties are durable (tensile strength of 250lbs) and suitable for robust applications, able to withstand high and low environment temperatures, light-weight and inexpensive when bought in bulk quantities. In addition, the ties eliminate the need for workers to carry additional equipment, such as wire cutting pliers, for the tie-down task. The zip-ties can be looped over a walk-board and scaffolding tube with the end of the tie pulled through the zip head and tightened. This method eliminates the need for high force and repetitive wrist motions observed in use of #9 gauge wire.

The second experiment assessed the use of existing coupler technology versus a proposed intervention for frame tube assembly and disassembly tasks. Current couplers include a nut and bolt mechanism requiring ratcheting. To eliminate the ratcheting action, a coupler design incorporating a ski-boot-like clamping mechanism was developed. This constituted a series of metal hooks on the lower-jaw of a coupler and a metal loop and lever swing arm mounted on the upper-jaw (see Figure 2).

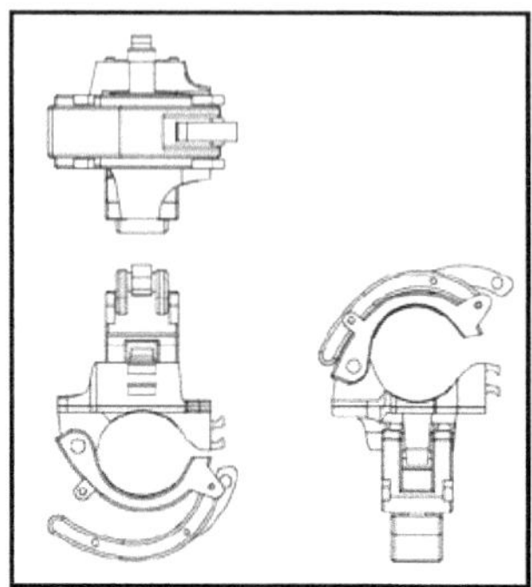

Figure 2: Design of proposed lever coupler incorporating "ski-boot" clamping mechanism.

When the wire is looped around a hook and the swing arm is pushed flush with the upper-jaw, the coupler is tightened around a frame tube. Subsequently, a cotter pin is inserted through the upper-jaw and swing arm to prevent inadvertent release.

The new coupler design was prepared using SolidWorks. Prototypes were built with a stereolithography machine and the models were used during the experiment. The new method of clamping was expected to eliminate any repetitive motion and reduce awkward postures when assembling scaffold frames with tubes.

For Experiment 1, a biaxial electro-goniometer (Model SG-150 from Biometrics) was used to measure flexion-extension and radial-ulnar deviation of the wrist. The device was positioned when the arm was in neutral posture with distal end placed over the third metacarpal and proximal end over the midline of forearm (see Figure 3). For Experiment 2, a biaxial electro-goniometer (Model SG-75 from Biometrics) and a single-axis torsiometer (Model T-110 from Biometrics) were used to record wrist motions and forearm pronation and supination, respectively, as observed in the tube coupling tasks. With the hand of a participant extended horizontally, the torsiometer was mounted with the distal end at the midpoint of the underside of the forearm and the proximal end on the interior portion of triceps. All three devices were connected to an eight channel analog to digital encoder unit (made by ProComp Infiniti) and integrated with the BioGraph Infiniti Software (Version 5). Data was sampled at 2048 Hz and recorded at 32Hz for analysis. Previously, electro-goniometry studies have used recording frequencies from 20Hz (Hansson et al., 1996) to 100Hz (McGorry et al., 2004). Digital cameras were also used during the experiments to record participant work behaviors.

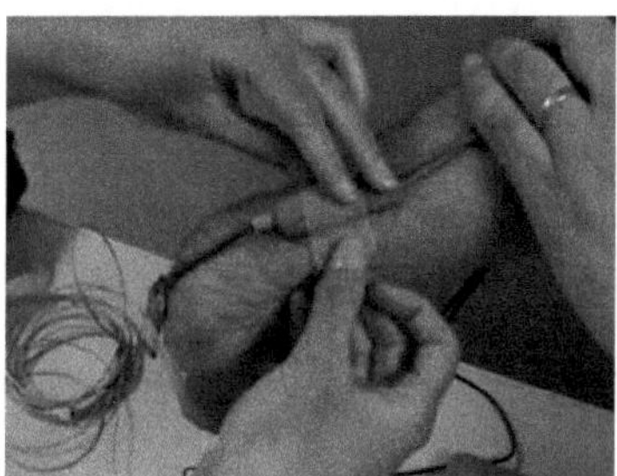

Figure 3: Electro-goniometer mounting for measurement of wrist flexion-extension and radial-ulnar deviation.

For Experiment 1, independent variables were: tie-down type with two levels (#9 gauge wire and plastic zip-ties), walk-board type with two levels (wood and metal), subtask with two levels (looping a tie around a walk-board and tightening it) and the position of the tie-down on the scaffold frame with two levels (upper and lower). The materials and the procedures were representative of real scaffolding operations. Consequently, the experiment followed a 2×2×2×2 randomized complete block design.

For Experiment 2, the independent variables were: coupler type with two levels (ratchet coupler and lever coupler), activity type with two levels (assembly and disassembly), and subtask with two levels (placement/removal of a coupler around tubes and clamping/unclamping of a coupler around tubes). The materials and the procedures were representative of real scaffolding operations. Consequently, the experiment followed a 2×2×2 randomized complete block design. The experiment conditions were presented in different orders across trials to control for learning and order effects.

Dependent variables were four joint movement angles measured with the electro-goniometers, including wrist flexion, extension, radial and ulnar deviation. The angles represented absolute posture positions during trials. Average time-to-task completion, determined from the videotapes, served as the performance measure. In Experiment 2, an additional variable was forearm pronation from a supine position.

Data collection spanned three days at a power utility training center. Participants were briefed on study goals and procedures, and provided with practice sessions for the proposed intervention trials. Each day was divided into three 2-hour sessions, with three 40-minute blocks in each session. For both experiments, each block consisted of four trials, approximately 4 minutes in duration. (Additional time was used for administrative tasks.) Three participants were involved in each test block serving as: (1) a test operator (T_O) performing all trials; (2) an operator's assistant (O_A) handing equipment to the T_O; and (3) a spotter (S) observing and completing surveys on the work activity and equipment designs. Roles were swapped at the end of each 40-minute block. Among the nine participants, there were three groups of operators that completed three different sessions in each experiment.

In Experiment 1, the pre-assembled frame included eight walk-boards with four of each type, and two work levels of 6" (lower) and 5' (upper) from the ground. Each trial required the T_O to tie-down a particular type of walk-board with a certain tie-type and to move among planks in a clockwise direction. Eight ties were completed in each trial. Consequently, a total of nine trial blocks were completed by each group with each operator completing 96 tie-downs.

In Experiment 2, a pre-assembled frame with eight couplers supporting eight vertical and horizontal tubes was created with one end open to adjustments. Each trial involved mounting one of two coupler types on the horizontal tubes, with four couplers mounted in clockwise order and disassembled in reverse.

It was hypothesized that with the use of the proposed interventions in both experiments, the wrist angles would be closer to neutral and performance time would be reduced due to elimination of unnecessary motion.

Data handling was standardized across both experiments with posture data synchronized with videos in order to conduct analyses on the specific conditions for each scaffold frame. SAS was used to determine the maximum and minimum values for each angle measurement for each condition set. A three-step data analysis approach was applied, including: (1) analysis of variance (ANOVA) for main effects testing and to identify significant learning effects across trials as well as potential outliers; (2) multivariate analysis of variance (MANOVA) to determine significant effects across angular responses due to their dependence; and (3) ANOVAs to identify specific significant effects on each response among the main effects and interactions revealed by the MANOVA.

3 RESULTS

Results from both experiments are presented in the following two sections. Initially, diagnostics were performed on statistical model residuals to test linear analysis assumptions, including constant variance and normality. Appropriate response transformations were applied in the case of any assumption violations.

3.1 Experiment 1

The MANOVA revealed six significant main effects and interactions across all angular responses: subject (p<0.0001), session (p<0.0001), tie (p<0.0001), level (p=0.0266), subtask (p<0.0001), and tie*subtask (p=0.0003). ANOVA results for tie-type revealed a significant impact on flexion (p=0.0341), extension (p<0.0001) and ulnar deviation (p<0.0001), but not for radial deviation (p=0.8865). For flexion, the mean angle with the existing tie was 42.8° and 26.8° for the zip-tie (reduction = 16°); for extension the angle was -79.7° with wire and -76.6° for zip-ties (reduction = 3.1°); for ulnar deviation the angle was 49.3° and 40.8° for wire and zip-ties, respectively (reduction = 8.5°). The tie*subtask interaction revealed significant differences in the angles of flexion, extension and radial deviation in looping and tightening. The interaction of tie-type with subtask revealed differences to occur in the tightening step (zip-ties produced a mean angle of 27.4° versus 55.7° for wire) but not in looping. The maximum wrist extension angle in walk-board tie-down was also reduced by 4% with use of zip-ties compared to #9 gauge wire. Further investigation of the interaction of tie*subtask revealed a 6.2° reduction in looping with zip ties (#9 gauge: -82.1°; plastic: -75.9°).

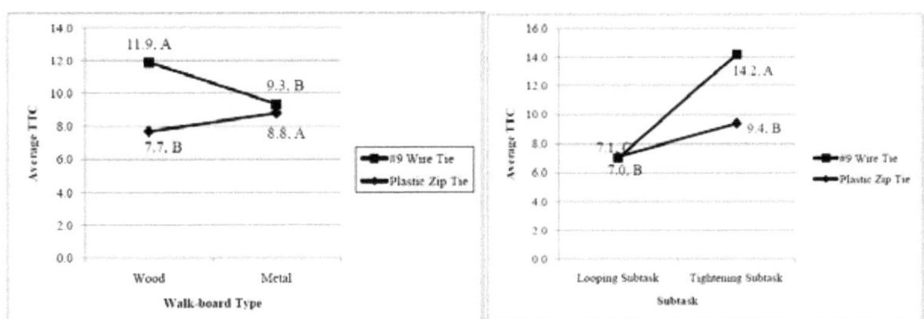

Figure 4, Left: TTC tie*walk-board interaction plot; Right: TTC tie*subtask interaction. (Note: Means with the same letter label are not significantly different at the p<0.05 level.)

An ANOVA on log (TTC) was conducted revealing tie-type to be significant (p=0.0016). The TTC for #9 gauge wire was 10.6s and 8.4s for plastic zip-ties yielding a 2.2s reduction. The interaction of tie*walk-board (p<0.0001) was also found to significantly affect TTC (see Figure 4, left side, for Tukey's post-hoc test results on Least Square mean values). The wire tie required 11.9s with wood boards, which was significantly longer (p<0.05) than the 9.3s for metal. Plastic zip-ties produced a TTC of 7.7s with wood boards, significantly shorter (p<0.05) than 8.8s for metal. Each tie-type was also significantly different from the other for each walk-board type.

The interaction of tie*subtask was also found to have a significant effect on TTC (p<0.0001) (see Figure 4, right side, for post-hoc test results). The #9 gauge wire required 7s for looping as compared to 14.2s for tightening (p<0.05). Plastic ties required 7.1s for looping, significantly lower than 9.4s for tightening. Each tie-type was also found to be significantly different from the other (p<0.05) for each level of subtask.

3.2 Experiment 2

The MANOVA analysis revealed five significant main effects and interactions for all four wrist angles: subject (p<0.0001), session (p<0.0001), subtask (p=0.0189), coupler*subtask (p=0.0423) and activity*subtask (p=0.0076). Although coupler and activity were not significant as main effects, their interactions were and so the main effects could not be eliminated from further analysis. ANOVA results on coupler-type revealed a marginally significant effect for ulnar deviation (p=0.068), with no effect on flexion, extension, radial deviation or pronation. The interaction of coupler*subtask was significant for flexion (p=0.0003) and radial (p=0.0132) and ulnar deviations (p=0.0002), but insignificant for extension (p=0.7759) and pronation (p=0.8586). The interaction plot for flexion angle (see Figure 5 for post-hoc test results) shows the lever coupler produced a significantly smaller mean angle (p<0.05) than the standard ratchet coupler in tightening. The coupler types were not significantly different for the placement subtask.

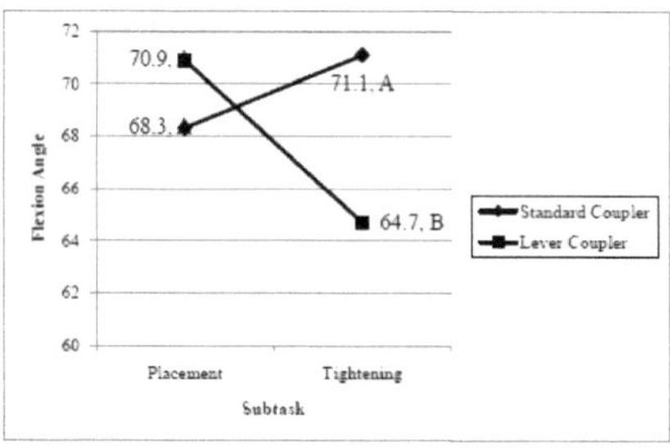

Figure 5: Flexion Angle coupler*subtask interaction plot. (Note: Means with the same letter label are not significantly different at the p<0.05 level.)

Concerning the same coupler*subtask interaction, for radial deviation (see Figure 9, left side for post-hoc test results), the angle was lower for tightening with the lever coupler than the standard coupler. For the placement subtask, there were no significant differences among couplers. For ulnar deviation, Figure 9 (right side) shows that the coupler types significantly differed for the tightening subtask, based on Tukey's test results. Lever couplers produced a lower mean deviation than ratcheting couplers.

ANOVA results also revealed coupler type to have a significant effect on TTC ($p<0.0001$). Mean TTC for existing couplers was 11s as compared to 12.5s for lever couplers. This result could have been due to operator limited familiarity with the new coupler technology. The assembly activity required a mean of 15.5s for each coupler as compared to 8s for disassembly. This difference in task time was due to the need to ensure proper coupler tightness during assembly.

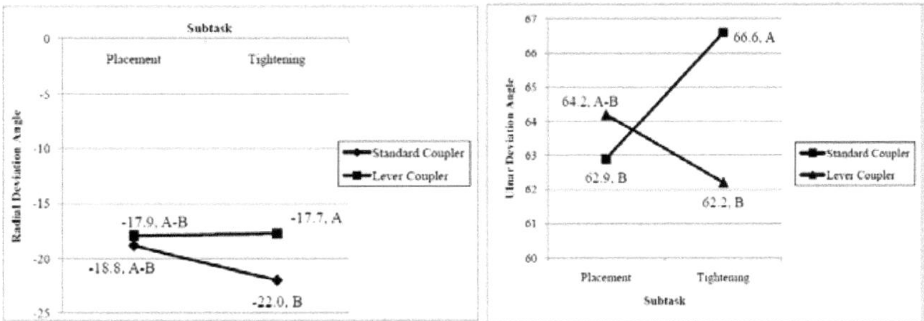

Figure 9, Left: Radial deviation angle coupler*subtask interaction plot; Right: Ulnar deviation angle coupler*subtask interaction plot. (Note: Means with the same letter label are not significantly different at the $p<0.05$ level.)

4 CONCLUSIONS

Our first hypothesis was that maximum flexion, extension, radial and ulnar deviation would be reduced when using plastic zip-ties. This expectation was supported by the positive effect of plastic zip-ties on all four wrist angle measures. From Experiment 1, wrist flexion angle was observed to be dramatically lower for zip-ties compared with #9 gauge wire. Use of plastic zip-ties also produced a 17% reduction in ulnar deviation (#9 gauge: 49.3°; plastic: 40.8°). Tie-type was not involved in interactions influencing ulnar deviation. For radial deviation, the tie*subtask interaction revealed no differences in for looping, but in tightening, the maximum angle for #9 gauge was substantially greater (-34.6°) than for plastic (-28.4°). These results can be primarily attributed to the elimination of wrist twisting motion observed while using #9 gauge wire during tightening versus a swift motion of pulling one end of a zip-tie through a zip-head. The wire ties required an average of 9.5s for tie-down while plastic ties required 7.9s, representing a 16.8% improvement. These results supported our second hypothesis of shorter task time in using the proposed interventions and maybe attributed to elimination of forceful repetitions in tightening #9 gauge wire.

With respect to Experiment 2, most surprisingly, the coupler type was not found to be statistically significant in influence on any of the response measures. However, the coupler*subtask interaction was highly significant, indicating that subtask type mediated the impact of the coupler design on motion behavior. For wrist flexion angle in the tightening subtask, the new lever coupler produced a reduction of 6.4°, or a 4% decrease in maximum angle, from the standard coupler with no difference observed in removal of couplers. These results may be attributed to the ratcheting action required with the current coupler as opposed to the swift single motion tightening achieved in using the lever coupler. With respect to radial deviation, a reduction of 19.5% in the maximum angle was observed for tightening with the lever coupler versus the ratchet coupler, with no reduction for looping. For ulnar deviation, a 6.7% reduction was observed in maximum angle with the lever coupler across subtasks due to elimination of the ratcheting motion. Thus, use of lever couplers had a positive effect on three of five joint angles and it was concluded that the new coupler design was beneficial for frame tube coupling. The second hypothesis tested the effect of coupler type on task completion time. The lever couplers actually took 12% longer to use than the ratchet couplers. Even with the experiment training session, operators may not have been as proficient with use of the lever clamping mechanism and locking pins as the nut and bolt mechanism of the ratchet coupler.

Consequently, both of our research objectives of conceptualizing interventions to reduce potential operator strain/sprain injuries, and conducting experiments to evaluate the impact on posture and performance, were achieved. Future research should involve field studies to gather data under a broader range of scaffold frame configurations (worker posture positions) and actual environmental conditions.

REFERENCES

Elders, L. A., & Burdorf, A. (2001). Interrelations of risk factors and low back pain in scaffolders. Occupational and Environmental Medicine, 597-603.

Hansson, G. A., Balogh, I., Ohlsson, K., Rylander, L., & Skerfving, S. (1996). Goniometer measurement and computer analysis of wrist angles and movements applied to occupational repetitive work. Journal of Electromyography and Kinesiology, 6, 23-35.

McAtamneya, L., & Corlett, E. N. (1993). RULA: a survey method for the investigation of work-related upper limb disorders. Applied Ergonomics, 24 (2), 91-99.

McGorry, R. W., Chien-Chi, C., & Dempsey, P. G. (2004). A technique for estimation of wrist angular displacement in radial/ulnar deviation and flexion/extension. International Journal of Industrial Ergonomics, 34 (1), 21-29.

Mital, A., & Channaveeraiah, C. (1988). Peak volitional torques for wrenches and screwdrivers. International Journal of Industrial Ergonomics , 3 (1), 41-64.

Moore, J. S., & Garg, A. (1995). The Strain Index - A Proposed Method to Analyze Jobs for Risk of Distal Upper Extremity Disorders. American Industrial Hygiene Association Journal, 56 (5), 443-458.

Van der Beek, A. J., Mathiassen, S. E., Windhorst, J., & Burdorf, A. (2005). An evaluation of methods assessing the physical demands of manual lifting in scaffolding. Applied Ergonomics , 213-222.

Vink, P., Urlings, I. J., & van der Molen, H. F. (1997). A participatory ergonomics approach to redesign work of scaffolders. Safety Science , 75-85.

Waters, T. R., Putz-Anderson, V., & Garg, A. (1994). Applications Manual The Revised NIOSH Lifting Equation. Cincinnati: U.S. Department of Health and Human Services.

CHAPTER 3

Development of Prototype System on Workers' Awareness Information Assessment for Safety Management in Petroleum Facilities

Shozo Tanuma, Toshiya Akasaka, Yusaku Okada
Keio University
Kanagawa, Japan

ABSTRACT

Potential risks threatening labor safety are increasing at many industrial plants in Japan. These risks are brought by such factors as changing age structure, fewer opportunities for manual operation due to advancing automation, and fewer opportunities for experiencing trouble. One urgent task facing the industry is to teach young workers so that they can be aware of potential hazards to safety. The task is urgent because the Japanese industry is about to see in a few years a massive retirement of Japanese baby boomers, who possess abundant experience and play a major role in training young workers. To succeed in the task, one possible solution is the one consisting of the following three steps.

1) Develop a prototype system for collecting and assessing information about what potential hazards they are aware of in their everyday work.

2) Identify some important features from the information gathered from experienced labor, and establish an education system based on that result.

3) Develop and implement a training system to enable workers to be aware of potential hazards.

This paper focuses on the first of the three agendas above. The system we developed is to gather information about potential hazards that workers are actually aware of, and to sort out and analyze that information. It is not a brand-new idea to collect information about not only the incidents that actually happened but also potential hazards that might cause an incident in the future. However, past attempts failed to formalize information, only providing open-ended questions asking writers to explain in their own terms what possible hazards they were aware of. This makes it difficult to make use of the information in an integrated way. Thus, they failed to collect information in the way that the information could contribute to resolving the second and third of the three agendas above.

Therefore, we study on how to gather the information potential risk in the workplace make a prototype of the evaluation method of collection results, intended to lead to safety education and training increase the level of awareness of the situation of workers, as shown 2), 3),

First, we sort out and categorized about 2,500 reports gathered in a petroleum plant for the past one year. The reports include information about potential hazards to safety, incidents that actually took place, and basic facts about incidents.

Next, we analyzed the information based on the past academic researches on labor safety and then summarize the list factors and countermeasures related to occupational safety. Thought the item was based on this list, we considering items described in the potential incident reports.

Keywords: potential incident, situation awareness

1 CLASSIFICATION OF THE REPORTS BY SRK MODEL

1.1 SRK model

As the first step in order to linking the improvement in ability to realized of the worker awareness and the information was collected potential incidents in the company, we were classified by SRK model for potential incident reports of about 2,400 worth of oil companies in the past year.

SRK model (Figure 1) was proposed by J. Rasmussen is divided into three categories.

1. Skill base: Action that can be executed without conscious as behavior patterns. For example, go step-by-step what to change in the eye, how to move the hands of a simple assembly operation. It's not a thinking process, such as workers were conscious attention or control. Skilled work can be said.

2. Rule-based: Action to be executed based on the (operating procedures, for example) rules, such as cultivated through education and training. Skills-based and have different points have been the formation of intent with its rules; its boundary is difficult to determine.

3. Knowledge-based: Action to perform, such as understanding in unfamiliar situations and with little or no experience, no experience in the past, and the state trial and error, and prediction of the effect. Briefly, due to the action based on the

information given, turned and thought, exploring the answer, it determines the behavior, thinking and judgment. In particular, workers are less skilled, this behavior accounts for a lot.

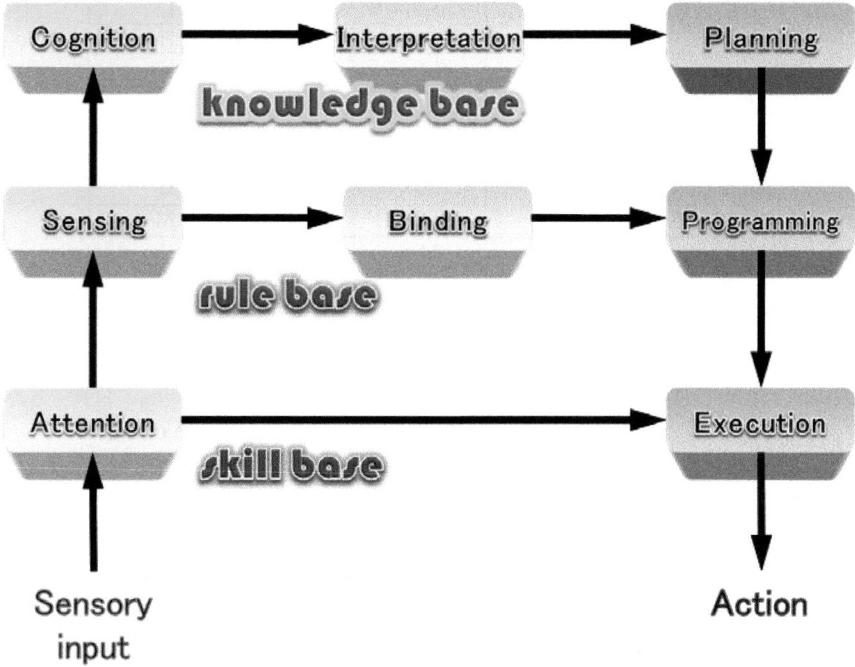

Figure 1 SRK model

1.2 Classification method

To account for the objectivity of the item, fewer omissions, and the relevance of the work proficiency, classification was used for the items "action", "age" of potential incident reports. "Action" is made up of 23 choices, and focused the difficulty of input (such as situational awareness and recognition), judgment (such as learning of the procedure and knowledge), output (such as operation and control) about the contents of 23"action" is shown in Table 1.

Table 1: The difficulty of the "action"

	Input	Judgment	Output
1walking	L	L	L
2bicycle operation	L	L	L
3bike operation	M	L	L
4motorists	M	L	L
5device operation	M	H	M
6equipment operation	H	H	H

7button operation	H	H	L
8valve operation	M	H	L
9equipment inspection	M	H	L
10equipment repair	H	H	H
11aerial work platforms	M	M	L
12transport	M	M	H
13aerial work platforms	M	M	H
14work in the machine	H	H	H
15rising-falling	L	L	L
16cleaning	L	M	L
17computer operation	H	H	L
18firefighting operation	M	M	M
19handling heavy	M	H	H
20sampling	H	H	M
21liquid handling	H	H	M
22construction/work attendance	M	M	M
23office work	M	H	L

Criteria L, M, and H

L: You can learn to work less than one year without the guidance of others

M: Takes about one to three years

H: Take more than three years

Based on the “age” and difficulty of the "action", perform the classification of SRK.

Action of (L, L, H) requires a high proficiency in work. So errors in this action, there may be an error resulting from a failure in operation and control.

→ Considered that proficiency of the work is not enough

→ Other than S

Action of (L, H, L) takes high level of knowledge and complex procedure, so errors in this actions due to the inability to master the procedure.

→ K

In this way, to classify other “action” to SRK and table 2 shows the classification criteria, including item of “age”.

Table 2: The classification criteria

	K	*KR*	*R*	*RS*	*S*
1walking					○
2bicycle operation					○
3bike operation					○
4motorists					○
5device operation	○				
6equipment operation	~27	28~34	35~41	42~48	49~
7button operation	~30				31~
8valve operation	○				

9equipment inspection	○				
10equipment repair	~27	28~34	35~41	42~48	49~
11aerial work platforms	~34	35~41	42~		
12transport	~34	35~41	42~		
13aerial work platforms	~34	35~41	42~		
14work in the machine	~27	28~34	35~41	42~48	49~
15rising-falling					○
16cleaning	○				
17computer operation	~30				31~
18firefighting operation	~27	28~34	35~41	42~48	49~
19handling heavy	~34	35~41	42~		
20sampling	~30				31~
21liquid handling	~30				31~
22construction/work attendance	~27	28~34	35~41	42~48	49~
23office work	○				

1.3 Classification result

Classification results based on the 2.2 is figure 2.

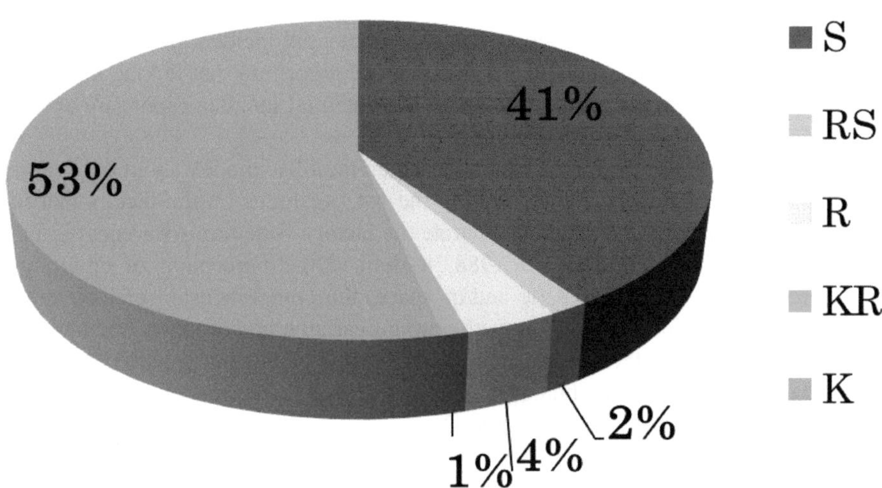

256 of 2361 data can't be classified because of blank for" action" or "age"
Figure 2 Classification results to SRK 5 stages

2 CLASSIFICATION TO THE FACTORS

2.1 Making a list of factors and items on the report

For linking the information about potential incidents to planning countermeasures and improvement in ability to realized system, we made the list of factors and items to collect these factors. We aimed classified to each process in the SRK; furthermore, potential incident factors not only SRK 5 stages. Figure 3 is flowchart on how to derive countermeasures.

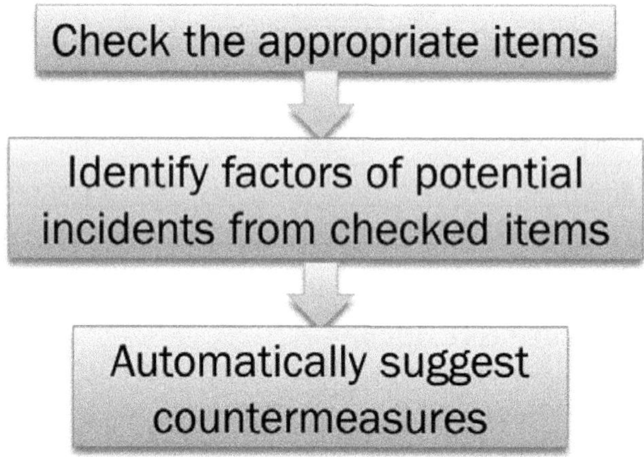

Figure 3 Flowchart on how to derive countermeasures

We made items on the report, factors of potential incident that are derived from the items, countermeasures for the factors, and algorithm that linking the items and countermeasures. So people who write the potential incident report are able to see countermeasures just check the items.

First, we made the list of factors based on the SRK model, about the factors that classification category. When making the list, we focus on the factors that inhibit the Situation Awareness, and enumerate the factors. Situation Awareness is a theory was proposed by Endsley in 1988, it defined the perception of environmental elements with respect to time and/or space, the comprehension of their meaning, and the projection of their status after some variable has changed, such as time. In the broader sense, there is a mental model of knowledge and a variety of situations is happening around the human. Level of situation awareness of workers is divided into three stages, as shown in Figure 3.

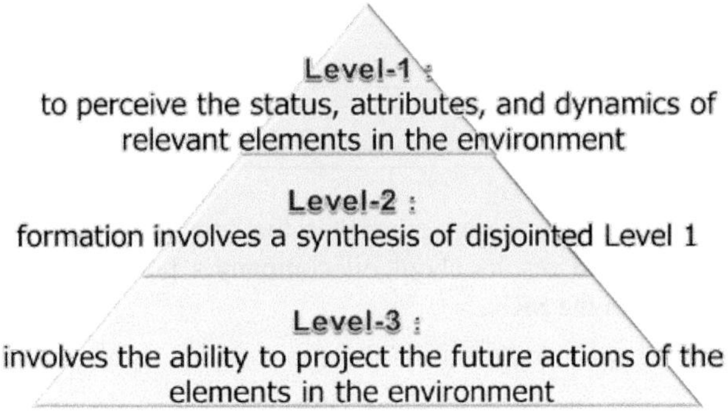

Level-1 :
to perceive the status, attributes, and dynamics of relevant elements in the environment

Level-2 :
formation involves a synthesis of disjointed Level 1

Level-3 :
involves the ability to project the future actions of the elements in the environment

Figure 4 Level of situation awareness

The purpose of this study is to step up the level of the situation awareness of workers.

Table 3 is part of the list of the factors.

Table 3 : The factors related to Attention

Factor		Description
Name	**Breakdown**	
Intervention of other work	By instructions from the human	Indication and another request comes in from work to others while working
	By non-human	Work of another may be necessary while working
Monotonic observed		Followed by a monotonous observation over a long period of time
Note interference environment		Environment such as temperature, humidity, noise and stench is bad
Excessive time margin		Too much time there is a deadline, easy to lose the attention

Table 3 is factors related to Attention of SRK process. We enumeration factors in the other SRK process like Cognition, Interpretation, Planning, Programming, and Execution. And considered the items that collected enumerates the factors from the potential incident report. Table 4 is the items for collecting factors related to Attention.

Table 4: The items for collect factors related to Attention

Factor		The corresponding item in the reference list
Name	**Breakdown**	
Intervention of other work	By instructions from the human	Intervention of other work Communication on the work
	By non-human	Intervention of other work
Monotonic observed		Monotonic observed

Note interference environment		Adverse environment
Excessive time margin		

We were also examined factors other than the Attention like this. Figure 4 is the tentatively classification results about oil company's potential incident reports of 120 to each process in the SRK.

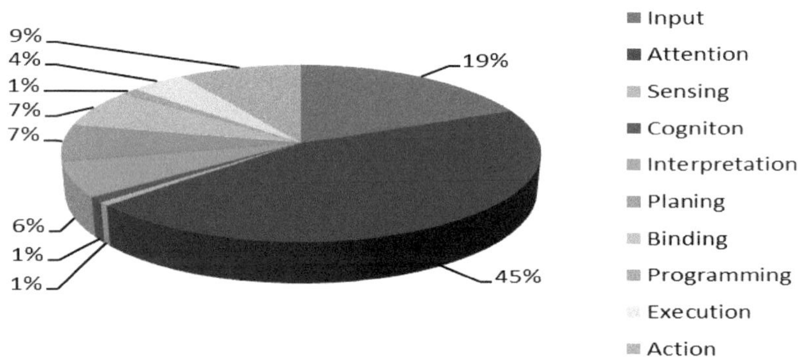

Figure 5 Classification results in the SRK process

Classified into potential incident data, SRK process and matching a list of countermeasures below, countermeasures policy presented automatically from potential incident information.

2.2 Making a list of countermeasures

We considered list of factors and items shown in the potential incident report in Chapter 2.1. Next, we made list of countermeasures when operate the potential incident reports at company, and gathered a number of factors, such as the advice can be presented. Table 5 is the list of countermeasures about factors for Attention.

Table 5: The list of countermeasures about factors for Attention

Education	Education of the importance of the consultation reports contact
	Education of how to make contact to lubricate the consultation reports (for management)
	Education of the possibility of error to increased dramatically that preparation, cleaning, etc. becomes insufficient
	Guidance so as not to manual disrespect

	Education of teaching methods to avoid trouble, the complaint in the workplace (for management)
Engineering	Ensure appropriate information sharing time
	Mechanisms absorb the information from the workers
	Review of the work burden (not to be too busy)
	Review of the storage location and reference method of the manual (To make it easier to workers)
	Periodic inspection of Preparation, cleaning, tidy situation
Enforcement	Carried out once a week at the morning meetings, the penetration of the overall purpose of improving "the safety and security of our customers"
	Site visits by the clerk safety guidance (Such as when the work atmosphere)
	Survey of the problems in cooperation and communication between workers
	Leadership system of the department length
Example	[Good practices that have been made to this factor group]
	Information sharing and understanding of in advance, thoroughness of well-known manual, insufficient preparation of the tool cause is not in trouble
Environment	Survey environment and improvement of the sound of the place to exchange information
	Ensure the storage location of the manual
	Review of the management system and storage location of the tool,
	Review of inspection system on the deficiencies of the tool

We made list of countermeasures also the other processes of SRK.

3 CONCLUSIONS

In this study, we classified into five categories using the SRK model, potential incidents reports have been corrected during the past year in an oil company. Then, based on the results of past research PSF, we examined the list of countermeasures compiled a list items for factors related to occupational safety, to collect further factors.

Future, we continue to promote considering of factors list and then check the validity of the factors list to evaluate the completeness of the factors enumerated and category balance. Also items shown in the potential incident report, we will investigate in consideration of objectivity, and completeness.

Then we scheduled to improve the evaluation system about ability to realize of the worker awareness, to expanded education and training in workplace safety.

REFERENCES

Rassmussen, J. 1986. Information Processing and Human-Machine Interaction. An Approach to Cognitive Engineering. Amsterdam: Elsevier, 1986.

Rassmussen, J.: Skills, rules, knowledge: signals, signs, and symbols and other distinctions in human performance models. IEEE Trans. Systems, Man, and Cybernetics. MSC-13, 257-267, 1983

Endsley, M. R.: Design and evaluation for situation awareness enhancement. Proceedings of the Human Factors Society 32nd Annual Meeting, Human Factors Society, Santa Monica, CA, 97-101, 988.

Endsley, M. and Garland, D. (2000) (eds.) Situation Awareness. Analysis and Measurement. Mahwah, NJ: LEA.

Banbury, S. and Tremblay, S. (2004) (eds.) Situation Awareness: A Cognitive Approach. Aldershot: Ashgate

Wickens, C. D.: Engineering Psychology and Human Performance (2nd ed.). New York: Harper Collins, 1992.

CHAPTER 4

Performance of New Raters Using an Observational Physical Ergonomic Risk Assessment Instrument

Kristin Streilein#, Ed Irwin*, Thomas Armstrong#*

Mercer Engineering Research Center*, University of Michigan#
Warner Robins, GA* & Ann Arbor, MI#, USA
kstreilein@merc-mercer.org, eirwin@merc-mercer.org, tja@umich.edu

ABSTRACT

This paper looks at the rating performance of raters using an observational physical ergonomic risk assessment instrument with which they had little or no experience. The physical stresses that are rated using this assessment instrument are: hand activity level, hand force, wrist flexion/extension, wrist radial/ulnar deviation, forearm rotation, elbow flexion, shoulder posture, neck posture, and back posture. Each stress is rated on a scale from 0 to 10 as described by W. A. Latko et al. (1997). Reliability and repeatability for this instrument for trained raters has previously been reported by Latko et al, but little is known about how individuals new to the instrument perform and how many jobs are needed to calibrate them.

80 novices rated 12 job videos that had previously been rated by faculty, staff, and graduate students having extensive experience using the risk assessment instrument. The consensus rating of the experienced raters was considered the "expert rating" and the assessments by the novice raters were compared to this standard.

Cronbach's (1955) error components were used to provide insight into which specific elements of the instrument the novice raters were having difficulty. The results of this analysis indicated no systematic bias. The participants were able to distinguish ergonomically risky jobs from those with low risk. They had significantly more trouble distinguishing between rating factors. The novice raters had the most difficulty in quantifying the specific level of risk at the body part level.

Analysis of observational ergonomic rating instruments using Cronbach's error components may help identify what level of assessment can realistically be expected of non-experts. The results imply that novice raters could be expected to successfully identify target jobs in need of further analysis, but are likely to misidentify the body parts at most risk. Interventions based on results with such limited specificity are unlikely to be successful. Negative outcomes, including continued costs of musculoskeletal disorders and wasted time and resources, can lead to reduction in management support and funding for ergonomic improvements.

Keywords: physical ergonomics, observational job analysis, error components

1 INTRODUCTION

The goal of this study was to examine the agreement of physical job stress ratings between new and experienced raters. Additionally we examined the effect on future rater performance of providing feedback to novice raters regarding the similarity of their rating to those of the experts.

2 BACKGROUND

This work is concerned with assessment by novice raters of physical job demands that are widely recognized as causal and aggravating factors of musculoskeletal disorders. Jobs can be assessed using a variety of methods ranging from instrumentation to observational methods. (Colombini, 1998; Fransson-Hall et al., 1995; Gilbreth & Gilbreth, 1920; Hignett & McAtamney, 2000; W. A. Latko et al., 1997; McAtamney & Nigel Corlett, 1993; McCormick, Jeanneret, & Mecham, 1972; Moore & Garg, 1995) Instrumentation methods are regarded as more accurate but also more costly, very time consuming, intrusive and impractical for large studies. Observational methods are frequently chosen since they are less costly, more time effective, and less disruptive to the worker than direct measurement methods (Li & Buckle, 1999). In many cases instrumentation is not available or it interferes with the work. Observations may be used in other cases as a starting point to determine if further assessment with instrumentation is merited.

In order for an observational assessment method to be successfully applied to the primary and secondary prevention of musculoskeletal disorders it needs to have good reliability, accuracy, and repeatability. Some studies have shown that "the self-report approach has a too low validity and reliability in relation to the needs for ergonomic interventions."(Li & Buckle, 1999) This work will help define a future study to see if a short training period or benchmark examples allow for an accurate and reliable assessment by a worker.

Little is known about the length of time it takes to learn an observational job assessment method. Most observational rating systems report inter- and intra-rater reliability only for experienced raters. "Few studies have been carried out on how to achieve good observation reliability, for example, through training format, training duration and the qualifications and requirements of observers"(Denis, Lortie, & Bruxelles, 2002). This work examines the performance of novice raters and the effect of feedback on novice rater performance. An on-line tool was chosen for this research, since using the web allows standardized tools to be made widely available – particularly to small employers and employers in remote sites who do not have immediate access to ergonomic resources.

3 METHODS

3.1 Experimental Design

A web-based training and assessment program was created. The program randomly assigned participants to one of the two feedback groups. Each group rated the same twelve jobs, selected to include a range of occupations, for ergonomic risk. The jobs came from a database that included job descriptions, video images of the job and physical stress ratings by an experienced job analyst. One group of novice raters was given feedback after each of the first 6 jobs; the second group got feedback after the second 6 jobs. Each job was rated while watching a web video which showed approximately 20 seconds of a representative cycle, which looped endlessly until the rating was completed.. The entire process took approximately one hour and participants were given a $30 Amazon.com gift certificate for their participation.

Each participant was given a sheet of all the rating scales to use as reference as they were rating. The rating scales they were given are described below. The ratings for each job were stored in a database for later analysis. Each participant also completed a short demographic survey to document their observational ergonomic assessment experience.

3.2 Physical Stresses Rated

The physical stresses that were rated included: hand activity level, hand force (peak & average), wrist flexion/extension, wrist radial/ulnar deviation, forearm rotation, elbow flexion, shoulder posture, neck posture, and back posture. Each stress was rated on a scale from 0 to 10 as described by W. A. Latko et al. (1997). Peak and average values were estimated from observations for each job, except for hand activity level, which included only an average value. For force estimation, 0 to 10 corresponded to 0 to 100% of maximum voluntary contraction (MVC). For postures, 0 corresponded to a neutral posture and 10 to the absolute value of the difference between neutral and the range-of-motion. The scales were created so that the posture score equals the absolute value of the joint angle divided by the neutral joint angle minus the maximum (minimum joint angle possible). For example, a score of zero for wrist flexion/extension implies that the wrist is neutral (straight) while a score of 10 implies that it is in maximum extension or flexion (which is approximately ±90°).

3.3 Descriptions of Jobs Rated

Twelve job videos were selected that covered a range of postures for each body part. The videos were selected from a variety of manufacturing, service, and office settings. The job videos had previously been rated by faculty, staff, and graduate students having extensive experience using the risk assessment instrument. The consensus rating of the experienced raters was considered the "expert rating" and the assessments by the novice raters were compared to this standard.

An attempt was made to include jobs that covered the greatest range possible for each posture, but in some cases it was not possible to obtain extreme values. For example, most people cannot sustain a high average shoulder posture, or an extreme neck posture. In other cases, extreme average values are possible, but not common. It is not uncommon for a worker to briefly flex or extend their wrist to a maximum or minimum value, but is uncommon for workers to maintain that position for extended periods of time.

4 RESULTS

4.1 Novice Rater Demographics

Eighty participants completed the entire survey. 44 were randomly assigned to group 0 (immediate feedback after first six jobs) and 36 to group 1 (immediate feedback after second six jobs). There were 47 male and 32 female participants (one person did not answer this question.) 6 people had technical training or some college. 17 people had a four year college degree and 57 people had attended graduate or professional school. The participants came from a variety of occupations including engineering (30), health and safety (14) and occupational therapy (7), and 28 with other jobs. Job tenure ranged from less than 6 months to over 10 years.

4.2 Ergonomics Experience

There was a variety of experience in ergonomics. 15 participants had no training in ergonomics while 37 reported they had taken a college course in ergonomics. 21 participants reported informal training, 24 reported attending ergonomics conferences, 15 reported continuing education classes, and 12 reported in-service training. Fewer participants had experience with observational job assessment methods. 27 participants said they had no training in observational job assessment methods, 20 reported informal training, 14 received info at ergonomic conferences, 8 reported continuing education classes, 8 reported in-service training, and 31 reported having taken at least one college course covering the topic.

34 participants reported that they never assess the ergonomic risk factors of jobs. 19 reported that they assess 1-2 jobs per year. 11 said they assess 1-2 jobs per month. 7 participants assess 1-2 jobs per week while the remaining 9 participants assess the ergonomic risk factors of jobs on a daily basis. Only 1 person reported extensive use of the ACGIH TLV for Mono-task Handwork (which includes the UM-Latko hand activity level rating scale). Most participants had never heard of the TLV (55 participants) while 17 have heard of it and 7 had basic training or used it occasionally. The NIOSH lifting equation had the greatest name recognition and use with only 31 participants saying they had never heard of it.

4.2 Ratings compared with Expert Raters

When the assessment tool was used by expert raters in the field, each rater rated the job on all of the risk factors. Ratings were compared once everyone had completed their individual rating and the rating assigned was discussed by the raters until everyone was within 1 point on the ten point scales. This produced the final consensus rating. The ratings done by the novice raters in this study were compared to the expert consensus rating to determine agreement. Table 1 shows the percentage of novice raters who were within 1 point on each of the scales while rating their 1st, 6th, & 12th jobs. For most body parts the agreement is not very close even for the 12th job where everyone had received several rounds of feedback on their ratings.

4.3 Error Components

Digging deeper into why the novice raters were so far off led to Cronbach's 1955 method for breaking down the source of errors in personnel ratings.(Cronbach, 1955) Cronbach's method breaks the overall error into 4 error components:

- E^2 is the elevation, which shows how much error is due to a systematic increase or decrease by the rater across all jobs and all scales; i.e.would the rater be significantly higher or lower than another rater across the board;
- DE^2 is the differential elevation, which shows how well the rater can rank-order the jobs across all scales; i.e. can they figure out which jobs are worst;
- SA^2 is the stereotype accuracy, which shows how well the rater does on different scales across all the jobs; i.e. can they figure out which body part to focus on;
- DA^2 is the differential accuracy, which is the rater's sensitivity to job differences on different scales; i.e. can raters make the fine-grained distinctions between jobs within each of the body parts (Murphy, Garcia, Kerkar, Martin, & Balzer, 1982).

Calculations were done on the data using formulae from Becker and Cardy (1986). The following graph represents the participant's error components.

Table 1 - Agreement with expert ratings - % of ratings within 1 point of the expert rating. (n=80 raters)

Ergonomic Stressor	% w/in 1 of expert rating		
	Trial 1	Trial 6	Trial 12
HAL	45	55	52.5
Force PEAK	43.8	47.5	60
Force AVERAGE	23.8	46.3	58.8
Wrist F/E PEAK	31.3	31.3	40
Wrist F/E AVERAGE	35	40	43.8
Wrist R/U PEAK	38.8	38.8	52.5
Wrist R/U AVERAGE	42.5	51.3	57.5
Forearm PEAK	33.8	43.8	47.5
Forearm AVERAGE	40	48.8	41.3
Elbow PEAK	68.8	71.3	61.3
Elbow AVERAGE	52.5	45	60
Shoulder PEAK	52.5	47.5	51.3
Shoulder AVERAGE	35	45	57.5
Neck PEAK	30	40	48.8
Neck AVERAGE	51.3	52.5	57.5
Back PEAK	47.5	42.5	53.8
Back AVERAGE	62.5	65	70

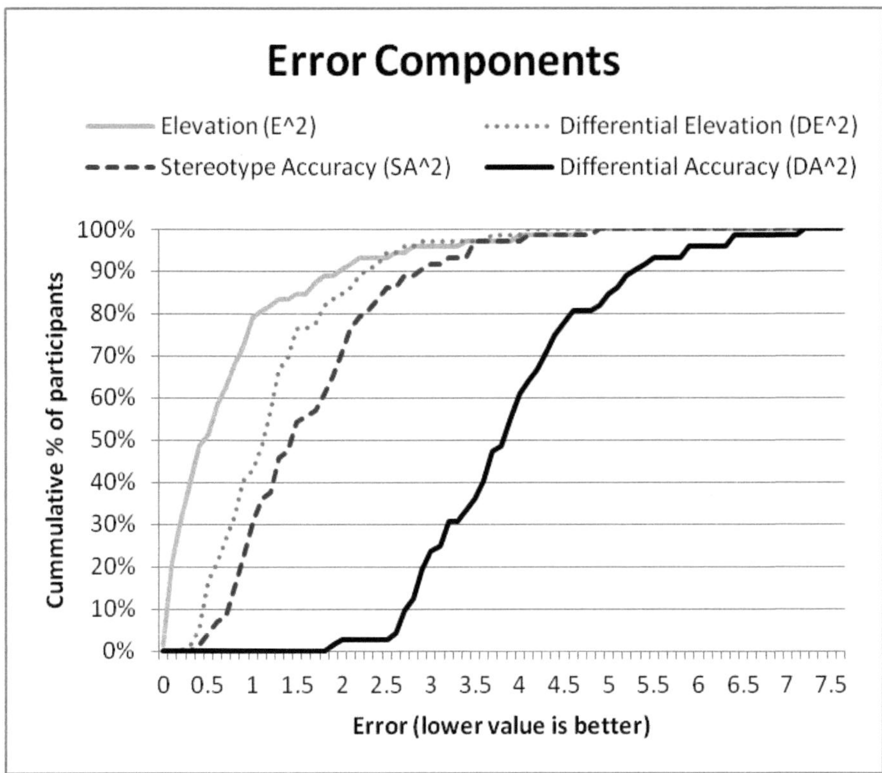

Figure 1: Cronbach's Error Components – Lower error is better. Note how the differential accuracy (DA^2) is significantly worse than the other error components.

5 DISCUSSION

5.1 Comparison with Expert Ratings

Table 1 shows that simply providing the rating scales or rating scales with feedback on ratings is not enough to calibrate novice raters, over the course of 12 observations. The lack of learning curve indicates there is something that is not intuitively obvious to novice raters that experienced ergonomists are doing when they are using observational rating scales.

5.2 Error Components

Using Cronbach's error components to analyze the data illuminates the implicit and inaccurate assumptions regarding the usability of rating scales by inexpert raters. This should be studied further in order to develop better training systems. The error elevation (E^2) line shows that the majority of individuals did not have a systematic shift in their ratings. There are few outliers that were systematically rating high or low indicating little bias due to rater demographics. The differential elevation (DE^2) line is a little worse, but shows that participants did a reasonably good job at rank ordering the ergonomic risk of the jobs. When asked to identify the worst jobs in a set of jobs, the participants would probably be able to rank the jobs in terms of global risk.

The stereotype accuracy (SA^2) is significantly worse. When asked to figure out which body parts represent the most risk, the participants start having problems. This is troubling, since figuring out which body parts are at risk directly corresponds to selecting appropriate technology or process changes to reduce the risks. Participants will also have trouble figuring out if proposed solutions address all the risk areas if they can't accurately identify the risk areas.

The differential accuracy (DA^2) line shows that once asked to make fine-grained distinctions across jobs, the participants aren't able to do it with any accuracy. DA^2 is less important when you only need to make gross distinctions between jobs like for grouping them into sets of high, medium, low risk jobs to prioritize funding for improvements but is very important for doing epidemiological studies where you need to be able to accurately quantify risk factors.

At the "gut level" the participants know which jobs are the worst but they have real issues applying the scales to determine the seriousness of the issue and exactly which body parts to focus on when trying to implement solutions.

5.3 Implications of Results

The results were based on a small subset of jobs and a small group of raters. If these finding are replicated in additional studies this has major implications for the development of ergonomic training programs and the execution of participatory ergonomic programs. These findings may start to explain why so many attempts to solve ergonomic problems fail if individuals do not consistently identify the body parts with the most risks and direct the solutions at the wrong body segments. This would also undermine the ability of train-the-trainer programs to be useful unless the programs focused on prioritizing jobs for evaluation by an ergonomic expert to identify the specific risks that need to be ameliorated. Participatory ergonomic programs would also need to make sure their teams were lead or guided by an ergonomic expert to ensure their efforts were focused on the specific risks present at the various high risk jobs they identified.

REFERENCES

Becker, B. E., & Cardy, R. L. (1986). Influence of halo error on appraisal effectiveness: A conceptual and empirical reconsideration. Journal of Applied Psychology, 71(4), 662-671. doi: 10.1037/0021-9010.71.4.662

Colombini, D. (1998). An observational method for classifying exposure to repetitive movements of the upper limbs. Ergonomics, 41(9), 1261-1289. doi: 10.1080/001401398186306

Cronbach, L. (1955). Processes affecting scores on "understanding of others" and "assumed similarity.". Psychol Bull, 52(3), 177-193. doi: 10.1037/h0044919

Denis, D., Lortie, M., & Bruxelles, M. (2002). Impact of observers' experience and training on reliability of observations for a manual handling task. Ergonomics, 45(6), 441-454. doi: 10.1080/00140130210136044

Fransson-Hall, C., Gloria, R., Kilbom, A., Winkel, J., Karlqvist, L., & Wiktorin, C. (1995). A portable ergonomic observation method (PEO) for computerized on-line recording of postures and manual handling. Appl Ergon, 26(2), 93-100.

Gilbreth, F. B., & Gilbreth, L. M. (1920). Motion study for the handicapped. London: G. Routledge.

Hignett, S., & McAtamney, L. (2000). Rapid entire body assessment (REBA). Appl Ergon, 31(2), 201-205.

Latko, W. A. (1997). Development and evaluation of an observational method for quantifying exposure to hand activity and other physical stressors in manual work. (University of Michigan Ph.D.), University of Michigan, United States -- Michigan.

Latko, W. A., Armstrong, T. J., Foulke, J. A., Herrin, G. D., Rabourn, R. A., & Ulin, S. S. (1997). Development and evaluation of an observational method for assessing repetition in hand tasks. American Industrial Hygiene Association Journal, 58(4), 278-285.

Li, G., & Buckle, P. (1999). Current techniques for assessing physical exposure to work-related musculoskeletal risks, with emphasis on posture-based methods. Ergonomics, 42(5), 674-695.

McAtamney, L., & Nigel Corlett, E. (1993). RULA: a survey method for the investigation of work-related upper limb disorders. Appl Ergon, 24(2), 91-99.

McCormick, E. J., Jeanneret, P. R., & Mecham, R. C. (1972). A study of job characteristics and job dimensions as based on the Position Analysis Questionnaire (PAQ). Journal of Applied Psychology, 56(4), 347-368. doi: 10.1037/h0033099

Moore, J. S., & Garg, A. (1995). The Strain Index: a proposed method to analyze jobs for risk of distal upper extremity disorders. American Industrial Hygiene Association Journal, 56(5), 443-458.

Murphy, K. R., Garcia, M., Kerkar, S., Martin, C., & Balzer, W. K. (1982). Relationship Between Observational Accuracy and Accuracy in Evaluating Performance. [Article]. Journal of Applied Psychology, 67(3), 320-325.

CHAPTER 5

Pragmatic Measures to Vitalize Railroad-Station-Service Employees' Activities That Can Reinforce Customers' Safety and Comfort

Naori TSUJI, Ryoko IKEDA and Yusaku OKADA

* Graduate School of Science & Technology, Keio University
Kohoku-ku Yokohama, Japan
fr071413@a2.keio.jp

ABSTRACT

Recently the purpose for the safety management in Japanese railroad companies is not only to prevent accidents but also to produce the high customers' satisfaction. However, the employees do not understand that the safety activities should include raising customers' comfort. Therefore, we discussed the method that can vitalize the employees' activities that aimed to reinforce customers' safety and comfort.

We developed a method for that purpose and tested this validity by applying it to railroad station service section. In addition, several new activities are introduced in station service section. To measure the effects of these activities, the survey on the safety activities was carried out for all employees, on 2009, before taking the activities, and on 2011, after taking the activities.

Keywords: Motivation for safety activities, Customers' Satisfaction, Human Error Management

1 BACKGROUND OF THE SAFETY ACTIVITIES IN JAPANESE RAILROAD COMPANIES

Recently the purpose for the safety management in Japanese railroad companies is not only to prevent accidents but also to produce the high customers' satisfaction. The safety of the railroad service is influenced by the subjectivity of customers who get the railroad service. In other words, customers' comfort becomes the important issues which have deeply influenced for the safety of the railroad companies. The issues should be solved immediately as soon as possible.

However, many employees think that safety activities are different from customer service. So, lots of safety activities which the safety managers planned for making customers feel safe are not executed in workplaces. It is assumed that the employees do not understand that the safety activities should include raising customers' comfort. Actually, near-miss incidents about customers' unpleasantness are much more unlikely to be reported than near-miss incidents about traffic.

Therefore, we surveyed a railroad-station-service and discuss the method that can vitalize the employees' activities that aimed to reinforce customers' safety and comfort. We developed a method for that purpose and tested this validity by applying it to railroad station service section.

2 SURVEY

Analysis of issues related human-error prevention action on a station service department

First, we surveyed the employees' understanding/motivation, which was held on the station service department with the object of the activities to prevent human-error, to clarify the issues to solve on this department. 447 employees were surveyed on 2009. The 20 questionnaires to survey the employees' basic understanding/motivation for human-error prevention action is shown in Table1. All questionnaires are answered on one-to-five scale; "Strongly Yes", "Yes", "N/A", "No", "Strongly No".

We analyzed the result of the survey by factor analysis, and we obtained following particulars. The contributing rates of factors in factor analysis are shown in Figure1. This figure suggests that factor 1 is major factor contributing to the employees' understanding and motivation for human-error prevention action.

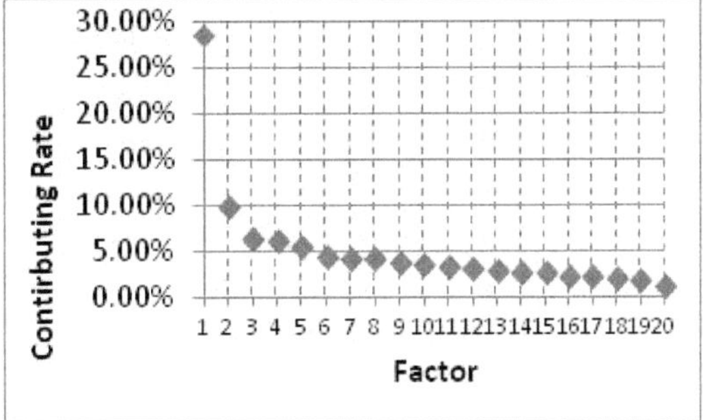

Figure 1 Contributing rate of factors (the result of the survey on basic understanding and motivation for human-error prevention action on 2009)

Table 1 Twenty questionnaires about basic understanding and motivation for human-error prevention action

Number	Contents
1	I think that I do not carry out human error.
2	I think that human error is carried out since skill is low.
3	If it can do, I will want to hide, when human error is started.
4	I think that the cause of human error results in the person who caused the error.
5	I think that the absent-minded act is the responsibility of a person concerned.
6	By efforts of the man, I think that human error is set to 0.
7	I think that the preventive measures of human error are the cultivation of men's ability which does not commit a mistake.
8	When human error is carried out, I think that fate was bad.
9	It is troublesome to write human error and a trouble report, and I write only necessary minimum.
10	I do not understand why the report of the phenomenon which has not been a trouble must also be written.
11	I do not know that there is a learning field which studies human error.
12	I have not received the lecture or short course about human error research.
13	The investigation about human error is the futility of time, and I do not want to carry out, if it can do.
14	I do not understand the purpose of analysis of human error.
15	The causative analysis of human error has not been carried out.
16	If only recurrence prevention is carried out, I will think that it is enough.
17	I think that the human error prevention activities performed now are enough.
18	It is not interested in the human error prevention activities of the other company.
19	I think that human error prevention activities are the problems which should think individually and should cope with it.
20	There is no necessity of managing human error as an organization etc. (Or it is hard-pressed)

So we focused attention on this factor. Factor loadings of each question numbers in factor 1 are shown in Figure 2. Figure 2 shows that the question number 13, 14 and from 16 to 20 is major consistuency of employees' basic understanding and motivation for human-error prevention action.

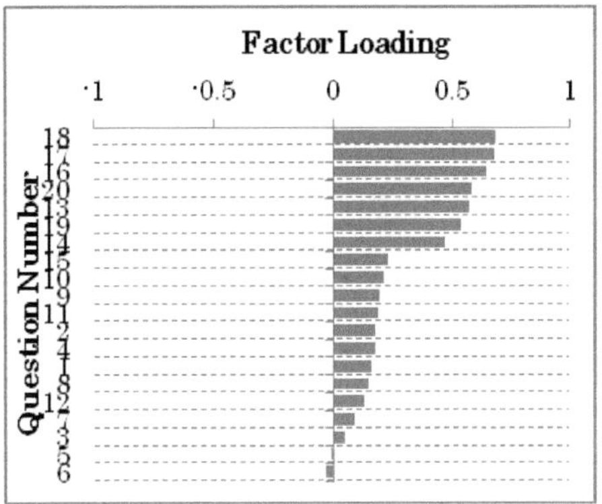

Figure 2 Factor loadings of each question in factor 1 (the result of the survey on basic understanding and motivation for human-error prevention action on 2009)

We analized the employees' understanding and motivation, furthermore, from another perspection which is the two type of groups; highly-motivated employees and lowly-motivated employees. Based on sum of score in 20 questionnaires (shown as Table 1), the employees are categorized into 5 groups (See figure3). Here, sum of score in 20 questionnaires is called as total score. In order to strengthen the interpretation of the result (shown as fig.3), these 5 groups are summarized into 2 groups; the first group is "lowly-motivated employees" who gets the total score less than 60, and the second group is "highly-motivated employees" who gets total score more than 60. The average score of each question in each group is shown in figure 4. We graphed the difference between the averages in each groups, too. According to figure 4, we considered from 11 to 14, 16, from 18 to 20 as factors which have a causual influence on the motivation and understanding gap.

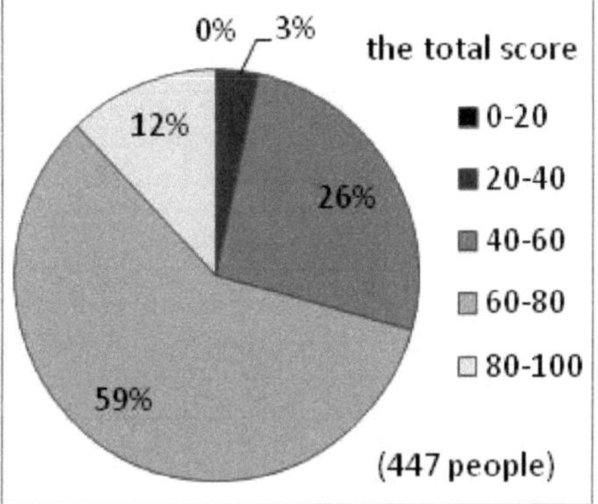

Figure 3 Five groups based on the total score of 20 questionnaires (the result of the survey on basic understanding and motivation for human-error prevention action on 2009)

These analysis from two perspectives indicates that question 13, 14, 16, 18, 19 and 20 are the issues related to human-error active prevention on this station service department. We selected these question number as important issues because these have high factor loadings and more than 1.0 deffernces between two groups. We produced the mutual relations between issues, latent causes of these issues and countermeasures against the causes in Figure 5 briefly, discuussing these issues with safety/service managers and employees in station service department. As will be noted from the figure 5, we decided the main purposes of activities to prevent human-error as "Construct the information sharing system" and "Improve the skill to deal with troubles".

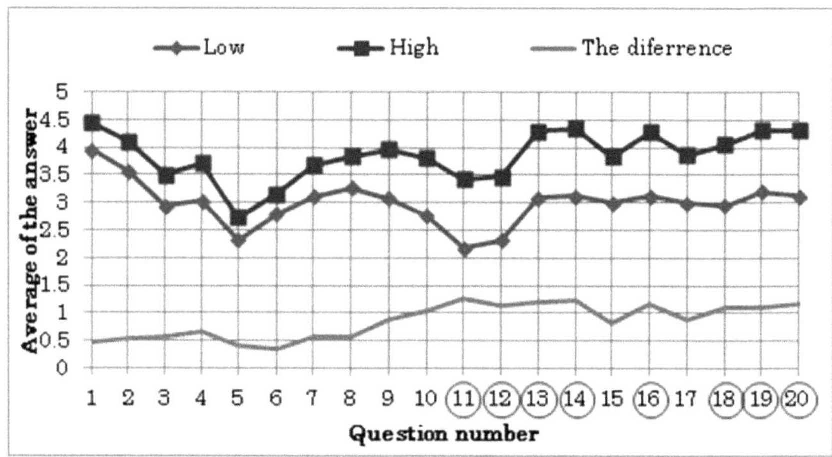

Figure 4 Average of the answers in each questions comparing highly-motivated group with lowly-motivated group (the result of the survey on basic understanding and motivation for human-error prevention action on 2009)

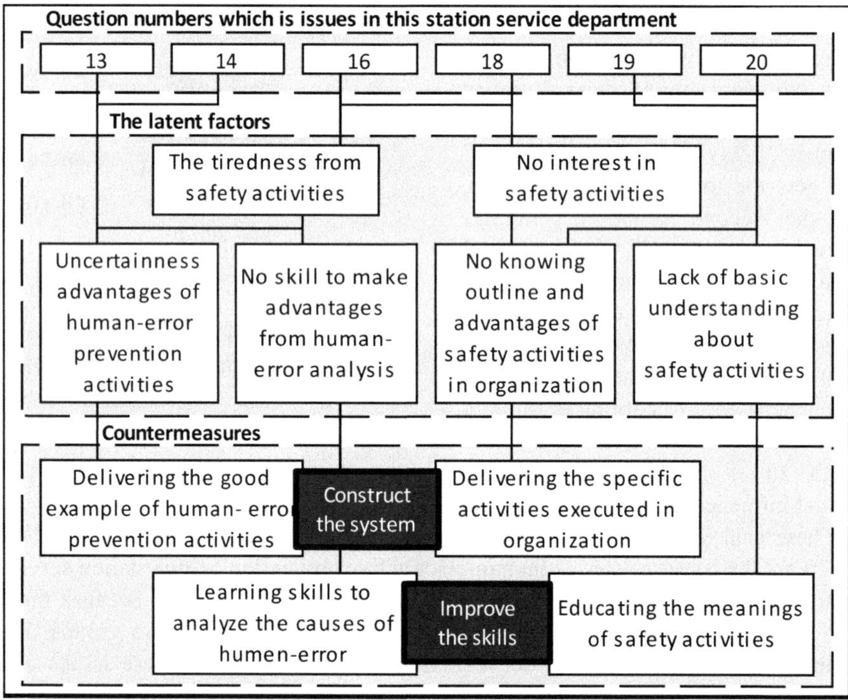

Figure 5 Conceivable latent factors and countermeasures against the issues in this station service department

3 ACTION

We considered the three primary purposes on the safety activities executed in station service department, based on the issues of human-error preventive activities and the target of improving customer satisfaction. The followings are them.

A) Project the more comfortable image of station service department
B) Construct the system to share information about safety activities, human-error prevention activities and troubles with customers
C) Improve the employs' skill to deal with customers' troubles and human-errors

The practical activities are as follows. These are executed in station service department for two years.

A-1) Check list on appearances/clothes of employees

Employees check their appearances, clothes and attentions to customers by themselves according to the checklist once a month. This checklist also can give employees' manager to a criterion for the caution about employees' appearances. It enables station service department to give comfortable and steady image for customers.

A-2) Refreshing greeting

In morning rush hours employees and managers give refresh greeting for customers together at ticket gate once a month. It aims not only to give comfortable image to customers but also to cultivate the employees' habit of greeting and to establish good communication between employees and managers. So this activity is related to purpose B and C, too.

A-3) Publication of "Subway News" for customers

This news delivers customers the information about employees' trainings and educations about customers' safety. It has been published intended to have customers feel more comfortable and relieved when they take trains.

B-1) Periodical meeting about troubles with station masters

These meetings are held to discuss about troubles, such as complaints and human-errors, which had been occurred in station service department once a month at least. The safety/service manager, station masters and some assistant station masters think and make out the way to deal with the troubles and recurrence preventions together.

It intends to share the detail information about troubles and make good communication between management level employees. And also it can help the station masters to improve their skill to deal with troubles, so it is related purpose C, too.

B-2) Publication of "Safety Service News" for employees

This news delivers employees the information not only about troubles occurred in station service department but also about the way to deal with the troubles and recurrence prevention which were decided in periodical meeting (B-1). It also provides useful information to improve customer service. The target for this news is employees not on high and middle management position.

It has been published intended employees to know the trouble information and the countermeasures and to lift employees' interest in safety activities.

C-1) The training session of human-error analysis

This session is aimed at station masters and some assistant station masters. In this training, they learn the way to analyze the latent causes of human-error and to make countermeasures. The countermeasures made in this session are like measures improving work environment, not like measures changing the consciousness of the employees.

Sometimes this session was held by experts who study human factors and who is not belong to station service department. Because it was expected that participants have come to be relieved of fixed idea by virtue of the outside instructors.

C-2) Studying session about the knowledges of other departments

This workshop is aimed at assistant station masters. The participants learn about other departments such as more detailed business content of crewman and the new function of the machines at wickets. This session is intended for the assistant station masters to get to appropriately provide direction for followers, even if the troubles concerned with other departments are happened.

4 RESULT & CONSIDERATION

Analysis of the efficient from twice surveys on the employees' basic understanding/motivation for human-error prevention action

To confirm the effect of safety activities described above, we surveyed the employees' basic motivation/understanding for human- error prevention action on 2011. The provided questionnaires were the same as which was provided on 2009.

First, we categorized the employees into 5 groups according to their total score of the 20 questionnaires and compared the result on 2011 with the result on 2009 in figure 6. As is obvious from the figure 6, the proportion of the middle level group decreased and the proportion of the very high level group increased instead. The proportion of the very low and low groups, however, was unchanged. It means that the safety activities have a very positive effect for middle and high groups but they have little effect for very low and low groups.

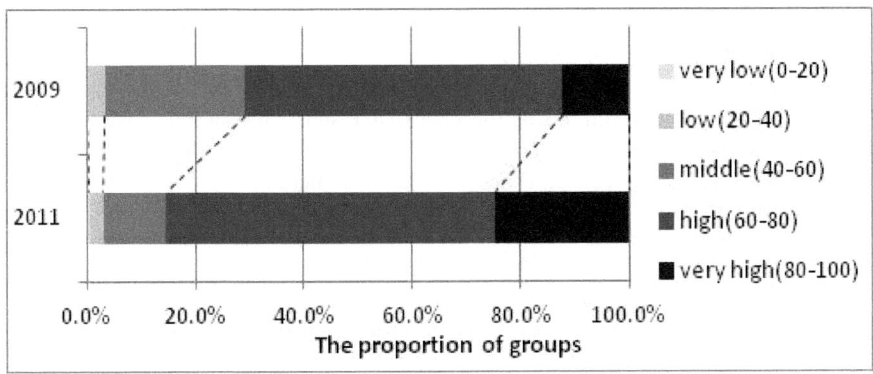

Figure 6 Proportion transitions of the five levels groups based on total score of 20 questionnaires which was haven on 2009 and 2011

Next, we compared the proportion of the group in each question which is selected as an issue at analysis phase in Figure 7. The employees are categorized into 5 groups according to his/her answer in each question. Figure 7 shows that the employees' understanding/motivation is improving in almost all groups. The number of employees who answered "Strongly Yes" in question 14 and 19, however, increased. It is inferable from this result that some employees understand the purpose of safety activities but can't understand the purpose of the human-error analysis and can't make good countermeasures without dependence on individual consciousness. So the training session of human-error analysis gets to be more important and the contents of the training should be more developed.

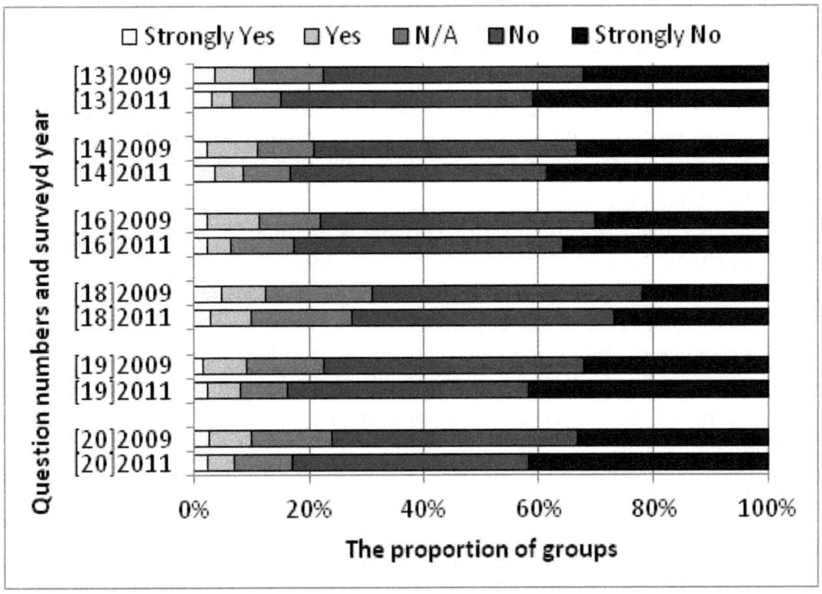

Figure 7 Proportion transitions of the each question numbers on 2009 and 2011

Analysis of the efficent from the result of the check list on appearances/clothes of employees

Figure 8 shows an example of the checklist result. The station masters give feedbacks, the result with his comments, to employees. From these result, it is revealed that employees' consciousness for the customers is improved. Almost all questions' result is improved.

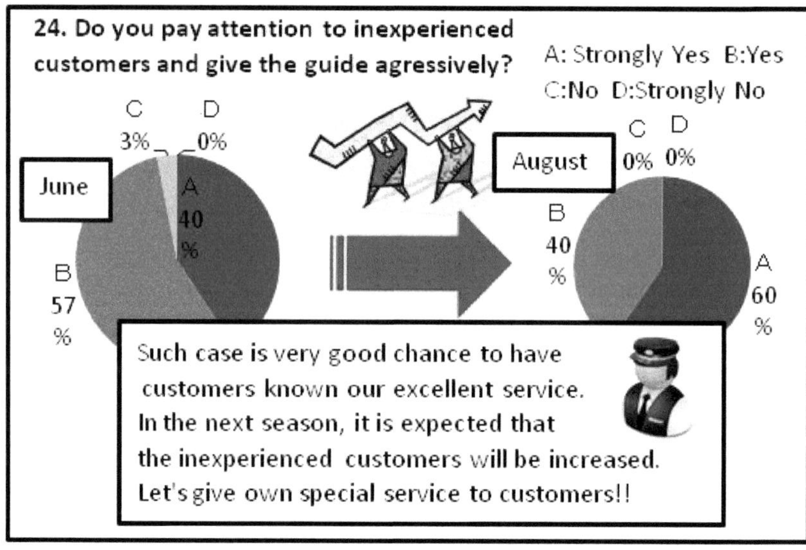

Figure 8 Example of the feedback on the result of the checklist

Analysis of the efficient from the number of complaints

The number of complaint which is related to handling customers is shown in Table2. According to Table2, the number of complaints about attitude and appearance decreased. It may be because of the effect of the safety activities such as checklist and refresh greeting.

On the other hand, the number of complains about wrong handling, negative attitude and announcement in the accident cases especially increased. We assumed that the latent causes of wrong handling have not been come out and managers have not been able to take efficient countermeasures. Combined with the increasing number of complaints about announcement in the accident case, we assumed that the way to response to customers appropriately has not been permeated in station service department. So safety activities related to purpose 'B' and 'C' get to be more important.

Referred to the safety activity related to purpose 'A', it is certain that the employees have developed their basic attitude for the customers (from the number of complaints about attitude and appearance). So it is better that the activity which enables the employees to achieve their aggressive attitude is executed.

Table2 Number of the complaints related to handling customers

		2009	2010	2011
Response to customers	Misguidance	3	5	4
	Lack of explanation	7	12	4
	Negative attitude	6	8	15
	Impolite language	2	4	4
	Wrong handling	23	25	35
	Lack of thanks	1	1	3
	Mishandling	3	8	3
Announcement	Announcement in the accident case	2	2	10
	Announcement at platform	4	7	7
Attitude	Chatting	1	2	3
	Arrogant attitude	7	5	1
	Dereliction of duty	14	9	4
Appearance	Hair	0	0	0
	Clothes	1	2	0

5 CONCLUSION

The survey for the employees' basic motivation/understanding for the human-error prevention activities revealed the issue on the station service department. We developed the safety activities, placing great value on the employees' consent. We have closely discussed about activities with managers and field workers on each section before the safety activities were executed. And then safety activities were executed for two years. The analyses confirmed the efficient. We were received the report there is another advantage such as the increasing number of near-miss incident report, the aggressive suggestion of project to improve environmental from field workers and the aggressive discussion of the safety activities in the working places. Next we intend to collect the customer's voice and to plan the activities.

REFERENCES

[1]Takeo YUKIMACHI, Reference List on Basis of GAP-W Concept and a Case Study, Human Factors, Vol. 9(1), 2004-07, pp. 46-62.

[2] Tomita Masahiro, Okada Yusaku, 2010, Developing a Method to Evaluate the Employee Satisfaction on Safety Management, AHFEI the 3rdd International Conference

[3] Okada, Y., 2005, Human Error Management of Performance Shaping Factors, International Conference on Computer-Aided Ergonomics, Human Factors and Safety

CHAPTER 6

Human Reliability of Chemicals Trucks Filling Process

Chih-Wei Lu, Chien-Yu Wu*

Department of Industrial & System Engineering,
Chung Yuan Christine University
Chun-Li City, 320 Taiwan

ABSTRACT

The human reliability is a major part to reduce industrial accidents. The goal of this study is to discover the construction and factors of relationship between human reliability and un-safety behaviors. The methods will be used are in-plant survey, job analysis, errors analysis and questionnaires. The results of this study shows 83.3% people known that miss would happened during the work processes. There were 43.3% workers thought they had made miss after they had done their jobs, 20% workers made miss during the last 3 months, and 80% workers had heard colleague talked about miss in the chemical factory. About what type of jobs were problem maker, 33.0% thought forklift transport process had highest miss rate, the next high (30.0%) was the operation of tanker processes, and 23.3% was the operation of pouring process.

Keywords: human reliability, chemical truck, filling process

1 INTRODUCTION

The human reliability is a major part to reduce industrial accidents and the proportion of human error has the tendency to rise gradually. Because the chemical industry occur injures with quite high severity degree in Taiwan and for every industrial country in the world, Especially, the factories don't attach great importance to human error in Taiwanese and only few references or information can be found, so we aim for relationship between human reliability and un-safety behaviors, to build the database of human errors information in chemical industry is necessarily and urgently.

The goal of this study is to discover the construction and factors of relationship between human reliability and un-safety behaviors. First, to investigate the operation of tanker with Allyl Alcohol and the operation of pouring Vinyl Acetate in the Dairen chemical factory, There are simple human machine interfaces in these operation zone, Including forklift transport objects, the operator controls and the movement of the operation of tanker.

2 METHODS

The methods will be used are in-plant survey, job analysis, errors analysis and questionnaires. First, working standard operation procedure, then use HTA (Hierarchical Task Analysis), pick up the error junction based on the SHERPA (The Systemic Human Error Reduction and Prediction Approach) that classified from the error taxonomy, then based on the Ramussen principle of motion levels of prescription.

3 RESULTS & CONCLUSIONS

The results of this study shows 83.3% people known that miss would happened during the work processes. There were 43.3% workers thought they had made miss after they had done their jobs, 20% workers made miss during the last 3 months, and 80% workers had heard colleague talked about miss in the chemical factory. About what type of jobs were problem maker, 33.0% thought forklift transport process had highest miss rate, the next high (30.0%) was the operation of tanker processes, and 23.3% was the operation of pouring process.

Based on the SHERPA that classified from the error taxonomy, for the operation of tanker with Allyl Alcohol, action errors has occupied a major part (39/50), the next is checking errors (8/50) and the last is selection errors (3/50). For the operation of pouring Vinyl Acetate, action errors has occupied a major part (20/24), the next is selection errors (4/24). For the both, action errors has occupied a major part (59/74), the next is checking errors (8/74) and the last is selection errors (7/74). Based on the Ramussen principle of motion levels of prescription, for the operation of tanker with Allyl Alcohol and the operation of pouring Vinyl Acetate, The major part is skill-based (34/74), the next is ruled-based (33/74) and the last is knowledge-based (7/74).

ACKNOWLEDGEMENTS

This study has been supported by national Science Council (NSC 99-2221-E-033-070-MY3), and College of Electrical Engineering and Computer Science Chang, Chung Yuan Christine University (CYCU-EECS-9901).

REFERENCES

Annett, J. (2003). "Handbook of Task Analysis in Human-Computer Interaction." Handbook of Task Analysis in Human-Computer Interaction,Lawrence Erlbaum Associates, Mahwah, NJ. Bell, J. and J. Holroyd (2009). "Review of Human Reliability Assessment Methods." Health and Safety Executive.

Embrey, D. (1986). SHERPA: A systematic human error reduction and prediction approach.

Ferry, T. S. (1982). Safety management planning, Merritt Co.(Santa Monica, Calif.).

Fishbein, M. (1979). "A theory of reasoned action: Some applications and implications." Nebraska Symposium on motivation 27: 65-116.

Harrison, M. (2004). Human error analysis and reliability assessment.

Heinrich, H. W. (1959). Industrial Accident Prevention. New York, MC Grew-Hill.

Hollands, J. G. and C. D. Wickens (1999). Engineering Psychology and Human Performance, Prentice Hall.

Hollnagel, E. (2005). "Human reliability assessment in context." Nuclear Engineering and Technology 37(2): 159.

Kirwan, B. (1996). "The validation of three human reliability quantification techniques--THERP, HEART and JHEDI: Part 1--technique descriptions and validation issues." Applied Ergonomics 27(6): 359-373.

Peters, G. (1966). "Human error: analysis and control." Journal of the ASSE 11(1).

Petersen, D. (1984). Human-error reduction and safety management. Deer Park, New York, Aloray.

Peterson, D. (1978). Techniques of safety management. New York, McGraw-Hill.

Ramsey, J. D. (1985). "Ergonomic factors in task analysis for consumer product safety." Journal of Occupational Accidents 7(2): 113-123.

Rasmussen, J. (1982). "Human errors. A taxonomy for describing human malfunction in industrial installations* 1." Journal of Occupational Accidents 4(2-4): 311-333.

Rasmussen, J. (1986). "Information Processing and Human-Machine Interaction. An Approach to Cognitive Engineering."

RBDM (2001). Risk-based Decision-making Guidelines: Applying Risk Assessment Tools-Chapter 12-Event Tree Analysis. Procedures for Assessing Risks, U.S.Department of Homeland Security

Reason, J., D. E. Broadbent, et al. (1990). Human factors in hazardous situations, Clarendon Press/Oxford University Press.

Rogers, A. E., W. T. Hwang, et al. (2004). "The working hours of hospital staff nurses and patient safety." Health Affairs 23(4): 202.

Ruckart, P. Z. and P. A. Burgess (2007). "Human error and time of occurrence in hazardous material events in mining and manufacturing." Journal of hazardous materials 142(3): 747-753.

Scott, L. D., A. E. Rogers, et al. (2006). "Effects of critical care nurses' work hours on vigilance and patients' safety." American Journal of Critical Care 15(1): 30.

Surry, J. (1968). Industrial Accident Research: A Human Engineering Approach, University of Toronto, Dept. of Industrial Engineering.

Swain, A. D. and H. E. Guttmann (1983). Handbook of human-reliability analysis with emphasis on nuclear power plant applications. Final report, Sandia National Labs., Albuquerque, NM (USA).

Tarrants, W. E. and A. S. o. S. Engineers (1988). Dictionary of Terms Used in the Safety Profession, American Society of Safety Engineers.

CHAPTER 7

An Improvement of Staffs' Motivation/Satisfaction for Safety Activities in a Railroad Operation Department

-Based upon investigation on staffs' consciousness /understanding about safety management for 3 years (2009-2011)-

Ryo Nihongi, Yusaku Okada

Graduate School of Science & Technology / Keio University
3-14-1 Hiyoshi Kohoku-ku Yokohama 223-8522 Japan
nicopo@z5.keio.jp

ABSTRACT

In railroad companies, to realize the ideal safety, human error management is critical issue that should not be neglected. Human error triggers various accidents, threatens the security of staffs, and blocks providing secure service.

However, it is so difficult to progress the human error management smoothly. One of important problems is a gap of the value sense on the safety activities between managers and staffs. When the gap grows larger, the staffs cannot understand the intension of several safety activities. Furthermore, the staffs' motivation become lower, and the effects of the activities are shrunk.

Therefore, based upon basic concept of PDCA management cycle, we aimed to propose strategies that can bridge a gap of the value sense in a railroad operation department. The contents of four stages in PDCA cycle to improve staffs' motivation/ satisfaction on the safety activities are considered.

In some case studies, we obtained following problems; insufficient communication in the team, vague knowledge about safety, 'Only I am all right' consciousness, too strong professional awareness, and so on. Against these problems, throughout discussion with a chief manager and a director in the railroad company, we proposed some strategies.

As a result of having executed some strategies, gaps of the value sense between managers and staffs became small, and various safety activities were activated. The effects of the proposed PDCA management cycle were also confirmed by plural questionnaires for staff's in an operation department. In the future, we will continue to investigate the effects of the actions, and will review the details of the actions.

Keywords: Human error management, Human reliability, Safety in Railroad

1 INTRODUCTION

In railroad companies, to realize the ideal safety, human error management is critical issue that should not be neglected. Human error triggers various accidents, threatens the security of staffs, and blocks providing secure service. Inside the organization, human error management leads to improvement of the staffs' value/quality. Outside the organization, human error management leads to improvement of social value/status.

However, it is so difficult to progress the human error management smoothly. A particularly important problem is a gap of the value sense for the safety activities between managers and staffs. When the gap grows larger, the staffs are not able to understand the intention of safety activities. Furthermore, the staffs' motivation for safety activities become lower, and the effects of the activities are shrunk.

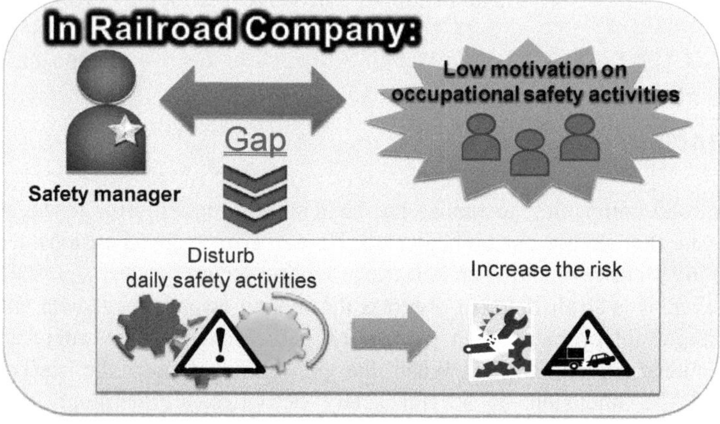

Figure 1 Influences by staffs' low motivation on occupational safety activities

Therefore, we aimed to propose strategies that can bridge a gap of the value sense for the safety activities between managers and staffs. In this study, an evaluation object is an operation control center (OCC) in a railroad.

2 METHOD

Based upon basic concept of PDCA (Plan, Do, Check, Action) cycle, we considered strategies that can improve staffs' motivation /satisfaction for the safety activities between managers and staffs.

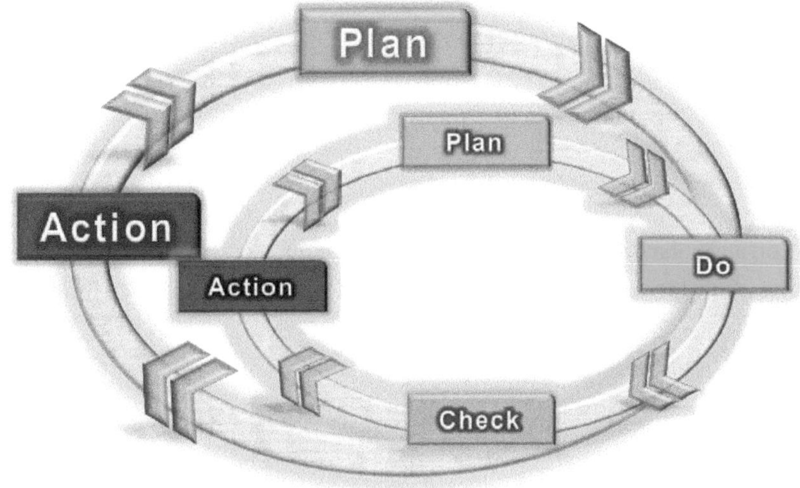

Figure 2 PDCA cycle to improve staffs' motivation/ satisfaction for the safety activities

2-1 PDCA CYCLE TO IMPROVE STAFFS' MOTIVATION/ SATISFACTION FOR THE SAFETY ACTIVITIES

The contents of four stages in PDCA cycle to improve staffs' motivation/ satisfaction for the safety activities are described as follows.

- *Plan-step*

In this step, the plan for safety activities is decided based on the result of investigation of staffs' consciousness/ understanding about safety management.
This investigation had been developed in our past study [1]. It uses a questionnaire consists of 55 question items. It investigates workers' understanding about human factors and current human error management in their organization. These question items evaluate condition of safety consciousness from 7 viewpoints as follows.

- ◆ "Fundamentals on human error" (question No.1 to No.20):

The depth of understanding on fundamental views about human error

- "Atmosphere" (question No.21 to No.25):
  The mental easiness to report when the person makes an error
- "Inspection system" (question No.26 to No.35):
  Devices on inspection system to prevent human error
- "Trouble research" (question No.36 to No.40):
  The contentment on trouble research
- "Prevention against reoccurrence of troubles" (question No.41 to No.45):
  The fulfillment on prevention against reoccurrence of troubles
- "Prevention system" (question No.46 to No.50):
  The completion of analyzing various incidents including potential incidents to prevent a future accident
- "Strategy" (question No.51 to No.55):
  The strength to confront human error as an organization

By investigating the reply for this questionnaire, we evaluated the respondent's attitude on each question items as a score out of 100 points. All people who belong to the organization should reply this questionnaire. To extract the organization's character from the gained scores, we defined a value called "low consciousness group score". This value evaluates the weight of the people who gave low conscious answer.

For example, the plan for safety activities is decided as against problems of whole organization based upon the investigation. The strategy of whole organization is defined based on the trend of whole organization. Also the strategies in each department are defined based on the trend of each department. Then the transparency of organization is evaluated with paying attention to a gap of the value sense between each job title based on the trend of each job title.

At this time, various factors which make a gap of the value sense for the safety activities are analyzed in each question in questionnaire's seat by root cause analysis (RCA). Then what to do to remove those factors is considered.

- *Do-step*

In this step, the plan for safety activities made in Plan-step is implemented by first-line staffs. It is important to use trial and error approach without considering the efficiency of the plan for safety activities and enable first-line staffs to review the plan for safety activities based on their subjective view. If the plan for safety activities didn't acquire staffs' satisfaction, it is necessary to take measure that remove factors which prevent the plan from acquiring staffs' satisfaction.

- *Check-step*

In this step, managers check and assess the efficiency of the plan for safety activities. To shrink the term of the plan for safety activities, the assessment system should be slim. And the method to announce the assessment should enable the result of this feedback to acquire staffs' satisfaction and understanding.

- *Action-step*

In this step, the investigation of staffs' consciousness/ understanding about safety management is retried by same questionnaire. Based on the trend of whole organization, the achievement of strategy in whole organization is assessed. And based on the trend of each department, the achievement of goal as each department is assessed. Then, the transparency of the organization is evaluated from result of each job title.

3 CASE STUDY

In order to verify the efficiency and validity of the proposed PDCA management cycle, we applied into an operation control center (of a Japanese railroad company). In this paper, we called the operation control center as OCC.

Here, we showed the safety management that was done in the target OCC, and observed the change of the staffs' consciousness on safety activities in OCC.

- *Plan-step*

We decided the plan for safety activities in 2011 based on the result of investigation of staffs' consciousness/ understanding about safety activities. Fig 3 shows the result of investigation implemented in 2009 and 2010.

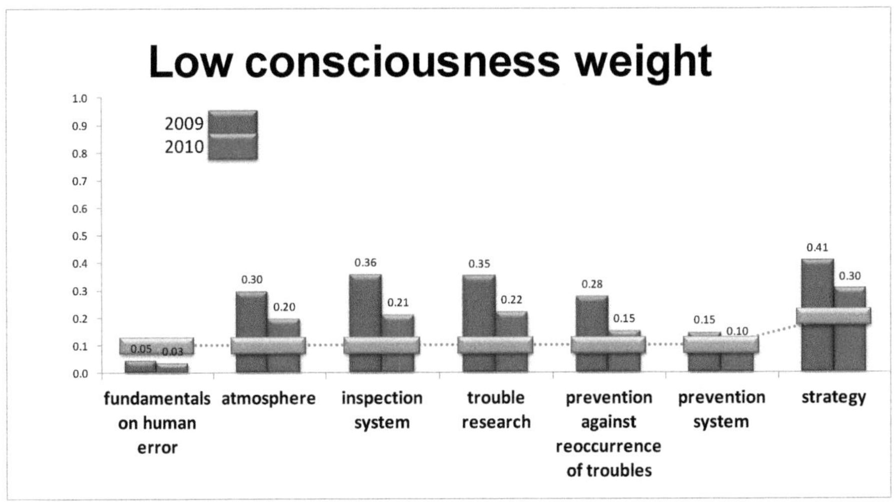

Figure 3 the result of investigation in 2009 and 2010

We focused attention on the low consciousness weight on viewpoint of "Fundamentals on human error" and "Atmosphere".

Then, by using RCA, we obtained following problems; insufficient communication in OCC, vague knowledge about safety activities, 'Only I am all right' consciousness, too strong professional awareness, and so on. Against these problems, we discussed with a chief manager and a director in a railroad company, and some strategies were proposed. Some strategies are as follow:

(1) "Publication of weekly safety newsletter"

A chief manager sends weekly newsletter about safety activities for all staffs. Newsletter gives various information and intention about safety activities to staffs.

(2) "Improvement of work environment based on staffs' requests"

Managers set up a questionnaire system that gleans information on staffs' various frustrations about work environment. Based on the information, managers improve the work environment.

(3) "Enforcement of a periodical group meeting"

Managers enforce a periodical group meeting that all staffs and managers participate in. In group meeting, participants discuss about safety activities and study by using passed accidents.

- *Do-step*

Those plan for safety activities were implemented by first line staff. In order to change staffs' 'Only I am all right' consciousness, the principle and the specific way of action in OCC are established. The principle and the specific way of action are voted from all staffs including managers in OCC. The established principle in OCC is "Cooperate mutually and operate exactly in emergency!" And the specific way of action is "Do thoroughly [Confirmation by pointing and calling] and [Reconfirmation by repetition]". Particulars are as follow:

[Confirmation by pointing and calling]

When you operate something, you should confirm the contents of your operation by finger pointing and verbal expression.

[Reconfirmation by repetition]

When you are prescribed some operation, you should reconfirm by repetition about the contents of your operation.

- *Check-step*

A consultant of human factors and managers discussed and assessed the contents and the efficiency of safety activities. Based on the assessment, the plan for safety activities were reviewed and adjusted appropriately to situation of OCC.

This questionnaire survey targeting all staff in OCC was carried out to verify the specific way of action had been executed. The result of this questionnaire survey showed that the specific way of action had become executed progressively. From this result, we gathered that staffs had become understand the importance of safety activities and staffs' motivation had become improved.

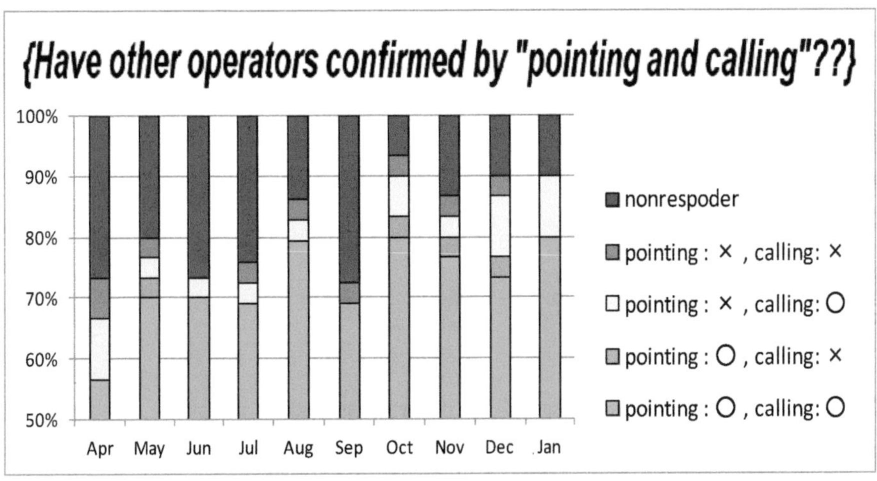

Figure 4 the result of investigation about "Confirmation by pointing and calling".

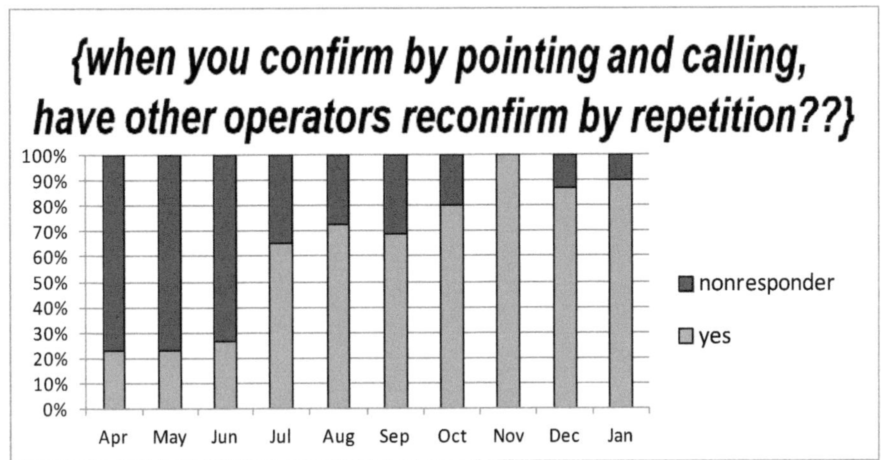

Figure 5 the result of investigation about "Reconfirmation by repetition".

- *Action-step*

As a result of having executed these actions in 2011, staffs' understanding and motivation about safety activities became improved. This graph shows the changes of value sense. In addition, a severe incident has not occurred since the management activity started.

These measures on improvement of staffs' motivation about safety activities have been executed for one year. Various effects were obtained. Additionally, other train operation sections in the railroad company have referred the measures that executed in OCC, and the motivation in other train operation sections is increasing.

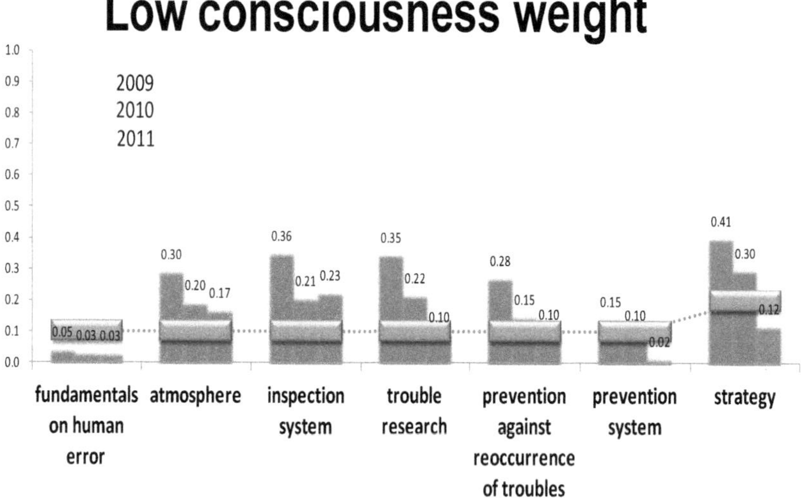

Figure 6 result of investigation in 2011

4 CONCLUSION

We proposed the safety management activities supporting method. As a result, a gap of the value sense for the safety activities between managers and staffs could be reduced, and various safety activities were activated. Especially, advanced strategies to increase human reliability of many operations in an OCC have introduced. The motivation about safety activities is increasing. In the future, we will continue to investigate the effects of the actions, and will review the details of the actions. Additionally, we wish to contribute to establish the high safety culture in not only OCC but also other train operation sections.

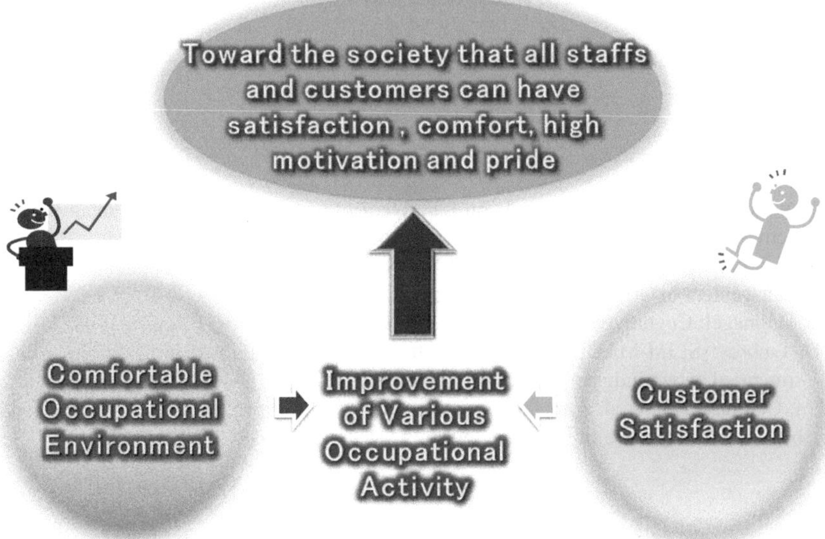

Figure 7 purpose of safety management on Human Factors

Figure OCC in target railroad company

REFERENCES

[1] R. Nihongi and Y. Okada, A Study on Improvement of Staffs' Satisfaction about Safety Activities in Medical Field, ICAP2010 in Korea, pp87-91, 2010

[2] A. Yokomizo, M. Tomita, Y. Mori and Y. Okada: A Study on Improvement of Safety Management in Organizations, Nordic Ergonomic Society, 2009

[3] E. Hollnagel: Cognitive reliability and error analysis methodology, Elsevier, 1998

[4] D. Embrey: SLIM-MAUD: An approach to assessing human error probabilities using structured expert judgment, NUREG/CR-3518, USNRC., 1984

[5] D.I. Gertman, D.I., et al: A method for estimating human error probabilities for decision-based errors. Reliability Engineering and System Safety, 35, 127-136, 1992

CHAPTER 8

New Forms of Education as Part of Effective Management of Occupational Safety and Health at Work

Karol Habina, Jan Donic, Andrea Lezovicova

BOZPO, s.r.o.
Prievidza, Slovakia
karol.habina@boz.sk

ABSTRACT

Education and training of employees is a fundamental principle of security measures. Trained and competent staff, sufficiently motivated to comply with OSH is a prerequisite for successful business. Organizations should have effective procedures for ensuring the competence and the competence of workers so that they are able to safely and properly organize and carry out their work.

Currently training in OSH carried out mainly by conventional traditional teaching by the characteristic structure of dialogues, lectures, seminars and typical technology such as projector and blackboard. New trends such as e-learning education form to wade into the foreground. Benefits of e-learning are:

- time saving
- costs saving
- the possibility of election to self-pace
- interactivity
- amount of visual perception, Example of images, sounds
- hyperlinks - links to other relevant information, legislation

The content of e-learning must be adapted to specific client requirements with regard to the internal regulations of the company. Individual work of participants is based on the study with e-learning modules through self-study. Participants have the opportunity to consult problematic issues and uncertainties with professional consultants. All approaches, signing and activities of students can be monitored on the basis of statistics.

People learn best when the subject matter lively, relevant to their needs and if it is consistent with the way how people learn._Active participation in the learning process in the interactive environment is important. E-learning form of education in OSH can have numerous negative effects to ensure the effective provision of information in this area. Negative effects can be very effectively eliminated by perfect knowledge of education possibilities for implementing e-learning format. If the service provider cannot do this elimination, this new trendy form of education may be replaced by citation:

"If people could not speak, they would found another way of harm"
P.F.

Keywords: health and safety, education, training, security measures, e-learning, new trends, workers, system of management

1. INTRODUCTION:

By improving security and health at work has a significant economic importance, because they address issues related to security and health protection, with creating favorable working conditions and working relations to optimize the work process and positive economic effect. It brings a reduction in losses, higher productivity, efficiency and quality of work– means greater prosperity of the company and thus also the whole society. The care of security and safety of health at work has also an important aspect of human, which presents cultural and social level of business and of state.

Industrial accidents, occupational diseases and inadequate working conditions are essentially due to improper organization of work. To ensure the continued prosperity of the company it is important, that the steering mechanism was introduced, which ensures proper functioning of the business entity. As in other areas of business management, and in area of safety and health at work it is necessary to establish an effective management system. As with the general management model, and also in the OSH management systems the starting point – formulation of corporate policy of OHS strategy.

The next steps are:

- planning
- organizational security (training, capacity and building of awareness)
- control and evaluation
- measures to improve

The last step of the cycle is the basis for formulating new corporate policy objectives of the OSH to a qualitatively higher level and graduation of the next cycle. That is the principle of continuous improvement.

2. EDUCATION

Education and training of employees is a fundamental principle of security measures. Trained and competent employee, sufficiently motivated observe the principles of OSH is a prerequisite for successful business. Organizations should have effective procedures to ensure competence and competence of staff, to be able to safely and correct to organize and carry out its work. General aims of education and other forms of education for OHS in general is to provide employees with the necessary knowledge and actual instruction, information, develop habits, skills and attitudes necessary for safe performance of activity and safe behavior. The given objectives are achieved by various forms of training, briefing, training, exercises, creating model situations, organizing campaigns, information activities (such as posters, brochures, posters and other). Within the education is necessary to paid the attention all aspects of work-related, for example and psycho-social relations, stress and so on.

2.1. Principles

Employees must be qualified to perform tasks related to OSH. Eligibility is subject to appropriate training, education, training, gaining experiences and skills.

Therefore monitoring capabilities of each member of work team and related education must be part of the OSH management system.

The system of education for OHS requires:

- perform a needs analysis for training,
- set goals and objectives,
- define the content part of (program) individual training,
- identify the appropriate procedures and methods,
- provide specialist teaching facilities,
- ensure didactic-technical, spatial, temporal and other conditions.

The training is conducted in accordance with pre-defined plan and according to established schedules. Training shall be documented in the form of records.

During the specification of the curriculum is essential to build on from:

- filling of work (with describing the activities with an increased risk),
- defining roles and responsibilities,
- regulations, rules and guidelines and principles of OHS,
- results of risk assessment,
- assessment of individual employees performances and neuropsychological stress,
- analysis of occupational accidents, occupational diseases and other damage to health,

- OSH policy,
- OSH management programs (incentive program, program against alcoholic, etc.).

2.2. Fundamental breakdown of educational activities

- admission (part of also includes a briefing on the job place),
- refill (at changes of workload, transfer to other work, introduction of new technologies),
- set by binding regulations (professional drivers, electricians, crane operators, welders and other.),
- repeated (renewal of knowledge gained in previous training and complete knowledge of new facts), repeat interval training is the set of selected professions and workplaces by special regulations, for other professions are required under the Act about OSH up to 2 - annual interval,
- particular, which have been determined as necessary, for example: on the basic of work accident,
- after the long-term absence of employee (especially after the injury and illness).
- additional training and participation in educational events (seminars, workshops and others.)

Breakdown under of target groups of employees /other people/

- heads of employees
- professional workers for OHS,
- workers in the professions, who must be trained in OSH

(electricians, crane operators, servicers of pressure systems, motor trucks etc.),

- representatives of employees,
- other staff,
- temporary workers, contract partners, suppliers, visitors, tours and like.

The effectiveness of training and the resulting level of competence should be evaluated. This may include evaluation as part of training exercise and / or appropriate checks "in the terrain", to the determine, whether achieve the required capacity, as employees apply their knowledge in practice. The results of evaluation of training effectiveness have lead to a possible revision of content, extent, if forms of education employees.

2.3. Inputs

- law regulations, norms, internal regulations, procedures, operating instructions and other requirements OHS;
- professional literature from relevant area
- policy of OHS, objectives and programs of OHS;
- definitions of roles and responsibilities;

- job descriptions (with describing the details of hazardous tasks, that is to be done);
- results of risks assessment;
- technical and technological documentation

2.4. Outputs

- required capacity for the performance of individual tasks;
- analysis needs training;
- training programs, plans for individual employees;
- range of training courses, products available for use within the organization;
- training records and performed verifications of knowledge;
- records from evaluation of training effectiveness.

3. NEW TRENDS

Currently training in OSH takes place mainly through the conventional traditional teaching, which are characteristic by structures of dialogues, lectures, seminars and typical technologies are projector and blackboard. However, at the forefront are plowed new trends namely distance learning at a distance and virtual learning. When deciding on a particular form of education (traditional teaching in the classroom, teaching at a distance using a fax, phone, email, teaching at a distance "face-to-face" (face to face) through videoconferencing system), is necessary to consider all aspects of the technologies available (their benefits, options, availability, features, price, quality) and under these circumstances (number of trainees, the aim and content of education, need for communication, availability of study materials) for a particular form of education and technology determine so, that learning objectives are met to the maximum satisfaction of lecturers and trainees. One of new trends is E – learning, which has considerable potential.

Benefits of e-learning education are:
saving time
saving of financial costs
the opportunity to choose their own pace
interactivity
the amount of visual perceptions, examples supplemented by pictures, sound effects,
hyperlinks - links to other relevant information, legislation.....
a new dimension to the evaluation - Monos test their skills, to determine their level of knowledge

The content of e learning education can be customized according to specific client requirements with regard to the internal regulations of company. The basis of study is individual work of parties training with e-learning modules through self-study. However, participants of course have possibility to consult problematic issues and uncertainties with consultants. The purpose of this course is to acquaint employees and heads of employees with the legislative and other regulations for compliance with health and safety at work. Most of the courses presented in the market is divided into several topics (modules) with an explanation and illustrative animation sand is complemented by related legislation with the possibility direct input from various parts of the text, each lecturer has access to the statistics of their courses and students. Based on statistics is possibly to watch all the approaches, logging and activities students. Shown also is the total time spent in e-learning system and also the time spent in a particular course - courses. Using statistical interface has a lecturer complete overview of the activities of his students. The curriculum is designed by the form of sequence. Each new chapter is accessible only after watching the previous. All chapters are completed by the control issue.

Questions used in courses:

a)yes / no

b)one correct answer from several options

c)more than one correct answer from several options

d)the coordinating issues –for example: assign the correct values - graphically processed – "pulling by the mouse "

e)graphic questions–for example: click on the correct designation

After correctly answering questions are the following chapters available according to specified order. Communication of students with lecturers is provided by sending internal reporting in system.

People learn best when, if is the subject matter alive, relevant to their needs and if is in accordance with the method, how to learn. Important is active participate on the learning process in the interactive environment. E - learning form of education in area of OSH, can have numerous negative effects in ensuring the effective provision of information in subject area. They may, however, very effectively eliminate, by perfect knowledge of the possible implementation of education through e-learning form. If the service provider cannot handle this elimination, can this new trend form of education move into position of fulfill the quote:

“If people could not speak, they certainly would have found another way, that they would hurt”

P.F.

Documents are processed by electronically and they provide clear information without interference:

-logical sequence of training by chapters

- active links to current legislation / direct links to a collection of laws /
- possibility to communication with supervisor by using the internal reporting in system
- ongoing verification of knowledge through the control questions
- use of effective tools, especially videos, animations, simulations, audio mix and text interpretation

Despite of all the positive, what provides an electronic form of training in the area OSH, is due to the specificity of this issue subject form of education should complemented by traditional methods directly on work places of employers. The target user group is although naturally limited by the necessity of technical equipment, but development of information in direct proportion it expands, thus the utility of this attractive form of education constantly continues to grow.

REFERENCES

National labor inspectorate - The rules of good practice OHS 2002, Management system for safety and health at work: The publication 2, April 2002

CHAPTER 9

BBS Program (Behavior Based Safety) – Way to Raising the Level of Health and Safety in Practice

Pacaiova, Hana., Nagyova Anna., Markulik, Stefan

Technical University of Kosice, Faculty of Mechanical Engineering
Department of Safety and Quality production,
Letna 9, 042 00 Kosice, Slovakia
hana.pacaiova@tuke.sk

ABSTRACT

The increase of the safety level depends on multiple factors. The legislative support, management awareness, staff's knowledge levels, and last but not least, the willingness and competencies of the organization's employees are indicators that significantly affect the successful implementation of new approaches. The creation of the environment where the occupational health and safety becomes matter-of-course is related to mutual relationships between management and employees. Therefore, stress should be laid not only on proper motivation but mainly on mutual communication. The functionality of numerous implemented management systems fails due to the underestimation of the quality and quantity of information crucial for the identification of dangers and threats. The BBS program provides for the efficient monitoring of dangerous activities for each profession. However, like other approaches, if misinterpreted, this approach shall become a burden without any effects on increasing the level of safety at work. This article reflects on the efficiency and functionality of standard OHS systems and it describes the approach that is being gradually applied in this area.

Keywords: safety, behavior, process

1 BBS AS AN IMRPOVEMENT TOOL

There exist numerous tools for the improvement of occupational health and safety in organizations. Companies make their own choice of tools they will use to ensure occupational health and safety for their staff. One of the tools focusing on this aspect is Behavior Based Safety (BBS) program, whose aim is to observe behavior that may lead to the occurrence of an undesired event. It is a paradox that the BBS does not have its roots in industrial manufacture where it is currently being implemented most frequently. The history of BBS started in the area of services, where causes of accidents and injuries were analyzed statistically.

BBS originated with the work of Herbert William Heinrich [Al-Hemoud, Al-Asfoor, May, 2006]. In the 1930s, Heinrich, who worked for Traveler's Insurance Company, reviewed thousands of accident reports completed by supervisors and from these drew the conclusion that most accidents, illnesses and injuries in the workplace are directly attributable to "man-failures", or the unsafe actions of workers. Of the reports Heinrich reviewed, 73% classified the accidents as "man-failures"; Heinrich himself reclassified another 15% into that category, arriving at the still-cited finding that 88% of all accidents, injuries and illnesses are caused by worker errors [Semcosh Fact Sheet, 2004].Heinrich's data does not tell why the person did what they did to cause the accident just that accident occurred. BBS programs delve into the acts that cause the accident. It delves into the workplace; environment, equipment, procedures and attitudes [Al-Hemoud, Al-Asfoor, May, 2006].

In order to use the tool, an organization has to define the goal it wants to reach by the implementation of BBS. In general, BBS is an algorithm or a procedure that involves observation of staff. The aim is to identify **unsafe behavior** and its cause, and to suggest measures that will prevent its future recurrence. Scientific publications define BBS as:

- philosophy or attitude,
- platform that forms the organization culture,
- extension of safety-oriented system,
- observation-based method,
- tool aiming at identifying **unsafe** behavior that may lead to an undesirable event.

An element common to all of these approaches is the basic tool of their implementation – regular observations that reveal activities with high probability of undesirable events occurrence.

"The aim of BBS is to achieve reduction in the occurrence of occupational injuries to the lowest possible level. BBS does not reach this reduction by changes in technology or production. BBS brings changes in the behavior of staff when performing their work-related activities."

2 UTILIZATION OF BBS IN PRACTICE

The difficulty of the implementation of the program itself in a particular organization depends on the level of its preparation. It is advantageous if the organization uses various tools of safe practice, safety management and occupational health and safety (OHS) or has implemented OHS management system.

BBS does not require provision of new machinery or equipment, but focusing on the human factor and human safety at work. One of the negative aspects that are currently observed in workplaces is stress, which has a negative influence on work behavior. In consequence of this factor, the staff can develop a type of "compulsory urge" to perform their duties even at the expense of breaching the prescribed safe behavior policies. The 2011 statistics in an observed mechanical engineering company shows that the highest proportion of injuries was caused by incautiousness of staff (Figure1).

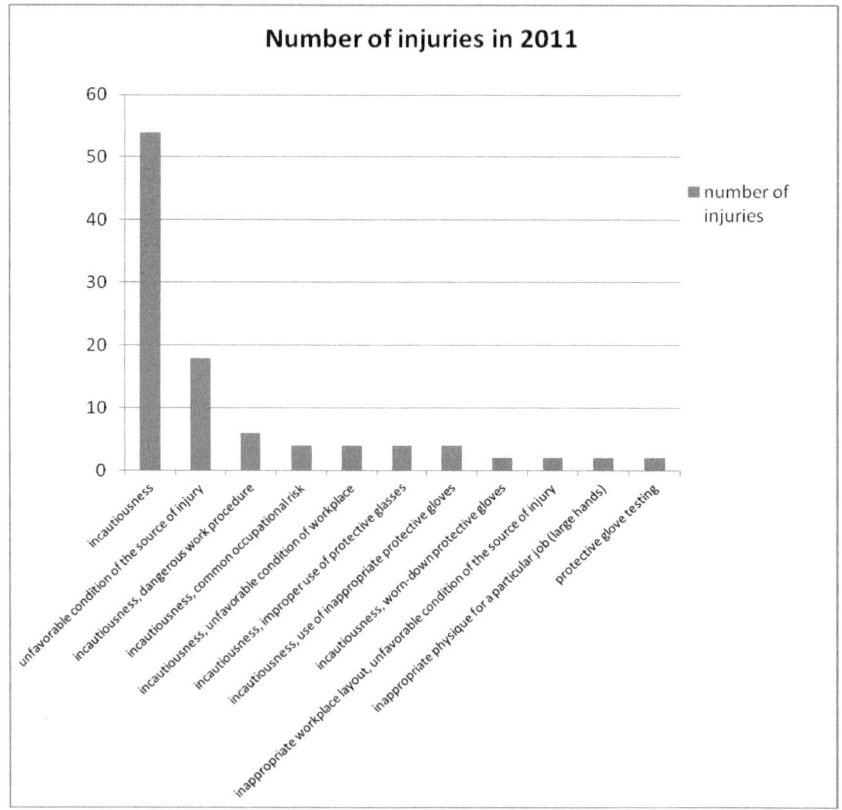

Figure 1 Number of injuries in a selected organization in 2011 [From Internal documentation of the industry company, Slovakia]

The injury statistics was performed in an organization that had implemented elements of safety management system, however, no safe behavior method or tool had been implemented in practice.

3 ALGORITHM OF IMPLEMENTATION BASED ON THE CONTINUOUS IMPROVEMENT PROCESS

Implementation of each system or method is based on a certain sequence of steps. The basic element of a system is a process, which is defined as a **set of activities using resources and management to enable the transformation of inputs into outputs.** During the transformation, certain resources are used in controlled conditions. However, the resources are not consumed in the process; they facilitate the process (e.g. people, technology). Factors that delimit the course of processes are known as regulators. These include laws, technical standards, regulations, internal documentation, etc.

The term "process" is often introduced in relation with "**process approach**". To understand this approach it is necessary to understand the interaction between different processes. The process approach is based on the principle of management and interaction of all processes of the organization to fulfill the specified purposes, such as safety, product quality.

The process approach involves identification, alignment and process control in such way that the output of one process is an input to the second process. This relation can be understood as a certain interconnection of processes (process chaining) (Figure 2).

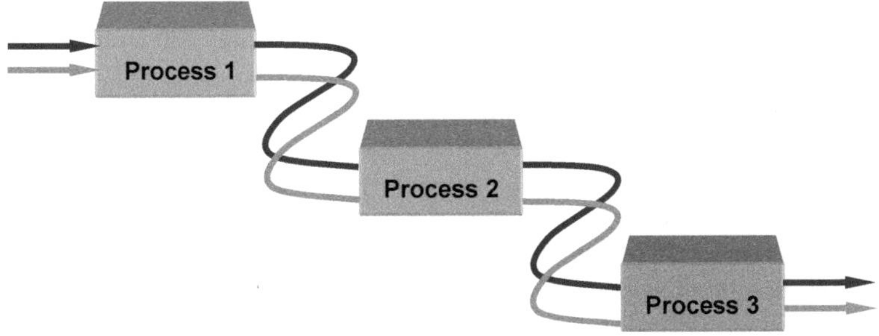

Figure 2 Illustration of the process approach (From Markulik, S., Nagyova, A.:Quality Management System, Faculty of mechanical Engineering, Technical university of Kosice, 2009. s. 79. ISBN 978-80-553-0306-2.)

The process approach gives more opportunities to achieve greater effect, since it focuses on the nature of processes. It can identify the key processes that are critical for the final effect, in accordance with business objectives. Due to its orientation on processes, the process approach eliminates stagnation and identifies opportunities for improvement.

If the BBS implementation is based on the process principle, it is possible to implement it according to the PDCA cycle – continuous improvement, which can be divided into 4 basic steps [Markulik, Nagyova, 2009] (Figure 3):

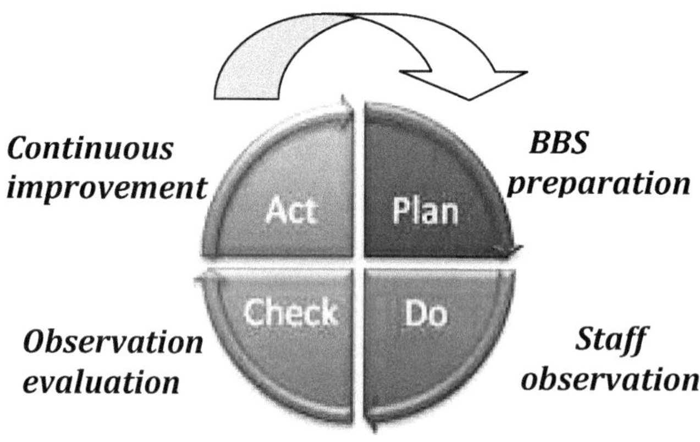

Figure 3 Basic phases of BBS program implementation (From Markulik, S., Nagyova, A.:Quality Management System, Faculty of mechanical Engineering, Technical university of Kosice, 2009. p. 79. ISBN 978-80-553-0306-2.)

3.1 BBS preparation

In the first phase it is necessary to identify aims and set the time schedule for the entire preparation (implementation), which will take into consideration the facts described in part 2.

The first step of planning is defining the work group that will be in charge of BBS implementation. The schedule for BBS implementation must be based on the analysis of the current state from the safety culture point of view. The result of the analysis has a great influence on the time span of the implementation. The basis of the BBS is the observation of staff and identification of unsafe activities. In order to make this observation objective and maintain its added value, the following need to be defined:

- work team of observers,
- number of observers,
- competences of observers (code),
- observation check lists,
- program (frequency and place) of observations.

3.2 Staff observation

The main aim of the observation is to concentrate on the behavior that increases the risk of a negative phenomenon occurrence. It should help the staff to think safe, so that their acts would always adhere to safe procedures. It encourages them to take personal responsibility for their actions that may affect their safety, as well as safety of or their workmates. The observation in an organization may be preceded by compiling a list of unsafe behaviors and their causes. The list should contain examples of unsafe behaviors, according to findings acquired from accident reports, interviews and observations. The list must be easily available and should express the iceberg principle. Its availability is crucial for its acceptance by the staff (Figure 4).

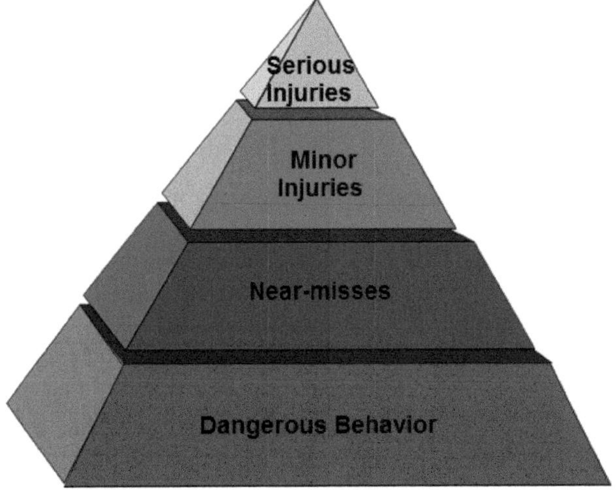

Figure 4 Iceberg principle (From Typcukova, D. The safety behaviour like a part of culture BOZP part on work place, Diploma Work, 2009).

The iceberg analysis identifies unsafe behaviors in order to recognize their causes. The data may be collected using information related to accident reports, occupational risks, interviews with employees or by brainstorming. In some cases, a combination of these tools may be used. The data may then serve as a basis for assembling a list or an algorithm of safe behavior, which should become a part of the operating procedure. Any deviations in the staff behavior during operation should be assessed by the observer from the viewpoint of safety. Should their behavior be considered risky, it is recorded in the checklist.

3.3 Observation evaluation

The observer must consult each case of unsafe behavior with the observed staff member. The employee must get a chance to explain their (unsafe) behavior. The

observer must follow the rules (as mentioned in 3.1) at observation or consultation with the employee in question. The interview provides the observer with immediate feedback, which makes it easier to identify the causes of the employee's unsafe behavior.
The feedback should be based on positive emotions (such as publicly expressed praise), which would lead the employees to adhere to the rules of safe behavior.

The evaluation of observation is the comparison of the actual and the expected behavior of staff. Operational procedures and instructions are very important in this process. If they are not appropriately elaborated, they can induce staff's unsafe behavior. One observation = no observation. Repeated observations provide a real picture of the level of safety (staff behavior) in a workplace. Multiple observations performed by several observers are also useful for elimination of observers' subjectiveness or prejudice against the staff. Summarization of the observation results is a step towards continuous improvement through defining corrective and preventative measures.

3.4 Continuous improvement

The feedback that occurs immediately after the observation between the employee and the observer can help to identify a problem – the cause of revealed unsafe staff behavior. It is still up to the observer to consider whether it was caused by individual failure or employee's nonobservance of instructions, or whether the cause has deeper roots. It is the analysis of the causes of staff unsafe behavior that allows for taking corrective and/or preventative measures.

"The BBS implementation requires especially a change in the staff's attitude, an ability to persuade them that BBS is a new tool that is not aimed at increasing their productivity, but a tool supposed to guarantee focus on safe behavior preventing injuries."

4 CONCLUSIONS

The implementation of BBS does not guarantee its effective functioning. An organization changes (develops) over the course of time – there are changes in technology, staff, operational procedures (instructions) - working conditions, which will influence the staff behavior. Continuous improvement of the system and reaching its aims requires regular observations and subsequent inspections of the changes introduced in the organization.

The advantage of BBS is that it does not have a prescribed implementation procedure. Instead, it can be customized to specific needs of an organization. The success factor for BBS is that the observers can dynamically use the knowledge of human behavior and keeping the assessment objective and impartial.

ACKNOWLEDGMENTS

The contribution is the result of the project implementation “Center for Research into control of technical, environmental and human risks for sustainable development of mechanical engineering production and products" (ITMS 26220120060), supported by the Research & Development Operational Program funded by the ERDF.

REFERENCES

Al-Hemoud, Ali M., Al-Asfoor, May M. (2006) "A behavior based safety approach at a Kuwait research institution." Journal of Safety Research, 37 (2) pp. 2001–2006.

Grencik, J., Pacaiova,H.: Fait with Maintenance crisis. In: Maintenance, 6, no. 1-2 (2008), p. 1,13-14,16. Internet: <www.udrzba.sk> ISSN 1336-2763.

IEC 60300-3-9: 2000, Dependability management - Part 3: Application guide - Section 9: Risk analysis of technological systems.

Internal documentation of the industry company, Slovakia.

ISO 31000:2009 Risk Management Principles and Guidelines.

Markulik, S., Nagyova, A.:Quality Management System, Faculty of Mechanical Engineering, Technical university of Kosice, 2009. s. 79. ISBN 978-80-553-0306-2.

McSween Terry, E. Value – Based Safety Process, Improving Your Safety Culture with Behaviour – Based Safety, Second Edition, 2003 a John Wiley & Sons, Inc., Publication, ISBN 0-471-22049-3.

Mikloš, V., Šolc, M.: Quality assurance by improving the human factor, Bezpečnosť - Kvalita - Spoľahlivosť : 5. International science conference: proceeding: Košice, 2011. - Košice : TU, 2011 S. 171-175. - ISBN 978-80-553-0612-4.

Pacaiova, H.: Maintenance Management 1.vid. Košice: TU v Košiciach, 2007, 127s. ISBN 978-80- 8073-751-1.

Pacaiova, H., Sinay, J., Glatz, J.: Safety and Risks of technical systems, SjF TU Košice, 2009, ISBN 978 - 80 -553 -0180-8.

Pacaiova, H., Sinay, J., Nagyova, A., Glatz, J., Balazikova, M.: Analyses of key performance indicators in gas industry Analyses of key performance indicators in gas industry In: New Technologies & Emerging Risks : 2nd iNTeg-Risk Conference : Dealing with multiple and interconnected emerging risks : June 14 - 18, 2010, Stuttgart, Germany - Stuttgart: Steinbeis-Edition, 2010. - ISBN 978-3-938062-33-3. - P. 459.

Semcosh Fact Sheet: Behavior Based Safety (2004).

Šolc, M., Girmanová, L., Petrík, J. : Process measurement performance of testing bar, In: Safety – Quality- Reliability: 5. International scientist conference Košice, 2011. - Košice : TU, 2011 S. 228-233. - ISBN 978-80-553-0612-4.

Typčuková, D.: The safety behaviour like a part of culture BOZP part on work place, Diploma Work, 2009.

CHAPTER 10

Using Fuzzy Set Theory to Model Resilience in Safe-Critical Organizations

Claudio H. S. Grecco[1], *Mario C. R. Vidal*[2], *Paulo V. R. Carvalho*[1,2]

[1]Instituto de Engenharia Nuclear - Comissão Nacional de Energia Nuclear
[2]Universidade Federal do Rio de Janeiro – COPPE/PEP
Rio de Janeiro, RJ, BRASIL
paulov@ien.gov.br

ABSTRACT

Resilience is the intrinsic ability of a system to adjust its functioning prior to, during, or following changes and disturbances, so that it can sustain required operations under expected and unexpected conditions. This definition focuses on the ability to function, rather than on being impervious to failure, and thereby overcomes the traditional conflict between productivity and safety. Resilience engineering has fast become recognized as a valuable complement to the established approaches to safety of complex socio-technical systems and methods to assess organizational resilience are needed. However, few, if any, comprehensive and systematic research studies focus on developing an objective, reliable and practical evaluation model for measuring organizational resilience. Most of methods cannot fully solve the subjectivity of resilience evaluation. To remedy this deficiency, the aim of this research is to adopt a Fuzzy Set Theory (FST) approach for resilience measurement in organizations based on leading safety performance indicators, defined according to the resilience engineering principles. The method uses FST concepts and properties to model the indicators and to assess the results of their application. To exemplify the method we performed an exploratory case study at the radiopharmaceuticals dispatch package sector of a radiopharmaceuticals production facility.

Keywords: fuzzy set theory, leading indicators, resilience engineering

1 INTRODUCTION

Resilience engineering emerges as a new paradigm of system safety, emphasizing that safe-critical organizations should be able to proactively evaluate work situations and human performance to keep safety under control avoiding accidents using a proactive approach that should be endorsed by all organizational organization levels. Recent research in safe-critical organizations findings indicates that safety emerges when an organization is willing and capable of working according to the demands of their tasks, and when they understand the changing vulnerabilities of their work environment (Hollnagel and Woods, 2006). Resilience performance indicators can be used to signal whether the system provides the capabilities for adaptation during the execution of work activities so that workers can handle the increased demand without sacrificing system safety. As the result, the organizations that use resilience indicators will be able to proactively evaluate and manage safety.

This paper presents a method for resilience assessment in organizations, based on: 1) the use of leading indicators in order to be able to monitor the effects of proactive safety work as well as anticipate vulnerabilities (instead of the normally used safety performance indicators based on lagging indicators, i.e., measuring outcomes of activities or things and events that have already happened); 2) the approach of resilience engineering in the development of indicators, which are based on six resilience engineering principles: top-level commitment, awareness, preparedness, flexibility, just culture and learning culture; 3) use ergonomic modeling techniques to fit the indicators according to the organizations activities; 4) the use of concepts and properties of fuzzy set theory to model the indicators and evaluation of results; 5) the development of a safety and resilience standard of a given organizational domain to serve as a baseline for assessments of safety and resilience of the domain. The method was applied for the identification potential problems in radiopharmaceuticals production process, indicating the level of resilience in some production phases, showing that it is a good tool to monitor the resilience of an organization.

2 MEASURING ORGANIZATIONAL RESILIENCE

The true challenge to measure resilience is to translate the trends of the six resilience principles listed above into observable actions – leading indicators – that can be monitored. The proposed method uses leading indicators related to top-level commitment, awareness, preparedness, flexibility, just culture and learning culture. The resilience measure must be made in direct relation with how a system performs, and how capable it is in monitoring and controlling performance throughout a given period. In this sense, Hollnagel and Woods (2006) claim that only the potential for resilience can be measured and not resilience itself.

Most of research on leading indicators is not directly related to measure organizational resilience. For instance, EPRI (1999) provides guidelines that helps organizational management to develop leading indicators predict possibilities of undesirable human performance according to nuclear industry safety culture criteria. Reiman and Pietikäinen (2010) present an overview on how to select and the effects of leading safety performance indicators in the domain of nuclear safety. Wreathall (2006) describes the importance to translate resilience engineering principles into leading indicators, but he does not say how.

A recent investigation of resilience in an electricity distributor in Brazil (Saurin, 2011) describes a method to identify sources of resilience and brittleness. The method adopted four resilience engineering principles (top management commitment, learning culture, flexibility, awareness), and it is based on the scoring systems and evaluation criteria adopted by the Brazilian Foundation for the National Quality. Although the method seems to be useful for auditing occupational health and safety management systems, the authors report some problems due the subjectivity of experts' opinions, scoring, and evaluations.

Our research aims to overcome some limitations of the methods described above. We propose a method that adopts leading safety performance indicators based on six resilience engineering principles to identify and measure sources of resilience. The method uses the Fuzzy Set Theory (FST) to cope with the subjectivity for attribution of indicators scores and its evaluation, and to calculate the consistency between experts' opinions to produce an ideal resilience pattern. This ideal resilience pattern will be compared with the workers evaluations, indicating how workers perceive the resilience during their daily/routine work.

2.1 FST for indicators modeling

FST provides an appropriate logical-mathematical framework to handle problems with the absence of sharply defined criteria, such as situations involving vague human concepts and reasoning, and an incomplete information because: 1) it deals with uncertainty and imprecision of reasoning processes and situations; 2) it allows the modeling of the heuristic knowledge that cannot be described by traditional mathematical equations and; 3) it allows the computation of linguistic information (Zadeh, 1996). FST is an extension of classical set theory where elements have degrees of membership. A generic element of the universe X, a fuzzy subset $\tilde{A}$, defined in X, is one set of the dual pairs: $\tilde{A} = \{(x, \mu\tilde{A}(x)) \mid x \in X\}$; where $\mu A(x)$ is the membership function or membership grade x in A. The membership function associates to each element x of X, a real number $\mu\tilde{A}(x)$, in the interval [0, 1].

FST deals with linguistic variables, a variable whose values are words or sentences in natural language, which can be represented as fuzzy sets that serve to describe vague reasoning, inherently imprecise and multidimensional concepts like safety and resilience. Furthermore, in the human discourse, variables are, normally, expressed by words, not by numbers. Thus, one advantage of using linguistic variables is that one can deal directly with semantic concepts of imprecise nature, with a consistent mathematical formulation. We can consider a leading indicator as a linguistic variable represented by set of four linguistic terms (Unimportant, Little

Important, Important and Very Important) to which correspond importance degrees used to assess the weigh this indicator by experts. These linguistic terms can be represented by triangular fuzzy numbers.

Typically the fuzzy reasoning model is performed on three stages. The first is a fuzzification stage, where crisp input data (either numerical or linguistic) is transformed in fuzzy sets. The second is the inference stage, where fuzzy rules are applied aggregating input data by means of fuzzy operators and fuzzy rules producing fuzzy output results. The deffuzzification is the last stage and it corresponds to the conversion of fuzzy results into crisp results to be presented to the decision makers.

We proposed a method to measure the organizational resilience based on the concepts of FST to cope the ambiguities in the evaluation of leading indicators. The first step of the method is the construction of a resilience ideal pattern for a given organization or organizational sector or department using experts' opinion. In this step, the leading indicators are linguistic variables represented by linguistic terms related to a set of linguistic terms represented by triangular fuzzy numbers. These triangular fuzzy numbers indicate the importance degree of each leading indicator. The process of expert opinions aggregation to build the resilience pattern uses the aggregation method developed by Hsu and Chen (1996). In the last step, the evaluation of leading indicators is performed by workers of the organization. The evaluation results are compared with the resilience pattern and then defuzzified using the center of area method. The result indicates the level of organizational resilience compared with the resilience ideal pattern. This method was applied in the dispatch package sector of the radiopharmaceuticals production service.

3. RADIOPHARMACEUTICAL PRODUCTION

The radiopharmaceuticals production service where the research was made is located in Rio de Janeiro, Brazil, and produces two types of radionuclides: ultrapure iodine-123 and fluorine-18, used as markers in several radiopharmaceuticals. Radiopharmaceuticals are radioactive product fulfilling all pharmaceutical requirements and intended for in vivo administration in humans with special purpose to carry out diagnostic or therapeutic procedure. Typical structure of a radiopharmaceutical has a carrier and a radionuclide. The carrier provides for the affinity to a special body tissue and radionuclide are intended to the purpose of detection only (diagnostics) or interaction with the tissue (therapeutics).

Radiopharmaceuticals production involves handling of large quantities of radioactive substances and chemical processing. The production includes two main processes: the production of the radionuclide, on which the pharmaceutical is based, and the preparation and packaging of the complete radiopharmaceutical as shown in figure 1.

Figure 1 Illustration of the main phases o radiopharmaceuticals' production. The irradiation in the isotope accelerator (no people allowed in the room), the manipulation of radionuclides in hot cells and finally de dispatch and monitoring of packages to be transported for hospitals radiotherapy services.

Before dispatching radiopharmaceuticals, each package must have a label indicating the maximum radiation level measured at its surface, according to the guidelines for radiation protection, to ensure that transport containers comply with regulatory requirements for transport of radiopharmaceuticals. Beside package and monitoring, the workers of the dispatching radiopharmaceuticals sector have to deliver the complete documentation for radiopharmaceuticals transportation. The team in the sector is composed by seven people (six men and one woman) with experience of over fifteen years in the activity of radiopharmaceutical dispatch. Sector activities are performed by three workers at weekly scales.

The activities at the radiopharmaceutical dispatch are dangerous due exposure and contamination risks to the workers while they handle radioactive substances. Moreover, the workers are faced with time pressure due to short-lifetime of the radiopharmaceuticals which ensures the least possible harm to the patient's body, but requires that it has to be used within a short period after production. An ergonomic work analysis (Grecco et al., 2010) indicated that the radiopharmaceuticals dispatch package has many sources of variability: seasonal variations of demand, variations resulting from production equipment's and radiation monitors' failures, different raw materials and inputs, changes in procedures, regulations and instructions. To cope with the situations variability and time pressure, workers' activities rely on ad hoc adaptations. In many situations, the violations of procedures were observed, however, most of time, they are viewed as appropriate actions, because they *make sense* to cope with time pressure and variability.

4. METHOD FOR RESILIENCE MEASUREMENT

The measurement of resilience using leading indicators and the FST concepts has the following main steps:

- Determination of the leading indicators framework based on six

resilience principles;
- Determination of a resilience ideal pattern;
- Evaluation of the resilience of the dispatch package radiopharmaceuticals sector based on resilience pattern.

4.1. Selection of leading indicators

The role of the leading indicators is to provide information on resilience and safety. Considering that there is no consensus about the best resilience indicator set, the selection of indicators was made according to the six resilience principles described in section 2, previous research in safe-critical domains (eg. EPRI 1999, 2000; Reiman and Pietikäinen, 2010), and based on ergonomic work analysis already made at the radiopharmaceuticals production (Greco et al., 2010). Table 1 list the some resilience principles and exemplify indicators and the metrics used to calculate their values.

Table 1 Resilience principles, indicators and metrics

Principles	Indicators	Evaluation (metric used)
1 Top-level commitment	1.1 Human Resources 1.2 ... 1.3 ...	1.1 ...availability of sufficient workforce is ensured ... ...
2 Learning culture	2.1 Info dissemination 2.2 ... 2.3...	2.1 ... adequate information dissemination on safety issues ... ...
3 Flexibility	3.1 Control the unexpected 3.2 ... 3.3...	3.1 ...personnel trained to cope with uncertainties ... ...
4 Awareness	4.1 Problems report	4.1 ... open atmosphere to report of errors... ...

4.2 Resilience Ideal Pattern

The aim is to obtain from experts in radiopharmaceuticals production and resilience engineering the degree of importance of each indicator of each theme, so that the radiopharmaceuticals dispatch package sector can be considered resilient, i.e., the relative importance an expert assign to each indicator for the sector achieve maximum (ideal) resilience. In this step the sector is not being evaluated, the aim is to discover the ideal of resilience level it should achieve.

A fuzzy model based on experts' assumptions increases its accuracy when increases the number of experts recognized by their knowledge and experiences in the domain (Ishikawa et al., 1993). The relative importance of the expert was calculated on the basis of subjective attributes (experience, knowledge of radiopharmaceuticals production safety and knowledge of the dispatch package radiopharmaceuticals). We used a questionnaire (Q) to identify the profile. Each questionnaire contains information of a single expert. The relative importance (RI) of expert Ei (i = 1, 2, 3, …, n) is a subset μi (k) € [0,1] defined by equation 1. Referring to equation 1, tQi, is the total score of the expert i.

$$RI_i = tQi / \Sigma tQ_i \tag{1}$$

Each leading indicator can be seen as a linguistic variable, related to a linguistic term set associated with membership functions. These linguistic terms are represented by triangular fuzzy numbers that indicate the importance degree of each indicator. It is suggested that the experts employ the linguistic terms, U (Unimportant), LI (Little Important), I (Important) and VI (Very Important) to evaluate the importance of the indicators. The importance degrees are assessed by requesting to each expert to weigh the indicators using their respective metrics.

The aggregation of the fuzzy opinions that combine the experts' opinions, which are represented by triangular fuzzy numbers, was made according to the similarity aggregation method proposed by Hsu and Chen (1996). The agreement degree (AD) between expert Ei and expert Ej is determined by the proportion of intersection area to total area of the membership functions. The agreement degree (AD) is defined by equation 2.

$$AD = \frac{\int_x \left(\min\left\{\mu_{\tilde{R}_i}(x), \mu_{\tilde{R}_j}(x)\right\}\right)dx}{\int_x \left(\max\left\{\mu_{\tilde{R}_i}(x), \mu_{\tilde{R}_j}(x)\right\}\right)dx} \tag{2}$$

If two experts have the same estimates AD =1. In this case, the two experts' estimates are consistent, and then the agreement degree between them is one. If two experts have completely different estimates, the agreement degree is zero. If the initial estimates of some experts have no intersection, then we use Delphi method to adjust the opinion of the experts and to get the common intersection at a fixed α – level cut (Hsu and Chen, 1996).The higher the percentage of overlap is the higher the agreement degree. After all the agreement degrees between the experts are calculated, we can construct an agreement matrix (AM), which give us insight into the agreement between the experts.

$$AM = \begin{bmatrix} 1 & AD_{12} & \cdots & AD_{1j} & \cdots & AD_{1n} \\ \vdots & \vdots & & \vdots & \vdots & \\ AD_{i1} & AD_{i2} & \cdots & AD_{ij} & \cdots & AD_{in} \\ \vdots & \vdots & & \vdots & \vdots & \\ AD_{n1} & AD_{n2} & \cdots & AD_{nj} & \cdots & 1 \end{bmatrix}$$

The Relative Agreement (RA) of expert Ei (i = 1, 2, 3, …, n) is given by equation 3.

$$RA_i = \sqrt{\frac{1}{n-1} \cdot \sum_{j=1}^{n} (AD_{ij})^2} \qquad (3)$$

Then we calculate the Relative Agreement Degree (RAD) of expert Ei (i = 1, 2, 3, …, n) by equation 4.

$$RAD_i = \frac{RA_i}{\sum_{i=1}^{n} RA_i} \qquad (4)$$

Now we can define the consensus coefficient (CC) of expert Ei (i = 1, 2, 3, …, n) by equation 5.

$$CC_i = \frac{RAD_i \cdot RI_i}{\sum_{i=1}^{n} (RAD_i \cdot RI_i)} \qquad (5)$$

A fuzzy number Ñ combining expert's opinions is the fuzzy value of each leading indicator which is also triangular fuzzy number. By definition of the consensus coefficient (CC) of expert Ei (i = 1, 2, 3, …, n), Ñ can be calculated by equation 6, where ñi is the triangular fuzzy number relating to the linguistic terms, U (Unimportant), LI (Little Important), I (Important) and VI (Very Important).

$$\tilde{N} = \sum_{i=1}^{n} (CC_i \cdot \tilde{n}_i) \qquad (6)$$

The resilience pattern as a reference for assessing organizational resilience is established by calculating the normalized importance degree (NID) of each leading indicator that make up each property relevant to resilient organizations. The normalized importance degree (NID) of each leading indicator is given by deffuzification of its triangular fuzzy number Ñ (ai, bi, ci), where bi represents the importance degree. Then, NID can be defined by equation 7.

$$NID_i = \frac{NID_i}{\text{highest value of } b_i} \qquad (7)$$

4.3. Evaluation of the radiopharmaceuticals dispatch package resilience

This last step in resilience evaluation is to obtain the attendance degree to the resilience pattern. Linguistic terms are used to assess the attendance degrees of each leading indicator at the radiopharmaceuticals dispatch package sector. The workers of the sector made the evaluation. It was suggested that the workers employ the linguistic terms, SD (Strongly Disagree), PD (Partially Disagree), NAND (Neither Agree Nor Disagree), PA (Partially Agree), SA (Strongly Agree).

Using center of area defuzzification method we calculate the attendance degree (AD) to the resilience pattern using equation 8. In equation 8, adj, is the attendance degree of the leading indicator j of the theme i in the dispatch package radiopharmaceuticals sector.

$$ADi = \frac{\sum_{j=1}^{k} NID_j . ad_j}{\sum_{j=1}^{k} NID_j} \tag{8}$$

5. RESULTS

The evaluation of the resilience of the radiopharmaceuticals dispatch package sector was performed by seven workers. In figure 3 we show the attendance degree of the top level commitment (similar figures were obtained for the other resilience principles). The difference in results is due to workers' perceptions of working conditions in the sector. We consider that all the workers have the same importance degree in this evaluation. Thus the average evaluate of the resilience was computed and showed in figure 3.

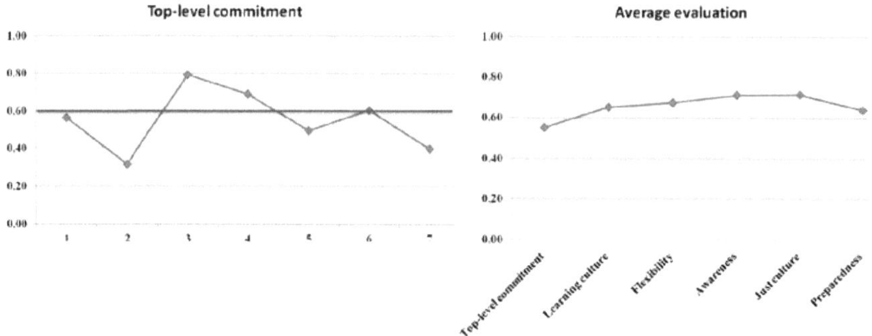

Figure 2 Attendance degrees of the top level commitment principle and average resilience evaluation.

For this first evaluation, we consider satisfactory an attendance degree greater than or equal to 0.6. The result of the average evaluation showed that the radiopharmaceuticals dispatch package sector presented satisfactory learning culture, flexibility awareness, just culture and preparedness. However, problems related to the top-level commitment appeared.

6. CONCLUSIONS

In this paper we described a case study in which a method for measurement of organizational resilience was used. We proposed a method that uses leading indicators and concepts and properties of FST. We develop a resilience pattern using a similarity aggregation method to aggregate fuzzy individual opinions, in which we consider the difference of importance of each expert. The case study in the radiopharmaceuticals production facility showed that this method offers interesting perspectives for the implementation of resilience engineering principles.

ACKNOWLEDGMENTS

The authors would like to acknowledge the Conselho Nacional de Pesquisas - CNPq and the Fundação de Amparo a Pesquisa do Rio de Janeiro - FAPERJ for support to this research.

REFERENCES

Electric Power Research Institute – EPRI, 1999. *Guidelines for Leading Indicators of Human Performance: Preliminary Guidance for Use of Workplace and Analytical Indicators of Human Performance.* Palo Alto, CA: EPRI.

Grecco C H. S., Vidal M. C. R., Bonfatti, R., 2010. Análise ergonômica do trabalho no Setor de Expedição de Radiofármacos de um Instituto de Pesquisas. *Proceedings of the XVI Congress of Brazilian Ergonomics Association – ABERGO 2010* (in Portuguese).

Hollnagel E., Woods D.D., Leveson N. editors, 2006. *Resilience Engineering. Concepts and Precepts*, Aldershot,UK: Ashgate.

Hsu H. M., Chen C. T., 1996. Aggregation of fuzzy opinions under group decision making. *Fuzzy Sets and Systems* 79: 279-285.

Ishikawa A., Amagasa M., Shiga T., Tomizawa G., Tatsuta R., Mieno H, 1993. The max-min delphi method and fuzzy delphi method via fuzzy integration. *Fuzzy Sets and Systems*, 55: 241-253.

Reiman T., Pietikäinen E., 2010. *Indicators of safety culture – selection and utilization of leading safety performance indicators.* Swedish Radiation Safety Authority, Report number 2010:07.

Saurin T. A., Carim Junior G. C., 2011. Evaluation and improvement of a method for assessing HSMS from the resilience engineering perspective: A case study of an electricity distributor. *Safety Science* 49: 355-368.

Wreathall J. Properties of Resilient Organizations: An Initial View, 2006. In: Hollnagel E., Woods, D. & Leveson, N. *Resilience Engineering: Concepts and Precepts.* Aldershot,UK: Ashgate.

Zadeh L. A. , 1996. Fuzzy Logic = Computing with words. *IEEE Transactions on Fuzzy Systems* 4: 103-111.

CHAPTER 11

Identification of High Risk Patients for Colonoscopy Surveillance from EMR Text

Eric Sherer

Division of Clinical Pharmacology, Indiana University School of Medicine
Health Services Research & Development, Roudebush VA Medical Center
Indianapolis, IN
Email address: ersherer@iupui.edu

Hsin-Ying Huang

Department of Industrial Engineering, Purdue University
West Lafayette, IN
Email address: huang230@purdue.edu

Thomas Imperiale

Division of Gastroenterology, Indiana University School of Medicine
Health Services Research & Development, Roudebush VA Medical Center
Regenstrief Institute, Inc.
Indianapolis, IN
Email address: timperia@iupui.edu

Gaurav Nanda

Department of Industrial Engineering, Purdue University
West Lafayette, IN
Email address: gnanda@purdue.edu

Mark Lehto

Department of Industrial Engineering, Purdue University
West Lafayette, IN
Email address: lehto@purdue.edu

ABSTRACT

In the United States, colorectal cancer (CRC) is third most common cancer diagnosed in both men and women. As such, screening tests such as colonoscopy are used to detect CRC. In addition, the removal of pre-cancerous adenomatous polyps via polypectomy during a colonoscopy is associated with lower lifetime incidence (Winawer et al., 1993) and mortality (Baxter et al., 2009). However, this risk is not uniform across the population, so a follow-up colonoscopy needs to be regularly scheduled especially for high-risk patients.

The purpose of this study is to calibrate and validate a computerized method to identify average-risk patients for screening colonoscopies, patients with inadequate preparation quality, and patents at high risk for colorectal cancer based on free text in the electronic medical record.

This study presents the use of naïve Bayes machine learning tool for prediction. The results showed the effectiveness of this automated method in terms of sensitivity (98%, 96%, and 100%) and positive predictive value (98%, 98%, and 85%), which lessens the effort of manual coding and improves its accuracy.

Keywords: coding accuracy, manual coding, computerized coding, text mining, and colonoscopy surveillance.

1 INTRODUCTION

Screening colonoscopy is recommended in the United States for average-risk adults (e.g. no family history of CRC, known genetic predisposition to CRC, or CRC-related symptoms) over age 50 as well as a diagnostic test in patients with gastrointestinal related symptoms such as chronic diarrhea or rectal bleeding. Following a screening colonoscopy, a patient's risk for CRC depends most strongly on the findings of the initial colonoscopy (Martinez et al. 2009); patients with an advanced or multiple adenoma are at higher-risk for subsequent CRC (Lieberman et al., 2007) while patients with no neoplasia are at lower risk (Imperiale et al. 2008). Thus, the guideline recommended surveillance intervals depend on the screening colonoscopy results and adequacy (Winawer et al., 2006).

The surveillance interval is shorter in patients who had multiple or advanced (i.e. ≥10 mm or with villous histology or high-grade dysplasia) adenomas removed at screening– and, therefore, are at higher risk for interval cancer. In addition, the recommended intervals are generally shortened from the guidelines in patients with inadequate preparation quality because of the potential that lesions may have been missed and could develop into interval cancer.

However, identifying the high-risk patients in an electronic database is difficult since the detailed colonoscopy results and adequacy are not coded (e.g. ICD-9 or CPT-4 codes). This information is available only in the free text of the colonoscopy procedure and pathology reports. Even though ICD or CPT codes are widely used in many health care facilities, the code accuracy has become an issue as it requires specially trained coders continuous repeats steps, making this manual coding process error-prone (O'Malley et al., 2005).

The objective of this study was to calibrate and validate an automated method to (1) identify patients who underwent an average-risk screening colonoscopy examination, (2) determine whether the procedure was of adequate quality, and (3) determine whether the patient at high-risk findings based on free text in the electronic medical record.

2 METHOD

2.1 Data collection

Free text from colonoscopy procedure and pathology reports for first-time colonoscopies performed from 2002-2008 at the Roudebush VAMC were extracted the electronic medical record using a previously validated extraction tool. The colonoscopy indication, preparation quality, and results of 3,713 of procedures were manually classified. The colonoscopy text and associated manual classifications were randomly assigned into a 80/20 mixture of calibration / validation sets. A naïve Bayesian technique for classification was calibrated and the sensitivity and positive predictive values were tested against the validation dataset for the identification of screening colonoscopies, inadequate preparation quality, and high-risk findings (see Figure 1).

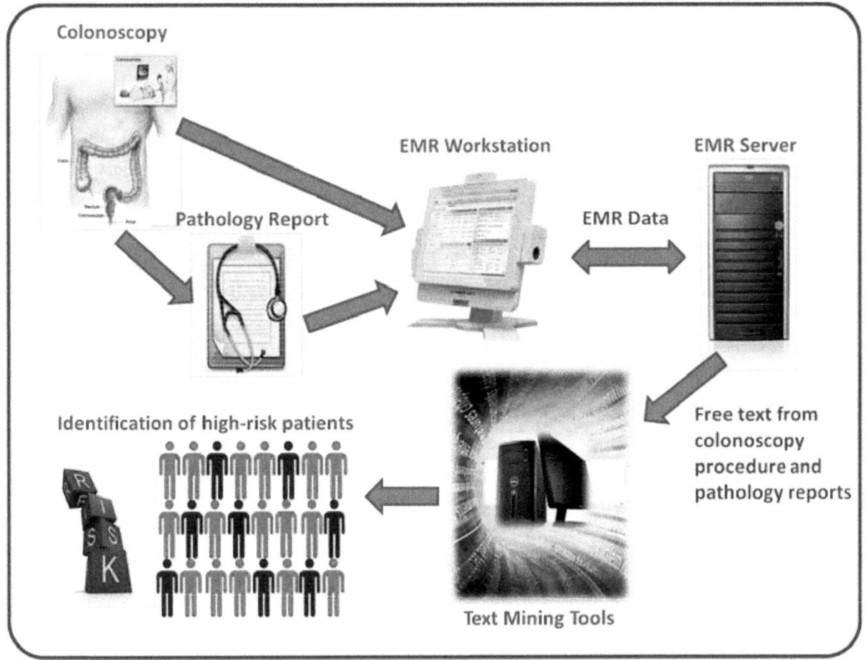

Figure 1 Data collection for text mining

2.2 Textminer: Bayes Classifier

To classify the free text data, we used the Textminer (Lehto, 2004) software program which uses a machine learning approach based on Naïve Bayesian classifier. First, the algorithm is calibrated to identify high risk cases from the manually classified narratives. The algorithm is then validated by systematically predicting the high risk cases from the new narratives and comparing with the manual classification. The Naïve Bayes classifier is based on the conditional independence assumption. The probability of assigning a particular event code category can be calculated as the expression shown in Equation (1) below:

$$P(Ei|n) = \prod_j \frac{P(n_j|E_i)P(E_i)}{P(n_j)} \quad (1)$$

where $P(E_i|n)$ is the probability of the ith event code category, E_i, given the set of n words in the narrative.

$P(n_j|E_i)$ is the probability of the jth word, n_j, given category E_i., $P(E_i)$ is the prior probability of category E_i, and $P(n_j)$ is the probability of the jth word, n_j, in the entire keyword list.

During the learning/calibration phase, the Naïve Bayesian model calculates the word probabilities from narratives that have been manually classified into high risk cases. Then it independently identifies high risk cases from colonoscopy procedure and pathology narratives. The model attempts to select the category which has the maximum probability for the set of keywords found in the narrative. This is an iterative process which includes assigning a probability for each category given each word in the narrative calculated from the learned dataset. Then the category most associated with a keyword (or keyword combination) is used to make the prediction. According to some studies (Wellman et al, 2004, Lehto et al., 1996), word combinations and word sequences are often accurate predictors.

In order to reduce the noise from the signal, a number of editing rules were applied on the text (see Figure 2). For example, we replaced all the different forms of words with same meaning and misspellings with a single word. Another editing rule which helped to reduce the noise was to delete very common words like "a", "an", "the" etc and words with very low frequency which do not help in predictions. In our case, some of the main predictor words were "polyps", "size", "10 (more) mm" etc. In addition, we identified cases where the size of polyps was mentioned to be 10 mm or more. Hence, we applied some editing rules on the free text to mark these words occurring in sequence together and identifiers were attached with those cases. We also used negative evidence for predictions by attaching separate identifiers for the absence of a particular predictor word.

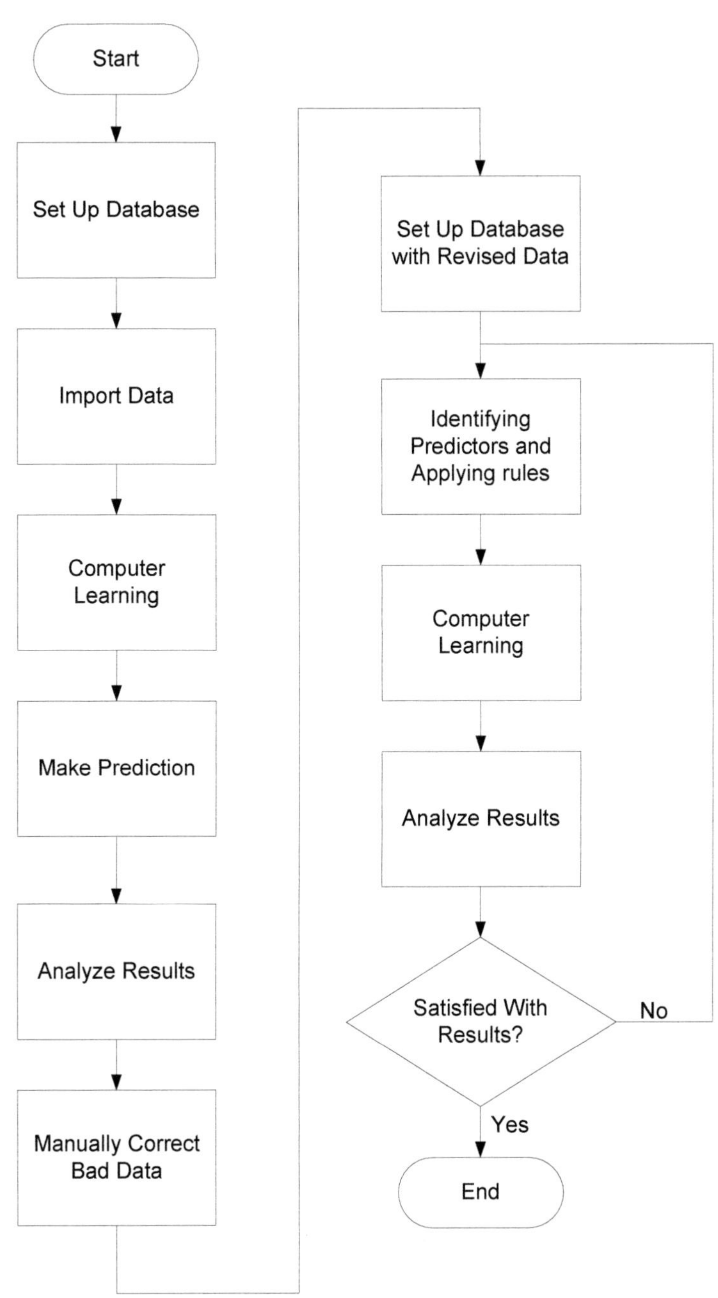

Figure 2 Flowchart – Process of Textminer Programming

3 RESULTS

The sensitivity and positive predictive value for identifying screening colonoscopies in the validation dataset were 0.98 and 0.98, respectively. The sensitivity and positive predictive value for patients with inadequate preparation quality in the validation dataset were 0.96 and 0.98, respectively. The sensitivity and positive predictive value for identifying all high-risk patients with multiple adenomas or advanced adenomas were 1.00 and 0.85 for validation cases. All cases which are not correctly predicted can be easily identified and manually corrected by examining those cases where the prediction strength is less than 95% from the naïve Bayes summary results.

4 CONCLUSIONS

A naïve Bayesian technique can accurately filter out screening colonoscopies from free text in the electronic medical record and identify patients at high-risk for colonoscopy surveillance. The textmining program helps reduce the burden of manual coding with improved accuracy and enables easy identification of cases from the wrong prediction for further manual correction. This study demonstrates the benefit of computerized coding which has been viewed as an effective way to improve the quality of care (Duszak et al., 2001).

REFERENCES

Baxter, N. N., M. A. Goldwasser, and L. F. Paszat, et al. 2009. Association of colonoscopy and death from colorectal cancer. *Annals of Internal Medicine* 150: 1–8.

Duszak Jr, R., D. Sacks, and J. Manowczak. 2001. CPT Coding by Interventional Radiologists: Accuracy and Implications. *Journal of Vascular and Interventional Radiology* 12(4): 447–454.

Imperiale, T. F., E. A. Glowinski, and C. Lin-Cooper, et al. 2008. Five-year risk of colorectal neoplasia after negative screening colonoscopy. *The New England Journal of Medicine* 359: 1218–1224.

Jemal, A., R. Siegel, and J. Xu et al. 2010. Cancer statistics, CA. *A Cancer Journal for Clinicians* 60(4): 277–300.

Lehto, M. 2004. TextMiner Manual (Ann Arbor, MI: ConsumerResearch, Inc.)

Lehto, M., and G. Sorock. 1996. Machine learning of motor vehicle accident categories from narrative data. *Methods Info. Med* 35(4-5): 309–316.

Lieberman, D. A., D. G. Weiss, and W. V. Harford, et al. 2007. Five-year colon surveillance after screening colonoscopy. *Gastroenterology* 133: 1077–1085.

Martinez, M. E., J. A. Baron, and D. A. Lieberman, et al. 2009. A polled analysis of advanced colorectal neoplasia diagnoses after colonoscopic polypectomy. *Gastroenterology* 136: 832–841.

O'Malley, K. J., K. F. Code, and M. D. Price, et al. 2005. Measuring Diagnoses: ICD Code Accuracy. *Health Services Research* 40: 1620–1639.

Wellman, H. M., M. R. Lehto, and G. S. Sorock, et al. 2004. Computerized coding of injury narrative data from the National Health Interview Survey. *Accident Analysis & Prevention* 36(2): 165–171.

Winawer, S. J., A. G. Zauber, and M. N. Ho, et al. 1993. Prevention of colorectal cancer by colonoscopic polypectomy. *The New England Journal of Medicine* 329: 1977–1981.

Winawer, S. J., A. G. Zauber, and R. H. Fletcher, et al. 2006. Guidelines for colonoscopy surveillance after polypectomy: A consensus update by the U.S. Multi-Society Task Force on colorectal cancer and the American Cancer Society. *Gastroenterology* 130: 1872–1885.

CHAPTER 12

The Effects of Repetitive Lifting on Heart Rate Responses and Perceived Exertion in a Young and Middle-aged Population

Mark G. Boocock and Grant A. Mawston

Health and Rehabilitation Research Institute,
Auckland University of Technology,
Auckland 1142, New Zealand
Email: mark.boocock@aut.ac.nz

ABSTRACT

Higher incidences of manual handling injuries have been identified in older adults compared to a younger population (Mackey et al., 2007, Sluiter and Frings-Dresen, 2007). Increased risk of injury may stem from a reduction in the functional capacity of older workers, which may lead to an inability to meet the physiological demands of repetitive manual handling activities (Mackey et al., 2007). However, to date, there has been few, if any, studies comparing the physiological and subjective responses of young and middle-aged workers involved in repetitive manual handling tasks. An understanding of differences in physiological and subjective responses in these groups of workers may provide an insight into potential mechanisms of fatigue-related injury within the ageing workforce.

Eleven young (19-36 years) and 11 middle-aged (40-55 years) participants with no recent history of low back pain, and no musculoskeletal, cardiovascular or respiratory disorder participated in the study. All participants performed an incremental cycle ergometer test to maximal effort where heart rate (HR) and ratings of perceived exertion (RPE) were recorded continuously and maximal aerobic capacity was ascertained. One week later, participants performed a repetitive lifting and lowering task (box weighing 13 kg) at a lifting rate of 10 lifts per min. The task was terminated when participants were either unable to continue lifting due to discomfort or fatigue, or had completed 20 minutes. HR was recorded continuously throughout the task and RPE were recorded every 30 s. Participants also reported any local body discomfort. Temporal profiles of HR and RPE for each age group were plotted using Lowess curves. A one way analysis of variance (ANOVA) with repeated measures was used to determine differences in mean HR and RPE at the sixth and final minute of the task.

Despite having significantly higher aerobic capacity, RPE of the younger participants was significantly higher than those of the middle-aged group throughout the lifting task. HR did not differ significantly between groups and displayed a gradual increase over time, indicting an inability to maintain a physiological steady state throughout the lifting task.

These findings suggest that irrespective of age, individuals are unable to maintain a physiological and psychophysical steady state when repetitive lifting and lowering a moderate workload. Overall, the middle-aged participants subjectively underestimated the physiological stress when compared to the younger group. A lower perception of the physiological stress could expose older workers to an increased risk of injury.

Keywords: low back injury, repetitive lifting, ageing, heart rate, ratings of perceived exertion

1 INTRODUCTION

Workers between the age of 45 and 55 have higher injury rates than those of a younger age (Mackey et al., 2007). In particular, these older workers show a higher prevalence and incidence of musculoskeletal injury in jobs requiring high physical work (Tuomi et al., 1991). It has been suggested that this increased risk of injury in older workers may be due to a decline in their level of fitness when compared to their younger counterparts (Mackey et al., 2007, Sluiter and Frings-Dresen, 2007). This is based on the premise that if an older worker's level of fitness is inadequate and unable to meet the demands of the task there is an increased likelihood of physiological fatigue and therefore, injury (Mackey et al., 2007).

The inability of an individual to meet task demands during repetitive physical work is often characterised by a continuous elevation in physiological measures, such as heart rate (HR). An increase in physiological demand is also associated with elevation in levels of perceived effort and localised muscular discomfort. Even when performing repetitive tasks at sub-maximal effort, elevations in physiological parameters and perceived effort over time can lead to eventual fatigue and exhaustion. Whilst comparisons have been made of the physiological and psychophysical measures in younger and older individuals when performing endurance exercises ((Heath et al., 1981)), there is limited information relating to the performance of repetitive manual handling tasks common to the workplace. Hence, the aim of this study was to investigate differences in HR and ratings of perceived exertion (RPE) in a group of young and middle-age adults when performing a repetitive lifting and lowering task.

2 METHODS

2.1 Participants

Eleven young (mean age = 24.6 years (yr) (standard deviation (SD) = 4.5 ± yr), mean stature = 1.8 m (SD = ±0.09 m) and 11 middle-aged (mean age = 46.5 yr (SD = ±3.3 yr), stature = 1.76 m (SD = ±0.06 m) participants took part in the study. Participants were excluded from the study if they had: a back complaint within the last six months; undergone any previous spinal surgery; any cardiovascular or neurological condition; or a musculoskeletal injury at the time of the study. Prior to the study, participants completed a health screening (Physical Activity Readiness) and Habitual Physical Activity Questionnaire. None of the participants were considered to be experienced in manual handling, i.e. undertook manual handling as part of a regular job. The AUT University ethics committee approved the study and participants were required to complete a written consent form.

2.2 Maximal excericse test

One week prior to the lifting task, participants performed a graded maximal exercise test on a cycle ergometer (ergobike medical8, daum elctronic gmbh, Germany) in order to measure maximal HR. Participants began the test by resting on the ergometer for one minute and then undertaking a warm-up for one minute by cycling at 60 revolutions per minute (RPM) at a power output of 20 W. Following the warm-up, participants performed an incremental maximal exercise test where power output was increased by 25 watts per minute at fixed cadency of 60 RPM until exhaustion, or until the participant was unable to maintain a cadence above 50 RPM. During the maximal exercise test, HR was recorded continually via a polar HR monitor (T34, Polar Electro, Finland) that transmitted data to a Fitmate Pro (COSMED srl, Italy) cardiopulmonary testing device. Participants were required to rate their perceived exertion (RPE) using Borg's 6-20 point rating scale (Borg, 1982) at the end of each 25 W increment. Maximal HR was determined by measuring the highest average HR over a 10 s period.

2.4 Lifting task

Participants were required to lift and lower a box weighing 13 kg at a frequency of ten times per minute. This handling workload was considered to be representative of repetitive handling tasks found in industry, e.g. as might be performed by baggage handling. The box (30 x 25 x 25.5 cm) was lifted from a shelf 15 cm above the floor to an upright standing position by holding onto two cylindrical handles that extended 6 cm from either side of the box, at a height of 17 cm above its base. The box was held at hip height prior to being lowered back onto the shelf. Throughout, participants were required to maintain a hold of the box handles. Participants commenced each lift and lower at the sound of a computer generated metronome operating at a frequency of 20 times per minute. They were not informed as to how long they would perform the lifting and lowering task, just that they should continue lifting until they became fatigued. The task was terminated if participants completed 20 min. Participants were verbally encouraged to continue and maintain the required rate of lifting throughout the task. A person was considered to be fatigued if they were unable to keep pace with the metronome or had a HR greater than that recorded during the maximal exercise test, or that the participant chose to stop lifting because of subjective fatigue or excessive discomfort. This was based on a modified version of the progressive iso-inertial lifting evaluation criteria proposed by Mayer et al. (1988). When lifting and lowering the box, participants were required to adopt a fixed foot position as close to the shelf as possible, but not touching it, and return the box to the same position each time. Participants were not instructions about the technique they should adopt when lifting or lowering.

HR was recorded continuously throughout the repetitive handling task using a polar T34 HR receiver (Polar Electro, Finland). This data was transmitted to a Fitmate Pro testing device (COSMED srl, Italy) which calculated mean HR every 15 s. Participants were asked to rate perceived exertion using Borg's 6-20 point rating scale (Borg, 1982) every 30 s. RPE were manually entered into the Fitmate Pro computer system.

2.6 Statistical analysis

HR was expressed as a percentage of maximum HR (%HRmax). Both %HRmax and RPE were plotted using Lowess curves (Dupont, 2002) to investigate temporal changes over 20 min. To determine differences between the beginning and the end of the repetitive task, the mean %HRmax and RPE were compared between the sixth and last minute. The sixth minute was chosen as a starting point for comparison because evidence reported in the literature suggests that during sub-maximal exercise at constant load, a steady state HR is reached within 3-6 min (Barstow et al., 1996). These comparisons were made using a 2 (young vs middle-aged) x 2 (six vs last min) analysis of variance (ANOVA). Statistical significance was set at 0.05 and all data analysis was conducted in the SPSS v16 software package (SPSS Inc., Chicago).

3 RESULTS

3.1 Task completion

Two participants from the younger group and three from the middle-aged group failed to complete the 20 min of repetitive lifting and lowering. Of those participants who were unable to continue lifting, both the younger and middle-aged adults stopped within approximately 14 min. Most participants stop lifting due to excessive discomfort in the lower back.

3.2 Maximum HR and changes in HR

When performing the maximum exercise test, there was found to be a significant difference in maximum HR (mean = 179, SD = ±12.1) when comparing between the young and middle-aged groups (mean = 165.7, SD = ±13.8) (p=0.026).

Figure 3 shows the temporal changes in %HRmax over the 20 min repetitive task for young and middle-aged participants. Lowess curves were found to provide the best fit to the data, and showed that both the young and middle-aged group displayed an initial mono exponential rise in %HRmax (first 3 mins) followed by a steady progressive increase in %HRmax over the duration of the task. Younger participants displayed slightly higher %HRmax than the middle-aged participants. The ANOVA showed that %HRmax significantly increased between the sixth and final minute of the task between groups (P = 0.01). However, there was no significant difference in %HRmax between the young and middle-aged group (Table 1).

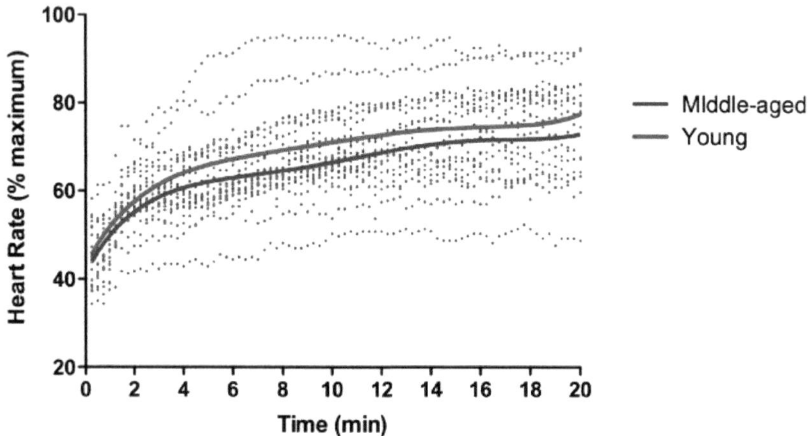

Figure 1. Individual (dotted lines) and lowess curves (solid lines) showing changes in %HRmax over the duration of the lifting and lowering task.

Table 1. The mean and standard deviation (SD) of the young and middle-aged participants' %HRmax and RPE for the sixth and last minute of the repetitive handling task, as well as the F-ratio and p values for the main effects of time and age, and their interaction.

Measure	Mean (SD)		Main Effects		Interaction
	6th min	Final min	Time	Age	Time-Age interaction
%HRmax					
Young (n=11)	67.9 (11.8)	74.8 (10.4)	F = 40.3	F = 0.85	F = 0.76
Mid-age (n=11)	62.9 (7.4)	72.1 (11.1)	*P* = 0.00	P = 0.37	P = 0.39
RPE					
Young (n=11)	15.7 (2.6)	18.6 (1.8)	F = 51.9	F = 8.28	F = 0.61
Mid-age (n=11)	12.3 (2.9)	15.9 (3.4)	P = 0.00	P = 0.01	P = 0.44

3.4 Changes in RPE

Figure 4 shows temporal changes in RPE over 20 mins. Similar to %HRmax, Lowess curves showed that both groups had an initial rapid rise in RPE followed by a slow increase in RPE for the remainder of the task. The ANOVA for RPE showed that RPE increased significantly between the sixth and final minute of the task for both groups ($P = 0.00$). There was also a main effect of age ($P = 0.01$), where older participant's RPE was significantly lower than that of the younger participants (Table 2).

4 DISCUSSION

This study showed that repetitive lifting of a constant load resulted in older participants displaying similar relative HR (%HRmax), but lower RPE than younger adults. Few studies, if any, have compared measures of HR for repetitive lifting in groups of young and older workers. However, data from non-lifting tasks would suggest that whilst middle-aged participants exhibit differences in absolute HR when compared to younger participants, there is no difference between groups when expressed as a percentage of their maximum HR (%HRmax) for a given workload (Tulppo et al., 1998).

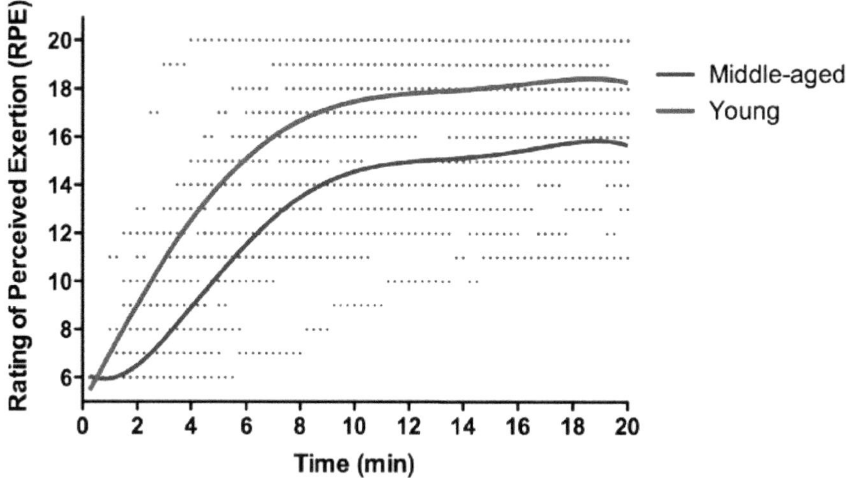

Figure 2. Individual (dotted lines) and Lowess curves (solid lines) showing changes in RPE over the duration of the lifting and lowering task.

Temporal changes in %HRmax and RPE were similar for both age groups. There was an initial rapid mono-exponential rise for the first 3 mins followed by a gradual increase until termination of the task. This suggests that participants had an inability to maintain physiological and subjective steady state. This pattern of a rapid rise followed by a slow component rise in HR has been documented during the performance of constant load exercise against heavy resistance above anaerobic threshold (Barstow and Mole, 1991, Herman et al., 2003). A slow component rise in HR has been associated with factors related to fatigue, such as rises in blood lactate, elevated intramuscular temperature and the recruitment of fast twitch muscle fibers (Barstow et al., 1996, Gaesser and Poole, 1996). The workload associated with the lifting and lowering task used in this experiment was considered to be representative of those found in industry.

Although both age groups showed similar temporal changes in RPE, middle-aged participants exhibited consistently lower RPE than younger participants throughout the duration of the task. Lower RPE have been found in older adults when compared to younger adults during the performance of static force production tasks, and it has been suggested that the subjective underestimation of physiological stress by older adults may be related to changes in the somatosensory system associated with aging (Pincivero, 2011).

5 CONCLUSIONS

This study found that when exposed to a 20 min repetitive manual handling task involving moderate workload, participants were unable to sustain a physiological and subjective steady state. Age did not significantly influence relative HR, but did influence participant's perception of effort, with older participants perceiving lower stress. This has possible implications for the risk of injury amongst older workers, placing them at greater risk than their younger counterparts.

REFERENCES

Barstow, T. J., Jones, A. M., Nguyen, P. H. & Casaburi, R. 1996. Influence of muscle fiber type and pedal frequency on oxygen uptake kinetics of heavy exercise. Journal of Applied Physiology, 81, 1642-50.

Barstow, T. J. & Mole, P. A. 1991. Linear and nonlinear characteristics of oxygen uptake kinetics during heavy exercise. Journal of Applied Physiology, 71, 2099-106.

Borg, G. 1982. Psychophysical bases of perceived exertion. Medicine and Science in Sports and Exercise, 14, 377-81.

Dupont, W. 2002. Statistical modelling for biomedical researchers. A simple introduction to the analysis of complex data, Cambridge, United Kingdom, Cambridge University Press.

Gaesser, G. A. & Poole, D. C. 1996. The slow component of oxygen uptake kinetics in humans. Exercise and Sport Sciences Reviews, 24, 35-71.

Heath, G. W., Hagberg, J. M., Ehsani, A. A. & Holloszy, J. O. 1981. A physiological comparison of young and older endurance athletes. J Appl Physiol, 51, 634-40.

Herman, C. W., Nagelkirk, P. R., Pivarnik, J. M. & Womack, C. J. 2003. Regulating oxygen uptake during high-intensity exercise using heart rate and rating of perceived exertion. Medicine and Science in Sports and Exercise, 35, 1751-4.

Mackey, M., Maher, C. G., Wong, T. & Collins, K. 2007. Study protocol: the effects of work-site exercise on the physical fitness and work-ability of older workers. BMC Musculoskelet Disord, 8, 9.

Mayer, T. G., Barnes, D., Kishino, N. D., Nichols, G., Gatchel, R. J., Mayer, H. & Mooney, V. 1988. Progressive Isoinertial Lifting Evaluation: I. A Standardized Protocol and Normative Database. Spine, 13, 993-997.

Pincivero, D. M. 2011. Older adults underestimate RPE and knee extensor torque as compared with young adults. Med Sci Sports Exerc, 43, 171-80.

Sluiter, J. K. & Frings-Dresen, M. H. 2007. What do we know about ageing at work? Evidence-based fitness for duty and health in fire fighters. Ergonomics, 50, 1897-913.

Tulppo, M. P., M√§Kikallio, T. H., Sepp√§Nen, T., Laukkanen, R. T. & Huikuri, H. V. 1998. Vagal modulation of heart rate during exercise: Effects of age and physical fitness. American Journal of Physiology - Heart and Circulatory Physiology, 274, H424-H429.

Tuomi, K., Ilmarinen, J., Eskelinen, L., Jarvinen, E., Toikkanen, J. & Klockars, M. 1991. Prevalence and incidence rates of diseases and work ability in different work categories of municipal occupations. Scandinavian Journal of Work, Environment and Health, 17, 67-74.

Section II

Work-related Musculoskeletal Disorders

CHAPTER 13

In vivo Investigation of the Effects of Force and Posture on Carpal Tunnel Pressure

McGorry RW[1], Fallentin N[1], Andersen JH[2], Keir PJ[3], Hansen TB[4], Lin JH[1], Pransky G[1].

[1]Liberty Mutual Research Institute for Safety, Hopkinton, MA, USA
Raymond.mcgorry@libertymutual.com
[2]McMaster University, Hamilton, Ontario, Canada
[3]Department of Occupational Medicine, Regional Hospital Herning, Denmark
[4]Orthopaedic Research Unit, Regional Hospital Holstebro, Denmark

ABSTRACT

Elevated hydrostatic pressures within the carpal tunnel have been associated in the literature with increased risk of carpal tunnel syndrome. A methodology and apparatus for in vivo measurement of pressures within the carpal tunnel is presented. A protocol to measure the response to the risk factors of varying wrist posture, applied grip forces and reactive wrist torque levels typical of industrial operations is described. Preliminary data will be presented.

Keywords: Carpal tunnel pressure, CTP, Carpal tunnel syndrome, CTS

1. INTRODUCTION

In addition to anatomic abnormalities or physiologic changes two other mechanisms have been associated with development of carpal tunnel syndrome (CTS). Mechanical compression of the Median nerve can occur either directly by application of external force or by internal constriction. Another potential mechanism is elevated hydrostatic pressures within the confined space of the carpal tunnel acting upon the Median nerve. In animal studies, brief exposure to carpal tunnel pressures (CTP) of 30 mmHg effect nerve conduction, and produce paresthesias (Lundborg et al., 1982), and sustained pressures of 30 mmHg or greater may affect nerve physiology, cause nerve injury and axon degeneration (Diao et al, 2005). Studies of patients with CTS conducted prior to surgery have shown elevated CTP, with pressures often returning to near normal levels following surgical release of the flexor retinaculum. Taken together the literature supports the hypothesis of an injury/disease pathway: pressure –> ischemia –> impaired nerve conduction (CTS). Elevated CTP could also exacerbate injuries where the primary agent was mechanical compression. The literature is fairly conclusive that wrist posture can affect CTP. Using wrist posture data collected coincident with CTP in healthy wrists, Keir et al. (2007) proposed guidelines for wrist postures based on a 25-30 mmHg threshold. However, the effect of wrist torque, with or without concurrent grip force has not been adequately studied. This report describes the experimental approach and preliminary results of a study designed to investigate the effect of the independent variables of wrist posture, wrist torque and grip force application, on the *in vivo* measurement of CTP. Few studies have measured tunnel pressures during gripping tasks typical of industrial operations. A systematic approach to evaluating these factors and their interactions could provide the basis for improving guidelines for upper extremity work, and job and hand tool design.

2. METHOD

2.1. Experimental apparatus

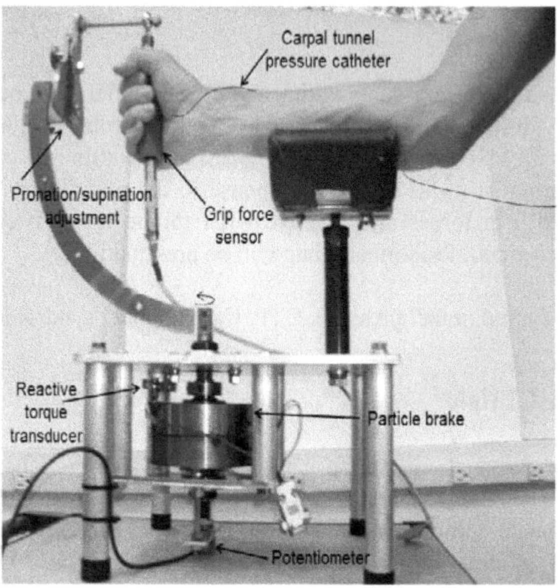

Figure 1. illustrates the main components of the experimental apparatus.

A system was developed to satisfy the following criteria: 1) simultaneously collect data in real time, under static and dynamic conditions, 2) control and record torques exerted in wrist flexion/extension, radial ulnar/deviation, and compound angles, 3) measure wrist angular displacement, 4) measure grip force (power grip or pinch grip), and 5) measure carpal tunnel pressure.

A magnetic particle brake had a controller that by varying the current supply provided a constant reactive torque to wrist motion. A load beam mounted between the particle brake and apparatus frame measured the reactive torque. A potentiometer aligned with the particle brake shaft and the participant's wrist recorded angular displacement of the wrist. Handles were mounted to custom-designed sensors for the Power grip and Pinch grip. A third device allowed torque production at the wrist without gripping, and was referred to as the No-grip condition. Carpal tunnel pressure was measured with a saline-filled 20 gauge nylon catheter connected to a continuous fluid pressure monitoring device (-50 to 300 mm Hg range).

2.2. Protocol

Fourteen right-handed male and female participants, 20 to 45 years of age, were recruited for a study involving a two-hour protocol. The catheter was inserted into the right carpal tunnel by a hand surgeon (TBH) at the distal volar crease. The position of the catheter tip was verified by ultrasonic imaging. Saline was administered by IV drip intermittently to ensure the catheter tip remained clear.

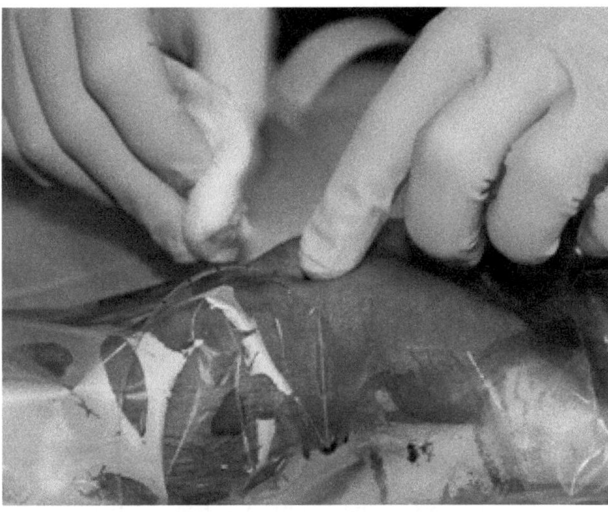

Figure 2. illustrates the surgical procedure for inserting the catheter used for the carpal pressure measurement.

The experimental protocol involved trials composed of two replications of full range of motion in four wrist postures: wrist flexion/extension (F/E), radial/ulnar deviation (R/UD) and two compound angles of F/E and R/UD. There were three grip conditions: Power grip, Pinch grip and No-grip. The Power grip condition was performed at three reactive torque levels: 0 Nm, 1 Nm, 2 Nm. The Pinch grip and No-grip conditions were performed at 0 Nm and 1 Nm levels. During the Power grip and Pinch grip trials participants were instructed to grip the handle firmly, but not as hard as possible. The dependent measures were carpal tunnel pressure, wrist angular displacement, wrist reactive torque and grip force.

3. RESULTS AND DISCUSSION

The experimental apparatus demonstrated the ability to monitor and record data on wrist and grip kinetics and kinematics simultaneously with carpal tunnel pressure at controlled levels of wrist moment and posture.

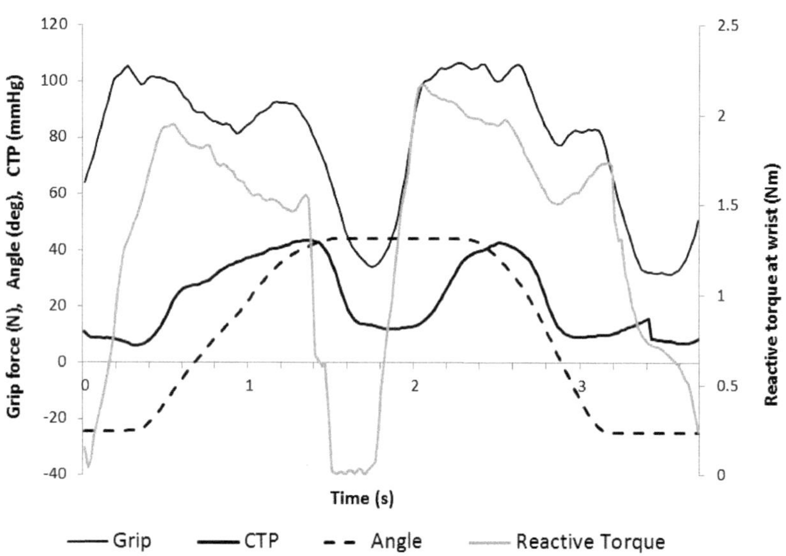

Figure 3. provides a sample output showing wrist angular displacement (+ flexion, - extension), carpal tunnel pressure, wrist reactive torque, and grip force collected during a cycle of wrist flexion-extension.

The resulting data should improve the knowledge base of the effect of these variables on CTP, and this information could be used in design of upper extremity work. Preliminary results will be presented.

REFERENCES

[1] Lundborg G, Gelberman RH, Minteer-Convery M, Lee YF, Hargens AR. Median nerve compression in the carpal tunnel-functional response to experimentally induced controlled pressure. J Hand Surg Am. 1982;7(3):252-9.

[2] Diao E, Shao F, Liebenberg E, Rempel D, Lotz JC. Carpal tunnel pressure alters median nerve function in a dose-dependent manner: a rabbit model for carpal tunnel syndrome. J Orthop Res. 2005;23(1):218-23.

[3] Keir PJ, Bach JM, Hudes M, Rempel DM. Guidelines for wrist posture based on carpal tunnel pressure thresholds. Hum Factors. 2007;49(1):88-99.

CHAPTER 14

Analysis and Evaluation of Muscular Strain While Working with Large Scale Touchscreens

Jennifer Bützler, Sebastian Vetter, Martin Kremer, Nicole Jochems, Christopher M. Schlick

Institute of Industrial Engineering and Ergonomics of RWTH Aachen University, Aachen, Germany
j.buetzler@iaw.rwth-aachen.de

ABSTRACT

As large scale touchscreens are used more frequently in different application areas, the ergonomic design of software applications for large touchscreens has to be considered. Large scale touchscreens are beneficial concerning the representation of complex information by reducing scrolling activities, but working with them also goes along with higher muscular strain than working on classical computer workstations. In this empirical study, we investigated the muscular activity for the eight most severely stressed muscles. Therefore an experimental setup with ten right-handed subjects aged between 21 and 32 years was used. A pointing task was chosen, since this is a characteristic task for touch input. In order to detect areas that are correlated with higher and lower muscular strain 14 target positions on the touchscreen were investigated. The results show significant effects of the target position on muscular activity. Thus, the ergonomic arrangement of buttons regarding muscular strain can be beneficial for the user. On the basis of our results we give practical recommendations for the ergonomic arrangement of interaction elements on large scale touchscreens.

Keywords: muscular strain, large scale touchscreen, ergonomics

1 INTRODUCTION

In many application areas such as manufacturing planning or computer-aided design, large scale touchscreens provide an alternative to conventional computer workstations. They allow the simultaneous representation of complex information and direct interaction and manipulation of objects by pointing movements and gestural input (Dietz & Leigh, 2001). Besides the advantages of this new interaction technique, ergonomically unfavorable body postures and high muscular strain might occur particularly when working in areas distal from the body. In particular, the muscles of the shoulder-arm system are subject to high muscular strain due to repetitive movements. This is critical especially for older users (Ahlström et al., 1992).

In order to analyze performance of repetitive work in terms of muscular activity electromyography (EMG) can be a useful tool (Marras, 2000). Regarding the applicability non-invasive surface electromyography (sEMG) is the preferred measurement technique.

The comparison of different input devices and the position of the input device were subject to ergonomic studies investigating the muscular activity of the shoulder-arm system (i.e. Cooper & Straker, 1998; Kumar & Kumar, 2008). Harvey & Peper (1997) for example investigated the effects of mouse and trackball position on the muscular activity. Regarding touch input, Shin & Zhu (2011) compared the muscular activity of the shoulder and the neck when working with desktop touchscreens versus standard keyboard typing and mouse input.

To our knowledge an elektromyographic evaluation of computer work with large scale touchscreens does not exist. One way of reducing muscular strain while working with large touchscreens, is to detect areas that are correlated with higher and lower muscular strain and to arrange buttons, menus and icons accordingly. Since pointing tasks are characteristic for touch input, in this study we investigated the muscular activity while performing pointing movements on a large touchscreen.

2 METHOD

The muscles and muscle groups investigated in the present study were selected on the basis of an empirical pre-study. The EMG system provides simultaneous measurement of the muscular activity with eight channels. Criteria for inclusion were the relevance of the muscle for the experimental task and the interference-free derivability via surface elektromyography (sEMG). For each relevant muscle or muscle group the maximum voluntary contraction (MVC) was determined. Then the average muscular activity during the execution of a pointing task was expressed relative to the MVC value in percent. By a comparison of the % MVC values the eight most strained muscles respective muscle groups when working with large scale touchscreens were identified: Forearm extensors, Forearm flexors, Brachioradialis, Triceps brachii cap., Deltoideus pars clav., Trapezius pars desc., Trapezius pars trans., Infraspinatus.

In order to analyze and compare the muscular strain while working with large scale touchscreens, the screen was divided into several sections, which were represented by 14 circular target objects and a home position. The sample of the main study consisted of ten right-handed subjects, five female and five male, aged between 21 and 32 years (M=26.8 years, SD=3.88 years) with an average body height of 1.74 m and a standard deviation of 0.12 m (M_{female}=1.65 m, SD_{female}=0.06 m, M_{male}=1.83 m, SD_{male}=0.09 m).

An 8-channel Noraxon TeleMyo (Noraxon, USA Inc.) system was used to record the electrical potentials of the examined muscle groups. The hardware used to register the pointing movements was the DiamondTouch screen developed by Circletwelve Inc. The DiamondTouch screen is a tabletop device (projection area 865 mm x 649 mm, 4:3 ratio) with a touch-sensitive surface of 1070 mm in diagonal. The images are projected from top via an LCD projector (1600 x 1200 pixels). Through capacitive coupling between a transmitter array embedded in the touch surface and separate receivers the subjects sit on, the attached control unit can distinguish multiple touch inputs.

Bipolar Ag/AgCl surface electrodes (Blue Sensor N, Ambu) were placed on the muscles with an inter-electrode distance of 22 mm. The skin was prepared according to the SENIAM Standard (Surface Electromyography for the Non-Invasive Assessment of Muscles, see Hermens 2011) recommendations. The electrode location on the muscle was determined based on Velamed (Shewman 2008). MVC normalization procedures were carried out in a pre-study before the main experiment. Therefore, each subject performed three trials of maximum isometric contraction for each muscle with duration of 5 seconds and 15 seconds of rest in between the trials.

The pointing task was carried out with the DiamondTouch screen lying on a table with a height of 755 mm. The subjects were seated on a chair in front of the screen with their legs bent (90 degrees) and their left arm resting beside the screen. The subjects were instructed to point the targets in a natural smooth movement. Accuracy of the pointing movement was emphasized over speed. The experimenter demonstrated and supervised a sample target block to familiarize the subject with the task and the test environment. The experimental task consisted of a typical pointing task which was carried out with the right index finger. Starting from a central home position (Ø = 31 mm) at the bottom of the touch screen each subject had to point the circular targets (Ø = 62 mm) which were arranged at various positions (Fig. 1). Reaching the target object, the participants had to perform the reverse movement back to the home position. Each target position was pointed four times. The target objects were presented in a random sequence. To minimize muscular fatigue a 5 seconds rest between each pointing movement was obligatory.

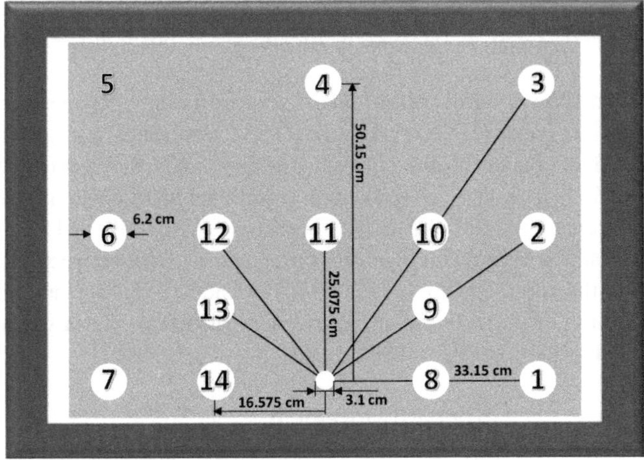

Figure 1 Position of the examined target objects

The resulting muscular activity was expressed as a percentage of the MVC value for each target position separately for the forward and backward movements. Due to the differences in muscular strength between men and women, the data was analyzed gender-specific. The muscular activity was analyzed by an ANOVA of repeated measures with gender as a between-group factor, and direction, target position and repetition as within-group factors. It was tested if the assumption of shericity has been met using Mauchly's test statistics. The significance level was set to p=0.05. The statistical software package PASW Version 18.0 was used to compute the descriptive and inferential statistics.

3 RESULTS

We analyzed effect of repetition for each muscle by ANOVA with repeated measures and could not detect significant effects for all eight muscles. Regarding gender, significant differences in muscular activity were found for some, but not all target positions. Hence men reached lower MVC values throughout all investigated conditions. The higher % MVC values of the female subjects can be attributed to the fact that men, compared to women naturally have higher physical strength. Furthermore the female group shows a smaller average body height than their male counterparts (M_{female}=1.65m, M_{male}=1.83) which may also influence the resulting muscular activity during pointing movements. Notwithstanding these gender differences the highest muscular strain (measured relative to the maximum voluntary muscle activation) was found for both men and women for the Deltoidus pars clav. and the Infraspinatus. Detailed results for these muscles are given in the following section.

The results for the Deltoideus pars clav. are depicted in figure 2 and 3. They show no significant main effects of repetition (F(2,16)=0.775, p=0.477). There were significant main effects for the gender ($F(1,8)=9.841$, $p=0.014$), for the movement direction ($F(1,8)=91.436$, $p<0.001$) and for the target position ($F(13,104)=36.428$, $p<0.001$).

Results also show a significant ordinal interaction effect between the movement direction and the gender ($F(1,8)=7.836$, $p=0.023$). Furthermore a significant hybrid interaction effect between the target position and the gender was found ($F(13,104)=2.363$, $p=0.008$) as well as a significant disordinal interaction effect between the repetition and the target position ($F(26,208)=1.615$, $p=0.036$). In addition, we found a significant disordinal interaction effect between the movement direction and the target position ($F(13,104)=31.664$, $p<0.001$) and a significant hybrid interaction effect between the movement direction, the target position and the gender ($F(13,104)=3.560$, $p<0.001$).

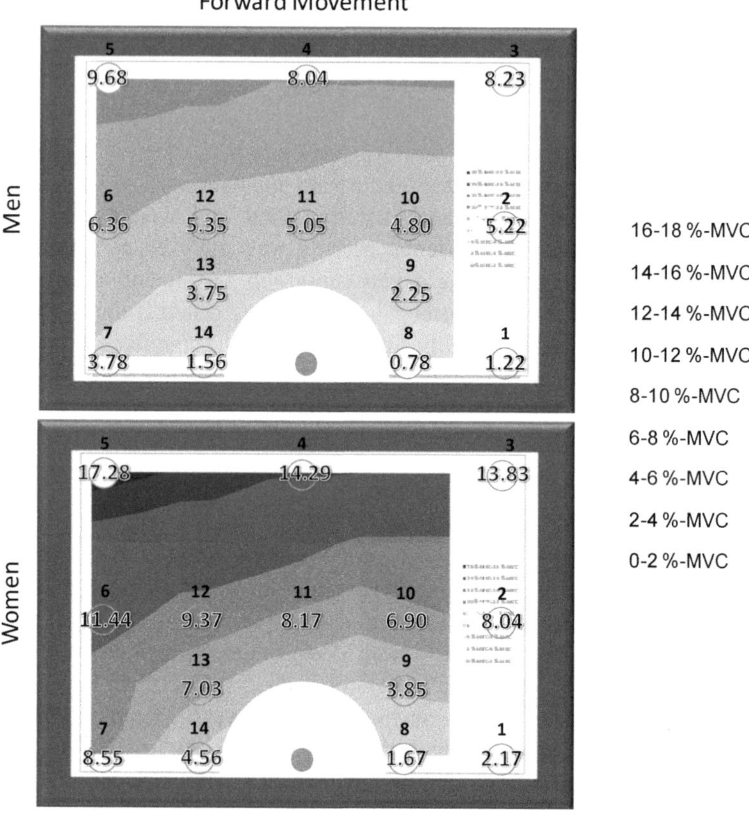

Figure 2 Muscle activity of the Deltoideus pars clav. in % MVC resulting from the forward movement for the male subjects (upper part) and the female subjects (lower part)

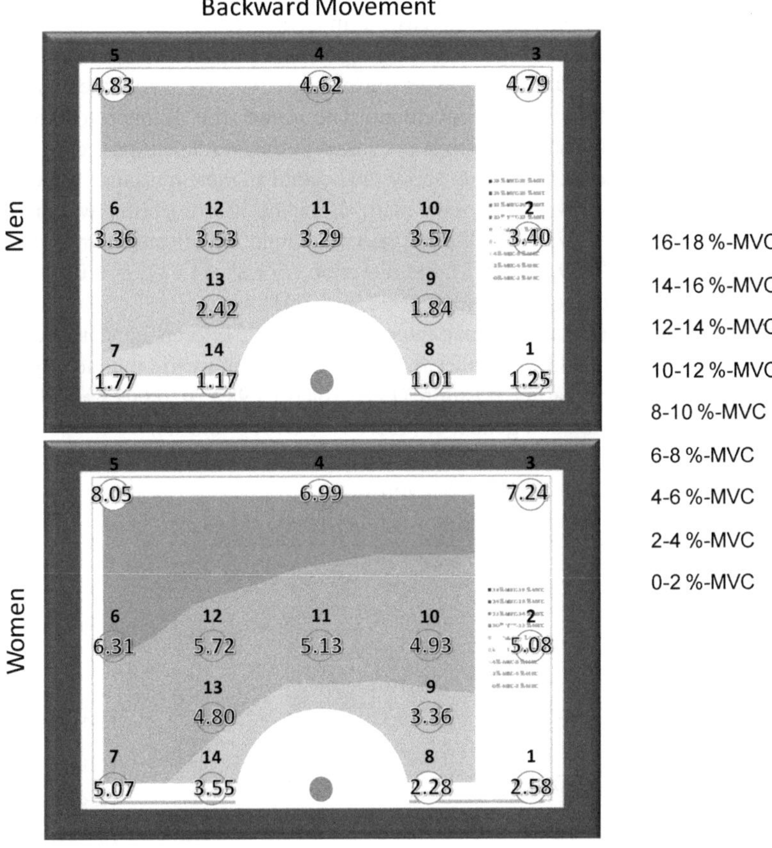

Figure 3: Muscle activation of the Deltoideus pars clav. in % MVC resulting from the backward movement for the male subjects (upper part) and the female subjects (lower part)

With regard to the interactions, we can interpret the main effect of the gender clearly. The muscular activity is significantly higher for the female subjects than for the male subjects. The main effect of the movement direction cannot be interpreted unambiguously (see Bortz, 2005). For movements to the target positions 1 and 8, the muscular activity is higher for the backward movement than for the forward movement. For movements to all other positions on the touchscreen, the muscular activity is higher for the forward movement. This can be explained by the overall low muscular activity in the Deltoideus when pointing to the right side. Moreover, the main effects of the target position cannot be interpreted clearly (see Bortz, 2005). Nevertheless, a trend which target position leads to higher and lower muscular activity can be seen.

As depicted in Figure 2 and 3 the highest muscular activity for the Deltoideus pars clav. was detected for the upper left touchscreen area (target position 5). For both gender-groups the highest muscle activation was found for the forward movement to target position 5. The female subjects show an average muscular activity of 17.28 % MVC value whereas the muscular activity of the male subjects is 9.68 % MVC value. Pointings to the target position 8 located at a position in front of the right shoulder required the lowest muscular activity for both male and female and for forward and backward movement. The higher the distance to this target position, the greater the strain which is provoked in the Deltoideus pars clav.

The results of the Infraspinatus show the second highest muscular activity and are depicted in figure 4 and 5. The main effect of the repetition was also not significant ($F(2,16)=0.073$, $p=0.930$). Again we found significant main effects for the gender ($F(1,8)=7.879$, $p=0.023$), the direction ($F(1,8)=24.137$, $p=0.001$) as well as for the target position ($F(13,104)=49.902$, $p<0.001$).

We found a significant ordinal interaction effect between gender and target position ($F(13,104)=2.843$, $p=0.002$) as well as significant hybrid interaction effects between direction and target position ($F(13,104)=13.982$, $p<0.001$) and between direction, target position and gender ($F(13,104)=3.379$, $p<0.001$).

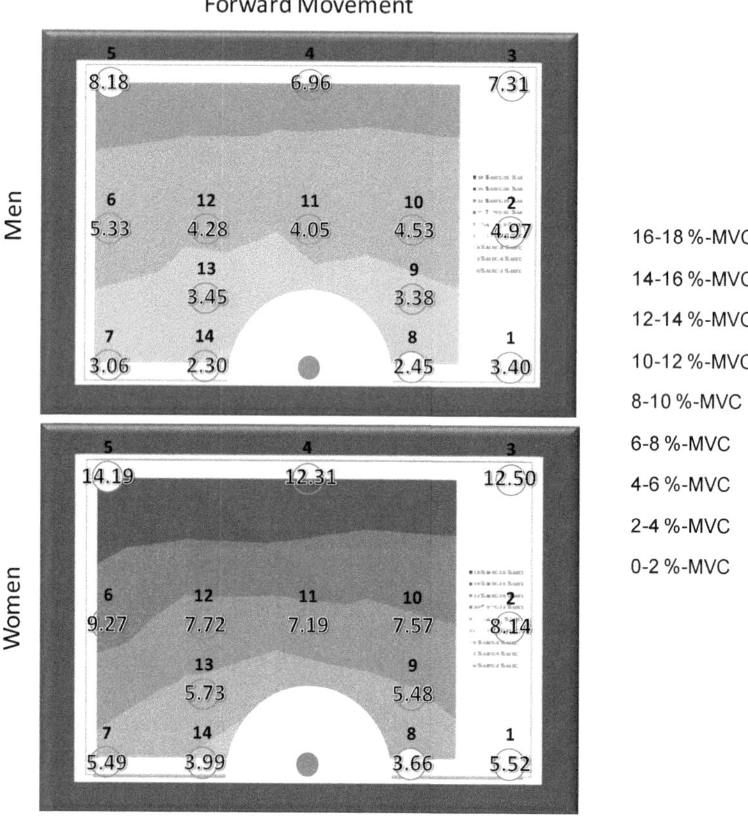

Figure 4 Muscle activation of the Infraspinatus in % MVC resulting from the forward movement for the male subjects (upper part) and the female subjects (lower part)

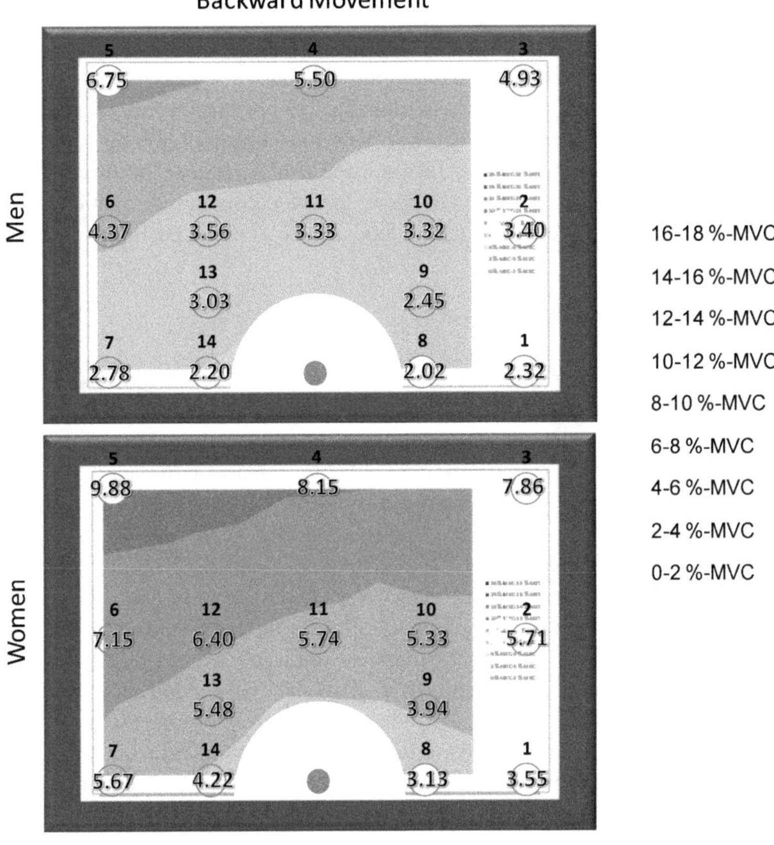

Figure 5 Muscle activation of the Infraspinatus in % MVC resulting from the backward movement for the male subjects (upper part) and the female subjects (lower part)

Considering the interactions, we can interpret the main effect of the gender clearly. The muscular activity is significantly higher for the female subjects than for the male subjects. The main effect of the movement direction cannot be interpreted unambiguously (see Bortz, 2005). For movements to the target positions 7 and 14 on the left side, in the female group the muscular activity is higher for backward movements than for forward movements, whereas for the male group when pointing to these positions the muscle activity is higher for the forward movement. For movements to all other positions on the touchscreen, the muscular activity is higher for the forward movement. Moreover, the main effects of the target position cannot be interpreted clearly (see Bortz, 2005). Nevertheless, a trend which target position leads to higher and lower muscular activity can be seen.

The areas that are correlated with high or low muscular activity in the Infraspinatus are similar to the results of the Deltoideus pars clav. (see Fig. 2, 3 and Fig. 4, 5). For both, female and male subjects, the muscular activity was highest for forward movements to target position 5 with 14.19 % of MVC value for the female and 8.18 % MVC for the male subjects. The target position with the lowest muscular activity is target position 8. The target positions 2 to 6 resulted in high muscular strain for forward and backward movements.

Altogether the results of the Brachioradialis, Triceps brachii cap., Trapezius pars desc., and Trapezius pars trans. show lower muscular activity but confirm the pattern found for the Deltoid pars clav. and the Infraspinatus. The forearm flexors and extensors were found not to be decisive for the ergonomic evaluation based on muscular strain as they were activated independently of target position.

4 CONCLUSION

The results show that pointing movements to the right, at the position of the right shoulder, require least muscular activity, whereas movements in the upper left area of the touchscreen result in the highest muscular activity. When performing pointing movements to target position 8, only the forearm is flexed and pronated. Further stretches, bends or rotations are not required for this sequence of motion. Target positions 1 and 9 as well as the starting point are also located in ergonomically favorable areas and can be reached with low strain. In contrary, pointings to target position 5 result in the highest muscular activity. Regardless of the body height, the user has to cover the longest distance to reach this position. Several muscles are involved in this movement sequence, as the user has to bend the upper body forward and fully extend the elbow joint. This is correlated with a high activation of the shoulder muscles, which stabilize and hold up the arm. In total, movements to all positions in the upper area of the touchscreen show similar unfavorable conditions for the user. Based on the results, recommendations for the ergonomic software development for large scale touchscreens can be given. An ergonomic input concept for right-handed persons should arrange frequently used buttons and interaction elements close to the position of the right shoulder. Buttons rather rarely used can be placed in more distal areas of the screen. Thus, the alignment of functions and menus at the top of the screen as often realized in conventional user interfaces should be avoided when possible.

The gender effects show that women experience higher muscular strain when working with large scale touchscreens. Consequently the time spent on working with large scale touchscreens should be shorter for women to reduce muscular fatigue and to prevent long-term disorders.

Since high muscular strain due to repetitive movements is especially critical for older users (Ahlström et al., 1992), future work should focus on the investigation of an older sample. A simple hardware-ergonomic approach to reduce the muscular strain particularly for target positions in the upper part of the screen is the variation of the tilt angle of the touchscreen (Vetter et al., 2010). Hence, the effect of different tilt angles on muscular activity while performing pointing movements to different target positions should be investigated accordingly.

ACKNOWLEDGEMENTS

This research was funded by the German Research Foundation according to the priority program no. 1184, Age-differentiated Work Systems. The authors extend their gratitude to all participants taking part in the experiment.

REFERENCES

Ahlström, B., S. Lehman, T. Marmolin. 1992. Overcoming touchscreen user fatigue by workplace design. *Proceedings of ACM Conference on Human Factors in Computing Systems CHI 92*: 101-102.

Bortz, J. 2005. Statistik für Sozialwissenschaftler (6. Auflage). Berlin: Springer.

Cooper, A. and L. Straker. 1998. Mouse versus keyboard use: A comparison of shoulder muscle load. *International Journal of Industrial Ergonomics* 22: 351-357.

Dietz, P. and D. Leigh. 2001. Diamondtouch: a multi-user touch technology. *Proceedings of the ACM Symposium on User interface Software and Technology* 14: 219-226.

Harvey, R. and E. Peper. 1997. Surface electromyography and mouse use position. *Ergonomics 40(8): 781-789.*

Hermens, H. J. 2011. SENIAM. Surface ElectroMyoGraphy for the Non-Invasive Assessment of Muscles. Accessed July 7, 2011, http://www.seniam.org.

Kumar, R. and S. Kumar. 2008. A Comparison of Muscular Activity Involved in the Use of Two Different Types of Computer Mouse. *International Journal of Occupational Safety and Ergonomics* 14 (3): 305–311.

Marras, W.S. 2000. Overview of Electromyography in Ergonomics. *Proceedings of the Human Factors and Ergonomics Society Annual Meeting* 44 (30): 5-534-5-536.

Shewman, T. 2008. Ableitpunkte für Oberflächen-EMG. Accessed November 28, 2011, http://www.velamed.com/downloads/SEMG-Muskel-Karte.pdf.

Shin, G. and X. Zhu. 2011. Ergonomic issues associated with the use of touchscreen desktop PC. *Proceedings of the Human Factors and Ergonomics Society Annual Meeting September* 55 (1): 949-953.

Vetter, S., J. Bützler, N. Jochems and C.M. Schlick. 2010. Ergonomic Workplace Design for the Elderly: Empirical Analysis and Biomechanical Simulation of Information Input on Large Touch Screens. In. *Advances in Understanding Human Performance: Neuroergonomics, Human Factors Design, and Special Populations*, eds. T. Marek, W. Karwowski and V. Rice. Boca Raton, USA.

CHAPTER 15

Shoulder Muscle Loading during Repetitive Reaching Task among Younger and Older Adults

Jin Qin[1, 2], *Jia-Hua Lin*[2], *Xu Xu*[2]

[1] University of Massachusetts, Lowell MA, USA
[2] Liberty Mutual Research Institute for Safety, Hopkinton MA, USA

ABSTRACT

Shoulder disorders and complaints constitute an important health problem in the working population. Research on shoulder muscle loading temporal change during low-intensity repetitive motion is limited. The goal of this paper was to evaluate the temporal pattern of shoulder muscle activities during a repetitive reaching task, and to compare the patterns observed in a younger and older group. Twenty participants completed an 80 min unloaded reaching task and the surface electromyography of upper trapezius and posterior deltoid muscles were recorded continuously. Both upper trapezius and posterior deltoid showed signs of localized muscle fatigue including increased amplitude and decreased median power frequency over the 80 min task. Muscle fatigue development was faster and greater in the older group than the younger group.

Keywords: shoulder, repetition, age, muscle fatigue

1 INTRODUCTION

Shoulder disorders and complaints constitute an important health problem in the working populations. In 2005, shoulder disorders amounted to 77, 800 or 6.3% of all nonfatal injuries in the private industry and 39% of all work-related musculoskeletal disorders (BLS, 2005). The prevalence of shoulder pain reported in the general population is 6%-11% under the age of 50 years, increasing to 16% - 25% in elderly people (van der Windt et al., 2000). Occupations that are most affected include computer users, assembly line, construction, food/meat processing, textile industry and packaging workers (Leclerc et al., 2004; Sommerich et al., 1993).

Repetitive movement often at low intensity is common in the workplace and is a known risk factor for WMSDs (NIOSH, 1997; Sommerich et al., 1993). Repetition is likely to result in muscle fatigue and many hypothesize that shoulder muscle fatigue is a precursor of shoulder complaints (e.g. Rempel et al., 1992; Takala, 2002). Quantitative assessment of the biomechanical loading and muscle fatigue and their temporal changes can help us understand the etiology of WMSDs and build up a database for dose-response analysis.

More research on low intensity repetitive motion during intermittent dynamic tasks is needed to study work-related risks on the upper extremity. De Looze et al. (2009) reviewed the studies on low-force work tasks and concluded that EMG manifestations of fatigue in the trapezius muscle do appear when the task intensity level is about 15-20% of maximum voluntary contraction. The injury mechanism may be different for isometric and dynamic muscle contractions. However, most studies used the isometric contractions at levels above 40% MVC to induce local muscle fatigue.

A rapid aging workforce in the US poses new public health challenge for preventing work-related injuries and illnesses. According to the US Census Bureau, the number of workers 55 and older will grow from 13% in 2000 to 20% in 2020. Age has been associated with increased risk of work-related musculoskeletal disorders (Miranda et al., 2008, Roquelaure et al, 2009). Despite the physiological and epidemiological evidence of correlation between age and WMSDs, data on how age modifies the effect of occupational tasks on biomechanical responses are limited.

This study was designed to evaluate the temporal pattern of shoulder muscle activities during a repetitive reaching task, and to compare the patterns observed in different age groups. We hypothesized that the 80 min reaching task without any external load will introduce localized muscle fatigue, and the temporal pattern will be different among younger and older adults.

2 METHODS

A repeated measures design laboratory experiment was conducted to answer the proposed research questions. Two age groups of female participants (ten in each group) with no history or current upper extremity MSDs were recruited. Participants' age, weight and height were shown in Table 1. The study was approved by the Internal Review Board of University of Massachusetts Lowell, and all participants had given their informed consent.

Table 1 Participants demographic information

Group	Age (yr)	Height (cm)	Weight (lb)
younger (n=10)	25.2 (3.9)	164.8 (5.2)	141.3 (24.0)
Older (n=10)	61.7 (4.3)	165.8 (9.8)	160.7 (39.2)

The task is designed to simulate low intensity repetitive upper extremity motions in the occupational settings. The task requires participants to reach and pick up washers and put them on a main board in front of them on a table. The washers were painted in five different colors and stored in bins with matching colors. The main board consists of a wooden base and 3 rows × 5 columns of hard plastic rods (height=65mm). The five columns of rods were also color-coded which match the colors of the bins and washers. The center of the table has a concave shape which has an inner edge and an outer edge (Figure 1). The edge of the board was aligned with the inner edge of the table. Participants performed the task while seated and their working shoulder aligned with the center of the main board and approximately two fists away from the inner edge of the table. The bins were placed along five lines drawn on the table 25° apart from each other (Figure 1). The center of the lines were the cross point of the outer edge of the table and a perpendicular line across the center of the main board. The heights of the table and the chair were adjusted so that participant's knee angle was 90° and the elbow height was similar to the table.

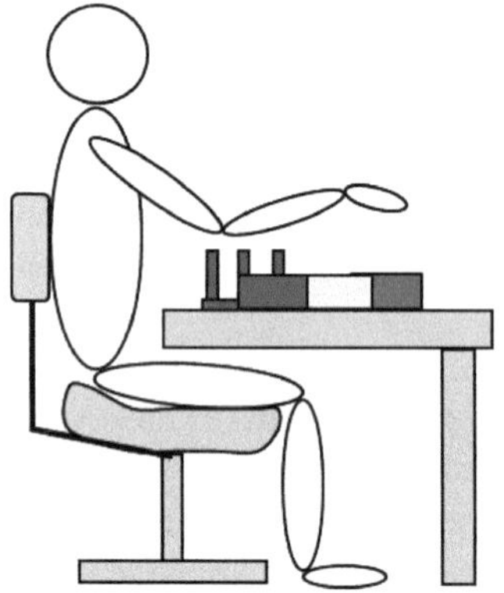

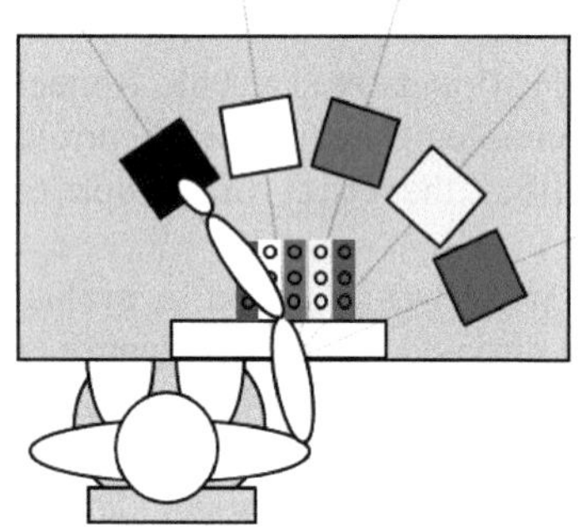

Figure 1 Experiment setup and layout

Participants were instructed to pick one washer at a time and put it through the rod with the same color from left to right (opposite for left-handed participants). Each reaching movement consists of steps of reach, grasp, back to main board, aiming and release. A metronome controlled the speed of reaching movement at 2 Hz. Participant went through three rows of rods repeated until all of them were full (~10min), and then the main board will be replaced with an empty one. Participants performed the task for a total of 80 min with short breaks (< 1 min) after each 10 min segment.

Four muscles were selected for electromyography (EMG) recordings: the descending part of upper trapezius and deltoid (anterior, middle and posterior). The myoelectric activity was recorded telemetrically by means of bipolar surface electrodes (Ag-AgCl, Noraxon, USA) with an inter electrode distance of 2.0 cm. The electrodes placement locations followed descriptions by SENIAM (surface EMG for non-invasive assessment of muscles) recommendations. The signals were amplified and filtered between 10 and 500 Hz. Data were acquired on a PC with a 16 bit analog to digital converter (National Instruments, Austin, TX) at a sampling rate of 1024 Hz.

In offline data processing, the EMG signals were digitally band-pass (10-300 Hz) and notch filtered. The root mean squared (RMS) value was calculated for epochs of 0.1 s. Power spectrum and median power frequency (MPF) were also derived. The EMG signals during dynamic reaching tasks were normalized to a reference voluntary electrical activity (RVE) obtained during an isometric submaximal reference voluntary contraction (RVC). RVC was recorded while subject seated with shoulder abducted 90° in plane of scapula, internally rotated and elbow extended (Boettcher et al., 2008) with a weight of 3 lb hanging over the distal side of the forearm. EMG during RVC was recorded for 15 s and the middle 10 s were averaged to obtain RVE.

Three one minute data were extracted from each of the 10 min segment (2-3. 5-6 and 8-9 min) for data analysis, resulting in total 24 (3×8 segments) data points throughout the whole recording period. Mixed effect model (proc mixed procedure in SAS 9.2) was used to evaluate the effect of age and time (repetition) on muscle EMG amplitude and power spectrum. For in the interest of length, we only presented the results of upper trapezius and posterior deltoid muscle in this paper.

3. RESULTS

Both MPF and RMS amplitude of the upper trapezius EMG showed significant change over time ($p<.0001$, Figure 1). MPF decreased while RMS increased indicating the development of localized muscle fatigue. The upper trapezius MPF ($p=0.03$) was greater and decreased faster (interaction=0.055) for the older group than the younger group. For additional analysis, the quadratic and cubic terms of the time variable were included in the model and both of the terms had significant effect on upper trapezius RMS amplitude ($p=0.04$ for t^2 and $p=0.004$ for t^3), as well as the interaction term of time (t) and age ($p=0.001$). The normalized RMS was consistently higher among the older group than the younger group however this difference was not statistical significant ($p=0.16$).

Similar to upper trapezius, posterior deltoid muscle MPF decreased and the RMS amplitude of the EMG increased over time ($p<.0001$, Figure 2). When the quadratic and cubic terms of the time variable were added into the model, all of them showed significant effect on RMS amplitude of the posterior deltoid EMG ($p=<.0001$ for t^2 and $p=0.0001$ for t^3), as well as the interaction terms of time ($p=0.046$) and time^2 ($p=0.0004$) with age. For MPF, the interaction between time and age (p=0.005) and between time² and age (p=0.043) were statistically significant. MPF decreased faster in the older group than the younger group.

Figure 1 Upper trapezius EMG median frequency and RMS amplitude during repetitive reaching task.

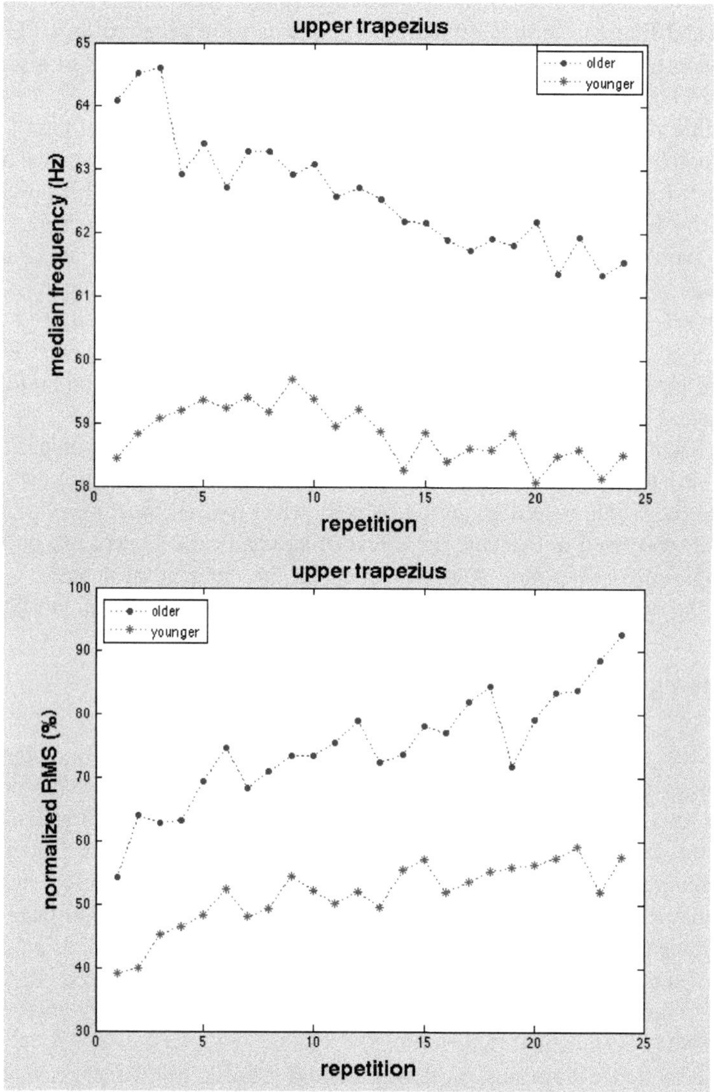

Figure 2 Posterior deltoid EMG median frequency and RMS amplitude during repetitive reaching task.

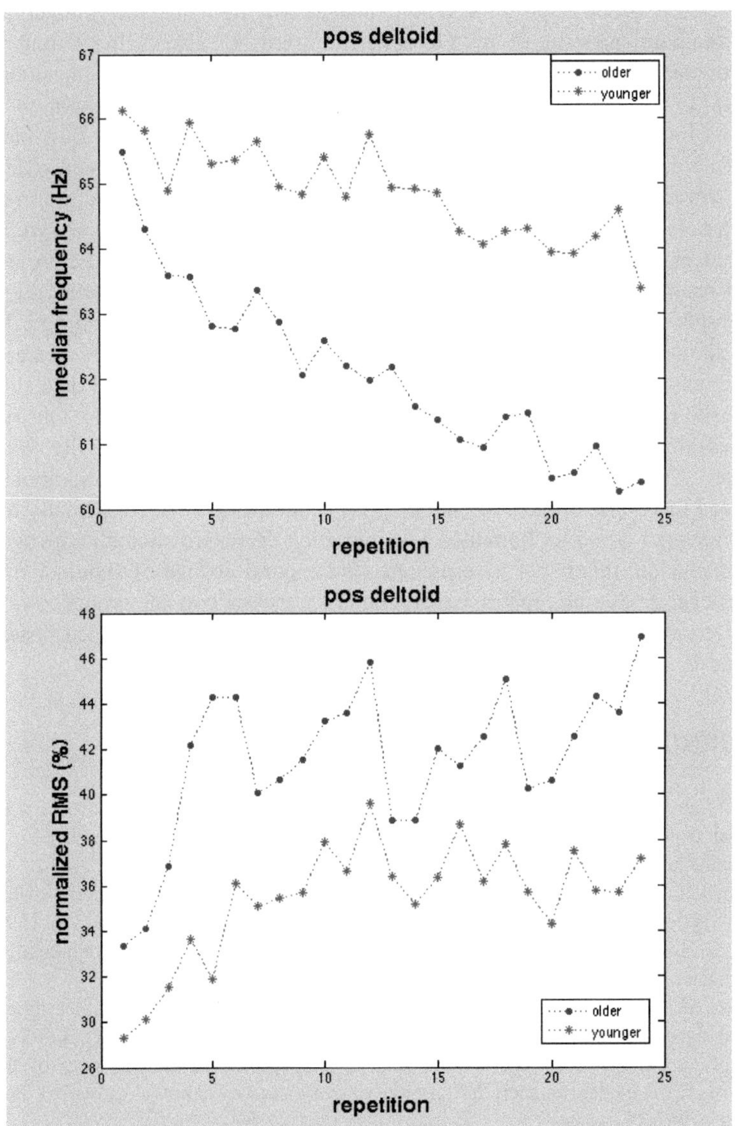

4 DISCUSSIONS AND CONCLUSION

This papers aims to evaluate the temporal patterns of shoulder muscle activities during prolonged repetitive motion among younger and older female adults. Shoulder muscle fatigue development was affected by repetition. Both upper trapezius and posterior deltoid muscles showed signs of localized muscle fatigue including increased amplitude and decreased median power frequency over the 80 min reaching task. Age group affected the temporal pattern of shoulder muscle activities. Older group often showed more change over time than the younger group.

The temporal patterns of muscle activities were not always linear but rather second or third order polynomial. This may be associated with the physiological responses to localized muscle fatigue, which needs further exploration. The significant interaction effects between age and the polynomial time terms indicate that muscle development over time were different between younger and older age groups. Muscle fatigue developed faster among the older group suggesting that older workers may be more susceptible and at higher risk of muscular symptoms and disorders than their younger counterparts. This is different from the previously reported results during isometric contractions. A review of the literature suggests mixed results on the association between skeletal muscle fatigue and age. Older individuals develop less muscle fatigue than the young in isometric contractions (Lanza et al., 2004; Yassierli et al., 2007a, 2007b). However limited data showed the opposite results during dynamic contractions (Christie et al., 2011). Our results confirmed that older individuals were less fatigue resistant than the younger individuals during low intensity dynamic shoulder motions. In addition, the normalized EMG amplitude values among the older group were consistently higher than the younger group for both muscles suggesting decreased strength with age.

In conclusion, this paper investigated the temporal change of shoulder muscle activities over an 80 min repetitive reaching task between two age groups. Evidence of localized muscle fatigue was observed and the older group developed faster and more muscle fatigue than the younger group.

REFERENCES

Bureau of Labor Statistics (BLS): Nonfatal occupational injuries and illnesses requiring days away from work, 2005. United States Department of Labor News, November 17, 2006. Available at http://stats.bls.gov/iif/oshwc/osh/case/osnr0027.pdf.

Boettcher, C.E., Ginn, K.A. & Cathers, I., 2008. Standard maximum isometric voluntary contraction tests for normalizing shoulder muscle emg. *J Orthop Res,* 26 (12), 1591-7

Christie, A., Snook, E.M. & Kent-Braun, J.A., 2011. Systematic review and meta-analysis of skeletal muscle fatigue in old age. *Med Sci Sports Exerc,* 43 (4), 568-77

De Looze, M., Bosch, T. & Van Dieen, J., 2009. Manifestations of shoulder fatigue in prolonged activities involving low-force contractions. *Ergonomics,* 52 (4), 428-37

Lanza, I.R., Russ, D.W. & Kent-Braun, J.A., 2004. Age-related enhancement of fatigue resistance is evident in men during both isometric and dynamic tasks. *J Appl Physiol,* 97 (3), 967-75

Leclerc, A., Chastang, J.F., Niedhammer, I., Landre, M.F. & Roquelaure, Y., 2004. Incidence of shoulder pain in repetitive work. *Occup Environ Med,* 61 (1), 39-44

Miranda, H., Punnett, L., Viikari-Juntura, E., Heliovaara, M. & Knekt, P., 2008. Physical work and chronic shoulder disorder. Results of a prospective population-based study. *Ann Rheum Dis,* 67 (2), 218-23

NIOSH 1997. Musculoskeletal Disorders and Workplace Factors.

Rempel, D.M., Harrison, R.J. & Barnhart, S., 1992. Work-related cumulative trauma disorders of the upper extremity. *JAMA,* 267 (6), 838-42

Roquelaure, Y., Ha, C., Rouillon, C., Fouquet, N., Leclerc, A., Descatha, A., Touranchet, A., Goldberg, M. & Imbernon, E., 2009. Risk factors for upper-extremity musculoskeletal disorders in the working population. Arthritis Rheum, 61. 1425-34.

Sommerich, C.M., Mcglothlin, J.D. & Marras, W.S., 1993. Occupational risk factors associated with soft tissue disorders of the shoulder: A review of recent investigations in the literature. *Ergonomics,* 36 (6), 697-717

Takala, E.P., 2002. Static muscular load, an increasing hazard in modern information technology. *Scand J Work Environ Health,* 28 (4), 211-3

Van Der Windt, D.A., Thomas, E., Pope, D.P., De Winter, A.F., Macfarlane, G.J., Bouter, L.M. & Silman, A.J., 2000. Occupational risk factors for shoulder pain: A systematic review. *Occup Environ Med,* 57 (7), 433-42

Yassierli & Nussbaum, M.A., 2007. Muscle fatigue during intermittent isokinetic shoulder abduction: Age effects and utility of electromyographic measures. *Ergonomics,* 50 (7), 1110-26

Yassierli, Nussbaum, M.A., Iridiastadi, H. & Wojcik, L.A., 2007. The influence of age on isometric endurance and fatigue is muscle dependent: A study of shoulder abduction and torso extension. Ergonomics, 50 (1), 26-45

CHAPTER 16

Whole-body Vibration When Operating Machinery

Svetla Fiserova

VŠB – Technical University of Ostrava
Faculty of Safety Engineering
svetla.fiserova@vsb.cz
Czech Republic

ABSTRACT

The article focuses on problems associated with the influence of whole-body vibration when operating machinery and deals with the evaluation of results of measurement of whole-body vibration in current real work conditions on selected machinery, i.e. four mobile machines and three heavy duty truck-mounted cranes. Vibration is caused by the resonance of parts of the body or tissues and the increased tonus of muscles maintaining the body in a stable position. The main negative effect is that on the operator's backbone. The obtained results confirm the necessity of paying, in the course of assessment, consistent attention to the prevention of exposure to whole-body vibration when operating machinery.

Keywords: Whole-body vibration, measurements, assessment, machinery, hygienic exposure

INTRODUCTION

To a human body, intensive vibration is transmitted most frequently from the oscillating parts of various machines and equipment, hand-held tools, means of transport, seats, work platforms, ant others. As vibration the motion of an elastic body or medium the individual points of which oscillate about an equilibrium position is regarded. When vibration acts on the human body, the degree of harmfulness is markedly influenced by the reaction of the organism, position of the body and limbs in relation to the direction of vibration, position and size of the area through which the vibration is transmitted to the human organism, and by forces produced by the human during exposure to vibration. It has been proved that vibration causes the unfavourable response of the human organism and that during long-term exposure, permanent damage to human health may occur.

Vibration acts on the human body in several directions and along several pathways. According to the type of vibration transmission, the following kinds of vibration are distinguished: **whole-body vibration transmitted to a sitting or standing person,** hand-arm transmitted vibration, whole-body vibration at a frequency less than1Hz, vibration transmitted in a special way and whole-body vibration in buildings.

Whole-body vibration transmitted to a sitting or standing person is assessed in a frequency range from 0.5 Hz to 80 Hz and is produced by a vibrating seat or platform in three axes of action.

As far as health effects are concerned, in a case of exposure to whole-body vibration, the effects are always system effects affecting the whole organism. Exposure to intensive whole-body vibration is connected with the unpleasant subjective perception of discomfort that can be assessed from the psychological and/or physiological point of view. Long-term exposure to whole-body vibration and shock in connection with a forced working position may result in backbone injury. Whole-body vibration at a very low frequency (less than 1 Hz) causes motion sickness. Generally, vibration induces general body fatigue leading to reduced attention, slowed and worsened perception, reduced motivation and decreased working efficiency. Vibration and shock effects on the human are observed with regard to ensuring comfort, working efficiency and health of exposed humans.

METHODS

Research conducted in current real work conditions on selected machinery, i.e. four mobile machines and three heavy duty truck-mounted cranes focuses on the verification of effects of whole-body vibration when operating machinery. The measurements were taken using standard methods with measuring instruments Bruel&Kjaer equipped with adequate software.

In the Czech Republic, methods of vibration exposure assessment even in the area of whole-body vibration are in accordance with legal rules of the European Union. The latest directive concerning vibration affecting humans was adopted on the 25th of June 2002, the Human Vibration Directive 2002/44/EC. It determines the minimum health and safety requirements regarding the exposure of workers to the risks arising from physical agents – vibration. Vibration is measured and evaluated according to standard methods, which we consider as methods contained in the Czech technical standard under conditions determined in the government Decree

No. 272/2011 Coll., on health protection against unfavourable noise and vibration effects. The government Decree prescribes, among other matters, limit values for occupational vibration exposure, determines both minimum measures leading to employee health protection and vibration risk assessment, states basic mathematical relations for exposure level calculation. The permissible limit value for exposure to whole-body vibration transmitted to employees for 8 hours, expressed as average weighted vibration acceleration level (RMS VTV recalculated to 8-hour working shift with reference to the duration of direct exposure) amounts to 114 dB, which corresponds to the acceleration value of 0.5 $m.s^{-2}$. The expert evaluation of whole-body vibration should be made in compliance with valid international standards, especially the following international standards: **ČSN ISO 5805,** *Mechanical vibration and shock – Human exposure – Vocabulary,* **ČSN ISO 2631-1,** *Mechanical vibration and shock – Evaluation of human exposure to whole-body vibration – Part 1: General requirements,* **ČSN EN 14253,** *Mechanical vibration – Measurement and calculation of occupational exposure to whole-body vibration with reference to health- Practical guidance.*

Before the whole-body vibration measurement itself, the workplace must be assessed and the activity of workers must be analysed. The analysis concentrates on the position of the workplace, the description of performed activities of workers, including the description of working position and the time image of actual exposure to whole-body vibration.

CHARACTERISTICS OF OCCUPATIONAL EXPOSURE EVALUATION

For the needs of hygiene evaluation of whole-body vibration, the vibration acceleration level L_a and the RMS vibration acceleration a_e are taken as determining quantities. The vibration acceleration level L_a is given by the following relation

$$L_a = 20 \log (a/a_0) \qquad [\text{dB}]$$

where
a is the instantaneous vibration acceleration in $m.s^{-2}$
a_0 is the reference vibration acceleration level in $m.s^{-2}$
a_0 is 10^{-6} $m.s^{-2}$.

The RMS vibration acceleration a_e is given by the following relation

$$a_e = \sqrt{\frac{1}{T}\int_0^T a^2(t)dt} \qquad [\text{m.s}^{-2}]$$

where
a(t) is the instantaneous acceleration in $m.s^{-2}$
T is the time for which the RMS acceleration in s is to be determined.

Vibration acceleration levels and RMS vibration accelerations are vibration quantities that are substitutable by each other. The weighted level of vibration acceleration L_{aeq} is such vibration acceleration level that corresponds to the frequency correction for the given way and conditions of transmission and the direction of vibration. It is expressed in dB. The total weighted level of vibration acceleration is given by the vector sum of weighted RMS accelerations along three orthogonal axes. In the case of whole-body vibration, vibration should be evaluated with regard to its effects on humans, in bands of frequency of 0.5 – 80 Hz. Whole-body vibration is measured and stated in the three directions of an orthogonal coordinate system. The total weighted level of vibration acceleration is after recalculation stated as RMS VTV (Vibration Total Value). RMS VTVs are recalculated to 480-minute, i.e. 8-hour working shift, when T = 8 hours (480 minutes).

REQUIREMENTS FOR WHOLE-BODY VIBRATION MEASUREMENT

When whole-body vibration acts on a human, the vibration spreads along three axes. **The x-axis** determines lateral vibration, i.e. vibration acting on a human from the side. This vibration causes swaying of the whole human body from side to side. Vibration acting on the front and rear parts of the human body is determined by the **y-axis**. When this vibration acts, the whole person sways back and forth. The **z-axis** determines vibration acting from below and from above, when the human turns from one side to the other. In the course of measurement of whole-body vibration, measurement must be carried out along the three above-mentioned axes, using a special sensor that is usually placed on a seat below the sitting operator of the machine, in accordance with relevant international standards. For the evaluation of total exposure it is necessary to determine in a standard way the actual period of vibration induced loading in a characteristic working shift.

STRATEGY AND MEASUREMENT METHOD

Whole-body vibration measurements were done in real conditions of work and in compliance with the long-term focus of the Laboratory of Occupational Safety and Ergonomics at the Faculty of Safety Engineering of VŠB – Technical University of Ostrava on physical factors of a working environment. The measurements were carried out in the years 2010 - 2011. All measurements were performed in accordance with relevant international standards for whole-body vibration measurement, after thorough preparation of measuring equipment, determination of actual exposure profile and duration, and after minimization of possible sources of undesirable effects on measurement results. Every measurement was repeated three times.

For all the measurements a Bruel&Kjaer device for vibration measurement and analysis VA 4447 (Fig.1) with a possibility of connection of a piezoelectric accelerometer of type 4520 with a seat adapter was used – Fig.1. The device VA 4447 is equipped with its own software Vibration Explorer. In all cases the measuring devices were calibrated using a calibrator B&K of the type 4294 before and after each measurement. Instrument equipment fulfils requirements ⅟₃foroctave band measurement that correspond to the requirements for screening measurement.

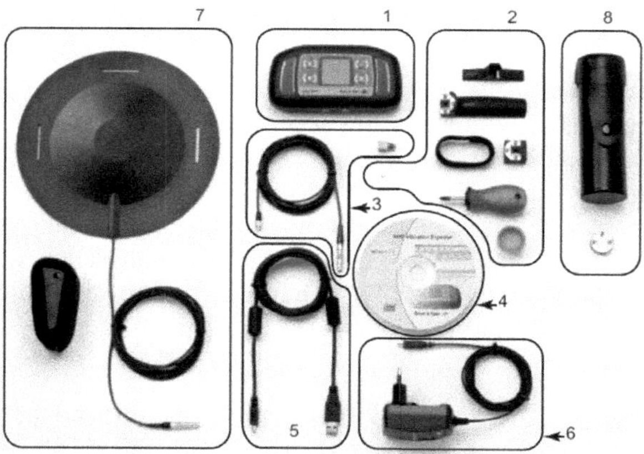

Fig. 1 1.Used measuring equipment 2.VA 4447 2.Fastening elements 3.Lemo cable, accelerometer of type 4520 4.Installation CD 5.USB cable fastening strap 6.Power adapter 7.Seat accelerometer adapter 8.Calibrator of type 4294

At first, test measurements were carried out. For the measurement of whole-body vibration, mobile machines used commonly in industry were selected. They were four mobile machines: Liebherr wheel loader (Fig.2), Linde fork-lift truck, Hangcha fork-lift truck (Fig.3) and GAZ open truck. All measurements were performed under the same conditions of transport in the framework of a large industrial company in outside storage areas, industrial roads, including a level crossing.

Fig.2 Liebherr wheel loader

Fig.3 Hangcha fork-lift truck

For the measurement of whole-body vibration from immobile all-terrain machines, real conditions in the course of routine work of operators (Fig. 4) of three heavy duty truck-mounted cranes of various types and capacities for lifting, lowering and moving a load – a 2500 kg weight (Fig.5) were selected. The following heavy duty truck-mounted cranes were used: TATRA AD28 (1989) crane with capacity of 28 000 kg, DEMAG AC40 (2000) with capacity of 40000 kg and DEMAG AC205 (1996) with capacity of 80.000 kg.

Fig.4 Seat in a truck-mounted crane cab

Fig.5 Example of moving a load

Fig.6 Illustrative photo of a heavy duty truck-mounted crane

In the case of a truck-mounted crane, there are two operator's places – the driver's seat in the undercarriage cab and the operator's place in the upper cab at the top of the slewing tower for controlling crane operations, as can be seen in an illustrative photo in Fig. 6. Heavy duty truck-mounted cranes are designed for use when the operating position and the mass of a load change frequently. Operations must be performed on consolidated and solid ground and the machine must be relevantly anchored by means of an installed support system of the machine.

RESULTS

In all measurements, a partial record was obtained for each measurement axis in both graphical and numerical form. At the same time, a record of recalculated MTVV values was received – maximum instantaneous RMS value of vibration, when the time of integration equals to 1s, Peak – average maximum value, RMS VTV for all three measurement axes. All obtained results were processed and evaluated using the software of the measuring device. In Figs. 8-9 there are examples of records of measurement in the x-axis and the z-axis for one of the measured heavy duty truck-mounted cranes and in Fig. 10-11 examples of record of recalculated measured values of exposure in other two heavy duty truck-mounted cranes.

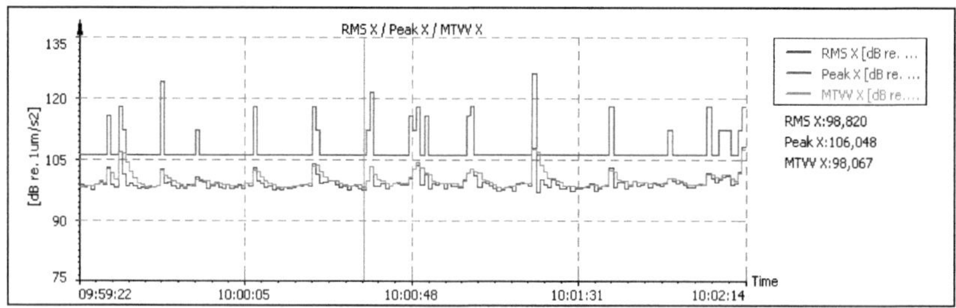

Fig.7 Record of measurement results for the machine TATRA AD28 for x-axis

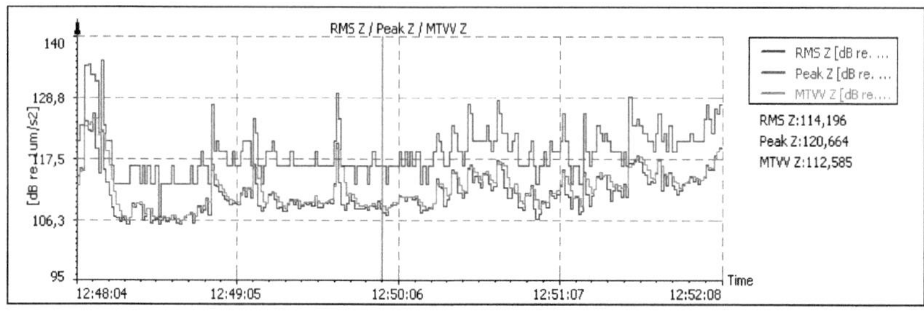

Fig.8 Record of measurement results for the machine TATRA AD28 for z-axis

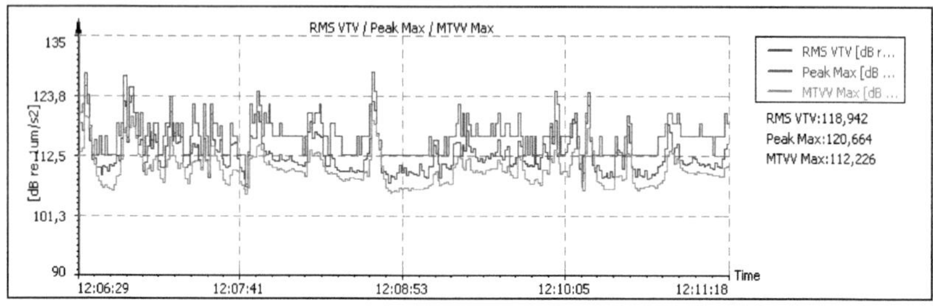

Fig.9 Record of recalculated measurement results for the machine DEMAG AC40

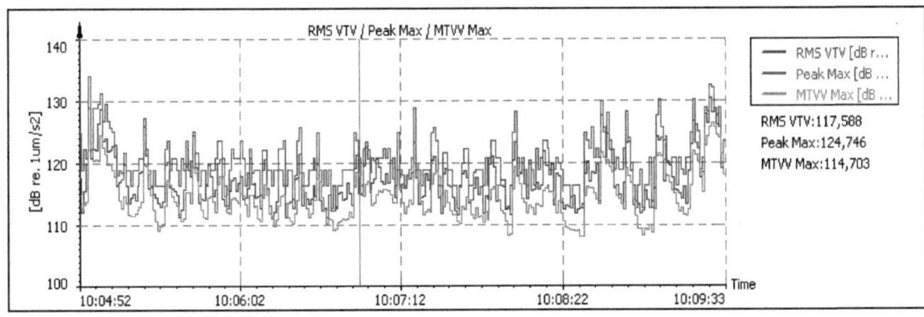

Fig.10 Record of recalculated measurement results for the machine DEMAG AC205

For actual exposure assessment, real time images of work with the determination of the real duration of exposure to whole-body vibration per shift should always be taken as a basis. In a case of the seven measured machines, when screening measurements were carried out in real conditions of work, recalculations of measured final RMS VTVs per eight-hour shift to A(8) [dB] were executed as if the exposure lasted for the whole shift. It is a case of calculation of the least favourable exposure duration in spite of the fact that this exposure per hour may be used as the basic item of data necessary for the calculation of optimum duration of exposure per shift to avoid exceeding the determined limit values. In Table 1 there is the final recalculation of RMS VTVs to A (8) [dB] for individual measured machines.

Table 1 Final recalculation of RMS VTVs to A (8) [dB]

	Linde I	Hangcha II	GAZ III	Liebherr IV	Tatra AD28 V	DEMAG AC40 VI	DEMAG AC205 VII
A(8) [dB]	95,4	97,4	92,5	95,3	111,0	**113,3**	**116,2**

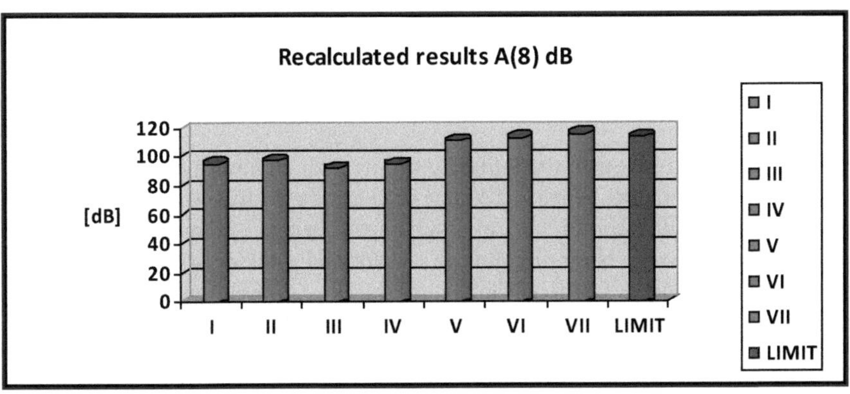

FIGURE 11 Final recalculation of RMS VTVs to A (8) [dB]

DISCUSSION

The source of vibration transmitted to the humans who operate mobile machines is a machine moving on roads on the surface and underground. The character of vibration is affected by the weight of the vehicle, its velocity, ground and road surface and technical characteristics of the vehicle; depends on the direction of vehicle travel, i.e. on the flat or on the slope, on braking and, on the contrary, acceleration of the vehicle. For the proper transmission to humans, above all seat parameters, vehicle suspension system, cab springing, vehicle dynamics with corrective actions of the driver and vehicle load are important.

Heavy duty truck-mounted cranes cannot be, from the point of view of work of the crane operator in a separate cab, considered to be mobile in the period of work, because the machines are all-terrain machines anchored firmly for operations performed. Whole-body vibration exposure is caused mainly by a vibrating crane jib in the course of moving large loads.

From the recalculated results presented in Table 1 and Fig. 11 it is clear that the hygienic limit of 114 dB, i.e. 0.5 $m.s^{-2}$, is kept even at the expected least favourable eight hour exposure A(8) in all mobile machines being examined. These results suggest that the technical condition as well as equipment of mobile machines being evaluated is, from the point of view of whole-body vibration transmission, favourable. Other conditions in terms of ground roughness can be causes of different levels of whole-body vibration exposure.

The results obtained during the operation of anchored heavy duty truck-mounted cranes when handling loads standardly suggest exposures that are, from the point of view of hygiene, severe. The worst results were obtained in a case of the heavy duty truck-mounted crane having the largest declared capacity. With reference to the fact that the time of work in the operator's cab on the anchored crane can be shortened merely by breaks for safety and hygienic reasons, the level of whole-body vibration exposure reaches or exceeds the hygienic limit. Prevention of possible unfavourable effects on health can be ensured above all by organizational measures. A system for compensation for health injuries caused by exposure to whole-body vibration is not introduced in the Czech Republic.

CONCLUSION

From the point of view of the influence of vibration on human health, the hand-arm transmitted vibration, which is a source of considerably frequent occurrences of occupational health injuries, is of high significance. However, repeated measurements of whole-body vibration in current real work conditions performed on selected machinery, i.e. four mobile machines and three heavy duty truck-mounted cranes have confirmed that the occurrence of whole-body vibration cannot be underestimated either. Heavy duty truck-mounted cranes are modern machines that are used in a wide range of applications. Occupational health protection must be ensured with reference to the characteristics of these machines in the light of whole-body vibration and must focus especially on keeping the adequate duration of exposure per shift.

REFERENCES

Fábry, V., & Fišerová, S., (2010), *Celotělové vibrace při obsluze mobilních strojů. (Whole-body vibration when operating mobile machines).* Master's thesis in the field of study Safety Engineering, Faculty of Safety Engineering, VŠB – Technical University of Ostrava, Czech Republic.

Náhlý, M.., & Fišerová, S., (2011), *Expozice hlukem a vibracemi při obsluze těžkých autojeřábů. (Noise and vibration exposure when operating heavy duty truck-mounted cranes).* Bachelor's thesis in the field of study Safety of Work and Processes, Faculty of Safety Engineering, VŠB – Technical University of Ostrava, Czech Republic.

Tureková, I. & Kuracina, R. & Rusko, M., *Manažment nebezpečných činností (Dangerous operation management)*, Trnava, AlumniPress, 2011, ISBN 978-80-8096-139-8

Smetana,C., (1998), *Hluk a vibrace. (Noise and vibration).* Sdělovací technika, Praha, Czech Republic

Sujová, E., *Princípy manažérskeho rozhodovania v podmienkach rizika.* In Aktuálne anažérske trendy v teórii a praxi (Actual manager's trends in theory and praxis) Žilina : Žilinská univerzita v Žiline, 2008, s. 163-168. ISBN 978-80-8070-966-2.

ČSN EN 14253/A1:2008 *Vibrace – Měření a výpočet expozice celkovým vibracím na pracovním místě s ohledem na zdraví-Praktický návod,* (harmonized with EN 14253+A1:2007 *Mechanical vibration – measurement and calculation of occupational exposure to whole-body vibration with reference to health – Practical guidance).*

Directive 2002/44/EC of the European Parliament and of the Council of 25 June 2002 on the minimum health and safety requirements regarding the exposure of workers to the risks arising from physical agents (vibration).

CHAPTER 17

An Assessment of Manual Material Handling and Postural Stress among the Warehouse Workers

Hema Bhatt, Promila Sharma

G.B. Pant University of Agriculture & Technology
Pantnagar, India
hemabhatt2000@gmail.com

ABSTRACT

Manual materials handling (MMH) is a component of many jobs and activities undertaken in life. Although this is an era of improved mechanization and technology but it has not eliminated manual materials handling (MMH). A significant problem associated with manual handling activities involving loading and unloading tasks is the fact that they are the primary cause of overexertion injuries. Loading and unloading tasks include diverse activities such as lifting, lowering, holding, pushing, pulling, carrying and turning of weights etc. Adoption of poor working posture in order to perform tasks could lead to a postural stress, fatigue and pain, which may in turn force the operator to stop work until the muscles recover. To prevent pain and injuries, the manual material handling tasks should be designed to take into account several risk factors related to the task being handled.This paper describes the results of an experimental study aimed at evaluating the postural discomfort during loading and unloading in warehouses through standard OWAS scale and a revised Nordic musculoskeletal questionnaire validated by Kuroinka *et al.* (1987). It was found through OWAS scale that the corrective measures were required immediately and as soon as possible in most of the activities in warehouse.

Keywords: workload, posture, loading, unloading, energy expenditure

1 THE MANUAL MATERIAL HANDLING TASK

Manual handling is a severe problem in developing and underdeveloped countries. Work related disorders has increased dramatically in these countries and workers are exposed to much worse conditions due to inadequate safety system, lack of awareness, lack of training of occupational safety, health and lack of ergonomic standards. A significant problem associated with manual handling activities involving loading and unloading tasks is the fact that they are the primary cause of overexertion injuries. Loading and unloading tasks include diverse activities such as lifting, lowering, holding, pushing, pulling, carrying and turning of weights etc. The types of back injuries most frequently reported are strains and sprains, dislocation (herniation) of the lumbar disc, fracture, joint inflammation (mostly L4/L5 and L5/S1; occasionally other joints such as the shoulder and hip), laceration of muscle tissue, contusion, and nerve (sciatic) involvement, often leading to activity limitation and workplace accidents.

Epidemiological and biomechanical studies have found that a combination of high external load and poor movement patterns cause a high internal load on the spinal structure and increases the risk of pain and injury. Poor movement patterns consist primarily of bending or twisting of the trunk, or both. Bending occurs during reaching and lifting of an object from a low to a high surface. Twisting of the trunk is mostly the result of inadequate workspace. Excessive bending and twisting of the trunk have been related to higher biomechanical and physiological costs and musculoskeletal injuries. The involvement of back and abdominal muscles in lifting activity has long been established. Several researches have shown that the application of ergonomic principles and programs in almost all workplaces result in increasing productivity and decreasing work related musculoskeletal disorders (WMSDs) (Saraji, J.N. *et al.* 2004). The moral and economic consequences that result from pain and injury made it necessary to study and, therefore, attempt to solve such a problem. Hence an effort in this research is made to ergonomically analyse the postural stress of the warehouse workers involved in manual loading and unloading work.

2 METHODOLOGY

The study population comprised the twenty warehouse workers employed at Food Corporation of India's unit at Rudrapur city in Uttarakhand. Personal interview method was used to elicit the information relating to individual details, type of work, musculoskeletal pain and discomfort with the locomotive organs by using the revised Nordic musculoskeletal and postural discomfort questionnaire. The already established OWAS (Ovako Work Assessment System) scale was used for work posture analysis.

3 RESULTS AND DISCUSSION

3.1 General profile of the respondents

The mean age of the respondents selected for the study was 36.75 years with standard deviation ± 5.99, height was 158.87 cm with standard deviation ±7.69 and weight was 68.42 kg with standard deviation ± 3.93 (Table 1).

Table 1. General profile of the respondents (N=20)

S. no.	Physical characteristic	Mean ± *S. D.*
1	Age (years)	36.75 ± 5.99
2	Height (cm)	158.87±7.69
3	Weight (kg)	68.42± 3.93

3.2 *OWAS- Scoring and determination of action category during loading and unloading*

OWAS analysis provides the opportunity to compare the job studies according to the number of postures which need to be corrected soon or immediately (Kivi P. and Mattila M.1991). The OWAS method collects observation information on worker postures on back, arms and legs. Each posture of the OWAS is determined by the four digit code in which the numbers indicates the postures of the back, the arms and the load needed. Each OWAS posture code is then analysed by using the individual OWAS classified posture combination to get the action category for each work phases. The classification for individual posture combination indicates the level of risk injury for the musculoskeletal system. If the risk for musculoskeletal disorder is high, then the action category indicates the need and urgency for corrective actions. So, accordingly the working postures of the respondents while performing various tasks involving loading and unloading activities were observed by the researcher and a code number was assigned to each posture by using the posture coding sheet of OWAS method. The position of back, upper limbs i.e. arms and lower limbs, legs as well as load of force used in carrying out activities were considered for analysis of posture.

1- *Loading grain bag on self:* Data presented in table 2 shows that maximum respondents kept their back bent with a forward flexion (75%), both arms are above shoulder level (100%), standing on one knees bent (65 %), and weight or force needed exceeded 20 kg. They were experiencing the pain in back arms, knees and neck.

2- *Carrying grain bag*: In carrying grain bag activity it was found that 60 % of respondents kept their back bent and twisted, both arms above shoulder level as reported by majority (75%) respondents, all the respondents were in a state of walking or moving (100%) while carrying grain bags from one place to other, and all respondents were carrying weight or force is exceeding 20kg.

3- *Unloading grain bag:* It was found that in unloading grain bag all respondents (100%) kept their back bent and twisted, with majority (50%) having arms at or above shoulder level, followed by respondents (30%) who adopted posture with one arm at or above shoulder level and few (20%) with both arm below shoulder level. Majority of the respondents (80%) adopted posture with on one or both leg kneeling , while 20% standing on one knee bent while unloading grain bags. In this activity too respondents carried load or force less than 10 kg.

4- *Loading grain bag on other:* In loading grain bag on other worker who will carry it, it was found that 50% of respondents kept their back bent while 30% bent & twisted. Both arms above shoulder level as reported by all of the respondents. Majority of respondents were standing on both knees bent, and all respondents were carrying weight or force is exceeding 20kg.

5- *Arranging grain bag*: In arranging grain bags 60 % respondents kept their back bent, all respondents kept their both arms below shoulder level, maximum (75 %) standing on both knees bent, weight and force was exceeding 20kg by all respondents.

3.3 Action level - Corrective measures needed for posture adopted during task involving loading and unloading

The codes assigned by the investigator to the postures adopted by the respondents while working on existing workplace were further analyzed to suggest action category for each adopted posture. The suggested action level categories were as follows:

1. *Loading grain bag on self*: Action level for adopted posture depicts that50% respondents need corrective measures immediately due to very poor posture and 35% respondents needed corrective measures as soon as possible followed by 15% due to less poor posture needed to be corrected in near future as shown in table 4.
2. *Carrying grain bag*: Data showed that only 15 percent respondents need corrective measures in the near future, 35 percent needed corrective measures as soon as possible and half of the respondents (50%) needs corrective measures immediately in their posture (table 4).
3. *Unloading grain bag*: It was found that 70 % respondents needed corrective measures immediately, while 30 % needed corrective measures as soon as possible.
4. *Loading grain bag on other*: Majority of the respondents (80%) posture needed corrective measures as soon as possible, with 20% indicating to be corrected immediately.
5. *Arranging grain bag*: It was found that only 35% respondents need corrective measures as soon as possible while 65% in the near future needed corrective measures.

Table 2. Posture adopted by respondents while performing various tasks involving loading and unloading work (N=20).

S. no.	Body posture and assigned Code no.	Action				
		LGS	CG	UG	LGO	AG
1.	**Back**					
	1. Straight	-	-	-	-	-
	2. Bent	15 (75)	8 (40)	-	10 (50)	12 (60)
	3. Twisted	-	-	-	4 (20)	4 (20)
	4. Bent & twisted	5 (25)	12 (60)	20 (100)	6 (30)	4 (20)
2.	**Arms/Upper limbs**					
	1. Both arms are below shoulder level	-	-	4 (20)	-	20 (100)
	2. One arm is at or above shoulder level	-	5 (25)	6 (30)	-	-
	3. Both arms at or above shoulder level	20 (100)	15 (75)	10 (50)	20 (100)	-
3.	**Legs**					
	1. Sitting	-	-	-	-	-
	2. Standing on both leg straight	-	-	-	7 (35)	-
	3. Standing on one straight leg	-	-	-	-	-

	4. Standing on both knees bent	7 (35)	-	-	13 (65)	15 (75)
	5. Standing on one knee bent	13 (65)	-	4 (20)	-	-
	6. Kneeling on one or both leg	-	-	16 (80)	-	5 (25)
	7. Walking or moving	-	20 (100)	-	-	-
4.	**Load/use of force**					
	1. Weight or force needed is 10 kg or less	-	-	-	-	-
	2. Weight or force needed exceeds 10 kg but is less than 20 kg.	-	-	-	-	-
	3. Weight or force needed exceeds 20 kg.	20 (100)	20 (100)	20 (100)	20 (100)	20 (100)

Values in parentheses indicate percentage.

LGS-Loading Grain Bag on Self
CG- Carrying Grain Bag,
UG- Unloading Grain Bag
LGO- Loading Grain Bag on Other
AG- Arranging Grain Bag

Table 3. Action level categories in OWAS method for work posture

S. No	Action Level categories	Posture
1	No corrective measures	Good posture
2	Corrective measures in the near future	Less poor posture
3	Corrective measures as soon as possible	Somewhat poor posture
4	Corrective measures immediately	Very poor posture

Table 4. Corrective measures needed for the posture adopted by respondents loading and unloading (N=20).

S. No.	Action category	No corrective measures (Good posture)	Corrective measures in the near future (Less poor posture)	Corrective measures as soon as possible (Somewhat poor posture)	Corrective measures immediately (Very poor posture)
1	Loading Grain Bag on Self	-	3 (15)	7 (35)	10 (50)
2	Carrying Grain Bag	-	-	6 (30)	14 (70)
3	Unloading Grain Bag		3 (15)	14 (70)	3 (15)
4	Loading Grain Bag on Other	-	-	16 (80)	4 (20)
5	Arranging Grain Bag		13 (65)	7 (35)	-

Values in parentheses indicate percentage.

Table 5. Action categories in OWAS method for work posture combination in loading and unloading work

BACKK	ARMS	1			2			3			4			5			6			7			LEGS
		1	2	3	1	2	3	1	2	3	1	2	3	1	2	3	1	2	3	1	2	3	USE OF FORCE
	1	1	1	1	1	1	1	1	1	1	2	2	2	2	2	2	1	1	1	1	1	1	
1	2	1	1	1	1	1	1	1	1	1	2	2	2	2	2	2	1	1	1	1	1	1	
	3	1	1	1	1	1	1	1	1	1	2	2	3	2	2	3	1	1	1	1	1	2	
	1	2	2	3	2	2	3	2	2	3	3	3	**3**	3	3	3	2	2	2	2	3	3	
2	2	2	2	3	2	2	3	2	3	3	3	4	4	3	4	4	3	3	4	2	3	4	
	3	3	3	4	2	2	3	3	3	3	3	4	**4**	4	4	**4**	4	4	4	2	3	4	
	1	1	1	1	1	1	1	1	1	2	3	3	3	4	4	4	1	1	1	1	1	1	
3	2	2	2	3	1	1	1	1	1	2	4	4	4	4	4	4	3	3	3	1	1	1	
	3	2	2	3	1	1	1	2	3	3	4	4	4	4	4	4	4	4	4	1	1	1	
	1	2	3	3	2	2	3	2	2	3	4	4	4	4	4	4	4	4	4	2	3	4	
4	2	3	3	4	2	3	4	3	3	4	4	4	4	4	4	4	4	4	4	2	3	4	
	3	4	4	4	2	3	4	3	3	4	4	4	4	4	4	4	4	4	**4**	2	3	**4**	

Table 6. Description of loading and unloading tasks and action level for adopted posture in different activities (N=20).

S.No.	Activities	OWAS posture codes				Action categories
		Back	Arms	Legs	Force	
1	Loading Grain Bag on Self	2	3	5	3	4 (Corrective measures immediately
2	Carrying Grain Bag	4	3	7	3	4 (Corrective measures immediately)
3	Unloading Grain Bag	4	3	6	3	4 (Corrective measures immediately)
4	Loading Grain Bag on Other	2	3	4	3	4 (Corrective measures immediately)
5	Arranging Grain Bag	2	1	4	3	3 (Corrective measures as soon as possible)

Postural stress on workers while doing loading and unloading work was calculated and was found to be 4 for the tasks like loading grain bag on self, carrying grain bag, unloading grain bag and loading grain bags on other, which means posture needs corrective measures immediately; 3 for the sub tasks like arranging grain bags, means posture needs corrective measures in near future (table 5 & 6).

4. Conclusion

A tremendous number of workers are routinely exposed to physical hazards and many of them develop one or more serious postural and musculoskeletal disorders during their working life time. Prevalence of these has increased dramatically in developing countries. So there is a need to address the inadequate safety system, lack of awareness, lack of training of occupational safety and health and lack of ergonomic standards and epidemiological studies.

ACKNOWLEDGMENTS

The authors would like to acknowledge from her heart to benevolent advisor Dr. Promila Sharma for valuable, cooperative and constant suggestions and benevolent criticism throughout the course of investigation and preparation of manuscript.

REFERENCES

Kurvonika, I., Jhonson, B., Kilbom, A., Vinterberg, H., Bier-Sorensen, F., Andersson, G. 1987. Stanardized Nordic Questrionnaire for the analysis of Musculoskeletal Symptoms. Applied Ergonomics, 18 (3): 233-237.

Saraji,J,N., Hassanzadeh, M.A., Pourmahabadian, M.and Shahtaheri, S. J. 2004. Evaluation of Musculoskeletal Disorders Risk Factors among the Crew of the Iranian Ports and Shipping Organication's Vessels. *Acta Medica Iranica*, 42(5): 350-354

CHAPTER 18

Effects of Different Inclination Angles on Biomechanical Loading of the Shoulder and Hand While Pushing Construction Carts

Yen-Hui Lin

Chung Shan Medical University, Taiwan, R.O.C.
yann@csmu.edu.tw

ABSTRACT

This study examines the effects of inclination angle, cart load, and cart type on muscular activities while pushing a construction cart. Twelve subjects pushed the cart on three different inclination angles: 0, 5, and 10 degrees. Cart loads were 30, 60, and 90 kg, and two types of construction cart (one-wheeled cart and two-wheeled) were used in experiments. Surface EMG (sEMG) system was used to measure muscle activity during the trials. Experimental results show that cart loads and inclination angles significantly affected muscular activities while pushing construction carts. Additionally, 90 kg card load and 10 degrees inclined surface also generated the highest muscle load comparing to the other situations. Muscular activities increased significantly in dominant hand with the one-wheeled cart when compared with the two-wheeled cart, suggesting that, in terms of muscle loads, the cart loads and inclination angles while pushing construction cart should be controlled to reduce the muscular loads

Keywords: Pushing task, construction cart, muscular activity, Surface EMG

1. INTRODUCTION

Working in the construction industry typically requires awkward postures, heavy lifting, and considerable exertion. Many workers performing such tasks complain of discomfort in their upper extremities and lower back over the course of a workday (Buchholz *et al.*, 1996; Jeong, 1998; Hoozemans *et al.*, 2001; Davis *et al.*, 2010). Meerding *et al.* (2005) reported that 59% of construction workers had musculoskeletal complaints, and 41% experienced low back pain in the preceding 6 months. Goldsheyder *et al.* (2002) identified a high prevalence of 82% for musculoskeletal disorders among stone masons.

Manual materials handling is common on construction sites, often involving lifting, carrying, and pulling or pushing heavy objects. Although lifting a load is generally considered hazardous and has been studied extensively, few data exist regarding the biomechanical load while pushing and pulling objects (Hoozemans *et al.*, 2001; Laursen and Schibye, 2002; Herring and Hallbeck, 2007). Frequent pushing and pulling has been observed as construction workers performed manual materials handling tasks (Hoozemans *et al.*, 2001). To minimize the load on the body during manual materials handling, wheeled carts have gradually replaced buckets, boxes, and other containers that were previously carried. Conventional wheeled constructional carts are one-wheeled or two-wheeled carts used to deliver masonry materials, such as cement, mortar, brick, and sand, to construct external and internal walls.

Construction carts are pushed and pulled on such surfaces as asphalt, flagstone, paving stone, gravel, grass, and occasionally soil. These surfaces have different resistances for cart movement. Significant differences in rolling resistance have been identified for carts pushed manually; soft surfaces have highest resistance (Al-Eisawi *et al.*, 1999). Such differences in rolling resistance may result in different magnitudes and directions of pushing or pulling force, and differences in working posture. Operating a construction cart should be considered in terms of problems associated with manual materials handling and, in particular, pushing and pulling activities. The objective of this study is to determine the task demands and loads on the shoulder and upper extremities under different task loads and inclination angles combinations, and to associate these demands with the strength of subjects. This study will provide evidence that supports ergonomic recommendations to promote workplace health by alleviating pain or fatigue of the shoulder and upper extremities while pushing carts.

2. MATERIALS AND METHODS

2.1 Subjects

Twelve college students (8 males and 4 females) were recruited and paid for their participation. Subject age ranged 20–23 (mean, 21.3). Average height was 167.6±8.6 cm and average weight was 66.2±15.8 kg. All subjects were in good health and had free of musculoskeletal and cardiovascular problems. All were right-handed and no subject had experience using construction carts. Prior to participation, all subjects signed an informed consent form indicating their participation was voluntary, and were informed of study objectives and procedures.

2.2 Apparatus

Two construction carts, a one-wheeled cart and a two-wheeled cart, were used in experiments (Fig. 1). The empty weight of the one-wheeled and two-wheeled carts was 14 kg and 15 kg with the force-measuring equipment, respectively. The cart wheels were made of slightly profiled hard rubber with a diameter of 25 cm and width of 8 cm. Handle height was 67cm in the vertical position. The cart was filled with 30kg, 60kg, and 90kg, and was pushed by subjects using both hands. A field study found that carts are most commonly used for transporting bricks, sand, concrete, and other construction materials on construction sites.

(a) (b)

Figure 1 Illustration of the cart in this study, (a) one-wheel cart; (b) two-wheel cart.

A surface EMG (sEMG) system was used to measure muscle activity via surface electrodes (Liu *et al.*, 2006). Four sEMG sensors were positioned based on the specific muscle location recommendations of Cram *et al.* (1998). These bipolar surface electrodes were attached bilaterally over the right and left biceps and trapezius muscle groups of subjects to record muscular activities. The sampling rating was 1000Hz per channel and data were analyzed using Viewlog software (Liu *et al.*, 2006). The subject's skin was abraded or shaved and cleaned with an alcohol pad when necessary. A ground electrode was placed over the lateral epicondyle. A series of calibrations were then performed to obtain individual baselines for maximal voluntary contraction (MVC) of each muscle group. The recorded sEMG data were subsequently utilized to normalize sEMG signals recorded during task performance by expressing these signals as a percentage of MVC (%MVC). All maximum contractions were performed three times, and the highest 1-s mean force was utilized.

2.3 Experimental design

The experiment had a three-factor design with repeated measures analysis of variance (ANOVA). Inclination angles (three levels), cart type (two levels), and cart load (three levels) were fixed factors. Subjects were the random factor. Three different inclination angles were used: level, 5^0 ramp, 10^0 ramp. The push trials were performed over a distance of approximately 10 m (*i.e.*, subjects pulled a cart forward for 10 m).

Two carts were tested, a one-wheeled cart, and a two-wheeled cart. The cart wheels were made of hard rubber and had a diameter of 25 cm. Cart load was 30, 60, and 90 kg. The experiments were performed on the outside, and only push forces were measured. During the experiment, each subject performed 18 trials (three different slope degrees with all three loads in the two carts). Task order was randomized across subjects. To present experimental data clearly, Table 1 lists the 18 experimental tasks in a fixed order.

Dependent variable was muscle activity (%MVC) measured from the sEMG for each of the four muscle groups.

Table 1 Eighteen experimental tasks used in this study

Experimental tasks	Inclination angles	Cart type	Cart load (kg)
Task 1	Level	One-wheel	30
Task 2	Level	Two-wheel	30
Task 3	Level	One-wheel	60
Task 4	Level	Two-wheel	60
Task 5	Level	One-wheel	90
Task 6	Level	Two-wheel	90
Task 7	5^0 ramp	One-wheel	30
Task 8	5^0 ramp	Two-wheel	30
Task 9	5^0 ramp	One-wheel	60
Task 10	5^0 ramp	Two-wheel	60
Task 11	5^0 ramp	One-wheel	90
Task 12	5^0 ramp	Two-wheel	90
Task 13	10^0 ramp	One-wheel	30
Task 14	10^0 ramp	Two-wheel	30
Task 15	10^0 ramp	One-wheel	60
Task 16	10^0 ramp	Two-wheel	60
Task 17	10^0 ramp	One-wheel	90
Task 18	10^0 ramp	Two-wheel	90

2.4 Experimental procedure

Prior to the experimental sessions, all subjects were informed of the study's purpose, procedures, and physical risks and informed consent forms were voluntarily signed. Experimentally significant anthropometric data were obtained, including body height, weight, and elbow height.

After anthropometric measurements were taken, the sEMG sensors were attached using double-sided tape collars. The sensors were then zeroed while a subject was in a relaxed standing position. Resting and set muscular activity measures were then recorded, such that sEMG data could be normalized during analysis. The EMG electrodes were placed on the forearm and upper back while a subject was in a pushing posture.

As mentioned, each subject participated in 18 experimental sessions. The experimental task was to push a construction cart in a realistic work situation. Subjects adopted a natural and comfortable stance to perform pushing tasks and were allowed to work at their own pace. Each session lasted approximately 5 min, and each subject performed no more than six trials on the same day.

All hand push forces were measured with carts with hard rubber wheels 25 cm in diameter on smooth asphalt. Subjects were given a 5-min break at minimum between trials to minimize muscle fatigue. This break was measured using a stopwatch. No subject practiced before the experiment. The order in which each subject performed each of the 18 trials was randomized.

2.5 Data analysis

All analyses used SPSS v 11.5.0 (SPSS, Inc., 2002). First, descriptive statistical analysis was conducted for all variables. Next, repeated-measures ANOVA was applied to each dependent variable to test whether it significantly affected any measure. *Post hoc* multiple-range tests were conducted to compare variable values when a factor was statistically significant at the $\alpha=0.05$ level.

3. RESULT AND DISCUSSION

Table 2 presents means and standard deviations of %MVC under all treatment conditions. Mean exertion force (%MVC) of the right trapezius (21.2 % MVC) and left trapezius (20.2 % MVC) was slightly lower than that of the right bicep (23.0% MVC) and left bicep (23.1% MVC). To identify factors impacting muscle loads, muscle activation levels of the four muscles were subjected to a three-factor design with repeated measures ANOVA (Table 3). The ANOVA results of sEMG measurements demonstrate that the main effects of the inclination angle and cart load, on the right trapezius, left trapezius, right biceps, and right biceps were significant difference ($p<0.05$). The interactive effect between inclination angle and cart load significantly influenced muscle activities of the right bicep ($F = 7.07$, $p<0.01$), and left bicep ($F = 11.56$, $p<0.01$), but not right trapezius and left trapezius. The interactive effect between the cart load and cart type significantly impacted the right trapezius ($F = 4.22$, $p<0.05$), while no interactive effects existed between inclination angle and cart type.

Table 2 Mean and standard deviations of Relative EMG signal activity (%MVC) in experimental tasks

Experimental tasks	Right trapezius (%MVC)	Left trapezius (%MVC)	Right biceps (%MVC)	Left biceps (%MVC)
Task 1	16.3	13.2	9.7	9.5
Task 2	15.5	13.7	6.7	10.0
Task 3	16.3	13.2	9.7	9.5
Task 4	15.8	13.3	13.3	14.5
Task 5	14.7	13.7	16.8	15.0
Task 6	15.8	13.3	13.3	14.5
Task 7	32.2	33.0	14.7	16.8
Task 8	34.8	31.2	12.8	15.3
Task 9	19.0	20.2	22.7	27.0
Task 10	14.0	14.0	13.0	12.0
Task 11	14.3	13.0	37.5	28.5
Task 12	19.7	20.3	36.5	29.8
Task 13	34.7	32.0	26.3	33.7
Task 14	21.2	21.2	24.5	29.2
Task 15	16.2	18.7	26.0	22.0
Task 16	21.2	21.2	23.0	22.7
Task 17	22.7	19.8	44.2	42.2
Task 18	17.5	19.3	41.8	41.7
Average	21.2	20.2	23.0	23.1

Table 3 Summary of ANOVA of relative EMG

Performance measures	Cart load	Cart type	Inclination angle	Load x angle	Load x type	Type x angle	Load x type x angle
Right biceps	25.77**	0.50	45.56**	7.07**	0.63	0.15	0.19
Left biceps	27.94**	1.96	41.44**	11.56**	2.95	0.07	0.81
Right trapezius	12.70**	2.20	34.03**	2.35	4.22*	0.80	0.66
Left trapezius	11.57**	0.08	13.26**	2.32	0.33	1.91	0.625

**p<0.01;*p<0.05.

The LSD multiple range test results indicate that mean %MVC of the right trapezius, left trapezius, right bicep, and left bicep under a cart load of 90 kg was significantly ($p<0.05$) higher than those under cart loads of 60 kg and 30 kg. The %MVC of the four muscle groups under the cart load of 60 kg was significantly ($p<0.05$) higher than those under the 30 kg cart load. The LSD multiple range test results for the effect of the three inclination angles on mean %MVC of the right trapezius indicate that mean %MVC while pushing on 10^0 ramp (22.3% MVC) and 5^0 ramp (22.3% MVC) was significantly ($p<0.05$) higher than that while pushing on level (15.7% MVC). Similar results also shown on left trapezius while pushing on 10^0 ramp and 5^0 ramp. Additionally, the LSD multiple range test results demonstrate that mean %MVC of the right bicep and left bicep were significantly ($p<0.05$) higher when pushing the cart on 10^0 ramp than when pushing on 5^0 ramp and level.

Although the hand and shoulder discomfort mechanisms remain unclear, forceful exertion, repetition, and static muscle load are significant risk factors for cumulative trauma disorders. Silverstein *et al.*, (1987) identified a correlation between repetitive tasks using high hand force and risk of hand tendonitis. In a study by Fennigkoh *et al.* (1999), a job requiring high force was defined as that requiring with >30% MVC, whereas a job requiring low force was defined as that requiring <10% MVC. In this study, muscular activity (*i.e.*, %MVC) increased over time from 6.7% MVC to 44.2% MVC during testing periods, ranging from an average of 20.2% MVC for the left trapezius to 23.1% MVC for the left bicep (Table 2); thus, pushing a construction cart was categorized as medium force. However, as the experiment task involved lifting plus holding a cart handle, and pushing a construction cart over a distance of approximately 24 m, this may have generated a highly static muscle load, resulting in fatigue, regardless of whether a subject's muscular activity was <30% MVC.

This study did not measure the coefficients of rolling friction for hard rubber wheels on different surfaces. Thus, resistance between rolling wheels and the surfaces was not measured. Future study is necessary, as noted by Al-Eisawi *et al.* (1999), to establish a database of coefficients of rolling friction for various wheel materials, tires, and surfaces that exist in industry.

4. CONCLUSIONS

This demonstrates that cart load and inclination angle affect muscular activities, while pushing construction carts. These muscle loads may increase risk for musculoskeletal disorders. Additionally, 90 kg card load and 10 degrees inclined surface also generated the highest muscle load comparing to the other situations. The significant increase in muscular activity while pushing a one-wheeled cart suggests that, in terms of muscle activities, the two-wheeled cart is better than the one-wheeled cart.

REFERENCES

Al-Eisawi, K.W., Kerk, C.J., Congleton, J.J., Amendola, A.A., Jenkins, O.C., Gaines, W., 1999. Factors affecting minimum push and pull forces of manual carts. Apply Ergonomics 30, 235-245.

Buchholz, B., Paquet, V., Punnett, L., Lee, D., Moir, S. 1996. PATH : A work sampling-based approach to ergonomic job analysis for construction and other non-repetitive work. Ergonomics 27, 177-187.

Cram, J.R., Kasman, G..S., Holtz, J., 1998. Introduction to surface electromyography. An ASPEN publication, Gaithersburg, Maryland, 317-329.

Davis, K.G., Kotowski, S.E., Albers, J., Marras, W.S., 2010. Investigating reduced bag weight as an effective risk mediator for mason tenders. Apply Ergonomics 41, 822-831.

Fennigkoh, L., Garg, A., Hart, B., 1999. Mediating effects of wrist reaction torque on grip force production. International Journal of Industrial Ergonomics 23, 293-306.

Goldsheyder, D., Nordin, W., Weiner, S.S., Hiebert, R., 2002. Musculoskeletal symptom survey among mason tenders. American Journal of Industrial Medicine 42, 384-396.

Herring, S., Hallbeck, M.S., 2007. The effects of distance and height on maximal isometric push and pull strength with reference to manual transmission truck drivers. International Journal of Industrial Ergonomics 37, 685-696.

Hoozemans, M.J.M., Van der Beek, A.J., Frings-Dresen, M.H.W., Van der Molen, H.F., 2001. Evaluation of methods to assess push/pull forces in a construction task. Apply Ergonomics 32, 509-516.

Jeong, B.Y., 1998. Occupational deaths and injuries in the construction industry. Applied Ergonomics 29, 355-360.

King, P. M., Finet, M., 2004. Determining the accuracy of the psychophysical approach to grip force measurement. Journal of Hand Therapy 17, 412-416.

Laursen, B., Schibye, B., 2002. The effect of different surfaces on biomechanical loading of shoulder and lumbar spine during pushing and pulling of two-wheeled containers. Applied Ergonomics 33, 167-174.

Liu, Y.P., Chen, H.C., Chen, C.Y., 2006. Multi-transducer data logger for worksite measurement of physical workload. Journal of Medical and Biological Engineering 26, 21-28.

Meerding, W.J., Ijzelenberg, W., Koopmanschap, M.A., Severens, J.L., Burdorf, A., 2005. Health problems lead to considerable productivity loss at work among workers with high physical load jobs. Journal of Clinical Epidemiology 58, 517-523.

Silverstein, B.A., Fine, L.J., Armstrong, T.A., 1987. Occupational factors and carpal tunnel syndrome. American Journal of Industrial Medicine 11, 343-358.

SPSS Institute, Inc., 2002, SPSS user's guide, Release 11.5.0.

CHAPTER 19

Effects of Restrictive Clothing on Lumbar Range of Motion in Adolescent Workers

Wichai Eungpinichpong[1], *Pattanasin Areeudomwong*[1], *Noppol Pramodhyakul*[1], *Vitsarut Buttagat*[1], *Manida Swangnetr*[2], *David Kaber*[3] & *Rungthip Puntumetakul*[1]

[1]Back, Neck and Other Joint Pain Research Group, Division of Physical Therapy, Faculty of Associated Medical Sciences, Khon Kaen University, Khon Kaen, 40002, Thailand
wiceun@yahoo.com
[2]Back, Neck and Other Joint Pain Research Group, Department of Production Technology, Faculty of Technology, Khon Kaen University, Khon Kaen, 40002, Thailand
[3]Edwards P. Fitts Department of Industrial and Systems Engineering, North Carolina State University, Raleigh, NC, 27695-7906, USA

ABSTRACT

The work force in Thailand currently includes a major adolescent segment. These persons seek to maintain a stylish appearance at work and tight tops and trousers are popular clothing choices. However, such clothing may lead to joint mobility restrictions and compensatory responses at the spine. Therefore, the objective of this study was to examine the effect of wearing tight pants on lumbar spine movement and low-back discomfort in simulations of manual material handling (MMH) tasks. Twenty young adults participated in the study with a balance for gender. During test trials, participants performed box lifting, liquid container handling while squatting, and forward reaching while sitting on a task chair when wearing tight vs. fit pants. Each task was repeated three times and video recordings were made as a basis for measuring lumbar range of motion (ROM).

This response was subsequently normalized based on baseline hip mobility. After each trial, participants rated low-back discomfort. Results revealed significant effects of both pant and task type on the normalized lumbar ROM and subjective ratings of back discomfort. The ROM was greater for participants when wearing tight pants, as compared to fit due, which was attributed to hip mobility restriction. Discomfort ratings were significantly higher for the tight pants than fit. These results provide an applicable guide for recommendations on work clothing fit in specific types of MMH activities in order to reduce potential of low-back pain among younger workers in industrial companies.

Keywords: work garments, hip mobility, normalized lumbar range of motion, low-back pain, manual material handling

1 INTRODUCTION

Increasing modern medical knowledge has identified cures for some spinal diseases. However, low-back pain (LBP) remains one of major health problems in modern society (Demoulin et al., 2007). Recent research has found LBP in adolescents to occur more frequently than previously suspected (Burton et al., 2006). Jones and Macfarlane (2005) projected the prevalence to be 70-80%. According to recent epidemiological studies of LBP in adolescents, prevalence rates in the USA and Switzerland were 36 and 51%, respectively (Burton et al., 1996; Costa et al., 2009; Demoulin et al., 2007; Thepdara et al., 2007).

The origins of LBP have been identified to include sustained and static loading of the spine (Althoff et al., 1992; Kanlayanaphotporn et al., 2003; Puntumetakul et al., 2009; Tyrrell et al., 1985), repetitive spinal movements (Healey et al., 2008; Rodacki et al., 2005; Tyrrell et al., 1985), vibratory forces on the back structures (Bonney et al., 2003), as well as limited hip range of motion (Wong and Lee, 2004). Adams and Hutton (1983) and Lee and Wong (2002) observed that altered movement patterns of the hips and spine may be a potential factor contributing to the development of LBP.

Tight pants, specifically sizes smaller than fit to a wearer's anthropometry, have become a popular clothing choice for Thai adolescents. Unfortunately, such clothing may restrict hip movement and alter spinal movement and trunk muscle activity during work tasks and leisure. According to prior research, hip movements are related to lumbar spine movements. Lee and Wong (2002) observed reduced hip mobility in association with increased spinal flexion and extension, as compared with unrestricted mobility conditions.

Historical workplace safety guidelines and regulations have addressed workers wearing loose clothing while operating machines and the potential for garments to become entangled in moving machinery parts (OSHA, 1992). Conversely, little, if any research, has focused on the impact of wearing excessively tight clothing at work, such as tight pants, on occupational illnesses for workers, such as LBP.

The objective of this research was to investigate the effect of wearing tight pants (smaller than fit sizes) on hip and lumbar spine movement in adolescents and perceived low-back discomfort. The study examined wearing tight garments as a cause of hip movement restriction in simulations of realistic work tasks in a lab environment. Specifically, we compared young adult lumbar range of motion during box lifting while standing or sitting in a chair and manual material handling (MMH) in a squatting position when wearing tight vs. fit (i.e., the correct size according to anthropometry) pants. This work was expected to complement prior research in the safety area by focusing on risks associated with wearing restrictive garments in work tasks vs. loose fitting clothing hazards.

2 METHODS

2.1 Subjects

A sample of twenty young adults was recruited from the Khon Kaen University community in Thailand. Inclusion criteria for the study were male or female aged between 18-25 years. The limited age range was selected to ensure subjects represented the target population for the research ("teenagers") but could still provide informed consent for participation (i.e., no minors under the age of 18 years). All subjects were also required to have a body mass index (BMI) between 18.7-25 kg/m^2. This range represents persons who are not considered to be "overweight" or "obese" (WHO expert consultation, 2004). The range was also limited to reduce the number of pairs of pants the research team had to purchase for subject participation.

Subjects were excluded from the sample for the experiment if they: (1) had previously experienced LBP, which required medication or consultation with a health professional, and/or had days away from work within the last 6 months caused by LBP; (2) were diagnosed with a medical condition that affected the musculoskeletal system, such as lumbar spondylosis, spondylolisthesis, lumbar herniated nucleus pulposus, ankylosing spondylitis or rheumatoid arthritis; (3) had recent lumbo-pelvic and/or abdominal surgery; (4) had symptoms, such as recent back pain, leg pain, or numbness in their back or legs prior to the experimental session; or (5) they were pregnant (O'Sullivan et al., 2006a).

2.2 Tasks

The tasks and equipment used in the test trials were as follow:

1) A forward reaching task required subjects to sit in a task chair with their feet flat on the floor with 90 degrees of flexion at the knee. They were then required to pick-up a box on a work table directly in front of them and place it at their maximum reach distance on the same table. Subjects were instructed to flex their spine and to extend their arms as much as possible (see Figure 1 (a))

2) A box lifting task required subjects to stand erect with their toes touching a tape line on the floor. They were then directed to bend at the knees and spine in order to pick-up a box on the floor and lift it to knuckle height. Subjects were then

permitted to bend at the knees with flexion of up to 90 degrees. (If maximum flexion exceeded 90 degrees, trials were repeated.) Subjects returned the box to the floor to complete the task (see Figure 1 (b)).

3) A squatting task, subjects were required to bend the knees with maximum flexion at the spine in order to pick up a package of drinking containers on the floor and to place it in a box (see Figure 1 (c)).

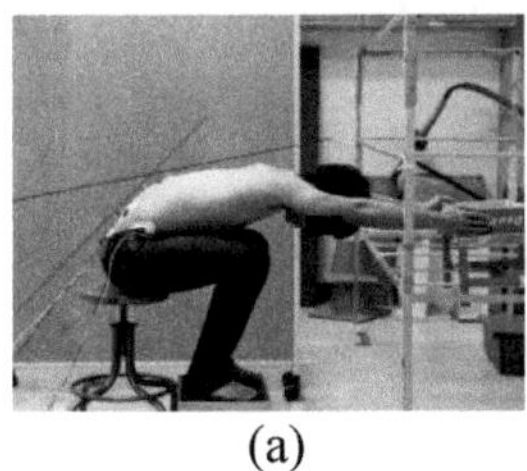

(a)

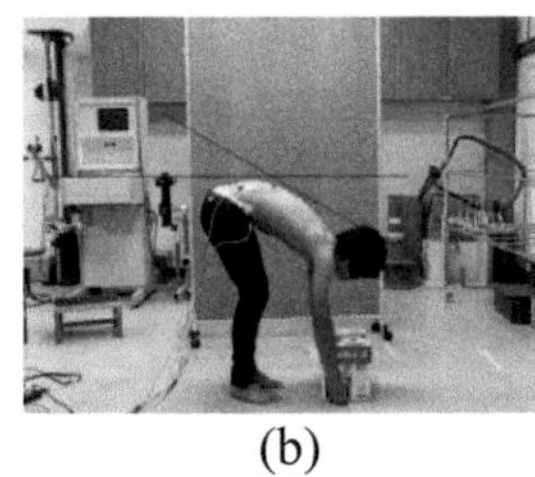

(b)

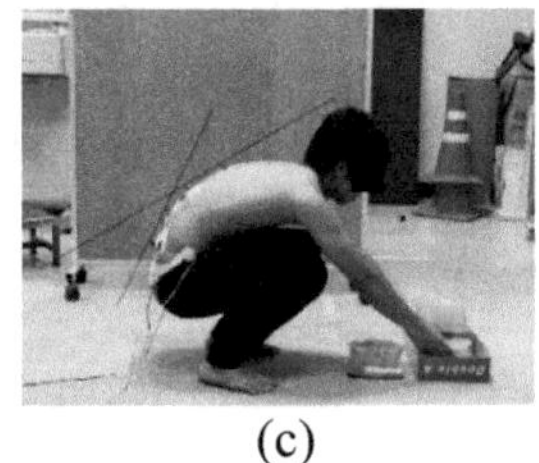

(c)

Figure 1 MMH tasks of (a) forward reaching, (b) box lifting, and (c) squatting.

2.3 Independent variables

The independent variables for the experiment were: (1) the type of pants with two levels, including tight (sizes too small for the wearer) and fit (the correct size according to anthropometry) long pants; and (2) the type of work task, including the maximal forward reaching while sitting, box lifting, and MMH while squatting.

Tight pants were objectively defined as pants with a waist size one size too small for the wearer. Beyond this, the pants were subjectively defined as causing subjects to feel "tightness" at the hip, buttock and thigh. They were assessed using a simple discomfort questionnaire in which subjects responded, "Yes" the pants feel tight at the hip, buttock or thigh, or "No" the pants do not feel tight in these areas.

Fit pants were objectively defined as pants for which the waist size was correct according to a subject's anthropometry. They were also subjectively defined as pants that did not cause subjects to feel tightness or restriction around the hip, buttock and thigh. The same subjective questionnaire used to assess the tight pants was used to assess the fit pants.

(The settings of the task type independent variable were described in the preceding section.)

2.4 Procedure

The present study followed a completely within-subjects design with all subjects being exposed to all combinations of garment fit and work task type. The procedure for the experiment involved randomly assigning subject to one of two groups, including those wearing tight pants first, followed by fit pants, and those wearing fit pants first followed by tight pants. Subsequent to donning a pair of pants a subject's hip range of motion was assessed in a supine posture position on a standard medical examination table. Directly following, subjects began performing the test trials for the various tasks. The order of task presentation was randomized within subject (i.e., subjects served as a blocking factor). As previously mentioned, each task was repeated three times by each subject with a single trial lasting approximately 5 seconds (for reaching, lifting or squatting). After completing three cycles of a task, subjects were asked to stand in an erect posture. Once a subject completed all trials in the first task, they moved on to the second task type and then the third. After having completed all nine test trials, subjects were permitted a 10 minute rest period while wearing a pair of fit shorts. At the close of the rest, they donned the alternate type of pants and resumed work task trials. At the completion of the second batch of test trials, subjects were debriefed and dismissed from the experiment. The entire experiment procedure took about 2 hours per subject.

2.5 Dependent variables

2.5.1 Lumbar Range of Motion

Regarding the lumbar range of motion (ROM) measurement, reflective markers were placed at vertebral joints along the spine, including T10, L2, L4 and S2. These joint locations have been previously observed to determine lumbar spine angle (O'Sullivan et al., 2006b). Videos of all test trials were recorded with a digital video camera with a frame rate of 30 Hz. All video files were imported into the VirtualDub software (video processing software) and individual frames were exported as image files. Image files were then inspected using VirtualDub for identification of the most extreme posture position for a specific task (i.e., a single frame for each test trial was used for analysis purposes). The spinal angle for a selected video frame was measured using the ImageJ analysis software. A line segment was constructed overlaying the marker images at T10 and L2 and another segment was constructed passing over the marker images for L4 and S2. The two line segments were extended to a point of intersection occurring approximately over the L3 joint. The angle between the two segments was automatically computed by the software. Related to this, all subject motion during task performance occurred in the sagittal plane, perpendicular to the line of view of the digital video camera. For data analysis purposes, the lumbar ROM was normalized based on the baseline hip ROM collected at the outset of the study for each subject.

2.5.2 Visual Analog Scale Discomfort Ratings

After each trial, participants rated their low-back discomfort using a visual analog scale (VAS). The scale included anchors of "none" to "intolerable". Subjects used pen and paper to complete the ratings by marking the location along the scale that best represented their feeling of discomfort after a trial. An analyst using a metric ruler subsequently measured the location and data was manually entered into a computer for analysis purposes.

2.6 Hypotheses

We expected that tight pants, typical of the style currently worn by Thai adolescents, would decrease hip mobility (flexion and extension) when compared to fit pants. Related to this, tight pants were hypothesized to increase lumbar ROM as a result of the restricted hip motion (Hypothesis (H)1). It is important to note that the lumbar ROM was not expected to decrease as a result of tight pants, as the waist line of test garments fell below the L5/S1 joint for all male and female subjects. The feeling of low-back discomfort as measured by the VAS was expected to increase as a result of tight pant use (H2). Beyond this, the task types were expected to differ in normalized lumbar ROM as well as perceived discomfort due to whole-body posture variations (H3). Specifically, box lifting was expected to generate the greatest ROM and discomfort ratings due to extreme spinal flexion compared to squatting MMH and reaching.

3 RESULTS AND DATA ANALYSIS

Diagnostics on the normalized ROM response and the VAS ratings revealed the data to be normally distributed and to have equal variances across the types of pants and tasks. A series of univariate ANOVA models were applied to the normalized ROM and perceived low-back discomfort responses. Models were initially constructed with factors, including: 1) subject; 2) pants type; 3) order of pants; 4) task type; and 5) order of task, as well as the interaction term of pant and task types. Both order of pants and tasks had no significant effect on the ROM or VAS responses; consequently, these factors were dropped from the statistical models.

With respect to normalized lumbar ROM, results revealed significant effects of both pant ($p=0.0004$) and task type ($p<0.0001$) (see Table 1). In line with hypothesis, the ROM was higher when wearing tight pants than fit (see Figure 2(a)). Regarding, task type, box lifting produced greater spinal flexion than squatting and reaching tasks (see Figure 2(b)). The ANOVA also revealed no significant interaction effect for the normalized ROM response.

Table 1 ANOVA Results on LROM Response

Effect	F Value	Pr> F
Subject	75.6284	<.0001*
Types of pants	17.7013	<.0001*
Task	12.3154	<.0001*
Type of pant * Task	0.6891	0.5046

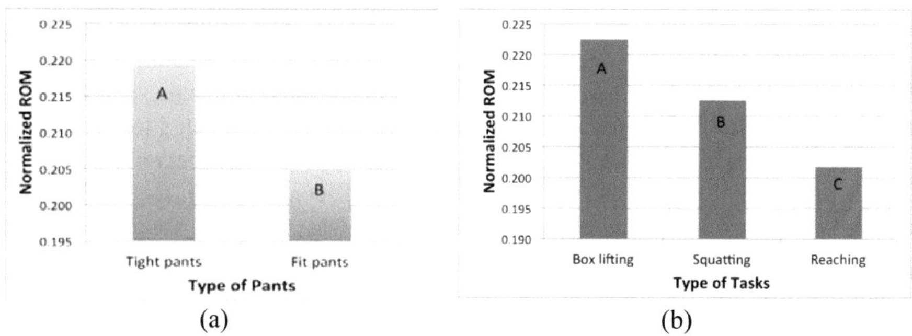

Figure 2 Mean normalized lumbar ROM for (a) type of pants and (b) type of tasks.

Significant effects for both pant (p<0.0001) and task type (p<0.0001) were also found on the subjective ratings of low-back discomfort (see Table 2). Results also revealed no significant interaction of the independent variables for low-back discomfort. Conforming with expectation, post-hoc analyses revealed the VAS ratings to be significantly higher for tight pants than fit (see Figure 3(a)). However, contrary to the hypothesis, subjects perceived squatting to cause the greatest discomfort followed by lifting and reaching, which were not different from each other (see Figure 3(b)).

Table 2 ANOVA Results on VAS Response

Effect	F Value	Pr> F
Subject	10.8418	<.0001*
Types of pants	437.6087	<.0001*
Task	5.8584	<.0001*
Type of pant * Task	0.3272	0.7217

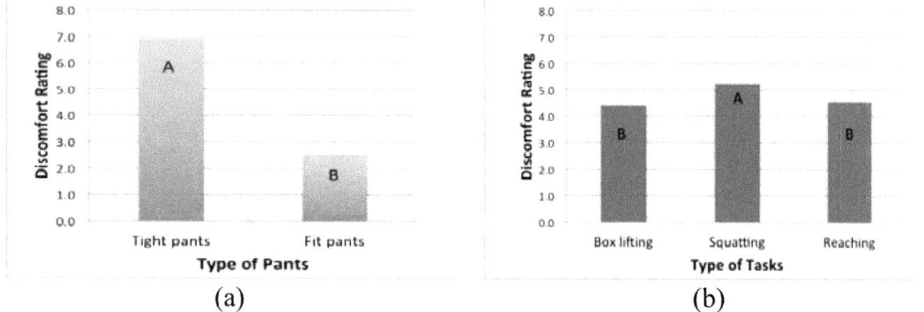

Figure 3 Mean discomfort ratings for (a) type of pants and (b) type of tasks.

4 DISCUSSION

The findings of this study revealed an increase in normalized lumbar ROM for participants when wearing tight pants during box lifting, MMH while squatting and maximal forward reaching while sitting. In baseline hip ROM measurement, we observed significant restriction in mobility when subjects wore tight pants. This restriction may have contributed to compensatory movement in the lumbopelvic region. Spine movements appeared to occur more frequently and with greater range during the experimental tasks requiring hip flexion. These observations were in line with expectation (H1) and prior research. Harris-Hayes et al. (2009) reported similar results but in observing different directions of hip movement that the present study. Other previous research (Esola et al., 1996; Wong and Lee, 2002) has found lumbar spine movement to have a relationship with hip movement, including forward, backward and lateral bending of the trunk. Therefore, alteration of either lumbar or hip movement, in general, appears to contribute to compensatory responses in the other movement.

Our results also showed that box lifting and squatting produced greater spinal flexion than reaching tasks. This finding was also in line with expectation (H3) and may have been due to each task requiring a different lumbar ROM for completion. For example, the forward reaching task produced less lumbar flexion than the box lifting and squatting. We observed that the arms and upper-back are key body segments to achieving this task as compared to the others. For this reason, the forward reaching task produced less lumbar ROM when compared with the box lifting and squatting.

Alteration of movement patterns of the lumbar spine and hip may be a possible factor contributing to the occurrence of low-back pain. Dolan and Adams (1993) reported that changes in spine and hip mobility would alter bending stresses on the spinal motion segment and, consequently, alter loads on facets of the posterior spinal ligaments (see Adams and Hutton, 1983). These outcomes could translate into pain experiences for workers. Our results on VAS ratings of low-back pain revealed tight pants to increase the feeling of discomfort, as compared to fit pants. This finding was also in line with expectation (H2). The experience of pain might also be due to the hip mobility restriction with the tight pants during task performance and leading to overloading of vertebral structures, such as the intervertebral discs and posterior spinal ligaments (Hansen et al., 1998; Pope et al., 2002).

Regarding task type, results also showed VAS ratings during squatting to be higher than box lifting and reaching. This was opposite to H3. This finding may be explained by the body position in each task. In a post-experiment interview, participants commented that the squatting posture required the greatest hip flexion, more so than box lifting and forward reaching. They also said they perceived the tight pants to restrict the hip movement most in the squatting posture (although the ROM response showed box lifting to be more restrictive) leading to a greater sense of discomfort than in the other tasks.

ANOVA results revealed no significant interaction effect for both ROM and subjective ratings. The lack of significant findings on the interaction term in all models might be due to the impact of pants overshadowing the task affect or tight and fit pants may have comparable effects on response measures across task types.

5 CONCLUSION

The findings of this study provide an applicable guide for recommendations on work clothing fit in specific types of MMH activities in order to reduce the potential for low-back pain among younger workers in industrial companies. For small to medium size companies (e.g., 10-100 workers) that may not provide workers with uniforms, companies should provide information on the potential for low-back pain associated with wearing tight pants and recommend fit or comfortable personal protective work clothing be worn during MMH tasks.

Regarding limitations of the present study, we did not examine lumbar ROM and back discomfort in more complex physical tasks, such as asymmetrical lifting. This type of lifting is common in industrial settings and reduces the generalizability of our findings. Second, we only investigated the effect of tight pants on young adult performance of MMH tasks. There is a need to look at the influence of restrictive garments on ROM and pain experiences in other working age groups, such as the elderly. Finally, we did not measure back muscle activity or compressive load at the lumbar spine for tight and fit pants in the various MMH tasks. Both of these responses may be important factors/predictors in low-back pain.

ACKNOWLEDGMENTS

This study was supported by grants from the Back Neck and Other Joint Pain Research Group, Khon Kaen University. David Kaber's work on the study was supported by a grant from the U.S. National Institute for Occupational Safety & Health (NIOSH) (No. 2 T42 OH008673-06). The opinions expressed in this paper are those of the authors and do not necessarily reflect the views of NIOSH.

REFERENCES

Adams, M.A. and Hutton, W.C. 1983. The mechanical function of the lumbar apophyseal joints. *Spine* 8 (3): 327–330.

Althoff, I., Brinckmann, P., Frobin, W., Sandover, J., and Burton, K. 1992. An improved method of stature measurement for quantitative determination of spinal loading.Application to sitting postures and whole body vibration. *Spine* 17(6): 682-693.

Bonney, R.A. and Corlett, E.N. 2003. Vibration and spinal lengthening in simulated vehicle driving. *Appl.Ergon.* 34(2):195-200.

Burton, A., Balague, F., Cardon, C., Erik sen, H., Henrotin, Y., Lahad, A.,Lecler, C.A., Muller, G., and van der Beek, A. 2006. European guidelines for prevent ion in low back pain. *Eur. Spine. J.* 15(Suppl2): 136-168.

Burton, A., Clarke, R., McClune, T., and Ti l lot son, K. 1996. The natural history of low back pain in adolescents. *Spine*21 (20): 2323-2328.

Demoulin, C., Distrée, V., Tomassella, M., Crielaard, J.M., and Vanderthommen, M. 2007. Lumbar functional instability: a critical appraisal of the literature. *Ann. Readapt.Med.Phys.* 50(8): 677-684.

Dolan, P. and Adams, M. D. 1993. Influence of lumbar and hip mobility on the bending stresses acting on the lumbar spine.*Clin.Biomech.*8 (4): 185–192.

Esola, M.A., McClure, P.W., Fitzgerald, G.K., and Siegler, S. 1996. Analysis of lumbar spine and hip motion during forward bending in subjects with and without a history of low back pain. *Spine* 21 (1): 71-78.

Hansen, L., Winkel, J., and Jorgensen, K. 1998. Significance of mat and shoe softness during prolonged work in upright position: based on measurements of low back muscle EMG, foot volume changes, discomfort and ground force reactions. *Appl. Ergon.* 29(3): 217-224.

Harris-Hayes, M., Sahrmann, S.A., and Van Dillen, L.R. 2009. Relationship between the hip and low back pain in athletes who participate in rotation-related sports. *J. Sport. Rehabil.* 18 (1): 60-75.

Healey, E.L., Burden, A.M., McEwan, I.M., and Fowler, N.E. 2008. Stature loss and recovery following a period of loading: effect of time of day and presence or absence of low back pain. *Clin.Biomech.* 23(6): 721-729.

Jones, G.T. and Macfarlane, G.J. 2005. Epidemiology of low back pain in children and adolescents. *Arch. Dis. Child.* 90 (3):312-316.

Kanlayanaphotporn, R., Trott, P., Williams, M., and Fulton, I. 2003. Effects of chronic low back pain, age and gender on vertical spinal creep. *Ergonomics* 46(6):561-573.

Lee, R.Y. and Wong, T.K. 2002. Relationship between the movements of the lumbar spine and hip.*Hum.Mov. Sci.* 21 (4): 481-494.

McLean, K.A., Chislett, M., Keith, M., Murphy, M., and Walton, P. 2003. The effect of head position, electrode site, movement and smoothing window in the determination of a reliable maximum voluntary activation of the upper trapezius muscle. *J.Electromyogr.Kinesiol.*13 (2):169–180.

OSHA. 1992. *Concepts & Techniques Of Machine Safeguarding* (Pub. 3067). Washington, D.C.: Author.

O'Sullivan, P.B., Dankaerts, W., Burnett, A.F., Farrell, G.T., Jefford, E., Naylor, C.S., and O'Sullivan, K.J. 2006a. Effect of different upright sitting postures on spinal-pelvic curvature and trunk muscle activation in a pain-free population. *Spine* 31 (19): 707-712.

O'Sullivan, P.B., Mitchell, T., Bulich, P., Waller, R., and Holte, J. 2006b. The relationship between posture and back muscle endurance in industrial workers with flexion-related low back pain. *Man. Ther.* 11 (4): 264-271.

Pope, M.H., Goh, K.L., and Magnusson, M.L. 2002. Spine ergonomics. *Annu. Rev. Biomed. Eng.*4: 49-68.

Puntumetakul, R., Trott, P., Williams, M., and Fulton, I. 2009. Effect of time of day on the vertical spinal creep response. *Appl.Ergon.*40(1): 33-38.

Rodacki, A.L.F., Fowler, N.E., Porvensi, C.L.G., Rodacki, C.L.N., and Dezan, V.H. 2005. Body mass as a factor in stature change. *Clin.Biomech.* 20(8):799-805.

Tyrrell, A.R., Reilly, T., and Troup, J.D. 1985. Circadian variation in stature and the effects of spinal loading. *Spine* 10(2): 161-164.

Wong, T.K. and Lee, R.Y. 2004. Effects of low back pain on the relationship between the movements of the lumbar spine and hip. *Hum.Mov.Sci.* 23 (1): 21-34.

WHO expert consultation. 2004. Appropriate body-mass index for Asian populations and its implications policy and intervention strategies. *Lancet.* 363(10): 157-163.

CHAPTER 20

Investigations on Eye Movement Activities in the Manually Controlled Rendezvous and Docking of Space Vehicles

Yu Tian, Shanguang Chen, Chunhui Wang, Qu Yan, Zheng Wang*

Science and Technology on Human Factors Engineering Laboratory, Astronaut Research and Training Center of China

Beijing, China
Shanguang_chen@126.com

ABSTRACT

Eye activity measures are increasingly employed to evaluate operators' mental states and the usability of interfaces. This paper focuses on the eye movement activities of operators performing the manually controlled rendezvous and docking (MCRVD) of two space vehicles during spaceflight. Ten male subjects participated in the simulated MCRVD experiments, in which the subject observed the image of the target space vehicle on the computer screen, and regulated the relative position and posture of the two space vehicles by manipulating the control handles. Each subject performed two consecutive RVD tasks of the same difficulty level. Eye movement data were collected by the ASL eye tracker H6. Eye activity measures, such as the blink rate, the blink duration, the Percent Eyelid Closure (PERCLOS), and the fixation related measures were extracted. As pervious research has revealed, manually controlled RVD can be divided into tracking control stage and accurate control stage, eye activity measures in the two stages were also calculated separately. Results show that the blink rate in the accurate control stage is significantly lower than that in the tracking control stage ($p < 0.05$), while both the PERCLOS and the blink duration in the two stages are low and show no significant difference. It indicates that the workload in the accurate control stage is higher than that in the tracking control stage, but no fatigue was found in the subjects during the experiments. For all of the ten subjects, longer control time corresponds to bigger PERCLOS, which means that there is a tendency for operators to be more tired as control time lasts longer, and validates that the PERCLOS is an effective index of mental fatigue. The fixation related measures show that human eyes were fixed on the image of the spacecraft nearly 80% of the time, while the numerical display of the spacecraft state were less fixed on, which validates that the numerical display only provides compensatory information. Among the numerical display areas, the velocity display area attracts more fixations and longer fixation dwell time, which implies that the velocity information is more important for subjects performing the RVD task (more fixations), and is more difficult to extract (longer fixation dwell time). Apparently if the velocity display is designed to be more noticeable, the overall visual workload of the operators may be reduced.

Keywords: eye movement activities, manually controlled rendezvous and docking, workload evaluation, interface evaluation

1 INTRODUCTION

There is a hypothesis that the eye movements of a person are mainly directed by his visual attention and are closely related to cognitive processes in the brain (Poole and Ball, 2005). Eye movements can be recorded acurrately by the eyetrackers. Measures of the eye activities, such as the fixations, the scanpaths, the blink rate, the pupil diameter, can be extrated from the eye tracking data recorded by the eyetrackers. Those eye activity measures can be applied in the evaluation of interface usablity and in the evaluation of the mental states of humans (Poole and Ball, 2005; Tsai, Viirre, and Strychacz, et al. 2007).

Manually controlled rendezvous and docking (MCRVD) of space vehicles is a complex human computer interaction (HCI) task in space. The MCRVD simulators provide facilities to train the astronauts to master necessary skills on the ground. Experiments with eye tracking of the operators in the MCRVD simulators were designed to evaluate the design of task interface and the mental states of operators in the task process.

2 METHODS

2.1 Subjects

Ten male technicians aged between 26 and 31 years from Astronaut Research and Training Center of China participated in the experiments. Each of them had education background with at least a bachelor's degree and work experience related to manned spaceflight projects.

2.2 The MCRVD Simulator and the Eye Tracker

Figure 1 shows the prototype of the MCRVD simulator on the ground (JIANG, WANG, and TIAN, et al. 2011). The vedio image of the target spacecraft obtained from the cameras is displayed on the monitoring interface, with the numerical data of the relative position and attitude of the two spacecafts obtained from the sensors overlaid on the corners of the interface. The operator observes the target and numerical data on monitoring interface and manipulates the controllers to control the chaser space vehicle to complete the RVD task (ZHANG, XU, and LI, et al. 2007). Performance data such as the control time and the fuel consumption are recorded by the system.

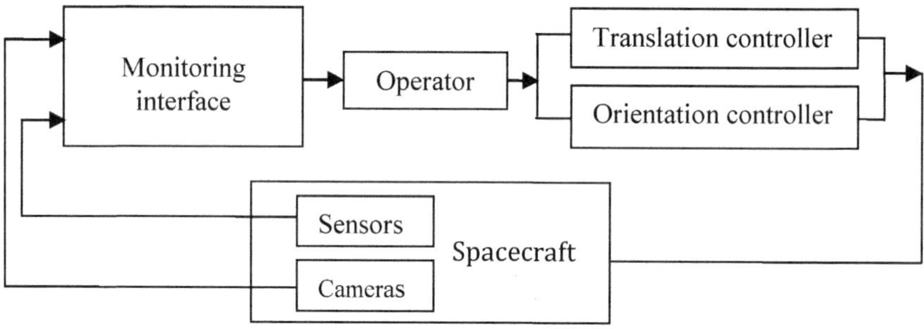

Figure 1 The prototype of the MCRVD simulator on the ground (modified from (JIANG, WANG, and TIAN, et al. 2011))

The eye movements of the operators in the MCRVD tasks are recorded by the H6 Head Mounted eye tracker produced by Applied Science Laboratories (ASL). The H6 Head Mounted eye tracker system records the location and diameter of the pupil at a sampling rate of 60 Hz. The system is designed to track gaze direction over approximately a 30-35 degree vertical visual angle and a 40-45 degree horizontal visual angle.

2.3 Experiment Design and Procedure

The subjects were trained twice to ensure that they were familiar enough with the MCRVD tasks. In the formal experiments with eye tracking, the eye traker was put on the head of the subject, and then calibration work was done for the eye tracking system. After that, each subject practiced once, and then performed two consecutive trails of MCRVD tasks of the same difficulty level, with around three minutes' rest in the middle. In the MCRVD tasks, the subject observed the monitoring interface, and regulated the relative position and posture of the two spacecrafts by manipulating the control handles. The initial distance of the two spacecrafts were set to be 100 meters, and the control time for one trail of the MCRVD task was mostly within 10 minutes.

3 RESULTS

3.1 Data Collected and Indices Calculated

In the experiments, the subjects' performance data in the MCRVD tasks and their eye tracking data in the task process were collected. The performance indices considered in this paper were control time and fuel consumption. Control time is the time from the beginning to the end of the MCRVD task. Fuel consumption is the fuel consumed (simulated by a series of rules) in the whole MCRVD task process.

Eye activity measures in the whole task process, such as the blink rate, the blink duration, the 80 Percent Eyelid Closure (PERCLOS) (Bergasa, Nuevo, and Sotelo, et al. 2006), the fixation distribution and dwell time in Areas of Interest (AOI) were calculated.

Previous research has revealed that MCRVD can be divided into tracking control stage and accurate control stage. Tracking control stage is from the initial distance to about 20 meters away from the target spacecraft. In this stage, due to the long distance, subjects mainly track the target by observing the shape features of the target spacecraft, and big deviation, attitude and translation of the two spacecrafts are eliminated in this stage. Accurate control stage is form 20 meter away from the target spacecraft to the end of the task. In this stage subjects have to narrow the deviation, attitude and translation of the two vehicles to meet the docking access requirements (JIANG, WANG, and TIAN, et al. 2011). Averagely, the tracking control stage counts for about 64% of the control time, and the accurate control stage counts for the remaining 36% of the control time. Eye activity measures in the two stages were also calculated separately by dividing the whole eye tracking data into the first 64% of the dataset and the remaining 36%.

3.2 Fixation Distribution and Dwell Time

Statistic data show that human eyes were fixed on the image of the spacecraft nearly 80% of the time, while fixations on the numerical display areas count for only about 20% of all the fixations. Among the numerical display areas, the velocity display area attracted about twice more fixations than the deviation display area, indicating that the velocity information may be more important for subjects performing the MCRVD tasks. The average fixation dwell time of the subjects on the velocity display area (0.92s) was much longer than the average fixation dwell time on the deviation display area (0.49s), indicating that the velocity information may be more difficult to extract.

3.3 Comparisons of Eye Status in the Two Stages

Results show that the blink rate in the accurate control stage is significantly lower than that in the tracking control stage (p value is 0.012, n=10). The decrease of eye blink rate in the accurate control stage indicates that there may be an increased demand involving visual attention in the accurate control stage (Wilson, 2002).

PERCLOS and blink duration are considered predictors of drowsiness and mental fatigue (Galley N, R. Schleicher, Galley L. 2004; Bergasa, Nuevo, and Sotelo, et al. 2006; Tsai, Viirre, and Strychacz, et al. 2007). Both the PERCLOS and the blink duration in the two stages were small and show no significant difference, indicating that the subjects were not drowsy at all during the experiments.

3.4 Eye Activity and Task Performance

Correclation of three eye activity measures with task performance was analysed (see Table 1). Blink rate shows a significant correlation with both control time and fuel consumption. This shows that the eye activities and the task propertities are interconnected: while the increase of task demand brings down the blink rate, the lower blink rate (may indicate higher concentration level) is acommplished by better task performance.

Table 1 Spearman's correlation of eye activity measures with performance indices of MCRVD tasks

performance indices	Blink Rate	PERCLOS	Bink Duration
Control Time	0.507*	-0.027	0.240
Fuel Comsumption	0.549*	-0.376	0.052

* at 0.05 level of significant difference.

Although the PERCLOS is not significantlycorrelated with the task performance, for each of the ten subjects who performed two trails of MCRVD tasks, task with longer control time was acommplished by bigger PERCLOS (see Figure 2). The results may indicate that there is a tendency of getting fatigue as control time increases, although they are not really drowsy. These data validate that PERCLOS is an effective predictor of mental fatigue.

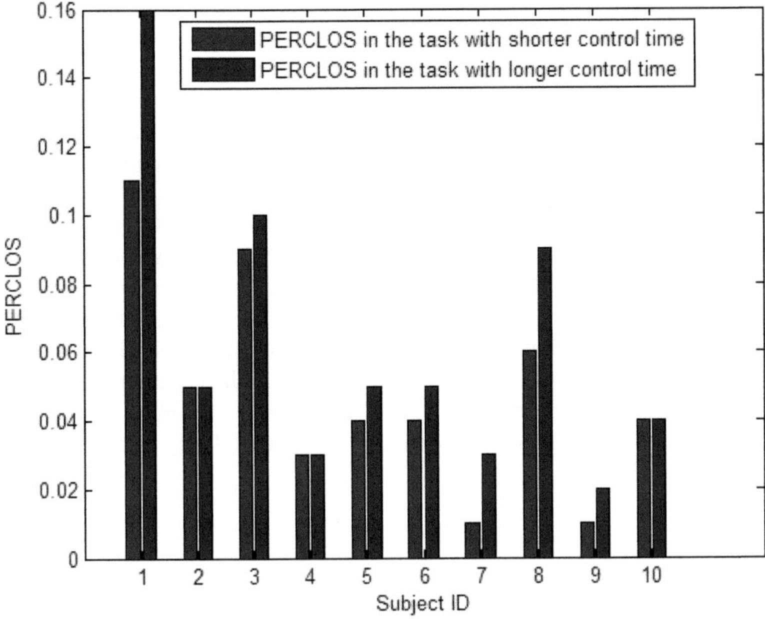

Figure 2 The PERCLOS of the subjects in the two trials of MCRVD tasks

4 DISCUSSION

The eye tracking data show that some area (such as the velocity display area out of the numerical displays) attracts more fixatioins than other areas. More fixations on a particular area indicate that this area is more important or more noticeable (Poole and Ball, 2005). As all the numerical displays are in a parallel relation, the velocity display area is not more noticeable, the only explanation is that the area is more important for the operators performing the MCRVD tasks.The velocity display area also attracts longer fixation dwell time, indicating that information presented in the area may be more difficult to extract. If velocity display is designed to be more noticeable, the overall visual workload of the operators may decrease. And from the decrease of the blink rate in the task process, it can be infered that the visual workload in the second stage of the task process (refered to as accurate control stage) may be significantly higher than the first stage (refered to as tracking control stage). So in the second stage, if numerical information is still needed, it should be brief and noticeale in order not to further increase the visual workload of the operator.

Two eye activity measures, the PERCLOS and the blink duration, do not show significant changes in the two control stages. These two measures are predictors of drowsiness (Galley N, R. Schleicher, Galley L. 2004; Bergasa, Nuevo, and Sotelo, et al. 2006; Tsai, Viirre, and Strychacz, et al. 2007), and the results may indicate that the operators are not drowsy at all in the task process. Although PERCLOS tends to increase in tasks with longer control time, the absolute value is still low, which shows that the operators are unlikely to fall fatigue in the MCRVD task.

Although the eye activities of the subjects show some similar characteristics, the individual differences in eye movements between participants on identical tasks were also found. For example the fixation patterns of the subjects were moderately different. So generally it is not proper to use a within-participants design in order to make valid performance comparisons (Poole and Ball, 2005). But some researches have illustrated that the eye activity measures may predict the task performance in a complicated way (Van Orden, Jung, and Makeig, 2002). In our study the blink rate shows a significant correlation with the performance of the MCRVD task. One possible explanation is that blink rate may be connected with concentration level of the operator, the more concentrated of the operator, the lower the blink rate tends to be. And when the operator is more concentrated, he is more likely to perform well in the task.

5 CONCLUSION

The present study demonstrated that the eye movement activities in the MCRVD task can provide useful information for interface evaluation and mental states monitoring. Eye tracking data and analysis demonstrate that the monitoring interface of MCRVD simulator provides sufficient information for the task, and the visual workload of the operator is acceptable. The velocity display area in the numerical displays is suggested to be designed more noticeable, as information in this area appears to be more important for the operators and is more difficult to extract. The visual workload in the accurate control stage of the MCRVD is higher that the tracking control stage, if numerical information is still needed in this stage, it should be brief and noticeale in order not to further increase the visual workload of the operator.

ACKNOWLEDGMENTS

This study is supported by the National Basic Research Program of China (No.2011CB711000).

REFERENCES

A. Poole, and L. J. Ball. 2005. Eye Tracking in Human-Computer Interaction and Usability Research: Current Status and Future Prospects. Chapter in C. Chaoui (Ed.): *Encyclopedia of HCI.* Pennsylvania: Idea Group, Inc.

Bergasa, L. M., J. Nuevo, Sotelo M. A., et al. 2006. Real-time system for monitoring driver vigilance. *IEEE Transactions on Intelligent Transportation Systems*, 7(1): 63-77.

Galley N., R. Schleicher, Galley L.. 2004. Blink parameters as indicators of driver's sleepiness–possibilities and limitations. *Vision in vehicles*. Amsterdam: Elsevier.

JIANG T., WANG C., TIAN Z., et al. Study on synthetic evaluation of human performance in manually con-trolled spacecraft rendezvous and docking tasks. *Digital Human Modeling - Third International Conference*, Orlando, USA, 2011.

Van Orden, K. F., T. P. Jung, and Makeig S. 2000. Combined eye activity measures accurately estimate changes in sustained visual task performance. *Biological Psychology*, 52(3): 221-240.

Wilson, G. F. 2002. An analysis of mental workload in pilots during flight using multiple psychophysiological measures. *The International Journal of Aviation Psychology*, 12(1): 3-18.

Y. Tsai, E. Viirre, C. Strychacz, et al. 2007. Task performance and eye activity: predicting behavior relating to cognitive workload. *Aviation, space, and environmental medicine* 78(Supplement 1): B176-B185.

ZHANG Y., XU Y., LI Z., et al. 2008. Influence of monitoring method and control complexity on operator performance in manually controlled spacecraft rendezvous and docking. *Tsinghua Science & Technology*, 13(5): 619-624.

CHAPTER 21

Upper Limb Strength Expression during Work Simulated Tasks in a Cohort of Black African Males

Davies SEH and Goba T

Cape Peninsula University of Technology
Cape Town, South Africa
daviess@cput.ac.za

ABSTRACT

Twenty black African males with a mean age of 22.7 years yrs (±2.8) voluntarily participated in a study that evaluated strength expression in common movements associated with manual work. To date there has been an absence and/or paucity of research and/or data on black Africans that describe upper limb strength expression in applied work tasks. Given the growing industrialization taking place in Africa it is necessary to provide a more informed understanding of the relevant performance attributes of persons who are likely to be exposed to associated manual tasks. Anthropometric measures of the participants indicated a mean stature of 1.73 m (±0.08) and mass of 73.1 kg (±10.35), along with a mean wrist and humerus width, respectively of 5.33 cm (±0.5) and 7.90 cm (±1.04). Strength expression means included a bench press of 61.1 kg (±18.2). A Lafayette dynamometer measured hand grip (right and left respectively) of 483.16 N (±57.4) and 462.5 N (±46.5). Isokinetic strength was measured using a Biodex System 3 PRO dynamometer, where maximal voluntary muscle contractions were recorded at testing speeds of 30, 60 and 90°/sec, peak torque (Nm). Mean values included wrist extension 7.15 Nm (±3.2), wrist flexion 5.4 Nm (±1.4), forearm supination of 11.2 Nm (±3.4), forearm pronation of 8.3 Nm ±2.4). The following work simulation strength tests revealed the following means: valve (36 cm diameter) clockwise turn (close) of 90.2 Nm

(±19.1) and anti-clockwise turn (open) of 87.61 Nm (±21.1); an 8 cm spherical grasp valve parallel grip utilizing pronation/supination forces resulted in means of rotate away (supination) 8.45 Nm (±2.1), rotate toward (pronation) 8.31 Nm (±2.6). Simulated screwdriver means included screw away (supination) 8.89Nm (±1.3), screw toward (pronation) 9.5 Nm (±2.7). An isokinetic test using a large wrench (horizontal) resulted in the following mean outputs isokinetic wrench forced away 325.39 Nm (±74.90), isokinetic wrench forced toward 290.4 Nm (±44.3). Isometric (static) fatigue tests using the wrench (mean maximal force per 30 second static load, with 10 second rest) produced the following outputs for episode 1: 337.6.2 Nm (±59), episode 2: 327.1 Nm (±70.8), and episode 3: 309 Nm (±74.5). Correlations were conducted on the variables, and it was interesting to note that bench press a standard measure of gross upper body strength was poorly correlated with pronation, supination,, extension and twisting actions associated with manual tasks using tools, and was significantly correlated only with isokinetic wrist flexion (0.64) and isokinetic wrench turning movements (0.55). These findings provide an indication of the strength expression in a cohort of young African males and should assist engineering design, as well as providing the foundation for a normative data base for assessing the readiness of individuals for work in physically demanding industrial situations.

Keywords: strength, isokinetic, African

1 INTRODUCTION

Pheasant and O'Neil (1975) make the point that the effectiveness of a large number of everyday activities depends ultimately on a person's capacity to grasp objects and turn them against resistance. It is also apparent that researchers in developed economies have contributed to the development of ergonomics data fundamental to the design of safe and usable products (Norris and Wilson, 1997). Examples of research that has developed and assimilated the most up-to-date anthropometric and physical strength data for countries around the world may be evidenced (*for example*) in publications relating to children (Norris and Wilson, 1995), adults (Pebbles and Norris, 1998) and older adults (Smith, Norris and Pebbles, 2000), along with research on torque exerted by women investigated by Bordett et al (1988), children Steenbreekers (1993), the elderly (Berns, 1981; Imhran Loo, 1989; Thomson, 1975).

However ergonomics research, notably regarding applied work task assessments, and the data generation regarding Black Africans in industrially developing countries (IDCs) is scarce (please note that South African Employment Equity Legislation identifies the following designated groups, Africans, Coloureds, Indians; along with Whites. For the purpose of clarity and understanding the term 'Black African' is equivalent to the term 'African', implicit in the use of these categorizations in South African Equity Legislation)

Many IDCs are characterised by high levels of unemployment, along with mediocre legislative and governmental oversight in terms of safety, occupational health provision or workers compensation. Workers in IDCs are therefore vulnerable and likely to be exploited, and they are likely to be subject to overt, and always indirect, pressure to suffer in silence rather than place their jobs at risk by complaining (Charteris and Scott, 2006). Thus it is reasonable to contend that in an environment where workers are vulnerable and potentially exposed to arduous physical work tasks that they are also at increased risk of injury. Charteris and Scott (2006) observed that people *(in Africa)* die of aids-related diseases, not of occupational back stress; consequently poor resources and more urgent priorities deflect attention away from the aleviating high levels of manual work and back stress that characterize the daily work experience for people in IDCs. However, whilst infrastructure inadequacies and limitations have been evident in many IDCs there are some promising indications that a number of countries on the African continent are gearing their economies towards a more sophisticated level of development.

The move towards a more complex and advanced industrial economy is apparent in a number of countries on the African continent with relatively high levels of GDP, notably South Africa, which has sought to implement health and safety legislation that is aligned with best practice world wide. An example of this has been the recognition of upper limb disorders, and related work induced illnesses and injuries; and the introduction of strategies and work based practices that militate against and/or eliminate exposing workers to tasks that may induce injury, such as legislation regarding Work-Related Upper Limb Disorders (WRULDs) Circular 180, which was introduced in 2002. The aim of this guideline is to give the Office of the Compensation Commissioner and health professionals dealing with work-related upper limb disorders (WRULDs) guidance on how to define, diagnose, manage and report these disorders. It also gives advice to employers regarding how preventative measures should be taken where such disorders in the workplace occurred and how to report these to the Department of Labour or the Department of Minerals and Energy. It is particularly pertinent to note that the preventative measures included in these draft guidelines relate to ergonomic interventions and systemic human factors management planning at the place of work.

The development of relevant and meaningful legislation in IDCs, along with the intention to improve work conditions, lessen the risk of injury, understand issues pertaining to absenteeism, as well as improving productivity is highly dependant on applied and relevant ergonomic research that provides data, which can guide improved industrial design, work practices, appropriate and responsive legislative/policy development that will (in this context) enhance our understanding of the capabilities of black Africans.

Adams (2006) contends that understanding the strength and motion capabilities of the hand and wrist are important in designing tasks to minimize the frequency and severity of work-related upper limb disorders. These cumulative injuries and functional losses, defined as cumulative trauma illnesses by the US Bureau of Labour Statistics account for 11% of all work related musuloskeletal disorders (illnesses) in the USA. Rorke (2002) observed that those physically well conditioned and able to meet task demands are less likely to suffer acute injury, or the negative effects of chronic degenerative disease. This is particularly relevant in industrially developing countries (IDCs) where there is a high prevalence of manual labour which is likely to cause more musculoskeletal disorders than those reported in advanced countries (Scott and Christie, 2002).

As indicated earlier in this paper, whilst there has been ergonomic research conducted with a view to developing normative data sets and standards with regard to isokinetic and torque parameters, such research initiatives have been scant in Africa, and have typically focussed on strength expression in sport (van Heerden, 1997 & 1998; Lategan, 2002). Ergonomic research that has been conducted in Africa has tended to investigate issues pertaining to occupations where the demands of the tasks associated cannot be adapted or ameliorated. Perhaps the best examples of such research were performed by Wyndham (1973) and Strydom et al (1975) in terms of categorisation of mine workers according to their tolerance of heat. More recently task evaluations and human performance parameters, again linked to pre-selection criteria for arduous occupations have been produced for Emergency Care Practitioners (paramedic rescue personnel) (Davies et al., 2008) and task demands when ECPs use the *Jaws of Life* (Parr and Davies, 2009).

2 METHODS

Twenty black African males between the ages of 18-24 years participated in this study. All participants were in good health and free from any upper extremity injuries or disorders. Details of the study procedures including benefits, confidentiality and voluntary participation were provided and written informed consent was obtained prior to testing.

Nine various upper extremity limbs were tested for strength and comprised seven isokinetic, one isometric contraction work load and a standard bench press. An isokinetic dynamometer, the Biodex System 3 Pro machine, was used for testing. Isokinetics has become an established method because of its popularity for validity, objectivity and repeatability in the evaluation of muscle strength (Perrin, 1993). All participants were subjected to anthropometric measurements, a warm up and a specific warm-up which consisted of four contractions at differing speeds on the test machine (Lategan, 2002). The dominant side was tested and a standardized encouragement routine was employed for all the tests simulated applied work tasks and considered grip and wrist action (Baltzopoulos and Brodie, 1989). The tests comprised wrist flexion and extension, forearm pronation and supination, valve head (36 cm diameter) turn clockwise and anti-clockwise, spherical valve head

(8cm diameter) rotations away and toward, isokinetic wrench pump action movement away and toward the centre of mass and finally isometric wrench.

During the wrist flexion/extension and forearm pronation/supination tests the participants were required to sit with their elbow flexed at 90^0, the dynamometer positioned at axis of rotation height (Biodex, 2006). The protocol used was isokinetic unilateral, as was the case for all tests with the exception of the isometric test, and comprised one set of five repetitions at a speed of 60^0/s. The other tests included turning a valve head (36cm diameter) clockwise and anti-clockwise. The test was conducted with the participant standing a comfortable distance from the dynamometer, allowing for a slight bend the arms/hands (36 cm valve) facing the dynamometer. A repetition was made up of a full turn clockwise and a full turn anticlockwise with the start position always at full pronation. The dynamometer orientation and tilt for clockwise and anticlock-wise was 90^0 and $+25^0$respectively. Similarly, one set of five repetitions at 60^0/s made up the protocol.

The dynamometer orientation, tilt and protocol for the screwdriver, spherical valve head (8cm diameter) rotation and isokinetic wrench tests was 90^0, 0^0 and one set of five reps at 30^0/s respectively. With the screwdriver and a spherical valve head (8cm diameter) rotation tests, the participants' hand was in line with the dynamometer attachment which meant the body was slightly but comfortably positioned to the left of Biodex.

Both the wrench tests required the participant to stand next to the dynamometer with the protruding part of the attachment (handle) in front of them at waist height. The participant was required to grip the handle with a slight bend in the arms and knees whilst performing the required action. With the isokinetic wrench away and toward, the participant was required to perform a pumping action with the start position at 0° and a full pump down and up constituting one repetition. With the isometric wrench the participant was required to maintain a body position similar to the isokinetic wrench. The protocol, however, consisted of three maximal isometric contractions episodes of 30s pulling/lifting wrench in an upward with a rest/recovery period of 10s between each exertional episode.

3 RESULTS

The following tables and figures provide an overview of means, along with standard deviations of the anthropometric characteristics of the partipants (Black African males n=20) along with their isokinetic and isometric (static) strength expressions during various protocols.

Table 1 Mean anthropometric characteristics of Black African Males (n=20), standard deviations are in parentheses

Anthropometric Measures	Males (n = 20)	
	Mean	SD
Age (years)	22.7	(±2.8)
Stature (metres)	1.73	(±0.08)
Mass (kg)	73.1	(±10.3)
% Body fat	13.4	(±7.1)
Wrist width (cm)	5.33	(±0.5)
Humerus width (cm)	7.9	(±1.04)
Bicep circumference (cm)	31.2	(± 3.5)

Table 2 Mean peak isometric performance of Black African Males (n=20), standard deviations are in parentheses

Mean Peak Isometric strength Measure(s) (N) During 30 second episodes: 1, 2 and 3	Males (n = 20)	
	Mean	SD
Episode 1: (30 secs of maximal static effort)	337.6	(±59)
Episode 2: (30 secs of maximal static effort)	327.1	(±70.8)
Episode 3: (30 secs of maximal static effort)	309.0	(±74.5)

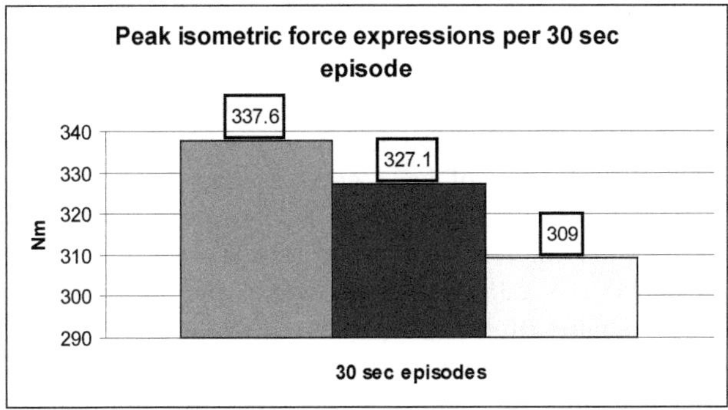

Figure 1 Illustration of the decline in peak isometric force expressions during three 30 second work episodes, separated by 10 second rest periods.

Table 3 Mean isokinetic and strength expression characteristics of Black African Males (n=20), standard deviations are in parentheses

Isokinetic (Nm) and Strength Measures (Kg)	Males (n = 20) Mean	SD
Bench press (kg)	61.1	(±18.2)
Hand grip (right)	483.1	(±57.4)
Hand grip (left)	462.5	(±46.5)
Wrist extension	7.1	(±3.2)
Wrist flexion	5.4	(±1.4)
Forearm supination	11.2	(±3.4)
Forearm pronation	8.3	(±2.4)
Valve (36 cm diameter) clockwise turn (close)	90.2	(±19.1)
Valve (36 cm diameter) anti-clockwise turn (open)	87.6	(±21.1)
Spherical grasp 8 cm diameter supination	8.45	(±2.1)
Spherical grasp 8 cm diameter pronation	8.31	(±2.6)
Screwdriver - screw away (supination)	8.89	(±1.3)
Screwdriver - screw toward (pronation)	9.5	(±2.7)
Wrench forced away (Nm)	325.3	(±74.9)
Wrench forced toward (Nm)	290	(±44.3)

2 DISCUSSION

The present research project that focused on a cohort of Black African males demonstrated that the established and reliable strength testing protocols, such as hand grip strength resulted in high levels of congruence with previous studies. Imrahn (2006) in his review of hand grip strength studies commented that one handed grip strengths have mostly fallen in the range of 450-600 N for adult males. It can be observed that the results from the present study for dominant mean hand grip strength of 483.1 N falls within this range, and also compares closely with results from other studies of 487.7 N (Adams, 2006); 409.8 N (Pebbles and Norris, 2002); 470 N (Bao and Silverstein, 2005). It was interesting to note that in terms of anthropometric measures only humerus width was well correlated with hand grip strength (0.5)

It was interesting to note that bench press a typical measure of gross upper body strength was poorly correlated with tasks found in the workplace that involve pronation, supination, extension and twisting actions of the hand/wrist when using tools such as screwdrivers, and interfacing with valves and levers/wrenches, and was significantly correlated only with isokinetic wrist flexion (0.64) and isokinetic wrench turning movements (0.55). It should also be noted that hand grip strength is often utilised as a predictor for upper body strength expression, however in this study it would appear that hand group was poorly correlated with pronation, supination, extension, flexion and twisting actions. Wrist width was reasonably correlated with forearm pronation (0.6) and supination (0.54), while bicep circumference was positively correlated with wrist extension (0.49) and flexion (0.49) as well bench press (0.63) and isokinetic wrench pull (0.49). The isometric wrench test during three consecutive work episodes of 30s (separated by 10s rest/recovery periods): 337.6, 327.1 and 309 Nm illustrated the stepwise peak force expression, along with the effects of fatigue on the operator (see Table 2 and Figure 1). Thus it is tentatively forwarded that selected anthropometric measures such as wrist width and bicep circumference appear to provide some insight into the predictive strength expression in some of the simulated work tasks characterised by forearm pronation/supination; and wrist extension/flexion, along with wrench operation.

2 CONCLUSIONS

The present study provides ergonomists, human factor specialists, engineers and designers a profile of anthropometric, strength and torque characteristics when applied to a number of simulated work tasks in a cohort of black African males. Whilst grip strength results from this study were consistent with previous reported, it is recognized that more research is needed in terms of replicating the other strength / torque testing protocols evidenced in this paper to ensure reliability and validity. The findings as they are presented provide an indication of the strength expression in a number of simulated work tasks by a cohort of young African males and as a consequence should assist engineering design, as well as provide a more informed foundation for a normative data base regarding the readiness of individuals for work in physically demanding industrial situations, and may be particularly relevant for industrially developing countries.

ACKNOWLEDGMENTS

The authors would like to acknowledge the Cape Peninsula University of Technology, University Research Funding Committee for their support of this project.

REFERENCES

Adams, S.K. (2006). Hand Grip and Pinch Strength. *International Encyclopedia of Ergonomics and Human Factors*, 2nd Edition, Taylor and Francis Group, Volume 1-3: 365-376.

Baltzopoulos, V. and Brodie, D.A. 1989. Isokinetic dynamometry: application and limitations. *Sports Medicine*, 8(2): 101-116.

Bao, S. and Silverstein, B. 2005. Estimation of Hand Force in Ergonomic Job Evaluations. *Ergonomics* 48(3): 288-301.

Bordett, H.M., Koppa, R.J. and Congelton, J.J 1988. Torque Required from Elderly Females to Operate Faucet Handles of various Shapes. *Human Factors* 30: 339-346.

Charteris, J. and Scott, P.A. 1999. Work-Hardening and Strength Expression: Effects on Isokinetic Curve Variability in a Manual Labour Cohort. *Journal of the Ergonomics Society of Southern Africa* 11(1): 20-25.

Davies, S.E.H,, Parr, B.M., and Naidoo, N. (2008). Physical performance characteristics of South African male and female Emergency Care Practitioners. *Journal of the Ergonomics Society of Southern Africa* 20(2): 3-14.

Imhran, S.N. and Loo, C.H. 1989. Trends in Finger Pinch Strength in Children, Adults and the Elderly. *Human Factors* 31(6): 689-701.

Imhran, S.N. 2006. Hand Grip Characteristics and Strength. *International Encyclopedia of Ergonomics and Human Factors*, 2nd Edition, Taylor and Francis Group, Volume 1-3, Chapter 85: 386-389.

Lategan L (2002). Normative Isokinetic Torque Values for Rehabilitation in South Africa. Unpublished PhD Thesis: University of Pretoria, Pretoria, South Africa.

Mital, A. Peak Volitational Torques for Wrenches and Screwdrivers. *International Journal of Industrial Ergonomics* 3(1): 41-64.

Norris, B.J. and Wilson, J.R. 1995. *CHILDDATA-The HandBook of Child Measurements and Caperbilities-Data for Design Saftey*. Department of Trade and Industry, London, UK.

Parr, B.M. and Davies, S.E. (2009). Physiological response of emergency care students during a stimulated extrication of a patient trapped in a light motor vehicle – implications for exercise training *Journal of the Ergonomics Society of Southern Africa* 21(1), 2009, 9 pp online

Pebbles, L. and Norris, B.J. 1998. ADULTDATA-The HandBook of Adult Anthropometric and Strength Measurements-Data for Design Safety. Department of Trade and Industry, London, UK.

Perrin DTT (1993). Isokinetic Exercise and Assessment. Champaign, Illinois: Human Kinetics Publishers.

Pheasant, S. and O'Neil, D. 1975. Performance in Gripping and Turning: A Study in Hand/Handle Effectiveness. *Applied Ergonomics* 6.4: 205-208.

Scott, P. and Christie, C. *Work-related upper limb disorders: Functional Anatomy*. Department of Human Kinetics and Ergonomics, Rhodes University, Grahams Town, South Africa, 2002.

Smith, S.A., Norris, B.J. and Pebbles, L. 2000. OLDER ADULTDATA-The HandBook of Measurement and Capabilities of the Older Adult- Data for Design Safety. Department of Trade and Industry, London, UK.

Strydom, N.B., Benade, A.J.S. and van Rensburg, A.J. 1975. The State of Hydration and the Physiological responses of Men during Work in the Heat. *Australian Journal of Sports Medicine* 7(2): 28-33.

Steenbreekers, L.P.A. 1993. Child Development, Design Implications and Acident Prevention. *Physical Ergonomcs Series*. Delft University Press, The Netherlands.

The Compensation Commissioner's Guideline for Occupational Health Practitioners & Employers to manage Work-Related Upper Limb Disorders (WRULDs) in terms of Circular Instruction 180. Dr Mmuso Ramantsi, Chief Medcal Officer Compensation Commissioner. 2 December 2002, Pretoria, South Africa.

Thompson, D. 1975. Ergonomics Data for Evaluation and Specification of Operating Devices on Components for use by the Elderly (*Loughborough Institute for Consumer Ergonomics*).

Van Heerden, H.J. and Viljoen, D. 1998. Preparticipation Evaluation in Young Rugby Players. *Medicine and Science in Sports and Exercise* 30(5):S231.

Van Heerden, H.J. 1997. Muscular Fitness and Intrinsic Injury Risk in Schoolboy Rugby. *South African Journal of Sports Medicine* 4(1):20-4.

Wyndham, C.H. 1973. The Physiology of Exercise under Heat Stress. *Annual Review of Physiology* 35: 193 – 220.

CHAPTER 22

A Work-Rest Allowance Model Based Individual Fatigue and Recovery Attributes

Liang MA, Zhanwu ZHANG

Department of Industrial Engineering, Tsinghua University
Beijing, P. R. China, 100084
Email:{liangma@tsinghua.edu.cn, jobbs.zhang@foxconn.com}

ABSTRACT

Appropriate work-rest scheduling is of great importance to reduce adverse effects of cumulative fatigue from physical work, to maintain workers physical conditions for manual handling operations, and to provide a safer working environment for the prevention of work-related musculoskeletal disorders (WMSD). It is believed that, to accomplish the same physical task, different individual workers may perform differently, and furthermore they have different fatigue progressions and post-work recoveries as well due to individual fatigue-related attributes. In this paper, a new work-rest allowance model is proposed based on a theoretical local muscle fatigue and recovery model. This new work-rest allowance model is compared with the other four existing allowance models, and effects from individual attributes on work-rest allowance are discussed. It is promising that the work rest allowance model enables us assign physical tasks for individual worker appropriately.

Keywords: work-rest allowance model; individual attributes; fatigue rate; recovery rate; individual factors

NOMENCLATURE

- *MVC* (unit: N): maximum voluntary contraction, maximum capacity of muscle, which equals to F_{max};
- $F_{cem}(t)$ (unit: N): current exertable maximum muscular strength at time instant t;
- $F_{load}(t)$ (unit: N): external muscular load at time instant t, the force which the muscle needs to generate;
- $F_{cem\ ini}$ (unit: N): the remained strength at the beginning of muscle recovery;
- k (unit: min^{-1}): individual fatigue rate, constant for a specific muscle group of an individual worker for a certain period;
- R (unit: min^{-1}): individual recovery rate, constant for a specific muscle group of an individual worker for a certain period;
- *%MVC*: percentage of the voluntary maximum contraction;
- f_{mvc} : relative load level $\%MVC/100$, $f_{mvc} = F_{load}/MVC$
- $p(\%)$: full recovery level, $F_{cem}(t) \times 100/F_{max}$
- $q(\%)$: fatigue level at the beginning of the recovery, $F_{cem\ ini}/F_{max} \times 100$.

1 INTRODUCTION

Cumulative muscle fatigue might lead to potential MSD risks (Chaffin et al., 2206), therefore appropriate work design is of great importance to prevent musculoskeletal disorders (MSD) resulting from physical tasks and to keep workers safe and health. Within work design, correct work-rest schedule should be assigned to avoid unwanted muscle fatigue. In order to schedule the work-rest appropriately, work-rest allowance models are often used in industry to assign work-rest paces.

The work-rest scheduling models are all developed based on the fatigue and recovery properties of physiological parameters. Fatigue is defined as "reduction of the functional capacity of an organ or organism" and recovery is defined as "increase of the functional capacity of an organ or organism, of which the functional capacity was reduced as a result of fatigue; recovery occurs by ending, reducing or changing the action which results in reduction of the functional capacity of an organ or of an organism." (Rohmert, 1973).

Conventional work-rest allowance models are often developed by using the actual holding time in comparison to maximum endurance time models (MET or maximum holding time, MHT) as a reference for fatigue/recovery status (El ahrache et al., 2006; El ahrache and Imbeau, 2009), and substantial discrepancies due to the inconsistencies of measurements have been reported by El ahrache and Imbeau (2009). Furthermore, individual attributes in fatigue and recovery have not been considered enough in the existing rest allowance models (Elfving et al., 2002). Those limitations result in lack of consideration in the work design stage and worker selection, and it may lead to overload for workers with weaker physical conditions.

Ma and his colleagues (Ma et al., 2009, 2011) have proposed a muscle fatigue and recovery model. The local muscle fatigue and recovery model is able to describe the fatigue and recovery in consideration of task parameters and individual fatigue rate and recovery rate. The fatigue and recovery model has been theoretically and experimentally validated, and it provides a new approach to decide the work rhythm by considering detailed progression in muscle fatigue and recovery.

In this paper, a work-rest allowance model is proposed to predict suitable work-rest allowance with consideration of individual fatigue and recovery attributes. This rest-allowance model is based on the muscle fatigue and recovery model (Ma et al., 2009, 2011). Individual attributes (fatigue rate and recovery rate) are integrated to predict the fatigue progression and recovery process during a static physical operation. Derived from this work-rest scheduling model, effects of individual fatigue attributes and recovery attributes are demonstrated, and the rest-allowance model is compared with the other models. Task parameters (such as, external load, posture, duration) can be combined with this model to predict the most suitable work-rest allowance.

2 MUSCLE FATIGUE AND RECOVERY MODEL

Muscle fatigue and recovery were proposed in Ma et al. (2009); Ma (2009); Ma et al. (2011), and they are introduced as below.

2.1 Local muscle fatigue model

Local muscle fatigue is described in Eq. 1.

$$\frac{dF_{cem}(t)}{dt} = -k\frac{F_{cem}(t)}{MVC}F_{load}(t) \tag{1}$$

The decrease of the local muscle strength can be described using Eq. 2 and the maximum endurance time is described by Eq. 3.

$$F_{cem}(t) = MVCe^{\int_0^t -k\frac{F_{load}(u)}{MVC}du} \tag{2}$$

$$t = MET = -\frac{ln\frac{F_{load}(t)}{MVC}}{k\frac{F_{load}(t)}{MVC}} = -\frac{ln(f_{mvc})}{k\,f_{mvc}} \tag{3}$$

2.2 Local muscle recovery model

Local muscle recovery model is formulated in Eq. 4.

$$\frac{\mathrm{d}F_{cem}(t)}{dt} = R(F_{max} - F_{cem}(t)) \tag{4}$$

The integration of Eq. 4 is Eq. 5, which indicates the recovery process after an operation.

$$F_{cem}(t) = F_{max} + (F_{cemini} - F_{max})e^{-Rt} = F_{cemini} + (F_{max} - F_{cemini})(1 - e^{-Rt})$$
(5)

Recovery time is the time necessary to restore the capacity to the full recovery level, and it depends on the selection of the full recovery level. The recovery time of a fatigued muscle from fatigue level q to recovery level p can be determined by Eq. 6.

$$t = -\frac{1}{R}\ln\left(\frac{pF_{max} - F_{max}}{F_{cemini} - F_{max}}\right) = -\frac{1}{R}\ln\left(\frac{p-1}{q-1}\right) \quad (6)$$

3 THE WORK-REST ALLOWANCE MODEL

Rest allowance (RA) in static work represents the resting time (RT) needed for adequate rest following a static exertion, and it is generally expressed as a percentage of holding time (HT) (RA% = 100 × RT /HT). In a static operation, the actual holding time (Eq. 7) can be expressed by the f_{HT} and the maximum endurance time derived from Eq. 3.

$$HT = f_{HT}\ MET = -f_{HT}\frac{\ln(f_{mvc})}{(k f_{mvc})} \quad (7)$$

The normalized remained capacity (fatigue level) at time instant HT is calculated by Eq. 8.

$$F_{cem}^{N} = q = \frac{F_{cem}}{MVC} = \exp(-k f_{mvc} * HT) = \exp(f_{HT}\ln f_{MVC}) = (f_{MVC})^{f_{HT}} \quad (8)$$

Therefore, according to the recovery model, the required recovery time to recovery level p is expressed in Eq. 9.

$$RT = \frac{-\ln\frac{p-1}{F_{cem}^{N}-1}}{R} = \frac{-\ln\frac{p-1}{q-1}}{R} \quad (9)$$

Then, according to the definition of rest-allowance, the RA is expressed in Eq. 10. Rest-allowance is a function of individual attributes (k, r) and working parameters (f_{HT}, f_{mvc}).

$$RA = \frac{RT}{HT} = RA(f_{HT}, f_{MVC}) = \frac{kf_{MVC}\ln\frac{p-1}{(f_{MVC})^{f_{HT}}-1}}{Rf_{HT}\ln f_{MVC}} \quad (10)$$

4 DATA ANALYSIS AND DISCUSSION

4.1 Holding time and recovery time

Holding time and recovery time are shown in Fig. 1. To understand the holding time and resting time could facilitate the understanding on our allowance model. Holding time is the monotonically increasing function of f_{HT}, while f_{mvc} keeps constant; it is also the monotonically decreasing function of f_{mvc}, while f_{HT} remains the same. The profile of the recovery time is calculated under recovery level p= 99.95%. Recovery time is the monotonically increasing function of f_{HT}, and it is the monotonically decreasing function of f_{mvc} as well. However, there are no substantial differences in recovery time from different fatigue levels to the high recovery level. As a result, the rest-allowance profile reaches to its lowest point, when f_{HT} approaches to 1 and f_{mvc} approaches to 0, since the endurance time is infinite and the recovery time is relative tiny. Theoretically, the highest point occurs while f_{HT} approaches to 0 and f_{mvc} approaches to 1. However, this case is extremely difficult to be reached in real manual operation.

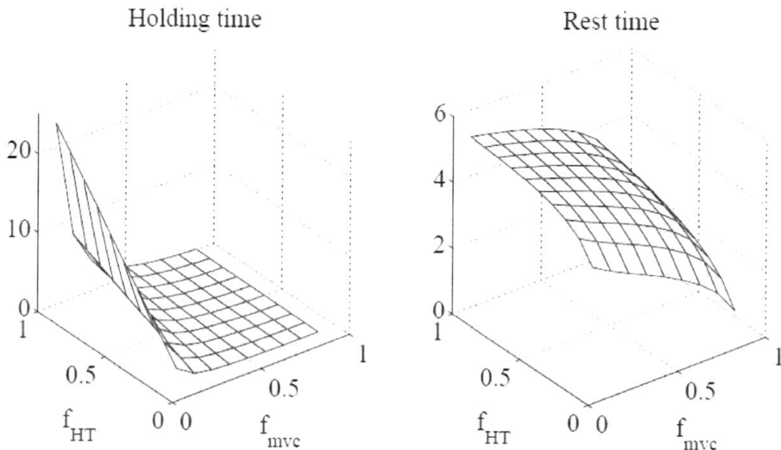

Figure 1: Holding time [min] and recovery time [min] using the theoretical approach

4.2 Effects of Individual Differences

In Eq. 10, k and R represent the individual factors on the rest allowance. The fatigue rate k determines globally the influence from each individual: larger fatigability k requires more time to recover when the other parameters remains the same, because holding time is shorter while the recovery time remains the same; larger recovery rate R results in shorter recovery time and therefore shorter rest allowance. The engagement of f_{HT} and f_{mvc} in determining RA is relative complicate, and it is graphically shown in Fig. 2. Suppose k = *1* and R = *1*, the profile shows the rest allowance for different force levels $f_{mvc} \in (0.1, 1.0)$ and different holding durations $f_{HT} \in (0.1, 1.0)$.

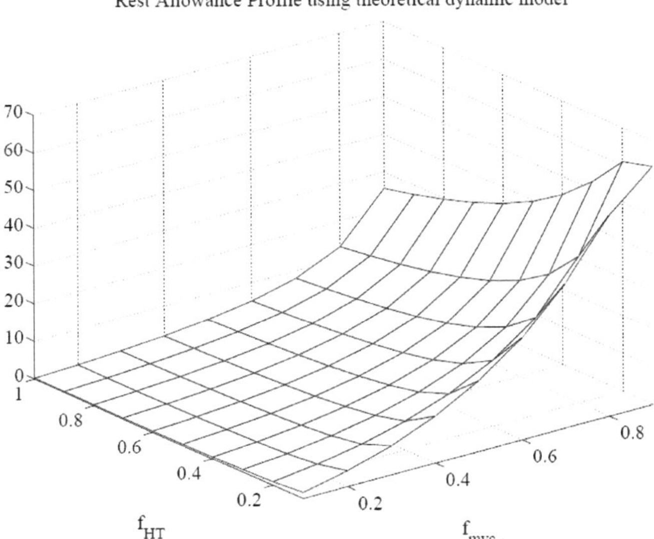

Figure 2: Rest allowance profiles using the theoretical approach

4.3 Comparison to Other Rest-allowance Models

Our model is compared to four rest allowance (RA) models summarized in El ahrache and Im- beau (2009) (see Table 1). Each RA model can be expressed as a function of f_{HT} and f_{mvc} according to El ahrache and Imbeau (2009).

Table 1: Rest allowance (RA) models (adapted from El ahrache and Imbeau, 2009)

Rest-allowance mode	RA(%)
Rohmert (1973)	$RA = 18 \times f_{HT}^{1.4}(f_{MVC} - 0.15)^{0.5} \times 100$
Milner et al. (1985)	$RA = 0.164 \times \left[4.61 + \ln\left(\frac{1}{100 - f_{HT}^{-1}}\right)\right]^{-1} \times 100$
Rose et al. (1992)	$RA = 3 \times MHT^{-1.52} \times 100$
Bystrom and Fransson-Hall (1994)	$RA = \left[\frac{\%MVC}{15} - 1\right]$

In Fig. 3, the predicted rest-allowance from our model is compared to the other models mentioned above. In our model, k and R, the fatigue rate and recovery rate are assigned as 0.94 and 0.96, respectively, which were mean values obtained from fatigue and recovery experiment of shoulder muscle group for 20 male Chinese subjects. In Fig. 3, it is observed that when f_{HT} is small, the differences between our model and the other models are relatively bigger; but when f_{HT} gets larger, the differences between our model and the other models get smaller. This indicates that our model could achieve almost the same pre- diction result as the others, when there is more fatigue due to longer holding time. However, when the holding time is relatively short, the prediction from this model still needs further verification. However, it should be noticed that in our model, k might be assigned individually. In this case, it is possible that the predicted value from existing models can be totally covered by using our model, such as the area between the two dashed-red lines with k= 0.5 and k= 1.5, respectively.

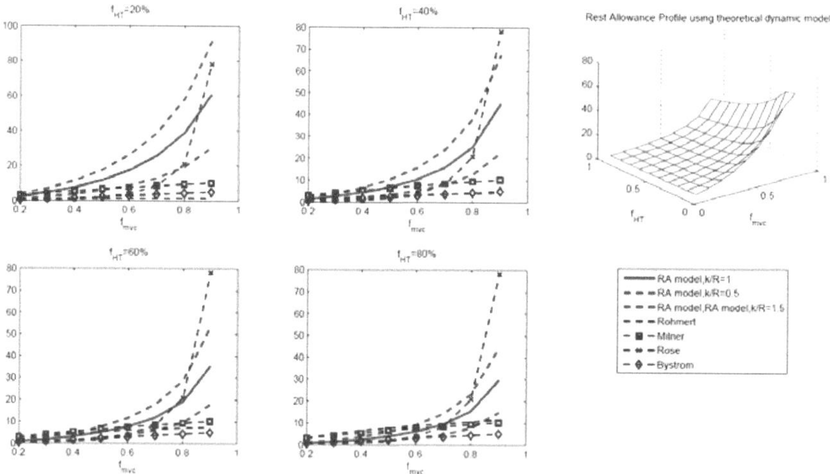

Figure 3: Rest allowance profiles using the existing RA models

It should be noticed that all the RA models in El ahrache and Imbeau (2009) are obtained from experimental data, and the substantial differences among these profiles indicate large differences in human recovery process. Although those differences can be explained by the different subjects participating in the experiment, methods to measure recovery, and modeling approach (El ahrache and Imbeau, 2009). Furthermore, it is believed that individual characteristics are the determinant factors. However, the personal factors are not considered enough in the models based on experiments.

In contrast, in our rest-allowance model, four parameters are used to calculate the suitable work schedule: k, R, f_{mvc}, and f_{HT}. Individual fatigue and recovery characteristics can be represented by k and R, and furthermore, both relative external load and relative work duration are also taken into consideration as traditional models. The RA model can be therefore generalized for industrial manual handling operations.

4.4 Discussion

In this paper, we proposed a simple rest-allowance model based on theoretical muscle fatigue and recovery model. In this way, it is promising to integrate individual physical attributes into work design. Fatigue rate (or fatigue resistance) and recovery rate are two representative terms within this model, and they are combined with task parameters to predict the fatigue progression. It stills remains a unexplored world behind this study, which is to find out factors determining the individual fatigue and recovery attributes. To accomplish this task, it still requires a great effort on it.

5 CONCLUSION

In this paper, a rest-allowance model based on a local muscle fatigue and recovery model is developed. In comparison to other rest allowance models, it is promising that this allowance model could be used to predict the work rest schedule for each individual appropriately. This model is computational efficient and it can be used in ergonomics application for work-rest allowance prediction and biomechanical applications.

ACKNOWLEDGEMENT

We appreciate the support for this study from the National Natural Science Foundation of China (NSFC, grant number: 71101079) and the grant from Foxconn, Yantai.

REFERENCES

Bystrom, S., Fransson-Hall, C., 1994. Acceptability of intermittent handgrip contractions based on physiological response. Human factors 36 (1), 158–171.

Chaffin, D., Andersson, G., Martin, B. 2006. Occupational Biomechanics. 4th Edition.Wiley-Interscience.

El ahrache, K., Imbeau, D., 2009. Comparison of rest allowance models for static muscular work. International Journal of Industrial Ergonomics 39 (1), 73–80.

El ahrache, K., Imbeau, D., Farbos, B., 2006. Percentile values for determining maximum endurance times for static muscular work. International Journal of Industrial Ergonomics 36 (2), 99–108.

Elfving, B., Liljequist, D., Dedering, ˚A., N´emeth, G., 2002. Recovery of elec - tromyograph median frequency after lumbar muscle fatigue analysed using an exponential time dependence model. European Journal of Applied Physiology 88 (1), 85–93.

Ma, L., 2009. Contributions pour l'analyse ergonomique de mannequins virtuels. Ph.D. thesis, Ecole Centrale de Nantes.

Ma, L., Chablat, D., Bennis, F., Zhang, W., 2009. A new simple dynamic muscle fatigue model and its validation. International Journal of Industrial Ergonomics 39 (1), 211–220.

Ma, L., Chablat, D., Bennis, F., Zhang, W., Hu, B., Guillaume, F., 2011. A novel approach for determining fatigue resistances of different muscle groups in static cases. International Journal of Industrial Ergonomics 41 (1), 10–18.

Milner, N., Corlett, E., O'Brien, C., 1985. Modeling fatigue and recovery in static postural exercise. Ph.D. thesis, University of Nottingham.

Rohmert, W., 1973. Problems in determining rest allowances Part 1: use of modern methods to evaluate stress and strain in static muscular work. Applied Ergonomics 4 (2), 91–5.

Rose, L., Ericson, M., Glimska¨r, B., Nordgren, B., 1992. Ergo-index- development of a model to determine pause needs after fatigue and pain reactions during work. Computer Applications in Ergonomics, Occupational Safety and Health, 461–468.

CHAPTER 23

Evaluation and Prevention of Work-related Musculoskeletal Disorders in Hungary

Gyula Szabó EUR.ERG.

Bánki Donát Faculty of Mechanical and Safety Engineering, Óbuda University
Budapest, Hungary
szabo.gyula@bgk.uni-obuda.hu

ABSTRACT

In spite of continued efforts the musculoskeletal disorders represent a major risk in the workplace. According to the European Working Conditions Survey 2010 on average 45.6% of the working population in Hungary reported that his / her work involves repetitive hand or arm movements.

Hazard identification, risk assessment and risk control in the workplace is required under Act No. 93 of 1993 concerning Occupational Safety and Health (OSH) with the principles of EU Council Directive 89/391/EEC on the introduction of measures to encourage improvements in the safety and health of workers at work. Employers are responsible for the realisation and occupational doctors and health and safety specialist with relevant qualification are permitted to carry a risk assessment. According to the law and the widely used Hungarian risk assessment guideline ergonomic risks are evaluated as part of the workplace risk assessment.

There is a wide range of ergonomic risk assessment tools available in the World, but only a few appears in the practice, used by people with general OSH knowledge but without specific ergonomic skills. The situation can be described by the fact published in the ESENER report by the European Agency for Safety and Health at Work that only 8% of establishments report using an ergonomics expert which is the second lowest value in Europe.

A possible way of ergonomic risk management is included in the harmonized European standard series "EN 1005 Safety of machinery. Human physical performance". Although the application of this standard is also obligatory it is often forgotten in workplace design or in case of machinery designed and produced in-house.

Although the risks of the manual handling, the force limits, body postures and movements, and high-frequency repetitive activities are separately determined by the EN 1005 standard, these and other amendments factors together, and mutually influencing determine the risk of work.

In the paper the contributing factors of physical risk are identified according to the EN 1005 standard, their context is analysed and an integrated work-related musculoskeletal disorders risk evaluation tools proposed to professionals with general OSH skills.

Key words: ergonomic tool, risk assessment, Hungary

1 INTRODUCTION

Since the 1990s with the expansion of automotive, electronics and machinery sectors, the rate of manual assembly and manual handling operations have increased in Hungary resulting high risk of work related musculoskeletal disorders. According to the European Working Conditions Survey 2010 on average 45.6% of the working population in Hungary reported that his / her work involves repetitive hand or arm movements, and 34.8% of the working population in Hungary reported that his / her work involve carrying or moving heavy loads.

The lack of a national language ergonomic screening and risk assessment tool is one reason why despite the risk management obligations for employers, the management of ergonomic risks is not solved.

2 HISTORY

During the 1970's era in Hungary, the central European country experienced a decade of ergonomics. Ergonomics was originally initiated by work psychologists. Doctors, health care professionals and representatives of different professions assisted the psychologists with their planning. This associated a strong mark in the development and application of ergonomics in Hungary.

The fact that these units lack the multidisciplinary approach to provide a professional group created a stronger barrier of development. In 1971 Hungary is connected to the ergonomics set up under the COMECON scientific-technical cooperation and soon after comprehended ISO standardization activities also.

A majority of products that were sold without design or ergonomics were accepted on the market due to there being no rivalry competition from vendors. But the suppression of COMECON was still a problem. The lack of social needs, receptive skills and the political priority still had an effect on ergonomics in society.

With time the system has declined, the laboratories started to close and the old cooperation began to weaken. In the 1990s, the regime changed dramatically, the old eastern orientation is replaced by a new western one, the old communist standards started to disappear. The health and safety research came to a holt because most of the attention is intended to change the regime, to the privatization, and to the decentralization of the R&D (Research and development).

The focus of the research was shifting to cognitive ergonomics.

At the same time may establish some progress in many areas of ergonomics:

- ✓ Compared to the previous, the Government subsidize/support financially more the developments.
- ✓ Broadening and strengthening of international research cooperation.
- ✓ Appear of the first naturalized ergonomics ISO standards.
- ✓ New relations in the EU.
- ✓ Creation of the MET (Hungarian Ergonomics Society)
- ✓ Connections to: IEA, CREE, FEES

3 PRESENT

As part of a state in the European Union (EU), Hungary enforces all safety requirements. The Hungarian occupational safety law is implements the EU OSHA frame directive requiring obligatory risk management by every employer including ergonomic risks. The requirements range from various policies such as use of work equipment, the directive on the minimum health and safety requirements for the manual handling of loads where there is a risk particularly of back injury to workers, the directive on the minimum safety and health requirements for work with display screen equipment.

Workplaces have to abide by rules and regulations that are put in place by government bodies.

European Agency for Safety and Health regulate all relevant law issues. However, it is an immense problem that only 10% of these standards are translated to Hungarian. According to the most recent data of the National Statistics Office (KSH) only 16.5% of Hungarians speak at least one foreign language fluently.

A range of information and downloadable data are only printed in English although, all Internet websites are able to be translated to any language that is required.

There is not any accurate statistics of accidents, injuries and diseases at work because of the characteristics of the occupational and primary health care system. There is also not any precise or correct data relating to work-related musculoskeletal disorders.

Since the acknowledgement of ergonomics, multinational companies become more aware of this issue and initiated training sessions that educated people about ergonomics. This was believed to help workers understand issues regarding health and safety while at work.

Despite being aware of ergonomic methods, workplaces have not improved essential safety standards that are required. It is alleged that workers that are not trained in REBA partake the questionnaire and are more likely to make mistakes compared to a REBA qualified individual. Based on a survey made in the

University of Óbuda of Budapest, 23 of 150 people use the REBA method. Others base their research on their own observations. As a result it is vital to obtain a risk assessment that is easy to use, Hungarian professionals are able to comprehend, and is able to be identified in Hungarian law. This will make a safer and happier working environment in workplaces and will reduce the amount of musculoskeletal disorders in the country.

4 COMPOSITE ERGONOMIC RISK ASSESSMENT (CERA)

The system-ergonomic approach were applied arranging the risk of the musculoskeletal disorders resulting from work affecting factors (Fig. 1.). The personal (operator), products (machines, tools, work objects) and environmental risk factors - that appear in the standard - are clearly identifiable. The remaining risk factors are the interaction variables, these has a high rates, because it gets the most attention by the evaluation of physical exertion. In our model the interaction variables of risk factors were classified into three groups, with the posture, strength and character of the motion variables.

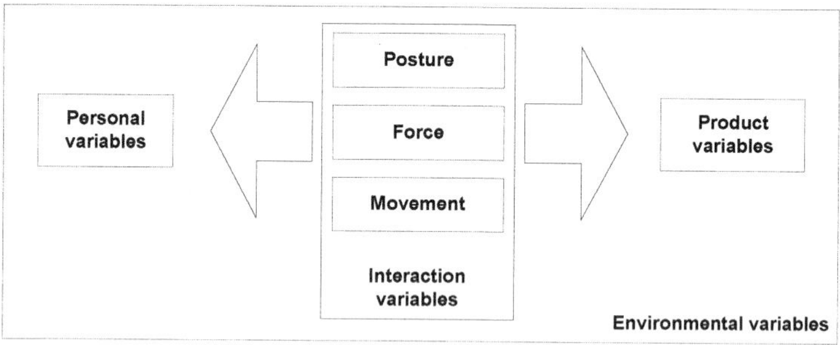

Figure 1: MSD Risk factors mentioned in the EN 1005 series

Risk factors that appear in EN 1005 standard series can be classified like:

- ✓ Risk factors with reference to other standards, or with the statement that attention needed in this regard,
- ✓ Risk factor with a recommended or required value or the expected state given,
- ✓ factors evaluated in detail in the standard, which several times also includes a multiplier factor.

The uneven distribution of appearing risk factors in each individual EN 1005 element refers to the primarily analysed risk. However risk factors belonging to other system elements are always present showing the strong interconnection among risk factors.

The Composite Ergonomic Risk Assessment (CERA) tool is an automated workbook with several datasheets. On the pictures some screenshots show parts of the CERA tool filled with data relating to a machinery manual assembly operation. Colour coding discriminates the obligatory or optional input fields and results of logical operations or calculations. When the evaluation includes sequence the result of the current step and the necessity of the following step is also colour coded.

Page A contains basic description of the operator, the workplace and working conditions (Fig. 2).

The purpose of Page B is the evaluation of manual handling operations according to EN 1005-2 which means the application of the modified NIOSH lifting equation.

The evaluation of forces limits on page C and posture on page D (Fig. 3a, 3b.) is relatively simple according to EN 1005-3 and EN 1005-4 accordingly thanks to the mostly separated factors.

A 1 **Ergonómiai kockázat értékelése az MSZ EN 1005 alapján**

Alap- és státuszadatok

	Munkahely neve, azonosítója:	Motorszerelő 13.				GM001020
	Lapok száma:					Lapszám:
I.	**A munkatevékenység:**				*Ha **Igen**, akkor minden tevékenységet külön kell értékelni.*	
1.	Összetett tevékenység:	**MO**		**I/N**		
2.	Jellege:	**MJ**				
3.	Álló munkavégzés	**MA**		**I/N**		
4.	Ülő munkavégzés	**MU**		**I/N**		
5.						
II.	**A munkakör jellemzői**					
6.	Megnevezése:	**KN**				
7.	Napi nettó munkaidő:	**KI**		**perc**		
8.	Pihenőidők:	**KP1,2,3**			perc	
9.						
III.	**A dolgozó(k) státusza**					
10.	Neve, azonosítója:	**DA**				
11.	Neme:	**DG**				
12.	Kora:	**DK**		**év**		
13.	Nemzetisége:	**DN**				
14.	Testmagassága:	**DM**		**cm**		
15.	Testtömege:	**DT**		**kg**		
16.	Egészségi állapota:	**DE**				

Figure 2: Example of our project, part of the CERA's status form

A 1 Értékelőlap a funkcionáli testhelyzetekből eredő kockázatokhoz

MSZ EN 1005-4:2005+A1:2009

Munkahely neve, azonosítója:	Motorszerelö 13.	GM001020

Törzshajlítás (előre-hátra) mértéke: **4** Megfelelő

1 2 3 4

Törzs fordításának vagy oldalra döntésének mértéke: **1** Megfelelő

1 2

A felkar függőleges irányú helyzete: **1** Megfelelő

1 2 3 4

Figure 3a: Example of our project, part of the CERA's posture evaluation form

B 1. Értékelőlap ipari gépek kiszolgálásával összefüggő emelési tevékenységekhez

MSZ EN 1005-2:2003+A1:2009

	Munkahely neve, azonosítója:	Motorszerelő 13.	GM001020	
I.	Lépés	*Kitöltendő? (ha M > 3 kg és L < 2 m)*	Igen	
1.	Vonatkozási tömeg:	Mref 15 kg	*Kiegészítő táblázat alapján kell megállapítani! (Mref munkalap)*	
	1. módszer: Átvilágítás a kritikus értékek átlagaival			
II.	Lépés	*Ha az **A** lap **V.29-37.** válasza: Igen*	Igen	
2.	1. eset: kritikus tömeg			
3.		Ha **M**< Mref 70 %-a	OK	IGAZ
4.		Ha **D** legfeljebb 25 cm	Nem	HAMIS
5.		Ha a mozgatás a csípő és a vállmagasság között van	Nem	
6.		Ha **A** = 0 és a törzs egyenes	OK	
7.		Ha a terhet a testhez közel tartják (**H**<25 cm)	OK	
8.		Ha **F**< 0,00333 Hz (1 emelés 5 percenként)	OK	
9.	2. eset: kritikus függőleges tömegelmozdulás			
10.		Ha **M**< Mref 60 %-a	OK	
11.		Ha a mozgatás a csípő és a vállmagasság között van (90-150 cm)	Nem	
12.		Ha **A** = 0 és a törzs egyenes	OK	
13.		Ha a terhet a testhez közel tartják (**H**<25 cm)	OK	
14.		Ha **F**< 0,00333 Hz (1 emelés 5 percenként)	OK	
15.	3. eset: kritikus gyakoriság			
16.		Ha **M**< Mref 30 %-a	Nem	
17.		Ha **D** legfeljebb 25 cm	Nem	

Figure 3b: Example of our project, part of the CERA's lifting form

The risk of repetitive movements according to EN 1005-5 is evaluated on page D. Since the detailed analysis required in this standard needs more knowledge than is expected from OSH professionals without ergonomic specialization the result shows only the verified red (risk) and green (no-risk) results. In case of yellow (in-between) risk zone an ergonomics specialist should be consulted and other tool should be applied.

There's a strong interest and expectation from OSH professionals and several field test will be done together with other evaluation methods and in the near future a validated easy-to-use Hungarian ergonomic evaluation tools will be available.

4 SUMMARY

A composite ergonomic risk assessment tool has been developed in Hungarian and is available in electronic datasheet format. This new understandable and usable tool increases the competency of the Hungarian OSH professionals with the evaluation ergonomic risks in the workplaces. The new Composite Ergonomic Risk Assessment (CERA) tool includes the evaluation of posture, manual handling, repetitive movement and force extension according to the harmonized European standard series EN 1005 Safety of machinery. Human physical performance.

ACKNOWLEDGEMENT

The project was realized through the assistance of the European Union, with the co-financing of the European Social Fund, project ID TÁMOP-4.2.1.B-11/2/KMR-2011-0001.

REFERENCES

EN 1005-1 Safety of machinery - Human physical performance - Part 1: Terms and definitions

EN 1005-2 Safety of machinery - Human physical performance - Part 2: Manual handling of machinery and component parts of machinery

EN 1005-3 Safety of machinery - Human physical performance - Part 3: Recommended force limits for machinery operation

EN 1005-4 Safety of machinery. Human physical performance. Part 4: Evaluation of working postures and movements in relation to machinery

EN 1005-5 Safety of machinery - Human physical performance - Part 5: Risk assessment for repetitive handling at high frequencyEuropean Agency for Safety and Health at Work. (2010). *European Directives.* Available: http://osha.europa.eu/en/legislation/directives. Last accessed 20th Feb 2012.

Eurofound. (2010). *Does your work involve repetitive hand or arm movements?*. Available: http://www.eurofound.europa.eu/surveys/smt/ewcs/ewcs2010_04_10.htm. Last accessed 2nd Feb 2012.

Hungarian Microcensus. (2005). *Population by knowedge of languages, age groups and sex*. Available: http://www.mikrocenzus.hu/mc2005_eng/volumes/02/tables/load2_1_7.html. Last accessed 2nd Feb 2012.

MET. (2010). *The tasks and objectives of the Hungarian Ergonomics Society*. Available: http://met.ergonomiavilaga.hu/subsites/index_eng.htm. Last accessed 20th Jan 2012.

Szabó, G., "A munkából eredő váz-izomrendszeri megbetegedések kockázatát befolyásoló tényezők", International Engineering Symposium at Bánki (IESB 2011), Budapest, Óbudai Egyetem, pp. 1-17, ISBN 978-615-5018-15-2, 2011.

CHAPTER 24

The Analysis of Complex Load of Machine Production Workers

Mária Kapustová, Ivana Tureková, Karol Balog

Faculty of Materials Science and Technology
Slovak University of Technology in Bratislava
Slovak Republic
maria.kapustova@stuba.sk
ivana.turekova@stuba.sk
karol.balog@stuba.sk

ABSTRACT

Ensure of thermal comfort in the workplace is an important factor for the safety and efficiency of employees. There is more important by work with hard physical work involving sources of excessive heat. An example of such work is forging press LZK 2500th. Blacksmith's work is characterized by high physical demands with the dynamic nature of muscle work. Workers are exposed to negative factor, which is excessive noise, too. Aim of this paper is to assess the complex worker heat stress and to design and validate new mathematical model to determine the human body complex workload.

Key words: heat stress, forging press, blacksmith, thermal balance, degree of the difficulty, thermal loading

1 INTRODUCTION

A satisfactory culture of work and working environment is one of the basic conditions of the healthy development of a modern human being and especially increasing of his living standard. Governments, including the social partners in the European Union, try to improve the qualitative value of the working environment over the minimal level determined by legislation. They realize that, in spite of existence of many EU directives and regulations aimed at improving the safety at workplaces, situation remains unsatisfactory (Jokl, 1984).

Notably unfavourable working conditions still remain in workplaces in engineering plants where working activities may cause unbearable workload to their workers during their working time. From the point of view of workers and in order to provide their health protection it is necessary for these plants to ensure the main economical and human pillars of their prosperity: a good working environment with quality of production and safety at work (Jokl, 1998). A mutual sequence of these system tools depicts Figure 1.

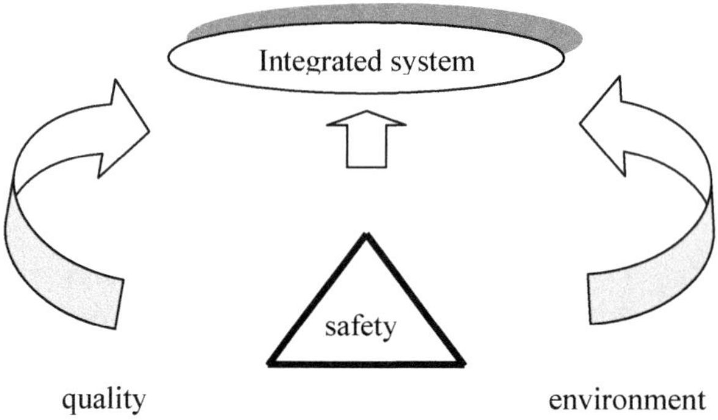

Fig. 1 Integration of system tools (Jokl, 1998)

2 COMMON PRINCIPLES OF WORKLOAD EVALUATION

The working environment is characterized by a complex system of negative factors that in every moment of the working process influence the human body with different intensity (Fišerová, 2011; Hettinger, 1990). The workload therefore represents the whole complex of external conditions and requirements in working system which have a disturbing influence on the health conditions of human body. Evaluation of stated load is commonly using an indirect quantity - factor of the difficulty Qdif which indicates how many times the real load is bigger than the allowable load (Cikrt and Málek, 1995; Raiskub and Hubač, 1980).

$$Q_{dif} = \frac{Z^r}{Z^p} \quad (1)$$

where Zr - value of real load, Zp – value of allowable load.

Note: Each workload has its evaluation criteria for calculation of coefficient of difficulty, e.g. the noise load evaluation criteria is the level of acoustic pressure

L_A[dB], the heat load criteria is the density of climatic heat flow q_{cl} [W.m^{-2}], the physical-dynamic load criteria is the expenditure of energy E_w [KJ.min^{-1}] , etc.

The analysis of current workload evaluation of employees shows that in practice the approach of partial workload evaluation predominates. This means that the effect of each influencing factor on worker's body during his working activity is evaluated individually, independently from other environmental factors present. The current way of evaluation is unable to consider the mutual synergism, i.e. the concurrence of all present negative factors of the working process on the human organism.(Raiskub and Hubač, 1980; Kapustová, 2002)

3 MATHEMATICAL MODEL FOR COMPLEX LOAD DETERMINATION

On the basis of actual need for practice, a new mathematical model was elaborated. This model enables us to express the total effect of negative environmental factors and it also enables us to evaluate the complex load of the human organism during monitored working time. Each working environment influences the human body during working time by several loading factors. The whole load Z is therefore the resultant of the individual loading factors impact Z_j, $j = 1, 2, ..., n$. It is commonly known that the individual loading factors have different rates within the whole loading. It is possible to assume that the loading factor Z_1 shares the whole loading by rate α_1, loading factor Z_2 by rate α_2 , etc. till loading factor Z_n by rate α_n , whereas it has to be valid (Kapustová, 2002):

$$\alpha_1 + \alpha_2 + ... + \alpha_n = 1 \qquad \alpha_j \in \text{(from 0 to 1)} \tag{2}$$

The coefficients values α_j, where $j = 1, 2, ..., n$ are called the importance coefficients of individual loading factors impact Z_j of the working environment. They characterize the rate of human organism load where the values α_j close to 0 express the reality that loading factor Z_j has a weak influence while values α_j close to 1 mean that influence of loading factor Z_j is very strong compared to other loading factors. The concrete values of the individual importance coeficients α_j are defined through the use of point method, which utilizes a point scale 0 - 10 and applies it to each sort of partial loading Z_j on separate organs of human organism T_i during concrete working activity. The principle of the point method as well as the way of determining coefficients of effect importance α_j for individual loading factors, is ilustrated in table 1.

The mathematical model described enables us to use a way of objective and also subjective evaluation of the factors´ influence while applying the point method. Assigning points by objective evaluation of the factor`s influence on human organism is often very time consuming, as it requires very close cooperation of experts from a wide fields of medicine, ergonomy, technical sciences, psychology etc., who evaluate the human work and its influence on health conditions from different points of view.

Table 1 Principle of Point Method (Kapustová, 2002; Kapustová 2005)

Loading / The body organ	Z1	Z2	Z3	...	Zn
T1	b11	b12	b13	...	b1n
T2	b21	b22	b23	...	b2n
T3	b31	b32	b33	...	b3n
...	...	...	...	...	...
Tm	bm1	bm2	bm3	...	bmn
$\sum_{i=1}^{m} b_{ij}$	$\sum_{i=1}^{m} b_{i1}$	$\sum_{i=1}^{m} b_{i2}$	$\sum_{i=1}^{m} b_{i3}$	...	$\sum_{i=1}^{m} b_{in}$

$$\alpha_j = \frac{\sum_{i=1}^{m} b_{ij}}{\sum_{j=1}^{n}\sum_{i=1}^{m} b_{ij}} \quad \alpha_1 = \frac{\sum_{i=1}^{m} b_{i1}}{\sum_{j=1}^{n}\sum_{i=1}^{m} b_{ij}} \quad \alpha_2 = \frac{\sum_{i=1}^{m} b_{i2}}{\sum_{j=1}^{n}\sum_{i=1}^{m} b_{ij}} \quad \alpha_3 = \frac{\sum_{i=1}^{m} b_{i3}}{\sum_{j=1}^{n}\sum_{i=1}^{m} b_{ij}} \quad \ldots\ldots$$

$$\alpha_n = \frac{\sum_{i=1}^{m} b_{in}}{\sum_{j=1}^{n}\sum_{i=1}^{m} b_{ij}} \tag{3}$$

For each working activity there are values of allowable loads known according to hygienic norms for particular loading factors Z_j. They are indicated by the sign Z_j^p and in each moment of working activity it is possible to determine also the real load Z_j^r. By means of these values and calculated values of weight coeficients α_j a degree of momentary complex load q_c is defined by relation:

$$q_c = \alpha_1 \frac{Z_1^r}{Z_1^p} + \alpha_2 \frac{Z_2^r}{Z_2^p} + \ldots.. + \alpha_n \frac{Z_n^r}{Z_n^p} = \sum_{j=1}^{n} \alpha_j \frac{Z_j^r}{Z_j^p} \tag{4}$$

where the expression $\alpha_j \frac{Z_j^r}{Z_j^p}$ defines the degree of real load of human organism by loading factor Z_j from the complex load by all loading factors Z_1, Z_2, ..., Z_n (Lehder, 2000; ISO 7933:1989).

4 THE PRACTICAL APPLICATION OF MATHEMATICAL MODEL

Detection and subsequent analysis of workloads were performed in a drop forge at the workplace of a forging crank press LZK 2500 where the blacksmith was observed during forging process of the drop forging „Flange“. This longitudinal drop forging with a very complicated shape and weight of 4.70 kg was produced by a forging crank press by means of a three-operating tool.

The scheme of the forging press workplace LZK 2500 is shown on figure 2 and the simplified shape of drop forging depicts figure 3. The detailed analysis and continuous observation of the blacksmith`s work enabled us to perform the selection of the work loadings which had a significant disturbing influence on employees fatigue.

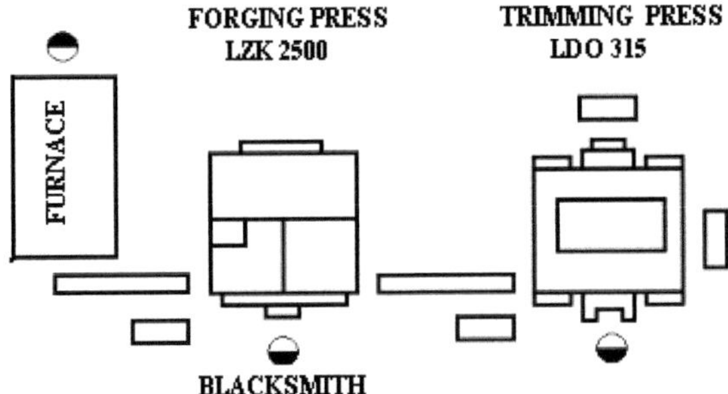

Fig. 2 Scheme of LZK 2500 press workplace

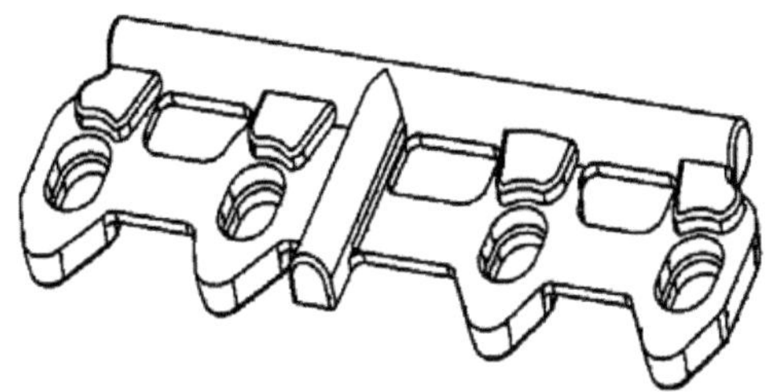

Fig. 3 Drop forging „Flange“

The scheme of the forging press workplace LZK 2500 is shown on figure 2 and the simplified shape of drop forging depicts figure 3. The detailed analysis and continuous observation of the blacksmith`s work enabled us to perform the selection of the work loadings which had a significant disturbing influence on employees fatigue.

The blacksmith´s body was mostly loaded by noise, excessive heat and physical work at the same time, while during the selection of three harmful substances the size and the period of their impact were crucial.

The influence of the three selected harmful substances (noise, heat and physical effort) on the worker was measured and evaluated during the first hour of the blacksmith`s work. Detected time responses of the noise, heat and dynamic loading of the worker in the workplace of the forging machine LZK 2500 during forging process for a period of 60 minutes are visually demonstrated by figure 4.

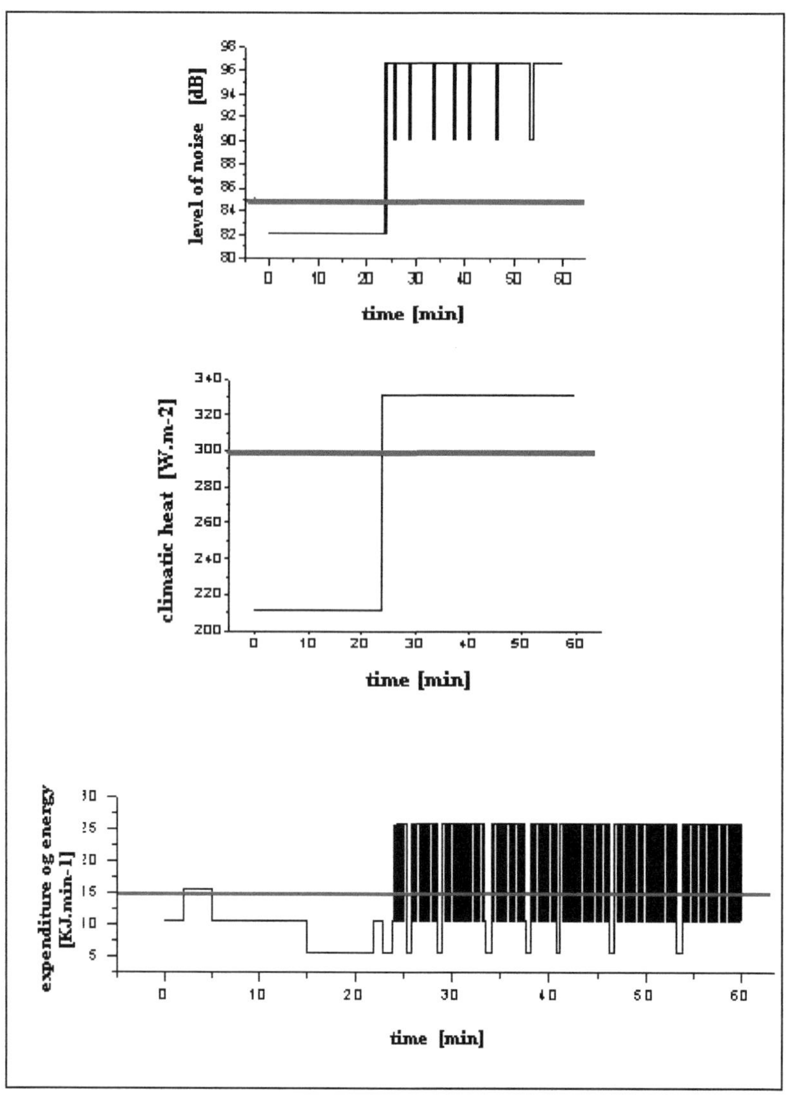

Fig. 4 Time responses of three selected loadin

The final aim of the model application is calculation of the total effect q_c of three loadings on the worker at any given moment of one-hour work according to relation (Fišerová, 2011). Important values of the weight coefficients α_j for individual types of loading Z_j (noise, physical, heat) were calculated using the point method by subjective evaluation. In table 2 are the concrete values of coefficients α_j gained by evaluating of separate loadings Z_j in examined workplace LZK 2500 completed by the worker, who expressed his feelings in a questionnaire.

Table 2 Values of weight coefficients α_j

Loading Z_j	α_j	Values
Physical Z1	α_1	0,439
Noise Z2	α_2	0,208
Heat Z3	α_3	0,353

The graphical explanation of the course of momentary complex load degree qc of worker while forging during the first 60 minutes of his work is demonstrated on the figure 5.

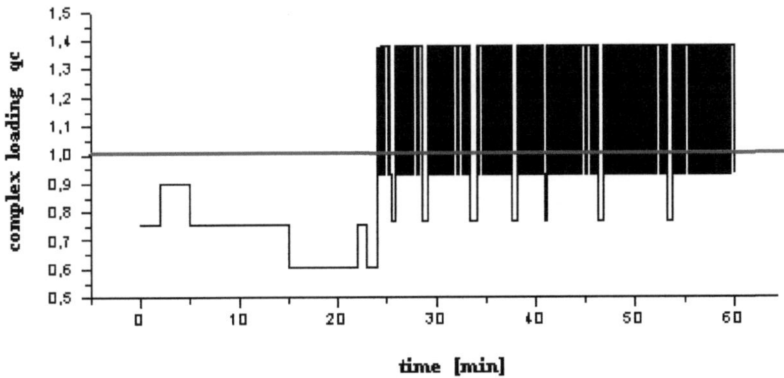

Fig. 5 The one-hour course of blacksmith`s complex load at work

5 THE IMPORTANCE OF COMPLEX LOAD FOR SAFE WORK

For considering the work environment quality as well as the rate of work safety at the workplace of forging press LZK, the course of complex load qc during the first 60 minutes of the blacksmith`s work (according figure 5) is very important. It is possible to find criteria which are determining comfort and eventually discomfort in the workplace on the basis of this graph. There are:

⇒ *rate, size* **of degree of momentary loading q_c**

⇒ *time of exposure*, **eventually** *period of duration* **of complex load q_c on blacksmith**

⇒ *frequency of repetition* **of complex load values q_c during the observing period**

Detecting the size and effect period of the total organism load qc is important for all environmental-technical or labour-medical fields that work together on the creation of a high level of safety and health protection of workers at work. Work hygiene recommends for practical needs to divide the areas of harmfulness (or the size of the health threat at work) in dependence of complex load degree value according to the table 3. On the basis of one-hour course analysis of worker´s complex load in the workplace LZK 2500, there occurred various working activities and blacksmith´s operations, which consider the real values of assessed working environment quality criteria in table 4.

Table 3 The influence of complex load qc on the health risk during work (Cikrt and Málek, 1995; Raiskub and Hubač, 1980; Lehder, 2000)

Complex load q_c	Areas of health risk at work
$q_c < 1$	Optimal area
$q_c = 1$	Allowable area with moderate harmful effect
$q_c > 1$	Area with high harmful effect

Table 4 Criteria of considering the work safety at the workplace LZK 2500 (Kapustová, 2002; Kapustová, 2005)

Number	Loading rate q_c	Type of working activity	Exposure time of loading [min]	Number of loading q_c repetitions
1	q_{c1} = 1,38	Forging process	24,2	121
2	q_{c2} = 0,93	Blasting, lubricating	8,47	121
3	q_{c3} = 0,89	Tool preparation	3	2
4	q_{c4} = 0,75	Waiting for semi-product	16,23	10
5	q_{c5} = 0,61	Walking	8	1
			Σ= 60 minutes	Σ= 255

It is evident from the graph in figure 6, that during one-hour blacksmith´s work on the forging press LZK 2500 the highest value of the complex load was qc1 = 1,38 and it corresponeded to the working activity of forging. It is a critical value of loading qc for the blacksmith`s organism, as this value according to table 3 belongs to the so called area with high harmful effect on health. From the aspect of safety and health protection at work is very important for this critical loading also the period of loading impact.

As it results from the figure 7, the blacksmith spent 24,2 minutes in the area of high harmful effect on health and during this time he forged out 121 forgings. For appraisal of the danger rate and work risk in the workplace of press LZK 2500 it is significant to observe the frequency of the complex load degree values repetition qc.

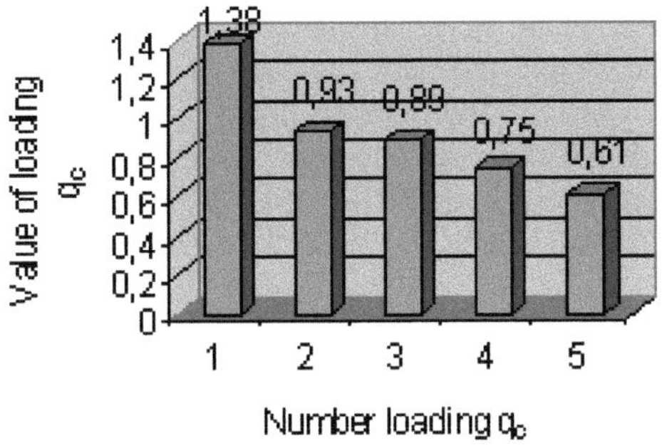

Fig. 6 Values of loading q_c

From the process of blacksmith`s one-hour work it was determined that the highest number of repetitions (i.e. 121- times in figure 8) of loading value during working process of forging was qc1 = 1,38. The same counting rate also has the loading value at tool blasting and lubricating qc2 = 0,93, but from the point of work risk it is very important to observe the counting rate of repetitions of complex load higher value qc1 = 1,38.

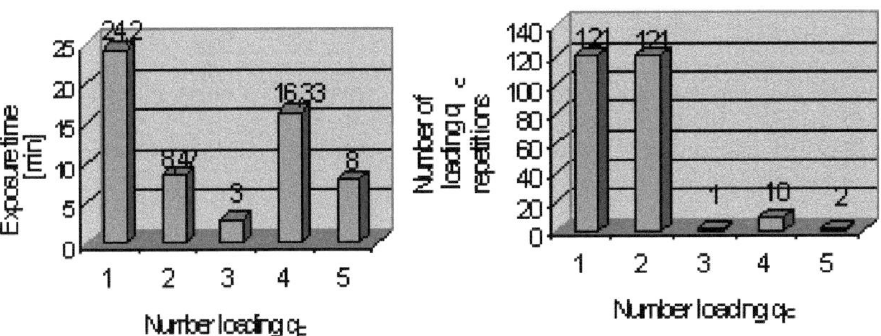

Fig. 7 Time period of load q_c impact

Fig. 8 Number of load q_c repetitions

6 CONCLUSION

The dynamic development of market economy and a strong competitive environment is constantly forcing the business subjects, especially in the field of engineering, to analyse and consequently optimize working conditions. Creating the optimal working conditions in the workplace is the main factor of stabilization of labour forces and the growth of work efficiency. The aim of this contribution was to mention the unbearable state of high workload which is frequent especially in workplaces such as hot machine plants, and to mention the new trends of evaluating the quality of working environment in order to increase working comfort in workplaces during active work. The calculated values and the courses of the immediate complex load enable the new approach at evaluation of comfort and eventually discomfort of working environments in machine plants.

The correctness of th the proposed model is clearly shown by results. These resulta and other standardized methods WGBT ISO 7243: 1989 and ISO 733: 1989 requirw another validification using real measurements based on other physical principles.

REFERENCES

JOKL, M. Optimization of physical conditions for human work, Prague: Labour, 1984.

JOKL, M. Energetic savings versus health. In Safe Work, 1998. N -2, pp. 9-14.

FIŠEROVÁ, S. Objectivization health risks for metal forming. In Integrovaná bezpečnosť 2011. Trnava : AlumniPress, 2011, pp. 33-40. ISBN 978-80-8096-153-4.

HETTINGER, TH. a kol. Hitzearbeit. Untersuchung an ausgewählten Arbeitsplätze der Eisen- und Stahlindustrie. ASER, 1990.

CIKRT, M. - MÁLEK, B.: Labour Medicine, Part I., Work Hygiene, Prague, CSPL, 1995.

RAISKUP, J.CH.- HUBAČ, M.: Professional - graphical Work Analysis, Bratislava, VUPL, 1980.

KAPUSTOVÁ, M. Ecologization of working places in engineering plants by means of complex load evaluation. Dissertation thesis. Bratislava. MtF STU, 2002.

KAPUSTOVÁ, M. The mathematical model for comfort at work determination in engineering production. In Acta Metallurgica Slovaca. 2005, Vol. 11, Number 1, pp.126-133.

Lehder, G. Working environment, work hygiene and ergonomics. Project P. TEMPUS-PHARE IB JEP13406-98. Modul 4.6. Kosice: TU, 2000.

ISO 7933:1989 Hot environments – analytical determination and interpretation of thermal stress using calculation of required sweat rates.

CHAPTER 25

A Design Proposal of Confirmation/Check Task to Ensure Workers' Active Will

Ryoko Ikeda,Naori Tsuji, Toshiya Akasaka, Yusaku Okada

Graduate School of Science & Technology / Keio University
Yokohama, Japan
Email: m6k6.u-u.wt.a6190@z2.keio.jp

ABSTRACT

Organizations have been increasingly interested in making their workers more safety-conscious, based on the premise that safety management are not put into action certainly if workers are not fully safety-conscious. However, being safety-conscious does not guarantee getting actively involved in safety management. Therefore, we aimed to propose the improvements for check/confirmation tasks by investigation of the workers' opinions and experiments. These results were integrated in the guideline. Several companies have been implementing our proposed guideline, and actually they highly appreciate our proposal. As a result, we obtained that workers were now more willing to conduct confirmation / check tasks than they had been before the improvements were made.

Keywords: Human Error, Safety, Confirmation / Check

1 INTRODUCTION

In many organizations safety management is keeping in very high level. However, as there is the possibility that the severe accidents happen, the safety management section in the organizations have been practicing the various countermeasures to prevent accidents. Especially, some organizations aim to grow up the customers' trust. This is based on the idea that establishment of safety management links to the customers' satisfaction. Here, the goal of the safety management is not only prevention of the accidents but also the improvement of the customers' satisfaction. In this strategy, the workers' motivation on safety activities/management is a very important key.

For example, the confirmation/check action is effective for recovering human error. However, if the workers' motivation on the confirmation/check action is low, the ability of recovery becomes so small.

Therefore, we investigate the workers' motivation on the safety activities to increase human reliability, and propose the suggestions to solve the issues that may shrink the workers' motivation. In this paper, the confirmation/check task is focused, and the human characteristics in the confirmation/check task are analyzed by the survey for the practical field and ergonomic experiment. In addition, by integrating the results, we proposed the guideline to growing up the workers' motivation and to make the potential possibility of human error so small.

2 SURVEY OF THE WORKERS' UNDERSTANDING LEVEL ABOUT SAFETY ACTIVITIES/HUMAN ERROR PREVENTION STRATEGIES

For 3 steel companies (15,000 workers), 4 railroad (20,000 workers), 3 medical institution (1,000 workers), 1 pharmaceutical company (500 workers), 1 oil refining company (500 workers), we made survey of the workers' understanding level about safety activities/human error prevention strategies for all employees.

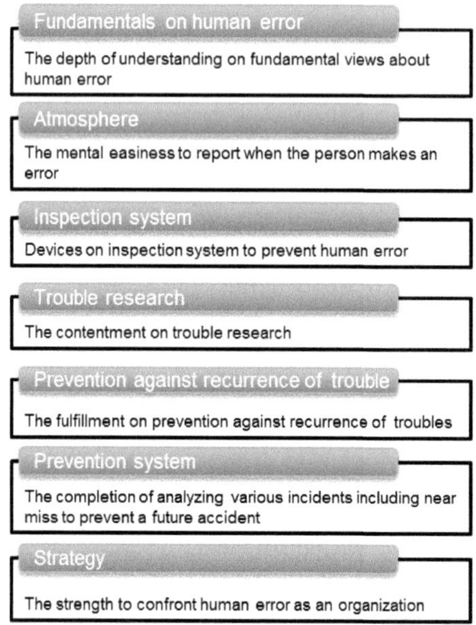

Fig.1 Seven Viewpoints in investigation for Workers' Active Will

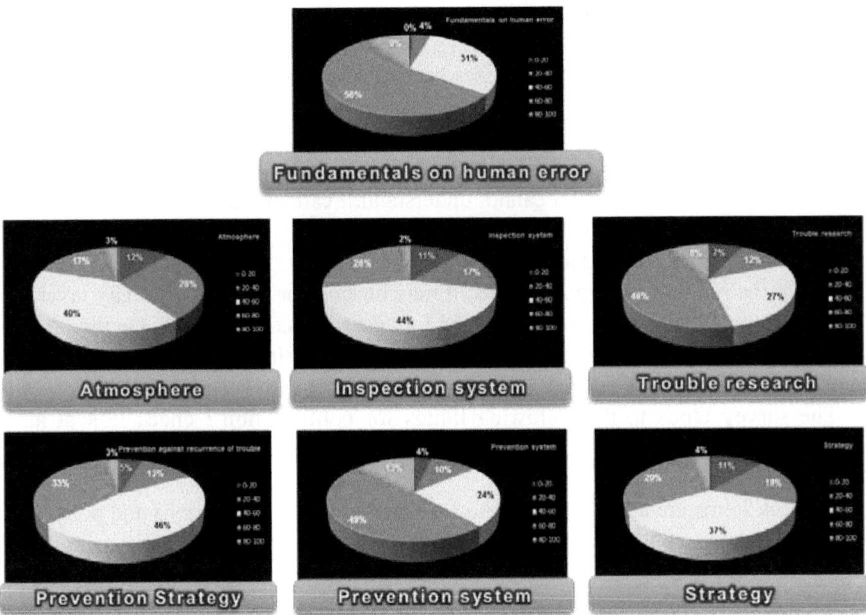

Fig.2 Ratio of 5 Layers on Workers' Motivation in 7 Evaluation Points

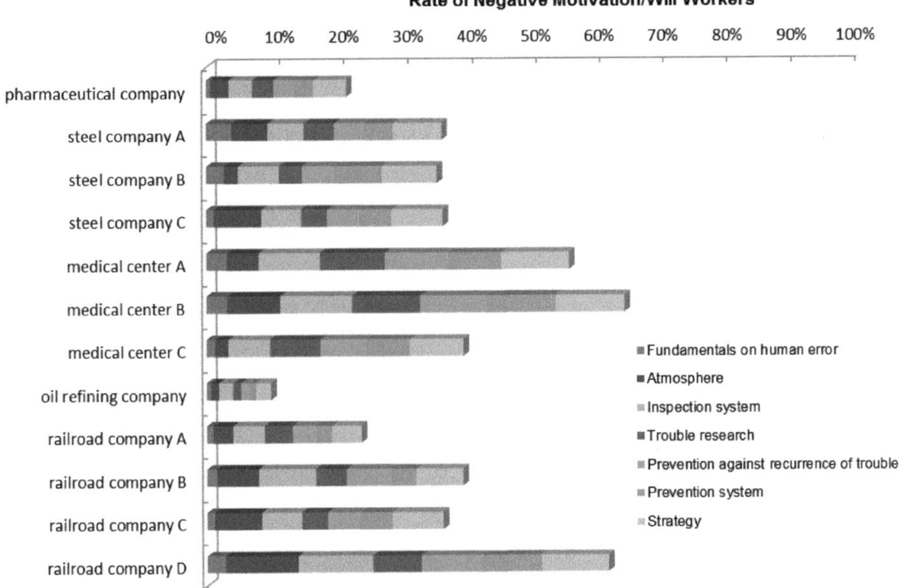

Fig.3 Rate of Negative Motivation Workers in Each Company

The investigation intends for all the employees in the object organization. We divide the result into 7 viewpoints (See Fig.1) and classify 5 layers (See Fig.2).

In these 5 layers, each layer expresses the following people.

Red layer; "I do not understand at all", "I cannot understand at all"
Orange layer; "I hardly understand", "I cannot understand"
Yellow layer; "If anything, I cannot understand, I cannot understand"
Green layer; "I understand enough"
Blue layer; "I understand and I agree with the idea".

As for the red and the orange layers, it may be said for "negative activity layer". And as for the yellow layer, it may be said for "passive activity layer". On the other hand, as for the green and the blue, it may be said for "positive activity layer".

The survey tends to the following things for confirmation / check task at all organizations.

If check items are added without the explanation of their intentions and possible benefits, workers are likely to have a negative sentiment, saying "here is yet another task which has nothing to do with us", "work is added which doesn't contribute to our safety". And workers often adjust their workload on their own to cope with the fixed volume of check items, leading them to conduct the task in a sloppy manner. On the other hand, confirmation / check tasks have hardly praised points because finding error sound like common sense. The sense increases to fail to mark or confirm check items. If they fail to mark or confirm check items, their managers can only rebuke them, saying "try harder", "be more careful", etc. Surely, directly measures correct nearly. But it is clear that the measures are not appropriate to investigate factors behind the failure. Gradually check function will decrease, if it must improve in the point which managers listen clearly to meanings, effects and importance of confirmation / check task for workers.

Especially, managers should not reproach confirmation / check task for missing problems with the idea of quiet task. It is important to enhance the motivation of workers through managers evaluate when workers found problems.

3 EXPERIMENT

In order to observe the human behavior in check/confirmation tasks, experiments were done. The object of the experiment is a simple check/confirmation work. Therefore, the flow of the operation is as follows.

Step 1: Check the state of the devices
Step 2: If the state is wrong, change the state
Step 3: Confirm the modified state of the devices

In such checking/confirming operation, the most important index is human error rate; omission error rate and commission error rate. However, in the experiments, because the task is so monotonous and the subjects' workload is so small, human error become hard to occur, and only human error rate would not be evaluated the human behavior enough. So the operational rhythm, which is the change of the progress speed, was added as the evaluation index.

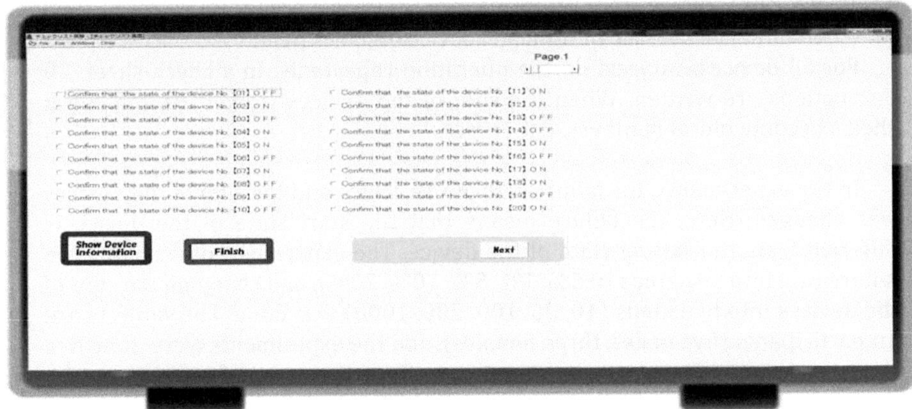

Fig.4-1 An Example of Checklist Page on Experimental Monitor

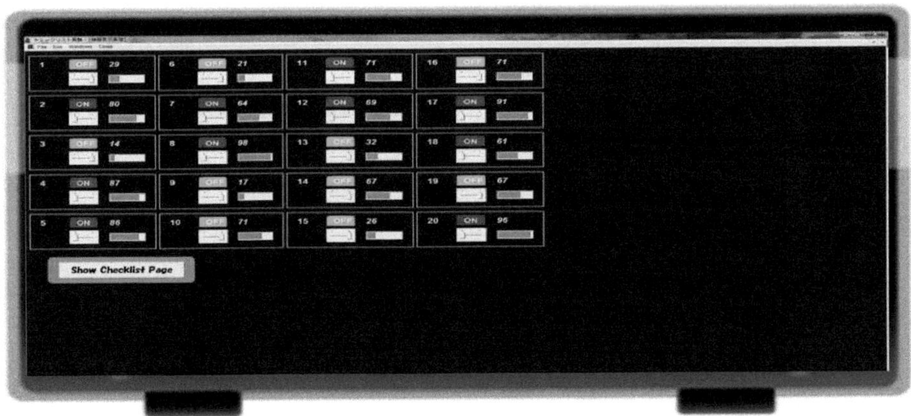

Fig.4-2 An Example of Device-information Page on Experimental Monitor

(1) Targeted Operation

Operation done in this experiment is as follows;

1) the subjects read check-sheet (See Fig.4-1), and recognize a instruction for the target device
2) the subjects memorize the setting state of the target device. Setting state is "ON" or "OFF".
3) the subjects observe the display (See Fig.4-2) that indicates the state of the devices, and confirm the state of the target device. If the state of the target device is wrong (the current state is different from the setting state written in the check-sheet), the subjects operate the switch of the target device, and confirm that the state of the target device becomes right.

For all devices, subjects do the operation repeatedly. In a check-sheet, 20 instructions are written. When the sum of the devices is over 20, the check sheets become plural numbers.

In the experiments, the failure rate of the device and the sum of the devices was changed. Here, the failure means that the start state of the device is different from the setting state of the device. The experiment of changing the failure rate into five kinds (0.5%, 1%, 5%, 10%, 20%), and changing the sum of the devices into five kinds (10, 20, 100, 200, 1000) was done. The subjects are 10 participants (five males, three females), and the experiments were done five times each. In addition, considering order effect, the order of the experiment was made random. The break time between experiments is over 2 hours.

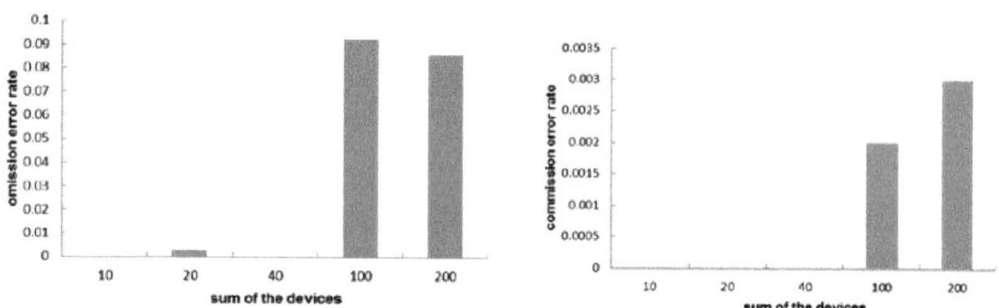

Fig.5-1 Relationship between Omission Error Rate and Sum of Devices

Fig.5-2 Relationship between Commission Error Rate and Sum of Devices

Fig.5 Relationship between Error Rate and Sum of Devices

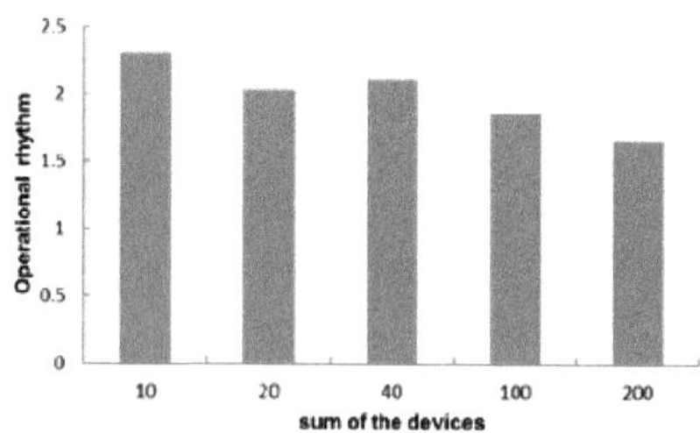

Fig.6 Relationship between Operational rhythm and Sum of Devices

(2) Relationship between the experiment result and the sum of the devices

For the evaluation indices, omission error rate, commission error rate and operation rhythm were used. Error rate is obtained by dividing [the number of error and commission error] by [the sum of the devices]. The operational rhythm is expressed as the change of the speed of check operation.

Results of experiments were shown as Fig.5 and Fig.6. In Fig.6, the results of error rate (omission error and commission error) were shown. The change of the error rate could be observed when the sum of devices is larger than 40. In the results of the operational rhythm, a little difference was observed. When the sum of devices is over 40, the check-speed become quick, and the disturbance of the speed become large.

So, in one check/confirmation task, it may be desirable that the sum of the devices is lower than 100. Especially, the monotonous of the check operation would be human behavior worse. So, not only simple check actions but also actions with thinking should be added in the task.

(3) Relationship between the experiment result and the personality

The subject answered a personality questionnaire, and the personality was analyzed. From our past studies, we obtained that when the "Consciousness" and "Neuroticism" in personality are high, the robustness of the monotonous work becomes strong. So, we investigated the relationship between the experiment result and the personality, but the particulars could not be obtained. From now, by collecting many subjects' data, we wish to clear the relation.

Table 1 Five primary factors and its subordinate factors on Personality

Main Factor	Subordinate	Main Factor	Subordinate
Extroversion	Extraverted	Neuroticism	Calm
	Energetic		Relax
	Talkative		At ease
	Bold		Not envious
	Active		Stable
	Assertive		Contented
	Adventurous		Unemotional
Agreeable-ness	Warm	Intellect	Intelligent
	Kind		Analytical
	Cooperative		Reflective
	Unselfish		Curious
	Agreeable		Imaginative
	Trustful		Creative
	Generous		Sophisticated
Concien-tiousness	Organized		
	Responsible		
	Conscientious		
	Practical		
	Thorough		
	Hardworking		
	Thrifty		

4 Guideline for Improvement of Check/Confirmation work

By combining the field investigation (in railroad companies, an oil refining company, a pharmaceutical company and medical centers) and the results of the experiment, we proposed the guideline to improve the check/confirmation work. The guideline aims to increase the operators' motivation in check/confirmation. The guideline consists of eleven items shown as follows.

** 1*

When the number of the check items must be increased, you explain the effect of additional intention and adding it in detail. Otherwise it is easy to come to invite the unfavorable consciousness, such as "that the work that is not the thing to contribute to safely of oneself increased", "the correspondence of the phenomenon irrelevant to oneself was increased".

** 2*

Set the number of the upper limits of the check item, and do not surpass it

** 3*

When the addition of the item is needed, review of the conventional item, and choose the items which can delete from the checklist.

** 4*

Do not have the excessive expectation for the error recovery by the check. Recognize the success of the recovery by checking/confirming as a good event, and commend a successful worker (do a prize). Do not scold that the recovery failed.

** 5*

When the checking/confirming task is failure, analyze the latent factors, and improve the work environments based upon the operators' centered design

** 6*

Use plural marking way. Avoid using only one-character marking way
For example, writing the value, marking the letter, and so on

** 7*

Clear the importance of the check items. Educate the relation between the importance and the risk.

** 8*

Decide a plan of a check/confirmation task (contents, time) in consideration with the worker's characteristics (such as personality, experience, skill, age, and so on)

** 9*

Collect the voices of the workers regularly (Only usual communication cannot collect enough information to improve latent problems)

** 10*

Construct a system to enhance the knowledge/technical tradition based on proud workers' idea

** 11*

Do not depend on only a check/confirmation task in human error prevention. When the countermeasures against human error is discussed, propose the plan except the check/confirmation task

5 VERIFICATION

In order to verify the validity of the proposed guideline for check/confirmation tasks, we practice the case study. The case studies have been executed in 12 sections of 3 companies. As the case studies started in November 2011, the meaningful result is not appeared yet. However, we obtained good comments from the safety managers; such as "the workers' motivation for check work is increasing", "the work team began to discuss the way to improve the efficiency/accuracy of the confirmation tasks", "the many workers come to do the check/confirmation task politely" and "the communication between the workers and the managers are revitalized". In the results of the investigation on workers' active will for safety management, the tendencies could be observed. (See Fig.7)

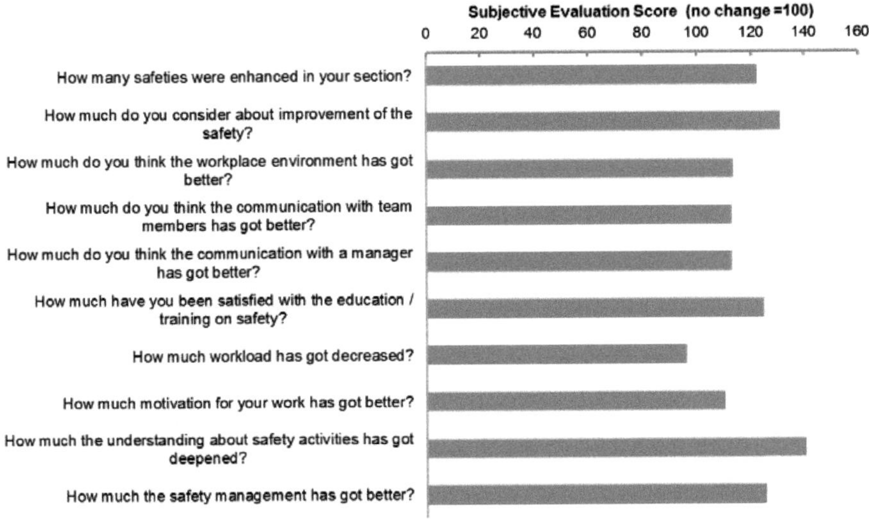

Fig.7 Subjective Evaluation Score about Safety Activities (Investigation by Questionnaires)

6 CONCLUSIONS

In this paper, we discuss the instruction way to ensure the workers' active will on human error prevention strategies. Especially, we focused on the check/confirmation tasks, and the problems were examined from field research and an experiment. From the results, we proposed the guideline for check/confirmation task. In same case studies, the efficiency of the instruction based on the proposed guideline is obtaining.

However, the bailers in practical fields were also found out. Therefore, through the discussion with the safety managers in the companies, we aim to produce the system that can support the safety-service management activities.

REFERENCES

[1] Takeo YUKIMACHI, Reference List on Basis of GAP-W Concept and a Case Study, Human Factors, Vol. 9(1), pp. 46-62, 2004

[2] Tomita Masahiro, Okada Yusaku, 2010, Developing a Method to Evaluate the Employee Satisfaction on Safety Management, AHFEI the 3rdd International Conference, 2009

[3] Okada Yusaku: Human Error Management of Performance Shaping Factors, International Conference on Computer-Aided Ergonomics, Human Factors and Safety, 2005

[4] Rasmussen, J: Information Processing and Human-Machine Interaction: An Approach to Cognitive Engineering. New York: Elsevier Science Publishing Co., Inc., 1986

CHAPTER 26

Enhancement of Workers' Capability to Analyze the Latent Factors in Troubles

Takumi Maeda, Akasaka Toshiya, Yusaku OKADA

Graduate School of Science & Technology, Keio University
3-14-1 Hiyoshi, Kohoku-ku, Yokohama, Japan
Maeda_t_23@yahoo.co.jp

ABSTRACT

In safety management in the organization, identifying the latent factors of the troubles is one of important tasks. However, almost of methods for identifying latent factors behind incidents are too difficult for normal workers to utilize.

Therefore, we intend to discuss the instruction issues that the workers can aware the latent factors in troubles easily as follows.

1. Reference Lists on PSFs (Performance Shaping Factors)
2. Analysis Strategy
3. Knowledge Sharing

These issues could be introduced from the survey of practical fields. In this paper, we intend to propose the instruction method to improve the quality of causes analysis based on these issues. Especially, we made a guide table to introduce latent factors. In addition, we produced the prototype of computer program that can support the identification of the latent factors in troubles. As a result of some experiments, the number of factors that the workers identified increases remarkably, the quality of the identified factors is also improved, and many effective countermeasures are produced. Thus some effects of our proposed instruction guide were obtained.

Keywords: Performance Shaping Factors, Trouble Analysis, Human Reliability, Human Error

1 INTRODUCTION

In safety management in the organization, identifying the latent factors of the troubles is one of important tasks. However, almost of methods for identifying latent factors behind incidents are too difficult for normal workers to utilize. The identification task was entrusted to a few special workers. So, the incidents, which are smaller than the accidents, were not analyzed enough, and the countermeasures against the future accidents were not discussed from different view. These problems happens the various troubles in usual work. Furthermore, in emergency, the influences of the problems become the threats that can spread the troubles.

Therefore, many organizations desire that more workers have insufficient capability to analyze/identify the latent factors of the trouble. The development of the capability depends on only workers' field experience on safety activities. But, it takes dozen of years before the workers have sufficient capability to analyze latent factors by this learning based upon the field experience. Thus, the effective instruction method is not established.

Therefore, in this paper, we intend to discuss the instruction issues that the workers can aware the latent factors in troubles.

2 SURVEY

In troubles caused by human error, next two viewpoints are important. The first viewpoint is the reducing the possibility of human error by improving the state of various 'performance shaping factors (PSFs)'. The second view point is the safety system to shut the effect of human error; such as fail safe.

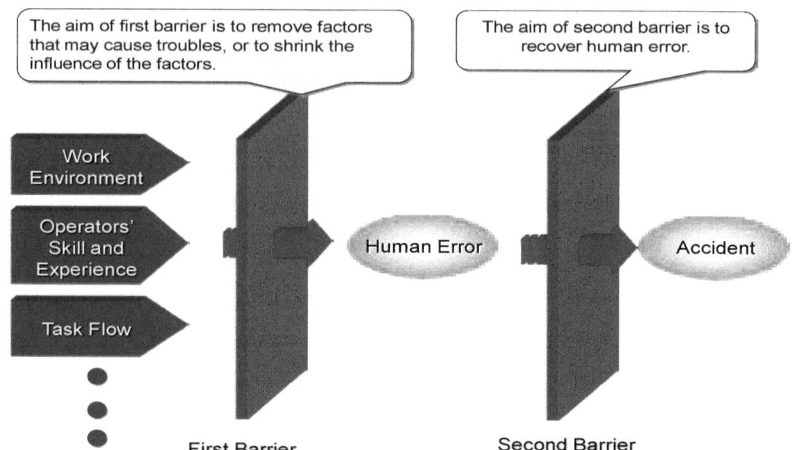

Fig.1 Fundamental Policy on Human Error Prevention Strategy

To prevent accidents the two viewpoints should be managed in detail. However, in many organizations, discussion on the first viewpoint is so insufficient. This is why latent factors of human behavior observed in the troubles could not be identified.

So we investigated the current state of trouble analysis in practical organizations. As a result, following problems were obtained.

1) The analyst cannot judge what the latent factors are.
   This is because the analyst does not have the knowledge about human factors. In other words, the minimum requirement for the analyst is the knowledge about special skill in the object work. So, in almost cases, lack of skill/knowledge/experience, shortage of education/training is picked up as the factors of human error. From only these factors the background of human error cannot be estimated.
   Performance shaping factors (PSF) is the factors that affect human behavior. Human error is also one of human behavior. Accordingly, PSFs in human error behavior are regarded as causes of human error.
   In order to prevent reoccurrence of human error, PSFs that causes human error should be extracted exhaustively, and the extracted PSFs should be removed or improved.
2) In many cases, the analyst group consists of the member in same section. So, the direction of the analysis tends to be biased. In order to expand the view of analysis, the member with different knowledge/experiences should add to the analysis team.
3) The many analysts decide the countermeasures against the troubles before the analysis starts. Accordingly, the factors are identified from the defined countermeasure. Of course, this cannot be called the analysis.

We intend to improve the three problems in trouble analysis.

3 METHOD

3.1 Identification of PSF

In order to obtain PSFs in target human behavior, the method that extracts PSFs is used. As PSFs have a hierarchical structure, PSFs that causes PSFs should be also investigated. From the PSFs structure whose summit is human error, the human error prevention strategy is planned. Here, the prevention strategy should have great effects on improvement of PSFs.

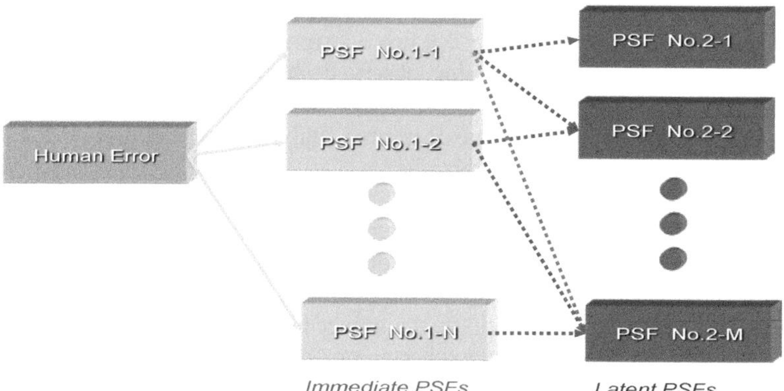

Fig.2 Hierarchy of Performance Shaping Factors (PSF)

Immediate PSFs: Improvement of immediate PSFs cannot be expected to prevent other trouble occurrence. The prevention countermeasures against the immediate PSFs tends to depend on the specification of object work. So the countermeasures cannot be applied for tasks in other departments, and possibility of occurrence of similarity troubles in other departments is not changed, that is, the possibility keeps high.

Latent PSFs: As latent PSFs exist in other departments, the prevention strategy against the latent PSFs is expected to reduce the possibility of human error occurrence. Only if the information of the strategy is transmitted to other departments, the similarity troubles will be prevented

However, these methods have only function that support to describe the extracted PSFs. So, if the subject who analyzes the PSF has few experiences of PSF analysis, sufficient results would not be obtained.

In particular, if latent PSFs that are background factors of human error were not extracted, the effect of the planned prevention strategy would not be expected.

For supporting extraction of latent PSFs, the following method, which has the reference such as PSF keyword table, is effective. There are a lot of cases that more than one human error exists in one trouble. So, with observing the operational sequence, the detail that leads the trouble should be cleared. Every human behavior that related to the trouble should be analyzed on PSFs.

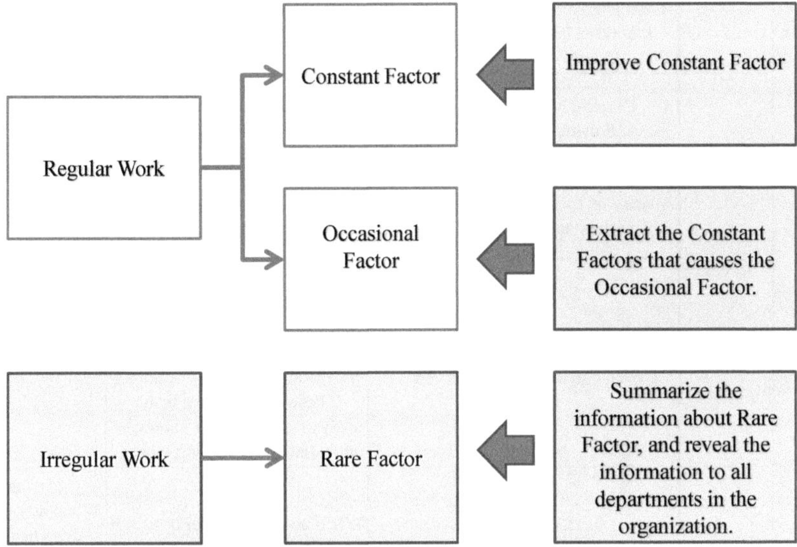

Fig.3 Categorization on dynamics of PSFs

In addition, it is important to examine the dynamics of PSF. Because dynamic PSF is occasional factor, there exist other factors that cause the dynamic factors.

From the points described above, we made a PSF Reference List to support the identification of the latent factors in troubles. The list is categorized into 6 levels.

Level 0: Superficial Factors; Only personal efforts can be selected as countermeasures.

Level 1: Direct Factors; Direction of the countermeasures is so wide.

Level 2: Direct Factors: Direction of the countermeasures is wide.

Level 3: Immediate Factors: Countermeasure are 'Review of work rule', 'Regulation of a prohibited operation', and so on. The countermeasures tends to depend on the specification of the object work.

Level 4: Latent Factors: The effect of the countermeasures is high.

Level 5: Latent Factors: The countermeasures can be expected to apply into various area, and to enhance the workers' motivation.

Table.1 Reference list

	Characteristics of identified laten factors	Maesures	Issues,Points to remember	Spread effect
Level0	Carelessly. Action slip. (illusion,Mishearing) A lapse of memory. Misunderstanding	Strengthen the sense of work. Do my best.	Telling "take care" as an action is not effective, because the person has already intended to take care. Additionally, it may become another stress and another factor to make error.	× Have an effect on persons that make a mistake
	Carelessness, Deceive oneself	Alerts. Worn.	Ofcourse it can be useful only if the person didn't check or didn't take care at all. But If not so, it will become a kind of the above useless actions. These actions have to be analysis of operating situations including the checking works.	▲
Level1	Lack of skill, Lack of experience	Education	When an organization does not educate workers, it is effective.However, if it dose, it is ineffective.	▲ When an organization does not educate workers,it is ○。But, orgnizations that do not have educatio system are not exist.
Level2	Lack of document The deficiency of manual Insufficient instructions Lack of communication	The review of the document. The review of the manual. Making instructions clear. Communication .	An orgnization do not have standards and establish standards.It do not provide a guideline.As a result, measures are ineffctive.	▲
Level3	Concrete problems in the work	The setting of the rule. Setting of the verboten.	It is necessary for a rule not to be broken.	○
Level4	Detailed piece of information of works	Work improvement. Environmental improvement. System improvement.	Continue the consideration from various angles based on m-shell, 4m, GAP-W, etc. And provide a standardized means of identifing factors.	◎
Level5	Detailed piece of information of works	The safe management in the organization. Take measures that improvement	In order to convince workers the necessity of measures,not only provide safety measures against PSFs,but also give the employees training on safety as necessary.	◎

Table.2 Epexegetical reference list

	An abbreviated designation	An example sentence
List of Level 4 and Level 5 PSFs	Luck of experience	Unfamiliarity with a situation which is potentially important but which only occurs infrequently or which is novel
		Operetor inexperience(e.g. a newly-qualified tradesman, but not an "expert")
	Feedback	Poor, ambiguous or illmatched system feedback
	Confirmation	No clear direct and timely confirmation of an intended acyion from the portion of the system over which control is to be exerted
	Skills	A need to unlearn a technique and apply one which requires the application of an opposing philosophy
	Information sharing	An impoverished quality of in formation conveyed by procedures and person/person interaction
	Displays and procedures	Inconsistency of meaning of displays and procedures
	Independent checking	Little or no independent checking or testing of output
	Environment	A poor or hostile environment
	Capabilities or experience	A need for absolute judgements which are beyond the capabilities or experience of an operator
	Physical condition	Evidence of ill-health amongst operatives, especially fever
	Cooperation	Unclear allocation of functon and responsibility
	Emotional atress	High-Level emotional stress
	Unsafety condition	An incentive to use other more dangerous procedures
	Identified hazard	A mismatch between perceived and real risk
	Infoglut	A channel capacity overload, particularly one caused by simultaneous presentation of nonredundant information
	Education system	A mismatch betwwen the educational achievement level of an individual and the requirements of the task
	Intervention of other work	Task pacing caused by the intervention of others
	Too many workers	Additional team members over and above those necessary to perform task normally and satisfactorily
	Age	Age of personnel performing perceptual tasks
	Performance standards	Ambiguity in the required performance standards
	Overconfidence to the past	Invest too much faith in workers
	Instrumentation	Unreliable instrumentation(enough that it is noticed)
		A mismatch berween an operator's model of the world and that imagined by a designer
	Transfer knowledge	The need to transfer specific knowledge from task to task without loss
	Too many tasks	Perform task with perfoming many other tasks
	Progress	No obvious way to keep track of progress during an activity
	Poor preparation	Woeful lack of preparations for works
	Unscheduled works	Unscheduled works and interruptions
	Dangerous task	Alutitude, high-voltage, highly-emissive, etc.
	Break rules	Break rules
	Complexity	Need to take a complicated process
	Correction	No obvious means of reversing an unitended action

In trouble analysis, if the analyst team refer the List (see Table 1,2) then the more effective factors can be identified. In addition, by learning about following issues, the number of the latent factors will become large, and the view of countermeasures will expand. The issues in learning are 'Relation between primary PSF and the troubles', 'PSF structures in the trouble', "Assumption of the background of the trouble', 'Personal characteristics of the parties in the trouble", 'Cognitive process of the parties in the trouble', 'Communication flow of the parties in the troubles', 'Detailed information sharing ', 'Value sense sharing', and so on.

3.2 Group Work

Trouble analysis should be done by team that consists of members with different knowledge/experience. In such a team, the discussion about background of the troubles tends to be activated. In addition, by interchange of the members' experience and knowledge, the members' idea is integrated, and the analysis result beyond than the expected will be obtained.

3.3 Guide to Identify Latent Factors in the Trouble

By doing the improvements, we expect that many latent factors are extracted and the effective countermeasures are introduced. So, we made a instruction guide that summarized the improvements (Reference List, Learning about PSF features, and the instruction of analysis by group work). This instruction guide was programmed in PC environment, and can be used in many work spot. The example pictures of the instruction guide are shown as Fig.4.

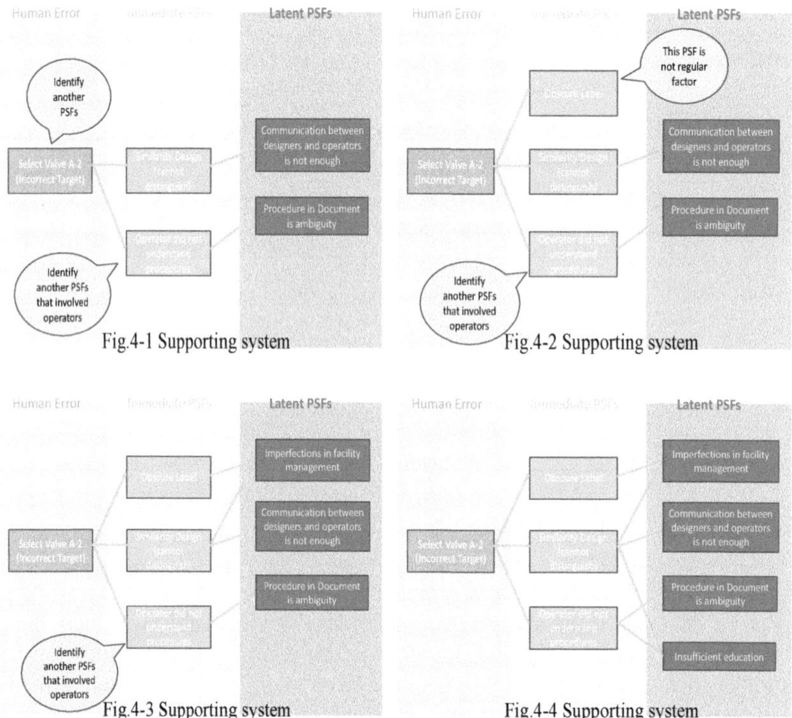

Fig.4-1 Supporting system

Fig.4-2 Supporting system

Fig.4-3 Supporting system

Fig.4-4 Supporting system

4 VERIFICATION OF THE INSTRUCTION GUIDE

In order to the validity of the instruction guide, we did some case studies. First, to observe the change by using of the instruction guide, we investigated the result in the conventional situation. The conventional situation means that the analysts were not educated the issues about human factors. After three months, for same analysts, we produced the instruction guide, the results of the analysis was obtained. The results were evaluated according to the level shown as Table.1. The evaluation results are shown as Fig.5.

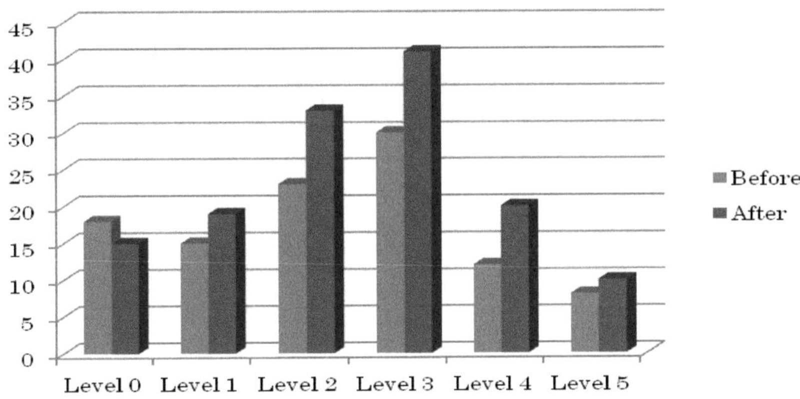

Fig.5 Number of identified factors

In this figure, we obtained that that the number of Level0 becomes small, and higher Level PSFs increased.

In other cases, we observed that the number of the Level 4 become large, and the effective countermeasures tended to be proposed. From the results of case studies, we could confirm the effectiveness and validity of the instruction guide.

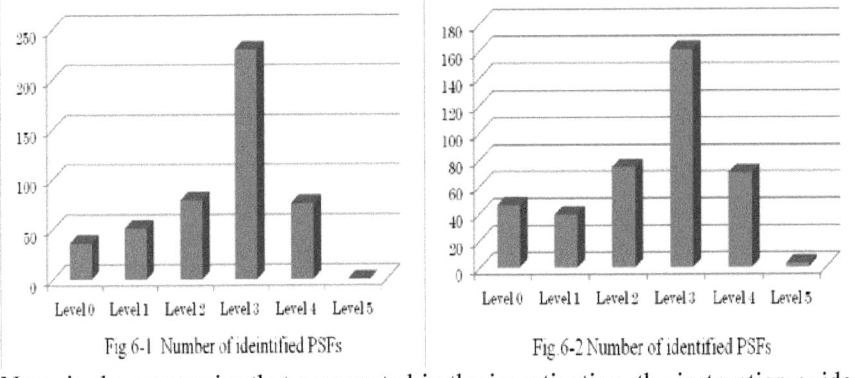

Fig.6-1 Number of ideintified PSFs

Fig.6-2 Number of identified PSFs

Now, in the companies that cooperated in the investigation, the instruction guide is introduced as formal education method.

5 CONCLUSIONS

In this paper, we discussed the instruction method that can support the identification of the latent factors in the troubles. Based upon the survey of the current in trouble analysis, we could obtain some improvements. Especially, for the analyst, to support the judgment whether or not the factor is the latent factor, we proposed the reference list of PSF. The list is categorized into 6 levels. The factors listed in low level class are not latent factors. In addition, some instruction issues were discussed. Based upon these issues, we proposed the instruction guide to support the identification of the latent factors in the troubles. Furthermore, the validity of our proposed instruction guide is confirmed in several case studies. In the companies that cooperated in the investigation, the instruction guide is added in a formal education system.

In the future, we will produce the detailed specification of the instruction method to enhance workers' capability to analyze the latent factors in troubles, and will decrease the workload of the education.

REFERENCES

Embrey, D.(1984) .SLIM-MAUD: An approach to assessing human error probabilities using structured expert judgment. NUREG/CR-3518, USNRC.

Gertman, D.I., et al.(1992) . INTENT: A method for estimating human error probabilities for decision-based errors. Reliability Engineering and System Safety, 35, 127-136.

Hollnagel, E, (1998). Cognitive reliability and error analysis methodology, Elsevier.

Kirwan, B. (1992). Human error identification in human reliability assessment, part 1: overview of approaches. Applied Ergonomics, 23, 299-318.

Okada, Y.(2003) An Analysis of Potential of Human Error in Hospital Work, 10th International Conference on Human Computer Interaction.

Miyata,K., Okada, Y., Koyanagi, A., Mita, K.(2003),Quality Management of Helicopter on Human Factors - A Proposal of Human Error Possibility Measurement Index -,15th Congress of the International Ergonomics Association, 2003.

Okada, Y., Watanabe,M, Shirai,M..(2003),An Analysis of Performance Shaping Factors in Nursing Work, 15th Congress of the International Ergonomics Association, 2003.

CHAPTER 27

Effects of Vibration Exposure on action Judgment: A Field Test for Professional Drivers

N. Costa [1,2*], *P. M. Arezes* [1,2] *and R. B. Melo* [3,4]

[1] Dept. Production and Systems, University of Minho, Guimarães, Portugal.
[2] CGIT - Research Center for Industrial and Technology Management, University of Minho, Guimarães, Portugal.
[3] Dept. Ergonomics, Human Kinetics Faculty, Technical University of Lisbon, Lisbon, Portugal.
[4] CIPER, Interdisciplinary Centre for the Study of Human Performance, Technical University of Lisbon, Lisbon, Portugal.
*ncosta@dps.uminho.pt

ABSTRACT

The aim of this study was to evaluate the vibration effects on cognitive and visual performance. With this purpose and to, hopefully, achieve innovative results this study included tests performed in a 'real' performing context. The considered sample was composed of 45 volunteers, from both genders, with ages between 21 and 62 years old (mean of 33.1±10 years old). In order to test the effects on a 'real' environment, the back of a 2.5 ton van was modified with the inclusion of two car seats and a platform to accommodate the 'Action Judgment Tester'. The used instrument was designed to examine the relation between the distribution of attention and the resultant reaction to ever-changing conditions. The movement of the van was performed in a closed circuit for each test and with a vehicle speed up to 30 Km/h. This circuit has two different pavements, one of asphalt and one of cobblestone. This feature allowed performing the 'Action Judgment Test' under three different conditions: (A) with the van halted, (B) with the van performing a circuit on asphalt, and (C) on cobblestone. Reaction judgment was done based on the degree of the effects of training and the total number of errors. The results show a degree of impairment as an effect of the vibration exposure level. The degree of impairment seems to be strong enough to obtain a poor result when vibration exposure levels are higher.

Keywords: Whole-body vibration exposure, performance, impairment.

1 INTRODUCTION

The occupational environment leads to numerous risk factors, among them there is the exposure to whole body vibrations (WBV).

The effect of exposure to WBV most often reported in the literature reviewed is the 'back pain'. Epidemiological studies are the basis of this statement. Bovenzi and Griffin are two authors who regularly perform this type of study, trying to systematize the profile of exposures and symptoms reported.

Magnusson et al. (1998) note the difficulty in establishing a clear link between exposure and effect, mainly due to the high number of disturbing factors or contributions that influence the risk associated with exposure to WBV.

Focusing attention on the possible change in the size of the vertebral column exposed to whole body vibrations, the volunteers used in the study of Bonney & Corlett (2003) underwent a simulated driving task. During this task, the subjects were exposed to WBV, according to the horizontal and vertical axis of the body, with a frequency close to the natural frequency of the human spine, 4 Hz. The results showed a statistically significant increase in the length of the backbone of subjects exposed. Cautiously, the authors point out the need to consider the revision of the model in which exposure to WBV overloads the spinal disc.

In 2006, Bovenzi et al. (2006) published the results of a study that included 598 professional drivers exposed to WBV. As a main conclusion the authors stated the fact that exposure to WBV and physical load factors (associated with manual handling and inappropriate postures) are two important components in the multifactorial origin of 'back pain' reported by professional drivers.

Although heavily focused on the effects on the back, the bibliography also refers to a wider range of effects caused by exposure to WBV. It seems, therefore, relevant to systematize some of these effects, since they help to better understand the possible effects on the cognitive level of operators exposed.

Following an approach related with the part of the body affected the study of Griffin & Hayward (1994) versed the effects of exposure to WBV on the reading. Regarding only in the horizontal component of the exhibition these authors found statistically significant reductions in speed of reading a newspaper notice, when the frequency of the vibrations was between 1.25 and 6.3 Hz and with higher magnitudes of vibration (1 and 1.25 m/s^2).

The visual system is also responsible for the perception of the motion of objects and was on the possible interference that WBV exposure might have on this mechanism that Peli & García-Pérez (2003) have addressed. These authors state that the image motion of objects on the retina caused by voluntary or reflex movements operator that whether he has received professional training or not, but he must be aware of relevant expertise.

The master site includes human-machine interaction equipment on the ground. It actively senses and recognizes the operator's movement intension, transmits it into the slave site after translating it into machine language. In addition it receives feedback information from the slave site, and rebuilds the remote environment for the operator on the ground.

The slave site usually means a dexterous space robot system interacting with target at remote site. It drives the robot to operate under the guide of the operator's decision, which result in full expression of the operator's intension. Thus it is not necessary for the space robot system to have intelligence. Besides that the slave site feed backward information acquired when interacting with targets.

2.2 Features of the system

The space operation system based on human-in-the-loop expands humans' action capacity by transferring human's intelligence to the remote site. The system has features as follows:

- The operator' body movements are actively sensed and automatically translated into machine language.
- The operator's movements are mapped into space robot's movements at the slave site. This would guide the robot to imitate the operator's behavior.
- The operator interacting directly with the mission target could concentrate on how to perform. This would improve operator's efficiency.
- The system could deal with unknown environment more effectively with the help of operator's expertise, intelligence, and experience.
- The space robot's DOF are controlled by operator's DOF so as to dexterously control complex multi-DOF robots.

2.3 Key technologies of the system

The kernel concept of the system presented in this article is applying humans' intelligent decision and judgment to dexterously control multi-DOF robots at the slave site so as to deal with unknown environment and non-cooperation targets. There exist four key technologies: behavior sensing, feature extracting, behavior reproducing and influence elimination of time delay.

A Behavior sensing technology

Behavior sensing technology is used to read the operator's operating behavior at the master site. The operator doesn't need to tell the robot by joysticks or mice which joint to move, in which direction and how long distance to move them, while facing with mission targets (the target here is reconstructed with feedback information). Instead the operator performs as if he is on the site. Meanwhile his operating intension is understood by the sensors via capturing his body's pose, tracing it continuously and describing it in the math coordinate system.

to keep the vehicle in motion at a constant speed during the visual and cognitive performance test, since no obstacles, pedestrians or other vehicles may interfere and disturb the driver.

The existence of two pavements created two different levels of exposure to WBV. In order to ensure statistically significant differences of these two exposure scenarios several preliminary tests with a vehicle with different speeds were undertaken. The characterization of exposure followed the methodology described in the standard ISO 2631:1997. The subsequent analysis of the various profiles generated by the different tests allowed the selection of WBV exposure conditions best suited to carry out the assessment of visual and cognitive performance, namely:

- Condition A, no exposure to WBV, halted vehicle;
- Condition B, exposure to WBV, vehicle moving at a speed of approximately 30 km/h on regular asphalt pavement;
- Condition C, exposure to WBV, vehicle moving at a speed of approximately 20 km/h on cobblestone pavement.

The selected speeds and the two test circuits used were chosen based on two different considerations. The first was related to the fact that the two profiles were representative of occupational exposure to different WBV conditions, a lower exposure level (average 0.20 ± 0.012 awz m/s^2) and higher exposure level (average of 0.54 ± 0.049 awz m/s^2). The second consideration was related to the fact that the trajectories and speeds made possible to maintain the test conditions during the assessment of visual and cognitive performance, thereby ensuring the reproducibility of the tests.

The back of a Citroen Jumper 2.5D (2.5 ton van) was modified in order to provide conditions to assess visual and cognitive performance, in a situation of real exposure WBV. This change included the introduction of two individual seats, and a support platform to accommodate the equipment to Action Judgment Test (AJT). On the seat of the test subject was placed the accelerometer, responsible for monitoring the acceleration values throughout the test. Thereafter, the values obtained for each test subject would create a profile of exposure and at the same time, ensure that all tests were performed under similar conditions.

The option of making the AJT inside a van aimed at obtaining real WBV exposure conditions.

The AJT (Action Judgment Test, Item No. 1105) was developed by Takei & Company, LTD. This instrument was designed to examine the relation between the distribution of attention and the resultant reaction to ever-changing conditions. The performance test tool/instrument is equipped with a synchronous motor to ensure a constant-speed of rotation for a disk that is marked with 16 red arrows and a peripheral red line. A subject is required to use a steering wheel to move two needles (on both the left and right side of his vision field) trying, from start to finish, not only to keep them clear of all the red arrows and the peripheral red line, but also to pass them on the side of each arrow neck and not on the arrow heads.

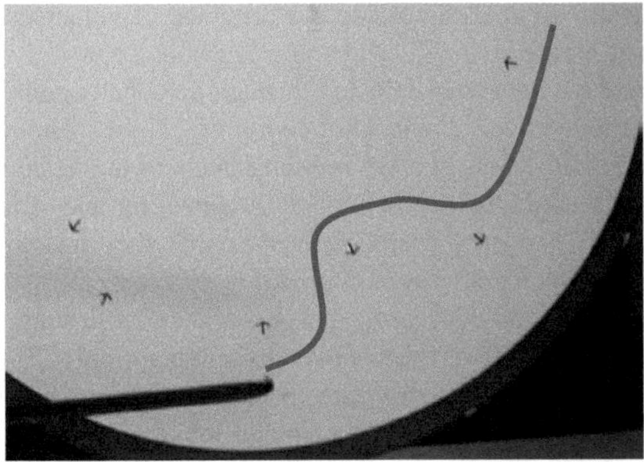

Figure 1 Path expected to avoid passing the needle to the front of the red arrows.

It is expected that subjects improve their performance while performing the test trial action, reaching a smaller number of errors in the last step, compared to the first. Thus, the authors considered relevant test to determine the 'training effect' (Takey and Company, 1996).

The value of 'training effect' is obtained according to Equation 1, where the letter L represents the value of the effect of training, the letter A represents the total number of errors recorded in the first minute and ten seconds, and the letter B represents the total number of errors recorded at three minutes and thirty seconds.

$$L = \frac{(A - B)}{A} \times 100 \qquad \text{Eq. 1}$$

The criterion of judgment is based on the value of the training effect (L) and the total number of errors (C) and can be obtained from the table 1 below.

Table 1 Criterion of judgment.

		Judgment	Total no. of errors (C)	Degree of the effect of training (L) %
	Excellent	+2	Below 69	Over 40.0
Eligible	Favorable	+1	70 – 90	20.0 – 39.0
	Good	0	91 – 104	0 – 19.0
Not	Unfavorable	-1	105 – 125	-20.0 – -1.0
eligible	Poor	-2	Over 126	Below -21.0

3 RESULTS AND DISCUSSION

The statistical analysis of the conditions of exposure to WBV undertaken by the volunteers during the AJT revealed significant differences ($p < 0.05$), legitimizing the claim of comparative analysis of the performance of volunteers over the three exposure conditions created.

The volunteers were randomly selected through personally addressed invitations addressed to the population of the University of Minho. The volunteer group included students from the first, second and third cycle of higher education, teachers from different Departments of the School of Engineering and Administrative and Social Services staff. The only prerequisite was that they had to have the legal authorization for driving cars. The sample of volunteers consisted of 25 males (56%), aged between 21 and 62 years', with experience of driving cars between 1-41 years. The female volunteers were fewer in numbers, only 20 (46%), and had ages between 21 and 46 years and driving experience of motor vehicles from 1 to 28 years. In general, the volunteers who participated in the evaluation of the effects of exposure to whole body vibration on the cognitive performance and visual had an average of 33.11 years' old (10.04 years standard deviation) and 13.24 years of experience and driving of motor vehicles (standard deviation 9.80 years).

The AJT results obtained by the 45 volunteers over the three WBV exposure conditions are summarized in Table 2.

Table 2 Summary of results obtained on the AJT.

	N	AJT according to the effect of training (AJ_L)				AJT according to the total number of errors (AJ_TE)			
		Mean	Min.	Max.	sd	Mean	Min.	Max.	sd
Condition A	45	0.47	-2	2	1.375	1.11	-2	2	1.352
Condition B	45	-0.11	-2	2	1.112	0.71	-2	2	1.471
Condition C	45	0.29	-2	2	1.141	0.62	-2	2	1.614

The results presented on Table 2 reveled interesting statistical significance, not only when comparing the values obtained after application of the criterion of the total number of errors (AJ_TE) but also when comparing the values obtained after administration of the criterion of the effect of training (AJ_L) . The analysis of the obtained AJ pairs value reveled significant differences ($p < 0.05$) between the values obtained for testing with the vehicle halted and the values obtained with the vehicle is moving, either over regular asphalt or uneven cobblestone pavement.

Based on the hypothesis presented and verified by Griffin & Hayward (1994), we can infer that the reduction of visual performance of subjects under the conditions of exposure to WBV adversely affect its performance in the AJT.

The number of errors made by the volunteers was another of the analysis parameters of the AJT. Statistical analysis of results obtained for this parameter showed a significant decrease in the action judgment value (AJ_TE) according to the worsening conditions of exposure to WBV. Namely, the average value of action judgment value (AJ_TE) decreased from 1.11 (condition A) to 0.71 (condition B, weighted vertical axis acceleration mean values of 0.20 m/s^2) and 0.62 (condition C, weighted vertical axis acceleration mean values of 0.54 m/s^2). The stated statistical significance confirms the hypothesis proposed by Newell & Mansfield (2008), that exposure to WBV adversely influences the reaction time.

4 CONCLUSIONS

The results obtained by applying the 'Action Judgment Test' showed a significant decrease in the performance of subjects under the conditions of exposure to WBV, by comparison with the condition of absence of exposure to vibrations. Regarding the 'Action Judgment Test' result, this decrease was recorded between the halted vehicle condition (condition A) and the condition of moving vehicle on a regular asphalt pavement (condition B) and between the halted vehicle condition (condition A) and the condition of moving vehicle on cobblestone pavement (condition C).

The results presented seem to indicate that there is an effect of exposure to whole body vibration on the visual and cognitive performance of the volunteers. However, this effect appears to be most noticeable on visual/motor coordination, put into evidence by the criterion of the total number of errors (AJ_TE). Assuming that using the criterion of the training effect (AJ_L) places greater emphasis on the cognitive/learning component, it was only possible to verify a significant difference in the performance condition between the halted vehicle and moving over regular asphalt pavement.

ACKNOWLEDGMENTS

This work was funded by Portuguese Funds through the Portuguese Foundation for Science and Technology (FCT) in the scope of the project PEst-OE/EME/UI0252/2011.

REFERENCES

Bonney, R. A., & Corlett, E. N. (2003). Vibration and spinal lengthening in simulated vehicle driving. Applied Ergonomics, 34(2), 195-200.

Bovenzi, M. (1996). Low back pain disorders and exposure to whole-body vibration in the workplace. Seminars in Perinatology, 20(1), 38-53.

Bovenzi, M., Rui, F., Negro, C., D'Agostin, F., Angotzi, G., Bianchi, S., et al. (2006). An epidemiological study of low back pain in professional drivers. Journal of Sound and Vibration, 298(3), 514-539.

Griffin, M. J. (1990). Handbook of human vibration. London: Academic Press.

Griffin, M. J., & Hayward, R. A. (1994). Effects of horizontal whole-body vibration on reading. Applied Ergonomics, 25(3), 165-169.

International Organization for Standardization, I. (1997). Mechanical vibration and shock -- Evaluation of human exposure to whole-body vibration -- Part 1: General requirements (ISO 2631-1:1997). Geneva, Switzerland.

Ljungberg, J., Neely, G., & Lundström, R. (2004). Cognitive performance and subjective experience during combined exposures to whole-body vibration and noise. International Archives of Occupational and Environmental Health, 77(3), 217-221.

Magnusson, M. L., Pope, M. H., Hulshof, C. T. J., & Bovenzi, M. (1998). Development of a protocol for epidemiological studies of whole-body vibration and musculoskeletal disorders of the lower back. Journal of Sound and Vibration, 215(4), 643-651.

Newell, G. S., & Mansfield, N. J. (2008). Evaluation of reaction time performance and subjective workload during whole-body vibration exposure while seated in upright and twisted postures with and without armrests. International Journal of Industrial Ergonomics, 38(5-6), 499-508.

Peli, E., & García-Pérez, M. A. (2003). Motion perception during involuntary eye vibration. Experimental Brain Research, 149(4), 431-438.

Schust, M., Blüthner, R., & Seidel, H. (2006). Examination of perceptions (intensity, seat comfort, effort) and reaction times (brake and accelerator) during low-frequency vibration in x- or y-direction and biaxial (xy-) vibration of driver seats with activated and deactivated suspension. Journal of Sound and Vibration, 298(3), 606-626.

Takey and Company, L. (1996). Item No. 1105, Action Judgment Tester, Operation Manual. Tokyo.

CHAPTER 28

An Investigation into Blood Pressure of Blue Collar Workers of Casting and Forging SMEs: A Study in India

Lakhwinder Pal Singh [1]

[1]Assistant Professor, Department of Industrial & Production Engineering, Dr B R Ambedkar National Institute of Technology, Jalandhar, Punjab, (India)-144011.

Corresponding Author: lakhi_16@yahoo.com,singhl@nitj.ac.in

ABSTRACT

The present study investigated the influence of workplace physical activities on HR & blood pressure of casting and forging industry workers. The study included a group of 132 male workers with mean age (SD) of 29.78 (7.96) years casting and forging industry and a control group of 50 male subjects with mean age (SD); 31.98 (8.87) years. Both the groups were assessed for heart rate (HR) and blood pressure parameters; systolic, diastolic and mean blood pressure. The industry subjects were found significantly lower HR (SD), systolic blood pressure (SBP) and mean blood pressure (MBP) parameters as compared to the control group subjects at $p < 0.05$. Workers who were engaged in transferring the hot work piece from furnace to forger are having significantly lower SBP as compare to the tool room workers and the control group subjects. At the same time the workers engaged in moulding and casting tasks were having lower SBP than the tool room workers. The study concluded that in Indian SMEs, the workers with work exposure of less than a decade, showed adaptive response to occupational noise (non annoyance to noise) and increased vascular control possibly due physical demand of work load as compared the control group subjects live a sedentary life style without adequate physical work.

Keywords: Blood pressure, SME Workers, Physical activity at work place.

1 INTRODUCTION

In a working population, occupational factors are believed to pose a threat to workers' cardiovascular health. Substantial research has shown that adverse physical and psychosocial work environments and working conditions, such as shift work and excessive workload, are related to cardiovascular diseases. The past researchers (Lang et al. 1992, Peter et al. 1993) have studied the effect of occupational noise annoyance and its combined effect with social support at work, nightshift work, and work satisfaction on blood pressure. A significant effect of noise annoyance on rising diastolic blood pressure (DBP) was observed (Peter et al. 1993). A long exposure to noise over 85 dB (A) could be a risk factor for high BP, contributing to an increase in BP in the population and possibly inducing major increases of BP among sensitive individuals (Lang et al. 1992). Ni Chun-Hui et al. (2007) found that both systolic blood pressure (SBP) and diastolic blood pressure (DBP) in the high frequency hearing loss group were significantly higher than those in the normal hearing group ($P < 0.05$). Kathleen et al. (1992), investigated, the effect of high job strain (defined as high psychological demands pins low decision latitude at work) on blood pressure was determined in 129 healthy, non-hypertensive men n=65) and women (n=64). Men with high job strain showed greater increases from screening to work, resulting in higher mean work blood pressure. Salonen et al. (1983), Khaw & Connor (1988) reported a association between blood pressure and dietary salt intake. Melamed et al. (1999) reported that chronic exposure to high noise level, objectively verified, was found to be associated with a significant increase in both SBP and DBP over a period of 2-4 years. Narlawar et al. (2006) reported that hypertension and hearing impairment were commoner in workers continuously exposed to high levels of occupational noise. Chronic exposure to noise levels typical of many workplaces was associated with excess risk for acute myocardial infarction death (Amelsvoort et al., 2000). It has been reported that a high prevalence of excess noise exposure at work, this association deserves further attention. Disturbance of the circadian pattern of cardiac autonomic control by working at night when the physiological system anticipates rest could explain part of the elevated cardiovascular risk in shift workers (Hugh et al, 2005). It has been suggested that an increased sympathetic dominance during a night shift sleep, indicating an inferior sleep quality (Ludovic et al. 2001). In Indian casting and forging SMEs, hazardous environmental conditions are inevitable. The occupational health and safety measures are ignored at the workplaces (Nelson et al., 2005). Developing countries like India lag far behind in implementing occupational health and safety programmes in their industries. Therefore present study assumes its significance for the assessment of occupational noise exposure and investigates the chronic effect of work load activities on cardiovascular autonomic control amongst casting and forging industry workers.

2 MATERIALS AND METHODS

The study included two groups of population, one group of 132 male subjects (workers) from casting and forging industries and another group of control (n = 50 male) subjects from non exposed population. These workers belonged to a common industrial environment where casting and forging activities were carried out, and the workers were engaged in different work area like; moulding, molten metal pouring, grinding, forging, punching, blanking, welding, gas cutting etc. The criterion of selection was agreement to participate in the study. The subjects of both the groups were brought to institute's Ergonomics laboratory for subject's demographic profile (age, height, weight, and work experience), details of alcohol intake. The measurements were carried out in a cabin inside the laboratory of the department. The room temperature inside the cabin was maintained at 24°C to 26 °C. Blood Pressure was measured in sitting position using a mercury sphygmomanometer and Stethoscope and Polyrite-D. Student's t-test was applied to statistically compare the various parameters of exposed and control group. One way ANOVA, ANOVA post hoc Tukey's analysis were applied for multiple comparisons among various sub groups, a p value of less than 0.05 was considered to be significant. The data were also analysed for influence of alcohol intake of alcohol, smoking and tobacco amongst exposed group subjects.

2 RESULTS

The industry workers showed significantly decreased resting heart rate with mean HR (SD) of 68.27 (10.23) beats per minute as compared to the control group with mean HR (SD) of 74.24 (10.26) ($p < 0.01$). In case of blood pressure (BP), the industry subjects were found significantly lower systolic blood pressure (SBP) and mean blood pressure (MBP) parameters as compared to the control group subjects at $p < 0.05$. The't' test results of Age, BMI, systolic blood pressure (SBP), diastolic blood pressure (DBP) and mean blood pressure (MBP) of both the groups are shown in Table. 3.1. The type of work condition was found to be associated with HR (SD) and BP as shown in Table 3.2. Workers engaged in moulding /casting section showed significantly reduced mean HR of as compared to control group. The post hoc Tukey's analysis with multiple comparison revealed that, the workers engaged in moulding/casting section showed significantly lower mean HR (SD) of 65.60 (9.84) as compared to the of control group subjects with mean HR (SD) 74.24 (10.3) at 'p' value less than 0.05.

2.1 Effect of alcohol, smoking and tobacco intake on HR and BP.

Within the exposed as well as control group, there was insignificant difference for HR and BP of alcoholic and non-alcoholic subjects (Table. 3.3). Similarly smoking and tobacco intake insignificantly influenced the HR and BP. At the same time, the industry subjects both smokers and non-smokers showed significantly lower values of mean HR (SD); 68.34 (10.57) and 67.46 (10.04) respectively than the control group (non-smoker subjects) with mean HR (SD); 74.24 (10.26) at 'p' value of less < 0.05

Table. 3.1 The't'-test results of exposed vs. control group for BP.

Parameter	Exposed Group (N = 132)	Control Group (N = 51)	Difference of Means	'p' value
	Mean (SD)			
Age (Yrs.)	29.65 (8.14)	31.98 (8.87)	-2.329	0.093
BMI	21.48 (3.78)	23.43 (2.62)	-1.94735	0.001
Mean HR	68.27 (10.23)	74.240 (10.261)	-5.969	0.001
Systolic BP	116.72 (9.62)	121.01 (6.15)	-4.2868	0.004
Diastolic BP	79.35 (7.82)	81.48 (4.93)	-2.1261	0.073
MBP	91.87 (7.98)	94.65 (5.04)	-2.7832	0.022

0.01> p value < 0.05; Significant at 95% confidence interval and p value ≤ 0.01; Significant at 99% confidence interval

4 DISCUSSION

The results of present study revealed that the industry workers showed adaptive response to the physical work load. The exposed group subjects showed significantly lower resting mean heart rate (HR) and BP as compared to control group. It has been agreed that lower heart rates are at least partially the result of increased parasympathetic tone (Yataco et al. 1997, Shin et al. 1997, Seals et al 1989 and Levy et al. 1998). Antelmi et al. (2004) also reported an inverse correlation of HRV with heart rate ($p < 0.001$).

In present study, the industry workers perform various activities at their work places like; lifting and carrying of heavy jobs, hot material handling, hot forging, moulding, pouring, grinding etc. *All these activities were the moderate-heavy category activities* (ACGIH, 2001). *The workers were performing such activities daily.* The workers generally work more than 8 hours/day i.e.10-12 hours per day. Probably the physical exercise comes as a part of their regular heavy work activities, which demonstrate a positive cardiovascular response to the body.

The industry workers who transfer the hot work piece from the oil fired furnace to the forger were found with significantly lower SBP & MBP as compare to the tool room workers and the control group subjects. These workers were doing their jobs at a high ambiance WBGT and therefore they were also under the sweat losses. At the same time the workers who were engaged in moulding/casting tasks were having significantly lower SBP than the tool room workers.

On the other hand control group subjects were physically less active and live a sedentary life style; therefore they showed significantly higher resting mean HR (SD) along with significantly higher BP as compared to the industry workers. The modern sedentary life style without regular physical activities (exercise) might have lowered the cardiac autonomic control amongst control group. This, physical inactivity may lead to coronary heart disease via increased adiposity, higher body mass, reduced cardiovascular fitness (Blair & Brodney, 1993) and raised blood pressure (Duncan et al. 1985). This alteration can be corrected by physical or yogic exercises (Sunkaria et al. 2010).

4.2 Cardiovascular Effects of Occupational Noise

A number of studies reported cardiovascular effects of occupational noise exposure. Chronic exposure to noise levels typical of many workplaces was associated with excess risk for acute myocardial infarction death (Hugh et al. 2005). Significant effects of noise annoyance on rising diastolic blood pressure (DBP) have been observed (Peter et al. 1993). A long exposure to noise over 85 dB (A) might be a risk factor for high BP, contributing to an increase in BP in the population and possibly inducing major increases of BP among sensitive individuals (Lang et al. 1992).

But in Indian environmental conditions of Casting & Forging industry, the long term exposure to noise lead to adaptability to noisy environment and noise induced hearing loss (NIHL). The industry workers also showed minimum level of occupational noise syndromes like; noise annoyance and anxiety etc.; rather they have accepted the occupational noise as an integral part of their job (Singh et al. 2010). At the same time the workers were not using hearing protection like; hearing plugs and ear muffs (Singh et al. 2010). During the adaptability to noise exposure and NIHL, the autonomic tone might have been affected. But the physical activities at the work place might have possibly conquered the effect of noise annoyance, and thus the HR and BP components; were significantly lowered. *However, hazardous effect of noise existed in the form of hearing loss and adaptability to the noisy environment. Whereas the risks of developing cardiovascular pathologies due occupational hazardous appear to possibly annulled by the heavy physical activities at work place. Thus an exposure of less than a decade to physically demanding even under hostile environmental factors showed improved cardiovascular control.* This is in common with Hamer et al. (2007), who also reported that exercise and physical fitness may act as a buffer to the detrimental effects of psychosocial stress exposure.

Table: 3.2 Effect of Occupation on BP parameters using ANOVAs results.

Parameter	Punching blanking (n=18)	Forgers (n=19)	Furnace to Hammer hot Job Handler (n=10)	Hammer Operator (n=9)	Grinding (n=44)	Moulding/ casting (n=19)	Tool Room/ Welding (n=13)	Control Group (n=51)	'p' value
Age(Yrs.)	25.50(7.56)	33.95(10.18)	25.50(4.55)	25.78(2.73)	29.11(6.19)	31.37(8.12)	34.31(11.00)	31.98(8.88)	**0.002**
BMI	20.07(3.14)	22.29(6.14)	19.65(2.59)	22.07(2.57)	21.65(3.21)	20.78(2.07)	23.69(4.35)	23.43(2.62)	**0.001**
Mean HR	**70.68(9.58)**	**66.92(11.95)**	**65.38(7.90)**	**69.11(9.42)**	**68.46(11.52)**	**65.60(9.84)**	**71.42(8.62)**	**74.24(10.3)**	**0.019**
Systolic BP	118.51(9.24)	117.68(11.99	**109.86(11.93)**	114.37(6.33)	116.97(9.12)	**114.28(6.65)**	**122.41(8.99)**	**121.01(6.15)**	**0.002**
Diastolic BP	78.59(9.46)	82.81(6.16)	75.86(8.78)	78.67(5.39)	79.53(8.01)	76.42(6.95)	82.18(7.165)	81.47(4.93)	0.068
Mean BP	91.90(8.35))	94.43(7.46)	**87.20(9.49)**	90.56(5.37)	92.01(8.11)	89.04(6.10)	**96.23(8.41)**	**94.65(5.04)**	**0.006**

1: Punching blanking, 2: Forgers, 3: Hot Job Handling Furnace to Forger, 4: Hammer Operator, 5: Grinding, 6: Moulding/ casting, 7: Welding/tool room, 8: Control Group. $0.01 > p$ value < 0.05; Significant at 95% confidence interval and p value ≤ 0.01; Significant at 99% confidence interval.

4.3 Effect of Alcohol, Smoking and Tobacco Intake on Cardiovascular Autonomic Control (HRV & BP)

In present study post hoc Tukey's analysis revealed that alcohol intake increased the gap of cardiovascular control between the exposed and control groups. The alcoholic subjects of industry showed significantly lower mean HR (SD) of 67.651 (10.754) as compared to alcoholic control subjects with mean HR (SD) of 76.333 (11.47). Habitual heavy drinking of alcohol is a risk factor for the development of hypertension (Makoto, et al 2000). Lang et al. (1987) also found a positive association between arterial hypertension and alcohol intake among workers of small and medium-sized companies in the Paris region. *In present study, alcohol intake insignificantly influenced the BP of industrial workers as well as control group. However the non alcoholic subjects of exposed group showed significantly lower SBP (at $p < 0.05$) as compared to the alcoholic subjects of control group. At same time smoking and tobacco intake did not influence both BP and HRV components. The same could be due to that the industry subjects were smoking 1-3 biddies or cigarettes/day; as such they were not chain smokers. Moreover, the heavy physical activities at work place seem to possibly overcome the effect of smoking.*

5 LIMITATIONS

In present study, investigation was limited to measure the combined effect of multiple hazardous on physiological parameters. But, the study lags in measuring the contribution of individual factors like; physical activity, heat stress on each parameter. The availability and willingness of the workers and the consent of the industry management to spare workers for the investigations put up a limitation on selecting the subjects randomly from the available population.

6 CONCLUSIONS

In Indian environmental conditions of Casting & Forging industry, the long term exposure lead to adaptability to working environment and showed significantly increased cardiovascular control. The study suggests that the cardiovascular vascular control was significantly better due to the regular heavy physical activities at the work place. Present study reports that, *the risks of developing cardiovascular pathologies due occupational hazardous are possibly annulled by the heavy physical activities at work place.* Thus hazardous effect of noise has been observed in the form of hearing loss and adaptability to the noisy environment.

ACKNOWLEDGMENTS

The authors; hereby acknowledge the co-operation and help extended by the management of casting and forging SME's. Authors are also thankful to the workers of these SMEs and the control subject, who have spared their valuable time and volunteering for the study

REFERENCES

Amelsvoort L G P M van, Schouten E G, Maan A C, Swenne C A, Kok F J (2000) Occupational determinants of heart rate variability. Int Arch Occup Environ Health. Vol. 73, pp: 255-262.

Heat Stress and Strain (2001), American Conference of Governmental Industrial Hygienists (ACGIH)

Antelmi I, De Paula R S, Shinzato A R, Peres C A, Mansur A J, Grupi C J (2004) Influence of age, gender, body mass index, and functional capacity on heart rate variability in a cohort of subjects without heart disease. Am J Cardiol Vol. 93, pp: 381–385.

Blair S N and Brodney S (1999) Effects of physical inactivity and obesity on morbidity and mortality: current evidence and research issues. Med Sci Sports Exerc; Vol. 31, pp: 646–662.

Duncan J J, Farr J E, Upton S J, Hagan R D, Oglesby M E and Blair S N (1985) The effects of aerobic exercise on plasma catecholamines and blood pressure in patients with mild essential hypertension. JAMA; Vol. 254, pp: 2609–2613.

Hamer M and Steptoe A (2007). Association between Physical Fitness, Parasympathetic Control, and Proinflammatory Responses to Mental Stress. Psychosomatic Medicine; Vol. 69, pp: 660–666.

Hugh W. Davies, Kay Teschke, Susan M. Kennedy, Murray R. Hodgson, Clyde Hertzman, and Paul A. Demers (2005) Occupational exposure to noise and mortality from acute myocardial infarction. Epidemiology Vol. 16, pp: 25–32.

Khaw KT and Connor E B (1988) The association between blood pressure, age, and dietary sodium and potassium: a population study. Circulation; Vol. (77), pp: 53-61.

Kathleen C. L, Turner J R, and Hinderliter A L (1992) Job Strain and Ambulatory Work Blood Pressure in Healthy Young Men and Women. Hypertension; Vol. 20, pp: 214-218.

Ludovic G.P.M. Amelsvoort V, Evert G. Schouten, Arie C. Maan, Kees A. Swenne and et al. (2001) 24-Hour Heart Rate Variability in Shift Workers: Impact of Shift Schedule. J Occup Health; Vol. 43, pp: 32–38.

Lang T, Fouriaud C, and Jacquinet-Salord MC (1992) Length of occupational noise exposure and blood pressure. Int Arch Occup Environ Health 63, 372-369.

Levy W C, Cerqueira M D, Harp G D, Johannessen K A, Abrass I B, Schwartz R S and Stratton J R (1998) Effect of endurance exercise training on heart rate variability at rest in healthy young and older men. Am J Cardiol; Vol. 82, pp: 1236–40.

Makoto T, Hisashi A, Yuji H, Yoshihisa F and Tsutomu I (2000) Association between alcohol intake and development of hypertension in Japanese normotensive men: 12-year follow-up study[*1]. American Journal of Hypertension; Vol.13 (5), pp: 482-487.

Melamed S, Boneh K, Froom P (1999) Industrial noise exposure and risk factors for cardiovascular disease: Findings from the CORDIS Study, Noise and Health; Vol.1 (4), PP: 49-56.

Ni CH, Chen ZY, Zhou Y, Zhou J W, Pan JJ, Liu N and et al. (2007) Associations of blood pressure and arterial compliance with occupational noise exposure in female workers of textile mill, Chinese Medical Journal; Vol. 120 (15), pp: 1309-1313.

Narlawar UW, Surjuse BG, Thakre SS. (2006), Hypertension and Hearing Impairment in workers of Iron and Steel Industry. Indian J Physiol Phaalrmacol; Vol. 50 (1), pp: 60-66.

Nelson DI, Nelson RY, Concha BM. Fingeruht M. (2005), The Global burden of Occupational noise induced hearing loss., Am J Ind Med; Vol. 48, pp: 446-58.

Peter L, Josef H, Walter WK (1993) Work noise annoyance and blood pressure: combined effects with stressful working conditions. Int Arch Occup Environ Health; 65:23-8.

Salonen J T, Tuomilehto J, and Tanskanen A (1983) Relation of blood pressure to reported intake of salt, saturated fats, and alcohol in healthy middle-aged population. Journal of Epidemiology and Community Health; Vol. 37, pp: 32-37.

Seals DR, Chase PB (1989) Influence of physical training on heart rate variability and baroreflex circulatory control. J Appl Physiol Vol. 66, pp: 1886–95.

Shin K, Minamitani H, Onishi S, Yamazaki H and Lee M (1997) Autonomic differences between athletes and nonathletes: spectral analysis approach. Med Sci Sports Exerc; Vol. 29, pp: 1482-90.

Sunkaria RK, Kumar V, Chandra S (2010) A comparative study on spectral parameters of HRV in yogic and non-yogic practitioners, International Journal of Medical Engineering and Informatics; Vol. 2 (1), pp: 1-14.

Singh L P, Bhardwaj A , Deepak K K, (2010) Occupational exposure in small and medium scale industry with specific reference to heat and noise. Noise & Health 12 (46), 37-48.

Singh L P, Bhardwaj A, Deepak KK and Sahu S (2010) Small & medium Scale Casting and Forging Industry in India: an Ergonomic study. Ergonomics SA 22(1), 36-56

CHAPTER 29

Ergonomics and Quality Interventions in Woodworking Technological Processes for Lightening the Workload

Henrijs Kalkis[1,3,4], *Valdis Kalkis*[2,3], *Zenija Roja*[2,3], *Valerijs Praude*[4], *Irina Rezepina*[1,4]

1Riga Stradins University, Faculty of European Studies, Latvia
2University of Latvia, Ergonomics Research Centre
3Latvian Ergonomics Society
4University of Latvia, Faculty of Economics and Management
Henrijs.Kalkis@lu.lv

ABSTRACT

The aim of the research is to analyze quality and ergonomics improvements in woodworking technological processes of furniture production after the ergonomics interventions for lightening the workload. The medium size woodworking enterprise was chosen and the investigation was done in three year period. Ergonomics risk assessment methods and quantitative process analysis tools were applied to analyze the human well-being, productivity and financial aspects of ergonomics interventions in process quality improvements. Results of ergonomics risk assessment KIM method showed that ergonomics interventions in the furniture production line lightens the workload for sorting, packaging and assembling operators. The cost-benefit analysis proved that an ergonomics intervention in furniture production line pays off in one year, while the benefits will continue to grow in next five years. Despite technical improvements, some work phases in furniture production line still requires hard manual handling, thus participatory ergonomics should be applied.

Keywords: Ergonomics, technology, quality, workload, woodworking

1 INTRODUCTION

Nowadays in Latvia we face rapid technological innovations, various kinds of organisations, changes in labour market, demanding needs from internal and external customers, step by step disappears difference between physical and intellectual work.

Quality management is a tool used to enrich processes of the company and to achieve customer requirements (Juran and Godfrey, 2000). Therefore the quality already has become one of prerequisites for enterprises to survive in today's business world. Not only the managers of enterprise, but also the employees understand the meaning of quality and its purpose to develop a framework which would ensure that every time a particular process is carried out using the specific information, methods, skills, also ensuring consequent controlling of the business processes (Besterfield, 2004). Thus the quality management provides definitions of clear requirements, information about the policy of quality and procedures assist to monitor processes and improve individual and team work performance (Dale, Van Der Wiele, and Iwarden, 2007).

Historically, the area of quality management constantly tried to find new ways to improve the organization's overall performance. Quality management techniques developed from statistical process control to quality circles (Juran, 1967) and widely used total quality management approach (George and Weimerskirch, 1994).

The success in today's businesses are ensured and measured by such approaches: Benchmarking, six sigma management, change and innovation management, Malcolm Baldrige National Quality Award (U.S.), European Foundation for Quality Management (EFQM) Business Excellence model (Europe), lean manufacturing, international quality standards (e.g. ISO 9001: 2008) a.o. (Bank, 1992; Freivalds, 2009). These new approaches are adapted to each type of business and include a number of process quality management techniques, including continuous process of quality improvement and management methodologies.

Therefore modern managers have to seek for new ways haw to improve organization performance, incl., process quality improvement, human resource development in safe, healthy and ergonomic work environment. Ergonomics solutions can help to select appropriate workload and optimal working conditions for workers, also improve production technology and have an enormous contribution in human resource development and increase of work abilities (Karwowski, 2005). There are several studies that verify ergonomics significance in process quality improvement. For example, J. Eklund and C. Botscha proved rapid improvements in the process quality if organization introduces not only the quality program, but also the ergonomic solutions. The authors indicate that such combination of implementation requires careful planning and investment (Botschka, 1996; Eklund, 1994). Studies have shown that costs for implementing the quality and ergonomic solutions results in the short and long term pays off, and it can be proved by economic cost-benefit analysis (Hendrick, 2003).

The aim of the research is to analyze quality and ergonomics improvements in woodworking technological processes of furniture production after the ergonomics interventions for lightening the workload.

The medium size woodworking enterprise was chosen and the investigation was done in three year period. In work process analysis sorting operators, packaging operators and assembling operators were studied and the following selection criteria were used: full-time workers, no acute musculoskeletal symptoms, work experience at least five years in the wood-processing industry, and full consent to participate. The all-male group consisted of woodworking operators (60). The average length of service was 8 ± 3 years. Background factors of the study group are shown in Table 1.

Table 1 Background factors of the study group

Variable	Workers (n = 60)	
	Mean ± SD	Range
Age (years)	37 ± 4	25–50
Height (cm)	180 ± 5	173–187
Weight (kg)	81 ± 9	65–97
Body mass index (BMI, kg/m^2)	25 ± 6	17–36

The research was carried out before (0-cycle) and after (Ergo-cycle) the ergonomics interventions in furniture production line. Ergonomics interventions were implemented with process quality improvement and actions were as follows: purchase of automatic lifting machines in packaging process, new machinery tools for sorting and assembling the furniture parts. Participatory ergonomics interventions involved job rotation, staff training, involvement in decision making about necessary improvements in work process.

2 METHODS

Data about quality and ergonomics problems was gained from plant workers with the help of interviews and checklists after the technological process improvements with ergonomics interventions.

For evaluation of workload severity was utilized the Key Item Indicator method (KIM) for estimation of ergonomics risk severity degree (Kalkis, 2008) and quick exposure check (QEC) method for quick identification and assessment of workload's influence on body parts (Brown and Li, 2003). In order to evaluate the process performance before and after the ergonomics interventions such methods were applied: Fault Tree Analysis (FTA) is a quantitative method that allows for establishing the unfavorable errors in the operation of equipment or machinery as well as faults in the technological and other processes and determine the causality of the unfavorable event (Dehlinger and Lutz, 2006) and Failure Mode and Effects Analysis (FMEA) for identifying and prevention of the problems with products or processes before the errors take place and the possible unfavorable effects occur (McDermott, Mikulak, and Beauregard, 1996). The computer software *FaultCat* (Burgess, 2011) and *ReliaSoft „Xfmea"* were applied. Financial aspects of process quality improvements with ergonomics interventions were calculated by Washington State Ergonomics Cost Benefit Calculator (WSECBC)[1].

3 RESULTS AND DISCUSSION

Workload severity risk degree Rd analyzed with KIM method for professions in furniture production line are summarized in Table 2.

Table 2 KIM method assessment scores and risk degree for professions in furniture production line

Furniture production line	Risk degree (Rd) (I – IV)
Before ergonomics interventions (0-cycle)	
Sorting operator	IV
Packaging operator	IV
Assembling operator	IV
After ergonomics interventions (Ergo-cycle)	
Sorting operator	II
Packaging operator	III
Assembling operator	II

Table 2 shows that the workload for sorting and assembling operators decreased from risk degree IV (very hard work) to II (moderate work), while the workload for packaging operator decreased to risk degree III (hard work) after the ergonomic interventions in process of furniture production. It can be explained that packaging operator despite the ergonomics improvements still has frequent arm movements and manual lifting and moving of finished products.

Assessing ergonomic risks by QEC method one can conclude that, when lifting and moving of furniture raw materials or complete furniture products are performed, the workload impacts the back, shoulder and arms, less loaded are joints of wrist/hand and neck. The investigated exposure scores by QEC before and after the ergonomics interventions in furniture production line are shown in Figure 1.

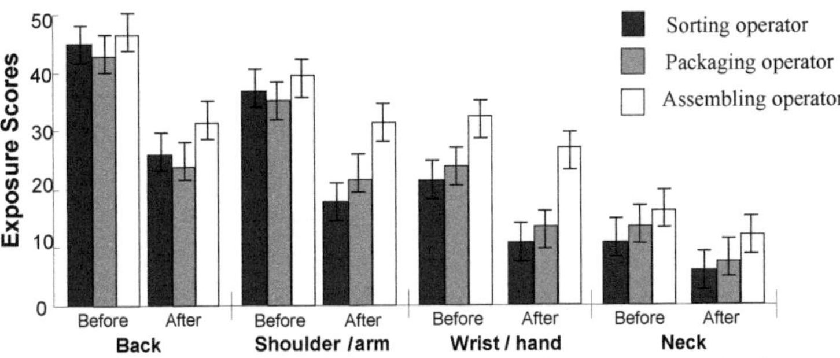

Figure 1 Exposure scores by QEC method before and after the ergonomics interventions in furniture production line

[1] Cost Benefit Analysis calculator, http://www.pshfes.org/cba.htm (*viewed 12th February 2012*)

Figure 1 shows that exposure scores for sorting, packaging and assembling operators are very high for back (44 ± 3), shoulders/arm (38 ± 4) and high for wrist/hand (34 ± 3) and moderate for neck (14 ± 2) accordingly to QEC method. After the ergonomics interventions the QEC results showed significant improvement in all exposure scores.

The FTA using the FaultCat software was made in furniture construction elements assembling and packaging stage. And the answer presented by the software after the ergonomics interventions was P=0.0094 – errors in the line due to stopping of the machinery and possible injury risk to the employees is likely to happen 3 times per year. In this case the main cause of the fault is still possible due mechanical errors in the equipment operation, including the external factor errors such as electricity interruption etc. After assessing the results of the fault tree analysis we must conclude that the likelihood of errors is significantly reduced by new technologies and ergonomic improvements, by lightening also the workload. FTA analysis was also confirmed with the FMEA results. FMEA was applied for analysis of equipment, ergonomic risks and products in furniture production line. The Risk Priority Number (RPN) decreased 3.31 times after the ergonomics interventions (Ergo-process) which led to the elimination of ergonomic risks and faults in furniture production line. Worth to mention that also the production volume increased and its quality improved. The results of FMEA analysis are summarized in the Table 3.

Table 3 Management of risks and probabilities by FMEA

		0-process				Ergo-process			
Process	**Likely kind of fault**	**Severity**	**Occurrence**	**Detection**	**RPN**	**Severity**	**Occurrence**	**Detection**	**RPN**
Furniture production line	**Machinery:**								
	Sorting errors	8	7	9	504	8	4	5	240
	Assembling errors	8	6	8	384	8	5	3	200
	Other faults	6	6	6	216	7	5	3	108
	Ergonomics risks:	8	8	9	576	8	6	5	160
	Muscles fatigue	8	7	7	392	8	5	5	200
	Heavy arm work	6	6	6	216	6	6	3	106
	Other faults								
	Products:								
	Dimension faults	8	10	9	720	8	4	5	160
	Design defects	8	8	8	512	8	6	4	192
	Other faults	6	6	6	216	6	6	3	108
Total RPN before technological improvements („0- process")					**8374**		RPN in „Ergo-process"		**2523**

FMEA results correspond with the cost benefit analysis indicating the ergonomic risk reduction of 70% that results also in process improvement (product developing efficiency) of 10%. The outcome of financial calculations about process improvements concerning ergonomics interventions are summarized in Table 4.

Table 4 Efficiency of ergonomics interventions

Efficiency	
The ergonomics effects	***Ergonomic risk reduction***
Eliminated the ergonomics risk of adverse effects	70%
Reduced the risk level of impact	40%
Reduced risk of exposure time	15%
Relieved heavy manual work	10%
Assessment of Productivity Improvement	
Level	***Process improvement***
High – accelerates the production process	10%
Moderate – reduces the waste motions	5%
Low – increases the comfort level	2,5%

The cost-benefit calculator WSECBC results shows that ergonomics interventions in furniture production line pays off in one year, while the benefits will grow every year, and the whole first year it sums up to 63 206 EUR, but in 5 years it is expected to benefit 319 030 EUR.

Hence the successful process improvement begins with analysis of the work requirements and the intervention of ergonomic basic principles in the technologies, i.e. introduce advanced technology to adjust human physical and psychical abilities, as well as enhance the safety of work environment. Such approach sets the opportunity for innovation, the ability to change and contribute to the sustainable development of the business in general. Undoubtedly ergonomics interventions require investments. Mostly the costs can be easily understood as fixed financial numbers, in order to improve the equipment, purchase of modern machinery or train employees for safer working methods, etc (Freivalds, 2009). While the benefits or ergonomics improvements are much more difficult to evaluate, because sometimes benefits are difficult to convert in monetary value, such as employee satisfaction, loyalty to organisation, satisfaction a.o. Managers often lack the understanding about benefits of ergonomics and therefore more researches are necessary in this field to prove ergonomics interventions in financial terms and convince managers that ergonomics opens the road for organizations to be more competitive.

4 CONCLUSIONS

Ergonomics risk assessment KIM method proved that ergonomics interventions in the furniture production line lightens the workload for sorting, packaging and assembling operators. Despite technical improvements, packaging operator's work still requires hard manual handling, thus participatory ergonomics should be applied. FTA and FMEA are recommended for process analysis for identification and prevention of problems, incl. ergonomics, before the faults take place and the unfavorable consequences arise. Ergonomics interventions should include not only improvement of machinery and working tools, but also employees training, job rotation, involvement in decision making a.o. The cost-benefit analysis is important to convince managers and stakeholders of the company about financial and less tangible benefits of ergonomics interventions. Hence the approach of combining ergonomics interventions in process improvement can result in effective identification and prevention of problems and enhance competitiveness of whole company.

REFERENCES

Bank, J. 1992. *The Essence of Total Quality Management*. London: Prentice Hall.

Besterfield, D. H. 2004. *Quality Control*. New Jersey: Pearson Education, 520 p.

Botschka, C. 1996. Participation in problem-solving a tool for improvement of working conditions? *Report: Division of Industrial Ergonomics and Centre for Studies of Humans*. Technology and Organisation, University of Linkoping, LiTH-IKP-R-921.

Brown, R. and Li, G. 2003. The development of action levels for the Quick Exposure Check (QEC) system. In: *Contemporary Ergonomics*, ed. P.T. McCabe, London: Taylor & Francis, pp. 41–46.

Burgess, M. Fault tree creation and analysis tool. *User manual*. http://www.iu.hio.no/FaultCat/manual.htm (viewed 15th February 2012).

Cost Benefit Analysis calculator, http://www.pshfes.org/cba.htm, (viewed 15th February 2012).

Dale, B. G., Van Der Wiele T., Van Iwarden J. 2007. *Managing Quality*. Malden: Blackwell Publishing, 280 p.

Dehlinger, J, Lutz, R., 2006. PLFaultCAT: A Product-Line Software Fault Tree Analysis Tool. *Automated Software Engineering*, 13: 169–193.

Eklund, J. 1994. Design for manufacturability – Consequences for quality and production personnel. *Proceedings of the 12th Triennial Congress of the International Ergonomics Association*, Canada, 4: 101–104.

Freivalds, A., Niebel, B. 2009. *Niebel's Methods, Standards, & Work Design*. 12th Edition, McGraw-Hill, 736 p.

George, S., Weimerskirch, A. 1994. *Total Quality Management*, John Willey & Sons, New York

Hendrick, H.W. 2003. Determining the cost-benefits of ergonomics projects and factors that lead to their success. *Applied Ergonomics*, 34: 419–427.

Juran, J. M., Godfrey, A. B. 2000. *Juran`s Quality Handbook*, 5th Edition, McGraw-Hill, 1872 p.

Juran, J. M. 1967. The QC circle phenomenon. *Industrial Quality Control.* 23: 329–336.

Karwowski, W. 2005. Ergonomics and human factors: the paradigms for science, engineering, design, technology and management of human-compatible system. Ergonomics, Vol. 48, 5: 436 – 463.

Kalkis, V. 2008. *Work Environment Risk Assessment Methods*. Riga: Fund of Latvian Education (in Latvian), 245 p.

McDermott, R., Mikulak, R., Beauregard, M. 1996. *The basics of FMEA*. USA: Productivity, New York: Quality Resources.

Nabitz U., Schramade M., Schippers, G. 2006. Evaluating treatment process redesign by applying the EFQM Excellence Model. *International Journal for Quality in Health Care*; Vol. 18, 5: 336–345.

XFMEA software, http://www.reliasoft.com/xfmea/ (viewed 15th February 2012).

Section III

Ergonomics and System Design

CHAPTER 30

Design Standards and Layout Methods of Multi-information Visual Interface Based on Eye Movement of Flight Simulator

Jun Hong 1 , Xiaoling Li 1, Keiichi Sato2, Lei Yao 1, Ying Jiang1

School of Mechanical Engineering, Xi'an Jiaotong University
Xi'an, China
Institute of Design, Illinois Institute of Technology
Chicago, USA
E-mail: xjtulxl@163.com

ABSTRACT

The goal of this research is to develop guidelines for optimizing the layout of instrument display system based on cognitive characteristics under the condition of multi-tasking with multi-information sources. Collecting eye movement data of take-off, level flight and landing based on multi-instrument display control test-bed. All these data were analyzed by the methods of statistical distribution, significantly different and data graphical processing. Our findings lead us to form some basic principles for better design of instrument panels. We put forward the following design suggestions and layout methods of instrument which are different from the traditional view: 1) The instruments with high attention conversion frequency should be placed together; 2) The instruments placed in core position should be much larger than those in non-core position; 3) Instruments should be placed in horizontal direction as far as possible. The conclusions of this research are applicable to aircrafts, automobiles, and other kinds of human-machine systems to improve the comfort and performance of operators.

Keywords: eye movement, flight simulation, visual interface, instrument layout, cognitive tasks

1 INTRODUCTION

In the process of flight simulation visual change is relatively large. Flight deck displays consist of GPWS、navigation display、integrated standby instrument system and so on. Therefore, the flight demands the pilot has better eyesight, which requires designers to research the visual interface of aircraft, while the aircraft instrument is the most concentrated expression of the visual interface. Instrument display system plays an important role in complex electromechanical equipment to contribute human-machine interactive function, monitor running condition of equipment and obtain important information about the external environment. If the instrument's display form, display quantity and display timing is designed unreasonably, they will cause the operator's visual fatigue, leading to more mistakes during operator working, and directly threatening the safety of personnel and equipment.

In the 1980s, Norman (1986) and Rasmussen (1986) respectively put forward the concept of cognitive engineering; it was also to be called interface science (Tamura, 1998). The research of cognitive engineering can reveal the reason of human judge mistake, the essence of disoperation and avoid the accidence happened.

About the response mechanism analysis of human visual system for visual interface, scholars have tried to use various methods to make display information best match with human visual characteristics. Amesbury and Schallhorn (2003) measured the minimum brightness contrast of required surface when human visual system identify different size objects, which based on contrast sensitivity index, and realized the quantitative check and analysis of human visual characteristics. Through the change of vibration frequency, vibration magnitude and vibration direction, Lewis (1980), Helander (1984) and Dupuis (1986) explored how these changes influence human visual recognition function. Liu Weiqi (1999) put forward the subjective color and brightness matching method, which was based on Munsell color system, and built the color brightness model of human eye subjective brightness between Munsell lightness values. In recent years, Liu Wei and Yuan Xiugan (2003) also built a simulated cockpit for synthetic ergonomic evaluation to research the cockpit layout, human-machine interface's comfort.

About the cognitive effect evaluation of human visual system for instrument relied more on subjective evaluation, fuzzy sets, gray system and the correct recognition rate in the process of design and evaluate instrument visual system (Bede Liu and David C.Munson, 1982; Keckler, Stadelman and Weiner, 1997; Kang Weiyong, Yuan Xiugan and Liu Zhongqi, 2008). Stark (2004) adopted subjective evaluation method, the main task of measurement method, auxiliary tasks measuring method and physiological measurement method to evaluate the cognitive effect of instrument. Fisher (1989) has shown that on average subjects are quicker to find a target option in a highlighted display than in a display without highlighting. Eui S.Jung et al (2000) put forward the visual cue is useful for visual display panel layout, which has made valuable research results. But most studies only from the view of subjective evaluation and objective measurement of physiological indexes to research the process of human cognitive, and did not take into account with the increase of information the physiological and psychological load will also increase.

Therefore, this research built a multi-instrument display control test-bed that enables flight simulation with the aircraft instrument interface and monitoring of eye movement of subjects. Eye movement can reflect eye sensitivity and delay of different places, size, color and the speed in multiple targets, multitasking situations. And fatigue of people can also be reflected from eye movement rules. The experiment required all subjects to perform a series of tasks including take-off, level flight and landing in particular routes. Take-off phase required subjects to fly to 5000ft height from the airstrip; Level flight phase required subjects to keep level flight time for at least 10min; landing phase required subjects to land on the airstrip safely from 5000ft height. Eye tracking system recorded the eye movement of the subjects on videos of flying scenarios, then the data of gaze fixation, the average pupil size, tracking speeds and saccade amplitudes can be collected, and we can draw a conclusion about the cognitive characteristics and layout methods of the multi-information visual interface.

2 METHOD

2.1 Subjects

In this study, 5 subjects were chosen. They were all college male students who had more than 100 hours of simulated flight experience. They were able to perform a variety of basic flight simulation tasks. In order to research cognitive characteristics in the condition of multi-task visual information, this research built test-bed that could simulate multi-task visual cognition. The test-bed is shown in Figure 1.

Adjustable seat system could provide safety and effective support for the driver. The adjusting mechanism of the seat allowed the seat to adjust the position before and after a distance of 0~±100 mm. The back of the seat could be adjusted 30° before and 90° back. Virtual instrument display system was mainly used to display visual information interface of different tasks. The height of the middle screen can be adjusted to the height range of 600~700mm and the both side two screens can be rotated freely outward 135°~360°. Visual system was used to simulate different scenarios of environments. In order to make the scene more realistic, this test-bed choose 54 inch LCD screen which has high color reproduction degree. Data of eye signal acquisition system is made of eye tracking device, and eye tracking system was the U.S. ASL company's mobile-eye series products.

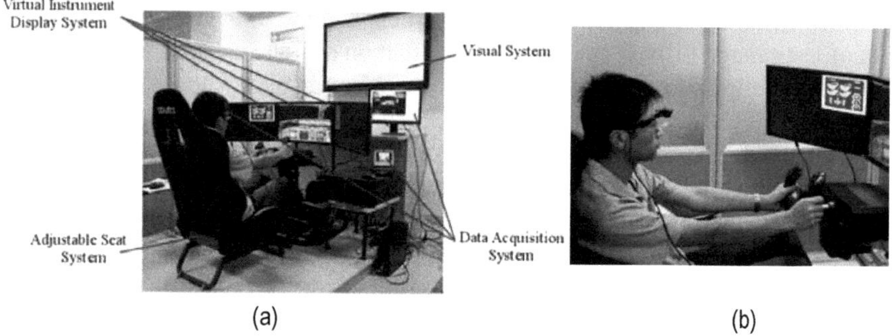

(a) (b)

Figure 1 Test-bed of multi-task visual cognition. (Figure1 (a) shows test-bed consisted of an adjustable seat system, virtual instrument display system, visual system and data acquisition system； (b) by eye tracking device to acquire eye movement data)

2.2 Experimental Procedure

All subjects were tested individually in a quiet, well-lit room. Each one calibrated the eye tracking system again before the formal experiment, and the calibration of eye tracking system procedure was as follows: 1)Rotate the eyeglass to make the three reflex points in the middle of the eye(Figure2(a)); 2)Swing the eyeglass to make the three reflex points more clearer. Subjects make some eye movement so that make sure we can find the location of the pupil (Figure2 (b));
3) Calibrate gaze fixation of eye and checking the calibration.

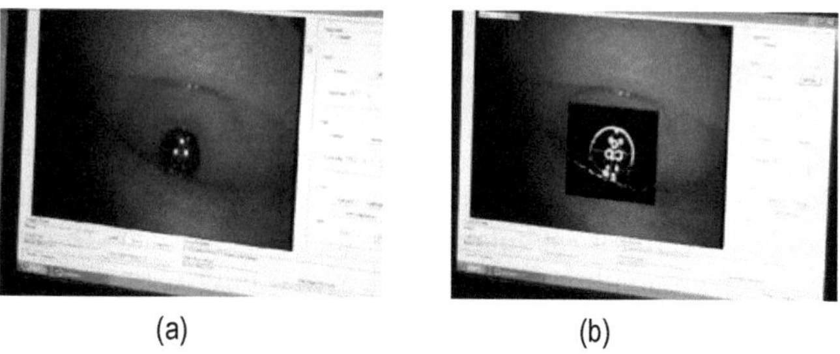

(a) (b)

Figure 2 Pictures of calibrate eye track system. (Figure2 (a) is to make the three reflex points in the middle of the eye; (b) is to make sure we can find the location of the pupil.)

After the calibration of eye tracking system was completed, the next procedure was ready to take-off; the take-off experiment process was as follows:

- Turn on the radio panel, input the heading of the target airstrip and activate the frequency;
- Open eye tracking system and eye movement data acquisition switch, push the accelerator pedal to the maximum and taxi to take-off. Experimental data were

collected by eye tracking system's software —Eye Vision. Gaze fixation of eye was shown in the picture when acquiring data(Figure3);

- Fly manually to 5000ft height and then adjust the aircraft heading to 52 degrees; Fly manually in the height of 5000ft height and heading of 52 degrees for 60s;
- Close eye movement data acquisition switch, turn on automatic pilot system and auto throttle. Keep the level flight for 3min and then open eye movement data acquisition switch;
- When the heading signal appears, turning on Approach Mode, intercepting the airstrip's magnetic heading and ready to landing. When landing completed, switch off the eye tracking device.

Figure 3 Distribution of gaze fixation point

The following was going to analysis the experimental data based on Excel, SPSS and MATLAB.

2.3 Analysis of Results

The data of gaze fixation, the average pupil size, tracking speeds and saccade amplitudes were collected in the experiment, and the percentage of gaze fixation and the average pupil size were analyzed by significantly different method. P value was an important index of significant test, and judging P value could get visual distribution of the subjects, which provides a theoretical basis for attention allocation rules, scanning principles and visual selective attention mechanism. And P value was gotten by T-test which was defined as the variance of two populations was unknown but the same used for significant different test of two averages.

Assuming that subjects gazed at scene that the pilot sees in the cockpit and instrument in different flight tasks were independent samples, and the number of gaze fixation, pupil size data were equidistant data. The samples obeyed the normal distribution.

Two populations mean were set to μ_1, μ_2. The hypothesis of T test was:

H0: Mean of two populations were equal: $\mu_1=\mu_2$

H1: Mean of two populations were not equal: $\mu_1\neq\mu_2$

Statistic calculation formula was:

$$t=\frac{\left|(\bar{X}_1-\bar{X}_2)-(\mu_1-\mu_2)\right|}{S_{\bar{X}_1-\bar{X}_2}}=\frac{\left|\bar{X}_1-\bar{X}_2\right|}{S_{\bar{X}_1-\bar{X}_2}},\nu=n_1+n_2-2$$

Among:

$$S_{\bar{X}_1-\bar{X}_2}=\sqrt{S_C^{\,2}(\frac{1}{n_1}+\frac{1}{\mathrm{n}_2})},\quad S_C^{\,2}=\frac{\sum X_1^2-\frac{(\sum X_1)^2}{n_1}+\sum X_2^2-\frac{(\sum X_2)^2}{n_2}}{n_1+n_2-2}$$

In the formula: $\overline{X_1},\overline{X_2}$ respectively were mean of two samples. ν was degree of two independent samples. n_1, n_2 were the number of two samples. Sc^2 was merger variance.

One side area of t value could be obtained from the formula. In SPSS, P value was defined as double area of one side area of t value. Judging P value could get visual distribution of the subjects, which provided a theoretical basis for attention allocation rules, and visual selective attention mechanism of human-machine interface design.

Tracking speeds and saccade amplitudes would expressed by the statistical distribution method in order to judge the change discipline of cognitive. According to human visual characteristics, eye elliptical structure almost caused the observation range is elliptical area. The statistical distribution of horizontal and vertical tracking speeds, as well as saccade amplitudes, can be described in MATLAB. The change trend of tracking speeds and saccade amplitudes could explain the different dimensions visual changes of the subjects.

3 RESULTS

The source of visual information was divided into two areas of interest (AOI, Area of Interest): Scene and Instrument. In this experiment, the percentage of gaze fixation, the average pupil size, tracking speeds and saccade amplitudes were collected. And all these data were analyzed by the methods of statistical distribution, significantly different and data graphical processing.

3.1 The Percentage of Gaze Fixation and Average Pupil Size

The percentage of gaze fixation is shown as Figure4 (a). Figure4 (a) shows the percentage of gaze fixation of scene is more than instrument. The times of gaze fixation in instrument in landing phase is more than take-off phase and level flight phase. When analyzing the percentage of gaze fixation in instrument region, the results of T test show take-off phase and level flight phase have no obvious difference($P=0.83>0.05$) by significantly different method. But they both have significant difference with landing phase($P_1=0.0032<0.05$, $P_2=0.0002<0.05$).

The average of pupil size shows as Figure4 (b). The average pupil size has no obvious difference in both scene and instrument ($P<0.05$) under different flight tasks by significantly different method. The pupil size on instrument panel is larger than on scene.

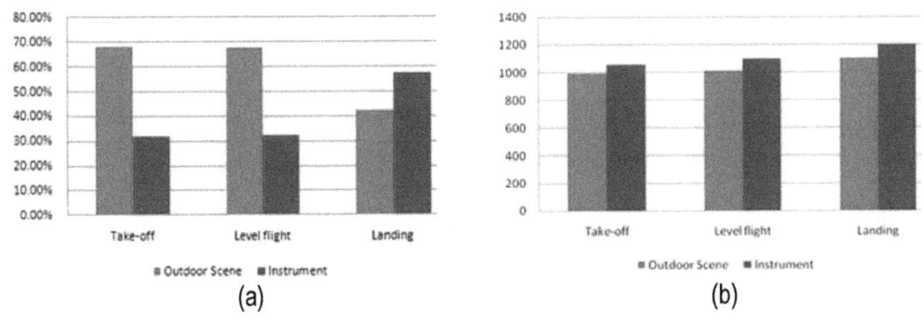

Figure4 Percentage of gaze fixation and the average of pupil size. (Figure4 (a) is the percentage of gaze fixation; (b) is the average of pupil size.)

3.2 Analysis of Scan Characteristics of Gaze Fixation

Tracking speeds and saccade amplitudes are main indexes of scan characteristics of gaze fixation. Tracking speeds in take-off phase and landing phase shows as Figure5.

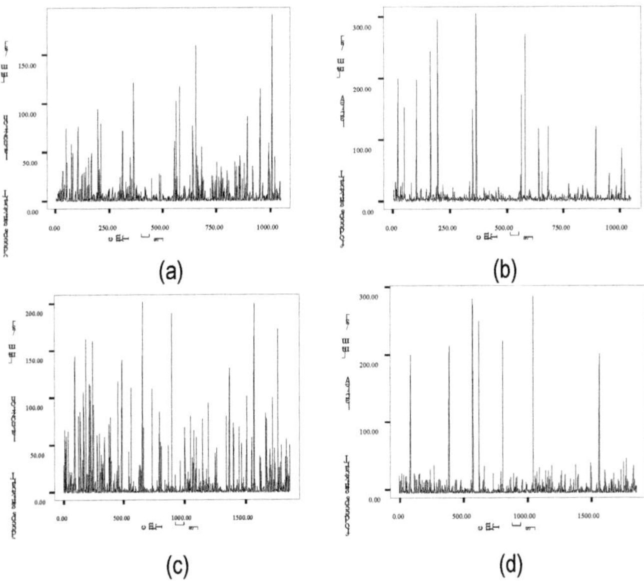

Figure5 Tracking speeds of different phases. (Figure5 (a) (b) is take-off phase horizontal tracking speeds and vertical tracking speeds; (c) (d) is landing phase horizontal tracking speeds and vertical tracking speeds)

From Figure 6, we can see saccadic amplitudes in take-off phase and landing phase has a significant jump time to time.

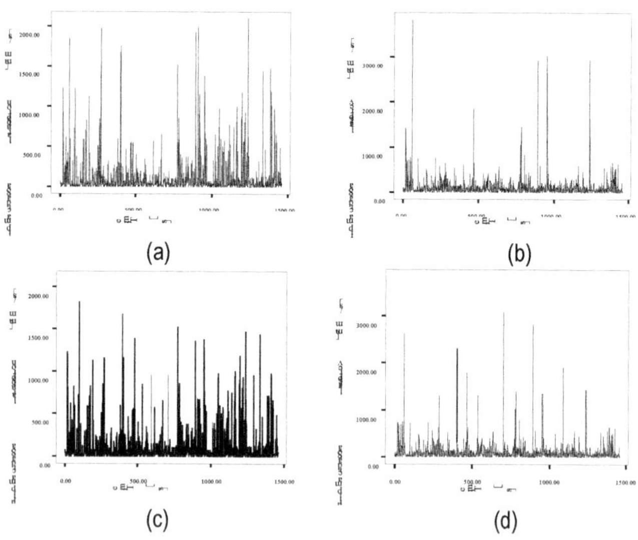

Figure 6 Saccadic amplitudes different phases. (Figure6 (a) (b) is take-off phase horizontal and vertical saccadic amplitudes; (c) (d) is landing phase horizontal and vertical saccadic amplitudes)

4 DISCUSSION

The number of gaze fixation in landing phase for instrument is more than scene. This is because there are more parameters in landing phase, and the subjects learn flying state by watching scene then confirm the information by watching instrument. There is more attention on the speed and height instrument in take-off phase. There is more attention on heading instrument in level flight phase, and most of the rest time is to watch scene. The differences in number of gaze fixation reflect the change of scan rule. It's generally thought that the number of gaze fixation expresses the degree of importance in the AOI, and by the degree of importance we can optimize the layout of instrument. When the subjects want to see a target clearly the pupil size will become larger, so the researchers believe that pupil size is one of sensitive indexes to measure the subjects' work load (Kang Weiyong et al, 2007). In the actual flight, the panel position is fixed and the function is clearly but there are still have lots of details, so the pilots' pupil size will get much larger than usual because they need watch carefully to obtain the whole flight information. However, the pupil size is less small when they watching scene because the pilots' visual field is wider and they feel more relax. Therefore, instrument should be designed in the best human vision field to avoid the average pupil size is too large. From above Figures of tracking speeds, whatever take-off or landing, we can see tracking speeds in horizontal direction is more than vertical direction, and then we can conclude that important parameters should be preferentially placed in the horizontal direction. From Figures of saccadic amplitudes we can see when the gaze fixation points are in the AOI, the pilots scan within a narrow range to conform the flight state.

Saccade amplitudes also identified most scan behavior occur in the horizontal direction.

The eye movement indexes can promote researchers to cognize visual interface. The percentage of gaze fixation can describe the transfer frequency of attention focus, which can understand the conservation, conversion and distribution of attention focus (Liu Zhongqi et al, 2006). According to the percentage of gaze fixation to keep the high attention frequency instrument in the center of the best sights, keep these high transfer frequency instruments layout together and according to the distribution of region attention. All these can enhance the readability of the information and reduce instruments watching time.

The eye movement indexes can measure the subjects' attention and express work load of different tasks and dimensions. The average pupil size is a sensitive index to reflect visual monitoring load and saccade characteristics with task difficulty's change. With increasing of task complexions, the increase of workload will causing visual tunnel effect and making the space of visual attention reduces. Therefore, the instruments laid out in core position should be much larger than the non-core position based on average pupil size in the process of design instrument, so that the operator's workload can be reduced. Instruments should be laid out in horizontal direction based on scan characteristics of gaze fixation points so that operators switch visual between various instruments in horizontal direction and confirm information in vertical direction.

5 CONCLUSIONS

This research analyzes distribution rules of attention, scanning principles and visual selective attention mechanism during the flight based on multi-instrument display test-bed. The study presents visual characteristics and cognitive characteristics of human under the condition of multi-tasking and multi-information dynamic instrument, so as to ensure the objectivity of evaluation and design of instrument display system. Human visual characteristics and cognitive characteristics have been obtained by the research of eye movement data, and the design standards and layout methods of instrument are shown as follows:

- Place the high attention frequency instrument in the center of the best sights;
- Place these high transfer frequency instruments together;
- According to the distribution of attention focus to divide regions for different kinds of instruments;
- The instruments laid out in core position should be much larger than the non-core position;
- Important instruments should been placed in horizontal direction.

The conclusions are applicable to aircrafts, automobiles, and other kinds of human-machine systems to improve the comfort and performance of operators.

ACKNOWLEDGMENTS

This research was funded by grants from National Natural Science Foundation of China (Grant No.61075069).

REFERENCES

Amesbury,E.C., and S.C.Schallhorn, 2003. Contrast sensitivity and Limits of Vision. *International Ophthalmology Clinics*, 43: 31–42.

Dupuis,H.,and G.Zerlett, 1986. *The effects of whole-body vibration*. New York: Springer-Verlag.

Fisher, D.L., and K.C.Tan, 1989. Visual displays: the highlighting paradox. *Human Factors*, 31: 7–30.

Helander,M.G., and B.A.Rupp, 1984. An Overview of Standards and Guidelines for Visual Display Terminals. *Applied Ergonomics*, 15: 185–195.

Jung, E.S., and Y.T.Shin, D.Y.Kee, 2000. Generation of Visual Fields for Ergonomic Design and Evaluation. *International Journal of Industrial Ergonomics*, 26: 445–456.

Kang, W.Y., X.G.Yuan, Z.Q.Liu et al. 2007. Analysis of Relations between Changes of Pupil and Mental Workloads. *Space Medicine & Medical Engineering*, 20: 364–366.

Kang W.Y.,and X.G.Yuan , and Z.Q.Liu, 2008. Optimization Design of Vision Display Interface in Plane Cockpit Based on Mental Workload. *Journal of Beijing University of Aeronautics and Astronautics*, 34: 782–785.

Keckler, A.D.,and D.L.Stadelman, D.D.Weiner,1997. Non-Gaussian Clutter Modeling and application to radar target detection. Proceedings of the 1997 IEEE National Radar Conference. Syracuse, NY, USA.

Lewis,C.H., and M.J. Griffin, 1980. Predicting the effects of vibration frequency and axis, and seating conditions on the reading of numeric displays. *Ergonomics*, 23: 485–499.

Liu, W.Q. , and R . Feng, F.K.Zhou, 1999. A Color Brightness Model Fitting Characteristics of the Human Vision. *ACTA OPTICA SINICA*, 19: 1426–1429.

Liu,W. ,and X.G.Yuan, D.M .Zhuang, et al. 2003. Development of a Simulated Cockpit for Synthetic Ergonomic Evaluation of Pilot. *China Safety Science Journal*, 13: 9–12.

Liu, B.D., and C.M .David, 1982. Generation of a Random Sequence Having a Jointly Specified Marginal Distribution and Autocovariance. *IEEE ASSP (S0096-3518)*, 30: 973–983.

Liu,Z.Q.， and X.G.Yuan , W.Liu, et al. 2006. Analysis on Eye Movement Indices Based on Simulated Flight Task. *China Safety Science Journal*, 16: 47–51.

Norman, D.A., and S.W.Draper, 1986. *User Centered System Design*. Boca Raton: CRC Press.

Rasmussen, J.， 1986. *Information Processing and Human-Machine Interaction*. New York: North-Holland.

Stark, J.M.， 2004. *Pilot performance and eye movement activity with varying levels of display integration in a synthetic vision cockpit*. America: Old Dominion University.

CHAPTER 31

A Study of Universal Design Principles Assessment

Moyen Mohammad Mustaquim

Department of Informatics and Media
Uppsala University
Uppsala, Sweden
moyen.mustaquim@im.uu.se

ABSTRACT

Goal in this pilot study is to explore the effect of universal design principles on the user's behavior as they use a system. It is found that the universal design principles are not really able to symbolize the system's attitude towards user's action on the system. This research result is a part of a larger and ongoing research effort to discover if a system is designed universally or not based on the design principles and thereby come up with new innovative universal design principles.

Keywords: Universal design, inclusive design, design principals.

1 INTRODUCTION

There has been debate among researchers about where great opportunities lay for innovating in the realm of human- computer interaction (Shneiderman and Maes, 1997). "Why should people have to adapt to systems, systems should adapt to people instead?" is a slogan that seems intuitively appealing to many users, as well as to researchers in the field of intelligent user interfaces (Höök, 1999). Considerable research has already been done and on progress in the human computer interaction fields within the area of design for all, inclusive design, universal design and assistive technology. As more research was conducted for improving product or user interface, more problems are found today associated with inclusive design. Different types of inputs in today's different types of device with different categories of user groups makes universal design concept challenging. Since it is a dynamic area of research which is rapidly changing, universal design is thus quite difficult to come up with. Universal design reflects a belief that the range of human abilities is normal and results in inclusion of people with disabilities in everyday activities(Center for Accessible Housing, 1995). The most significant benefit to the proliferation of universal design practice is that all consumers will have more products to choose from that are more usable, more readily available, and more affordable(Center for Accessible Housing , 1995). In this paper the universal design principles are studied under experimental condition with the users of an information system. The objective for user was to do their daily task in the industry setup where they work and record their activities successively for 7 working days. They did this by filling up set questionnaires at the end of work before they leave the work place. The objective of this study here is to understand how universal design principles are effecting on users work task in an information system setup. This work also shares the possibilities of other attempts to come up with the new universal design principles which can be useful and developed.

2 UNIVERSAL DESIGN

Universal design can be defined as the design of products and environments that can be used and experienced by people of all ages and abilities, to the greatest extent possible, without adaptation(Center for Accessible Housing , 1995). The philosophy underlying inclusive design specifically extends the definition of users to include people who are excluded by rapidly changing technology, especially the elderly and ageing, and prioritizes the role and value of extreme user groups in innovation and new product and service development. It also prioritizes the context of use, both physical and psychological, and the complexity of interactions between products, services and interfaces in specific contexts of use, such as independent living(Langdon et al., 2007). To make technology useful to and usable by older adults, a challenge for the research and design community is to ‘‘know thy user’’ and better understand the needs, preferences and abilities of older people(Czaja and Lee, 2007). It is fairly well established that many technology products and systems are not easily accessible to older people. There are of course a myriad of reasons for this, such as cost, lack of access to training programs, etc. However, to a large extent lack of accessibility is due to the fact that designers are unaware of the needs of users with varying abilities, or do not know how to accommodate their needs in the design process(Clarkson and Coleman, 2003).

2.1 The principles of universal design

The original set of universal design principles, described below was developed by a group of U.S. designers and design educators from five organizations in 1997 (Center for Accessible Housing , 1995). The principles are copyrighted to the Center for Universal Design. The principles are used internationally, though with variations in number and specifics analogy.

- Equitable Use: The design does not disadvantage or stigmatize any group of users.

- Flexibility in Use: The design accommodates a wide range of individual preferences and abilities.
- Simple, Intuitive Use: Use of the design is easy to understand, regardless of the user's experience, knowledge, language skills, or current concentration level.
- Perceptible Information: The design communicates necessary information effectively to the user, regardless of ambient conditions or the user's sensory abilities.
- Tolerance for Error: The design minimizes hazards and the adverse consequences of accidental or unintended actions.
- Low Physical Effort: The design can be used efficiently and comfortably, and with a minimum of fatigue.
- Size and Space for Approach & Use: Appropriate size and space is provided for approach, reach, manipulation, and use, regardless of the user's body size, posture, or mobility.

Although universal design is a concept that has more and more support around the world, there are still habits, ignorance and wrong priorities that need to be overcome. The British Standards Institute (British Standard , 2005) defines inclusive design as "The design of mainstream products and/or services that are accessible to, and usable by, as many people as reasonably possible ... without the need for special adaptation or specialized design." By meeting the needs of those who are excluded from product use, inclusive design improves product experience across a broad range of users. Put simply, inclusive design is better design which should be embedded within the design and development process, resulting in better designed mainstream products that are desirable to own and gratifying to use. In Europe, the term Design for All has a similar meaning to universal design. However, the term inclusive design also includes the concept of reasonable in the definition. Erlandson's organization of principles (NSF, 2006) addresses two vulnerabilities in universal design. First, as noted above, too often people assume that universal design is a synonym for barrier-free or accessible design. Many people think of universal design as focused primarily on the same conditions as barrier-free: people with mobility limitations especially wheelchair users and people who are blind. The second vulnerability is that it appears to promise too much. Too broad, it becomes insignificant – more concept than design strategy. The Erlandson scheme (NSF, 2006) of adopting three broad categories of human function captures the vast majority of conditions. It makes it easier to understand that barrier-free/accessibility is a floor upon which to build universal design. It makes it easier for a client to appreciate the difference between barrier-fee, accessibility and universal design. It points to the reality that we need more research and innovation to expand a repertoire of design solutions.

3 METHOD

The principles of universal design are intended to be verified with a factor analysis, reliability analysis, correlation matrix and multiple regression analysis of the data from a quantitative field experiment on a garments industry setup in Bangladesh. These methods are chosen since they will explain the relationship and dependencies of the variables and the analysis will be performed using the software package "SPSS"

The basis for the validation is questionnaire responses. The questionnaire items were designed to capture the variables of universal design principles. All scales used in the research were prior to the fieldwork tested and optimized for face validity with senior researchers and qualitatively tested with respondents demographically similar to the final field work respondents.

3.1 Procedure

On December 2010 a 3 weeks research visit to Bangladesh initiated this study. The garments factory that has been selected to run the research has more than 1800 employee in different sections. 100 workers selected for this study that are able to read and understand English properly. However the questions were also provided in the subject's mother tongue Bengali. For seven successive working days these workers were requested to fill up the questionnaires before they leave the work. The workers that were selected are the users of different systems on the same industrial location. This includes machine operator of different types, management department, information technology department, human resource department and accounting department. The machine operators are well trained for running fully automatic machines for garments production and packing. The rest of the department's workers are using customized design software for their respective department. The subjects were of mixed gender group and none of them suffer with severe physical disabilities of any kind. On April 2011 the same research was conducted, this time with 100 different employees of the same factory. The person, who monitored the test and collected data this time, was appointed by the researcher. On May 2011 the representative from the company in Bangladesh visited the researcher and handed over the new data. The statistical operations are then conducted based on the data of 200 users.

4 RESULT

The exploratory factor analysis (with factor loading value 0.40) shows that the data set leads to a three factor solutions (Figure 1). These three factors represent the three principles of universal design: flexibility of use, simple intuitive use and low physical effort. The reliability analysis shows that the data set for these three principles are reliable (Figure 2) since the value of chronbach's alpha is more than 0.7 for these three data sets. Other variables representing the four principles showed the corrected item-total correlation value to be less than or close to 0.3 which means each items are correlated with the total score with low degree of values which makes the data less reliable or unreliable. A bivariate correlation option from SPSS was run to find out the Pearson and Spearmann results for testing which variables are in regression with which variables. The result shows that flexibility in use variable can work as a dependent variable in a regression with simple intuitive use, low physical effort and error tolerance as independent variables. However, the variable error tolerance has a tendency of showing less confidence in the correlation matrix (Figure 3) which proves that there is no correlation between low physical effort and simple intuitive use. Finally a multiple regression was run where flexibility in use was used as a dependent variable and simple intuitive use, low physical effort and error tolerance were used as independent variables. The R square value shows that 39.4% of the variance (Figure 4) in the dependent variable, flexibility in use is explained by the model including other three variables. The sig value of the variable error tolerance is higher than 0.05 on the coefficient table (Figure 5) meaning that this is not a significant variable. Finally the part value for simple intuitive use on coefficient table (Figure 6) is 0.305. If we square this value we get an indication of the contribution of this variable to the total R square which is 9.3%.

Total Variance Explained

	Initial Eigenvalues		
Factor	Total	% of Variance	Cumulative %
1	8.456	56.376	56.376
2	1.761	11.741	68.117
3	.933	6.219	74.336
4	.609	4.060	78.396
5	.473	3.152	81.548
6	.403	2.684	84.232
7	.397	2.650	86.882
8	.364	2.427	89.309
9	.312	2.081	91.389
10	.280	1.869	93.259
11	.257	1.710	94.969
12	.225	1.498	96.467
13	.211	1.408	97.874
14	.179	1.192	99.066
15	.140	.934	100.000

Extraction Method: Maximum Likelihood.

Figure 1 Almost a three factor solution of the data

Reliability Statistics

Cronbach's Alpha	Cronbach's Alpha Based on Standardized Items	N of Items
,936	,936	5

Figure 2 Reliability analysis of flexibility of use principle

Correlations

		Flexibility s1-s5	Simple Use s39-43	Physical Effort s51+s52	Error Tolerance
Flexibility s1-s5	Pearson Correlation				
	Sig. (2-tailed)				
	N	[illegible]			
Simple Use s39-43	Pearson Correlation	,574**			
	Sig. (2-tailed)	,000			
	N	301	[illegible]		
Physical Effort s51+s52	Pearson Correlation	,548**	,605**		
	Sig. (2-tailed)	,000	,000		
	N	295	300	[illegible]	
Error Tolerance	Pearson Correlation	,002	-,030	-,059	
	Sig. (2-tailed)	,972	,609	,315	
	N	297	298	294	[illegible]

**. Correlation is significant at the 0.01 level (2-tailed).

Figure 3 Correlation matrixes to analyze variables in regression with each other

Model Summary[b]

Model	R	R Square	Adjusted R Square	Std. Error of the Estimate
1	,628[a]	,394	,388	1,14623

a. Predictors: (Constant), Simple Use, Physical Effort s39-43, Error Tolerance s51+s52

b. Dependent Variable: Flexibility s1-s5

Figure 4 Variance of the flexibility in use principle

Coefficients[a]

Model		Standardized Coefficients	t	Sig.	95,0% Confidence Interval for B
		Beta			Lower Bound
1	(Constant)		3,014	,003	,291
	Simple Use s39-43	,383	6,663	,000	,302
	Physical Effort s51+s52	,318	5,533	,000	,204
	Error Tolerance	,032	,702	,483	-,019

Figure 5 Insignificancy in error tolerance principle

Coefficients[a]

Model		95,0% Confidence Interval for B	Correlations		
		Upper Bound	Zero-order	Partial	Part
1	(Constant)	1,385			
	Simple Use s39-43	,555	,574	,364	,305
	Physical Efforts 51s51+s52	,430	,548	,309	,253
	Error Tolerance	,040	,002	,041	,032

Figure 6 Contribution of simple, intuitive use principle to R square is 9.3%

5 FINDINGS AND DISCUSSION

We found that flexibility of using a system could be measured by measuring how simple the system is to use and the level of physical efforts required for using the system. However, it is also found that while users are using a system they did not bother considering the factor how tolerant the system is against error. The analysis of data shows that 39.4% of the whole model is dominated by how flexible the system is to use and it is important to users that the system be simple and intuitive to use since this variable explains 9.3% of the model. The regression analysis explained that although a system is simple, intuitive to use that does not necessarily mean that it takes less physical effort to use that system and vice versa. But the users though that, the flexibility of a system is dependent on both how the physical effort is being given and how simple the system is to use. A three factor solution of data also dominates three design principles to be remarkably noticeable and other design principles to have less impact on the user's psychology, while they were using a system. The subjects of this study were not disabled but still did not feel the effect or need of other design principles which is interesting. This leads the argument of defining universal design or universal design principles from a new dimension. It should be strictly mentioned whether or not the keyword 'universal design' and 'universal design principles' are solely meant for design for handicapped or special need groups of people or not. If not, then probably it is time to alter the universal design principles and write them in a new way based on further research and other important factors that are dictating the design of today's information systems.

6 FURTHER RESEARCH

A path analysis of the data will be able to show the strength of the dependencies among the variables found in this study. Such analysis will also predict the direction of the dependencies. Partial Least Square (PLS) method can be used for doing path analysis. PLS has high probability of finding latent variables and is statistically strong. Scenarios like workers of a factory where they are supposed to do specific tasks in a sequential and ordered manner, PLS has greater statistical robustness than structural equation modelling. Making a PLS-analysis of the data from this study will show the degree and strength of dependency and interdependencies of these variables (principles) and the way they do or do not impact attitude, cognitive ability and behavioural change of users while they use a system. This study will be performed during the winter 2012.

7 CONCLUSIONS

In this study we found that universal design principles are not necessarily noticeable by the users of a system which satisfies those principles. Also it is noticed that, sometimes users are unaware of the effects of some design principles.

A product or system designed for disabled peoples, following universal design principles would probably return similar results, that users are unaware of some factors of the design. At the same time while usability test can be performed on a system or product, it can also be evaluated by the result of this research to see which design principles are dominating the system. The flaws of other design principles can thereby be found and improved if necessary.

REFERENCES

NSF 2006 Engineering Senior Design Projects to Aid Persons with Disabilities.

1995. Center for Accessible Housing. Accessible environments: Toward universal design. Raleigh :North Carolina State University.

2005. British Standard 7000-6:2005. Design management systems - Managing inclusive design – Guide.

CLARKSON, J. & COLEMAN, R., KEATES, S., LEBBON, C. (EDS.) 2003. Inclusive Design: Design for the Whole Population. 88-102.

CZAJA, S. & LEE, C. 2007. The impact of aging on access to technology. Universal Access in the Information Society, 5, 341-349.

HÖÖK, K. 1999. Designing and evaluating intelligent user interfaces. Proceedings of the 4th international conference on Intelligent user interfaces. Los Angeles, California, United States: ACM.

LANGDON, P., CLARKSON, J. & ROBINSON, P. 2007. Designing accessible technology. Universal Access in the Information Society, 6, 117-118.

SHNEIDERMAN, B. & MAES, P. 1997. Direct manipulation vs. interface agents. interactions, 4, 42-61.

CHAPTER 32

Hand/Handle Interface Affects Bi-manual Pushing Strength

Jia-Hua Lin, Raymond W. McGorry, and Chien-Chi Chang

Liberty Mutual Research Institute for Safety
Hopkinton, MA, USA
Jia-hua.lin@libertymutual.com

ABSTRACT

Hand-handle interface, which induces varied wrist posture, is seldom considered in most of the current upper limb biomechanical analyses of pushing and pulling strength. In our laboratory a study was developed to examine effects of handle rotation in the frontal plane (0°-horizontal, 45°, and 90°-vertical), anterior tilt (0°-parallel to the frontal plane, and 15°), and distance between two handles (31 and 48.6 cm) on pushing strength. A special testing station was constructed to measure upper limb push exertions with minimal contribution from the torso and legs. Both the horizontal (forward pushing) and vertical components of the pushing forces were measured within the station. Thirty-one study participants were recruited for the seated two-hand pushing strength tests. Referencing to the horizontal, straight, and a 31 cm between-handle distance handle configuration, the 45°-rotated and tilted handles with a 31cm between-handle distance allowed 6.7% more pushing output, while the horizontal and tilted handles with a 31cm between-handle distance resulted in 2.8% less. Subjective preference was correlated with normalized pushing strength (r=0.89). Tilted handles, at 45°-rotated and vertical positions received highest subjective ratings of preference among all handle configurations. Women exerted the greatest strength when the handles were the 31-cm apart, while men exerted more pushing strength with the 48.6-cm handle distance. The results demonstrated that pushing capacity was affected by handle rotation and tilt angles, and therefore such design should be taken into consideration when evaluating pushing tasks.

CHAPTER 33

Research on Evaluation and Optimization Design Method of Product Sensory Quality

Fu Guo, Yu Sun, Mingjian Zhu, Lei Zhao, Na Xu

Northeastern University of China
Shenyang, China
fguo@mail.neu.edu.cn

ABSTRACT

With the increasingly fierce market competition, users tend to pay more attention to the product sensory quality. Only the designer do grasp the consumer's requirements on the product sensory quality and translate their perceptual demands to the design elements can the designer make the product to meet consumer's demands.

This thesis makes a study of evaluation and optimization on product sensory quality. In this paper, the first step is to thoroughly analyze the current domestic and foreign research status on product sensory quality. Secondly, it makes a research on the significance of products and human perceptual cognition, where it introduces sensory quality evaluation factors, namely perceptual factor and introduces indicators to measure the level of sensory quality, namely the feeling discrepancy (FD) and feeling ambiguity (FA). Based on this, this paper then establishes the evaluation methods and steps of the product sensory quality. Thirdly, it presents an optimized design model for product sensory quality based on the analysis of the QFD method and Taguchi Method, and proposes a new method to optimize the design of product sensory quality, through which the optimization subject and emotion object of design are determined, the key appearance design variables are found by QFD method, and then the variables are parametric designed by Taguchi method. Finally, an example is given to demonstrate the effectiveness and validity of the proposed method on the optimization of the sensory quality. The result shows that the feeling discrepancy (FD) and feeling ambiguity (FA) of the optimized car are reduced by 35.92% and 46.52% respectively, which provides a reference for the other products.

Keywords: sensory quality, quality function development, Taguchi Method, perceptual factor, optimization design

1 INTRODUCTION

With the constant development of socioeconomic, consumers increasingly focus on the satisfaction of the products on the demand for psychological (mental). So however to improve sensory quality of products has become more and more important. Sensory quality (Hsin-Hsi Lai, Yu-Ming Chang and Hua-Cheng Chang, 2005) refers to design products which satisfy the users' personal emotional needs and expectations. The related study of sensory quality abroad dates back to 1982. Holbrook and Hirschman (1982) put forward that when a consumer decide whether or not to purchase a product, they not only place importance on a product's physical quality, but also employ their sentimental responses. Kolter (1992) emphasized that personal characteristics has an influence on products' purchasing. Study of Norretranders (1998) showed that vision stimulated consumption behavior mostly, and appearance, package, advertisement, board, and website of the products are the most important factor to vision, so they are dominant in design. Not until 2005, did Hsin-Hsi Lai, Yu-Ming Chang and Hua-Cheng Chang present the real concept of products sensory quality. Then they made a robust design approach for enhancing the sensory quality of products. From then on, the study on sensory quality of products also caused more and more domestic scholars' attention. From the perspective of consumers, Jianning Su and Heqi Li（2005）had studied on the feeling feature of materials in industrial design. They derived the consumers' preference image from the product shape material intention survey. By analyzing this, they concluded the corresponding relationship between consumers' preference image and product design materials. Daqing Li (2006) pointed out that studying on sensory quality in China is valid and reasonable. He made an elemental investigation on sensory quality evaluation method and improvement method. At the same time, he analyzed the main research directions for sensory quality. Shihu Xu and Donghai Ye (2006) discussed that color had an influence on people's physiology and psychology, analyzed that environment, custom and other factors affecting the understanding of products' color, and finally proved that colors of industrial products were strongly linked to the fashionable colors.

Literature research above shows that the research on sensory quality of products is still in the initial stage. It lacks a scientific and complete research system. While this paper presents a products sensory quality evaluation method and proposes an optimization design method based on sense factors, which provides the technical support for product designers.

2 A PRODUCT SENSORY QUALITY EVALUATION METHOD

(1) Founding sensory quality evaluation indexes, namely sensory factors. At first through various means to collect Kansei words about products extensively. In the process of product representative samples selection, pictures are firstly classified according to similarity through questionnaire, and then, representative samples are found by multi-scaling, cluster analysis, and other mathematical methods. The next step is to establish the sensory quality evaluation scales. Selected Kansei words and representative samples should be combined in a questionnaire by semantic differential method to investigate the respondents; researchers should analyze the resulting data by correlation coefficient matrix analysis, project analysis and factor analysis; delete invalid words; merge some high interrelated data; and finally set up a sensory quality evaluation scale. The last step is confirming products' sensory factors which are determined by factor analysis on sensory quality evaluation scale utilizing the investigated data and establishing the model which reflecting overall preferences of customers. Sensory factors are founded by means of regression analysis between them.

(2) Measure index of products sensory quality. In order to evaluate sensory quality of the products better, the authors introduce two quantitative indicators, namely feeling discrepancy (FD) and feeling ambiguity (FA) (Lai, Chang & Chang, 2005, pp.447-449). And then make some improvement on them.

(3) Evaluating sensory quality. Defining the target feelings is the first step. Target feelings are the values which the designed products should be achieved and they are regulated by designers. The second step is to evaluate feelings. With the help of nine-mark or other image scale, researchers should employ defined sensory factors and the evaluated object pictures to combine a questionnaire by semantic differential method, investigate the respondents, and then process the investigated data. The value of customer overall preference, which is called the value of sensory quality evaluation is finally calculated through the model reflecting customer overall preference and sensory factors. The assessment result should be analyzed in the end. The main step is to analyze the sensory quality evaluation for products by feeling discrepancy (FD) and feeling ambiguity (FA).

3 AN OPTIMIZATION DESIGN METHOD OF PRODUCTS SENSORY QUALITY BASED ON INTEGRATED MODEL OF THE QFD AND TAGUCHI METHOD

As the highlight of the QFD method is to translate customer needs into the design variables, which there is a lack of technical support tools in later design. However, the Taguchi method focuses on engineering design, which ignores user demands and lacks of technical support tools during the system design phase. Based on this, this paper proposes an optimization design method of products sensory quality based on integrated model of the QFD and Taguchi Method, as shown in Figure 1. The model's optimization processes are as follows:

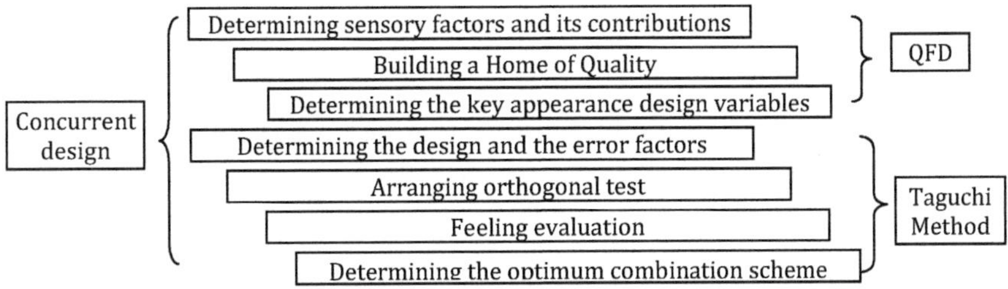

Figure 1 Optimization Design model of quality based on QFD and Taguchi method

(1) Ascertaining optimized object and target feeling of optimal design. A target feeling is used to be expressed by one-nine image scale, the lowest value is 1 and the highest value is 9. For example, if quality factor is the sensory factor of the products, when a designer defining the target feeling is 9 in optimizing the design of products' sense quality, which means designers want to optimize the design so that customer's preferences towards quality could reach the highest level.

(2) Utilizing the QFD method to confirm product's key appearance design variables. Firstly, ascertaining the contribution of various sensory factors *Ki* (i=1,2, ···, m). In this part of research, according to the determined sensory factor of products above, the contribution of various sensory factors to products' sensory quality can be worked out by regression model between general preferences of customers and sensory factors. The following is founding a Home of Quality. When determining appearance design variables of products' sensory quality, researchers mainly adopt the method of expert investigation. Employ questionnaires to determine the relation between sensory factors and appearance design variables *rij* (i =1,2,······, m; j =1,2,······, n), set up a relation matrix between sensory factors and design variables, also install a correlation matrix about design variables, and eventually, construct a Home of Quality with the above two matrix. Ultimately, the key appearance design variables are confirmed. In the research on sensory quality, the key appearance design variables are also determined by the design variables weighted importance degree *hj*.

$$h_j = \sum_{i=1}^{m} k_i r_{ij} \qquad (1)$$

If design variables of item no.*j* are closely related to a number of sensory factors, and these factors have a very high contribution degree (*Ki* larger), the value of *hj* is high, in other words, this design variable is very important, and it should be as a key appearance design variable.

(3) Designing the parameters of product's key appearance design variables by the Taguchi Method. In this part, design factors are those product's key appearance design variables which determined above. The levels of design factors are confirmed by the existing study on design parameters of the corresponding parts of the products. Error factors in this study basically point to the personal characteristics which is a direct bearing on the success or failure of the tests. Then

choosing the appropriate inner and outer orthogonal arrays, and arranging the rational orthogonal test. Selections of inner and outer orthogonal arrays are respectively according to the number and level of design factors and error factors. After that, sensory quality evaluation methods which proposed in this article should be made use of to estimate the feeling. Then, evaluating and analyzing the received data, and utilizing the model between general preferences of customers and sensory factors which has mentioned above to find the general preferences of customers for each product. The last step is to fix on the optimum combination scheme. The "smaller-the-better" S/N ratio of each scheme, which is an index to distinguish whether sensory quality of the products is stable or not can be calculated by the means of sensory quality evaluation value of products. The "smaller-the-better" S/N, η, can be given by

$$\eta = S/N \qquad (2)$$

Where S is the main effect of the factor and N is the error effect. Afterwards, figure out different level of S/N ratio in all design factors by orthogonal array and S /N ratio of each scheme, where the higher the value of S/N ratio is the better. A high S/N ratio of a design variable in one parameter level indicates that this variable is good in this parameter level. In this condition designers can design an excellent product with a higher sensory quality. So, combining the highest S/N ratio of a design variable in one parameter level is the optimum combination scheme.

4 CASE STUDY

With a sedan as the optimized object, by the use of the optimization designing model of products sensory quality which is established in this article, this case aims to optimize the sedan.

(1) Ascertaining sensory quality expression tools of the car-sensory factors.

This case quotes the determined sensory factors of a sedan, including temperament factor, harmony and personality factor in Fu Guo and Gaiyun Liu's study of customer-oriented study on design support technology based on perception of car style, and it also references the relational model between general preferences of customers and sensory factors they have worked out:

Y=0.996+1.427 (temperament factor) +0.620 (harmony factor) +0.516 (personality factor) (3)

Temperament factor is the sedan's overall styling features directly, harmony factor mainly refers to whether the appearance of the sedan is harmonious and the overall line feeling feature, and personality factor mainly points to the styling feeling feature of the sedan compared with other sedans.

(2) Confirming optimized object and optimal design's target feeling.

In this case, the authors just optimize the design of two-dimension outline in the optimal design process (see Fig.2). The optimum design target value of harmonious factor, personality factor and temperament factor are respectively 9.

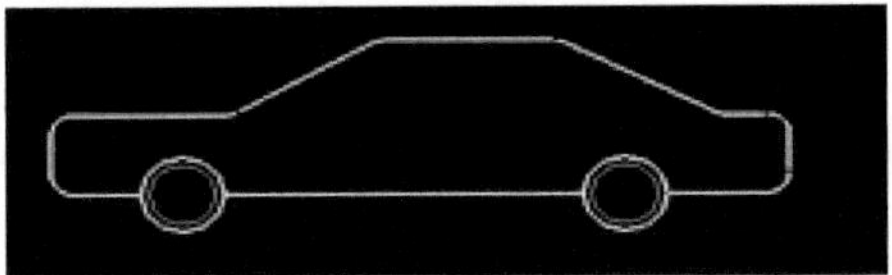

Figure 2 The sidewise two-dimensional contour of a sedan

(3) Making use of the QFD method to determine sedan's key appearance design variables. Through the model between general preferences of customers and sensory factors concludes that contributions of temperament factor, harmonious factor, and personality factor to sedans' sensory quality are 1.427, 0.620, and 0.516 respectively. 15 sedan design variables (see table.1) are determined by expert investigation and literature review. Employ questionnaires to found the relation matrix between sensory factors and design variables and the correlation matrix about design variables. There was a total of 60 questionnaires sent out, where 53 effective questionnaires were collected. In Fig 3, the degrees of relation in design variables are the average value marked by evaluators, and the degrees of correlation in appearance design variables are got equally. Eventually, the key design variables can be determined by design variables weighted importance degree hj.

Table 1 sedan's appearance design variables

number	Design Variables	number	Design Variables
1	Car's length	9	Rear region's length
2	Car's height	10	The gradient of fore windshield
3	Chassis's height	11	The gradient of rear windshield
4	Wheel axle's height	12	The gradient of rear bonnet
5	Front overhang' length	13	The gradient of fore bonnet
6	Rear overhang' length	14	The gradient of fore fender's bottom
7	Fore region's length	15	The gradient of rear fender's bottom
8	The height of car's face		

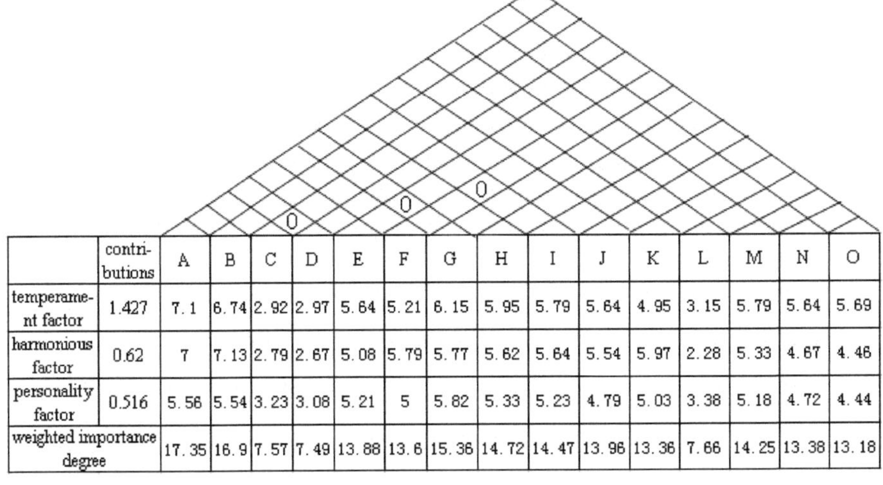

	contributions	A	B	C	D	E	F	G	H	I	J	K	L	M	N	O
temperament factor	1.427	7.1	6.74	2.92	2.97	5.64	5.21	6.15	5.95	5.79	5.64	4.95	3.15	5.79	5.64	5.69
harmonious factor	0.62	7	7.13	2.79	2.67	5.08	5.79	5.77	5.62	5.64	5.54	5.97	2.28	5.33	4.67	4.46
personality factor	0.516	5.56	5.54	3.23	3.08	5.21	5	5.82	5.33	5.23	4.79	5.03	3.38	5.18	4.72	4.44
weighted importance degree		17.35	16.9	7.57	7.49	13.88	13.6	15.36	14.72	14.47	13.96	13.36	7.66	14.25	13.38	13.18

Figure 3 HOQ of sedan sensory factor and design variable

The calculation results obviously indicate that weighted importance degree of C, D, and L are 7.57, 7.49 and 7.66 respectively, which are significantly lower than other variables'. By querying the sedan manual, the three variables are not important considered variables in the process of appearance design. Therefore, this case determines 12 key appearance design variables which are A, B, E, F, G, H, I, J, K, M, N, and O.

（4）Adopting the Taguchi Method to optimize the key appearance design variables of cars.

Taguchi design factors are the key appearance design variables for sedans determined by QFD. Three parameter levels of 12 key appearance design variables (see table 2) are determined by measurement tool MB-Ruler and literature review method.

Personal trait and product involvement are chosen as error factors in this paper. In order to measure personality (w), the authors adopt e scale of the Eysenck Personality Questionnaire-Revised Short Scale for Chinese (EPQ-RSC; Eysenck, 1975), and this factor including two-levels namely inward and outward. Product involvement (x) refers to the consumer expresses concern for, or participates in situations on the basis of inherent needs, worth, and interest. The measurement of product involvement adopts the modified Personal Involvement Inventory (PII) by Zaichkowsky(1994) . The factors also include both high and low levels.

Table 2 sedan's appearance design variables

Level	Level 1	Level 2	Level 3	Level	Level 1	Level 2	Level 3
A/m	4.9-5.5	4.5-4.9	4.0-4.5	I/m	1.48-1.73	1.24-1.48	0.99-1.24
B/m	1.48-1.53	1.43-1.48	1.38-1.43	J/°	36-43	30-36	23-30
E/m	0.95-1.11	0.78-0.95	0.61-0.78	K/°	49-60	37-49	26-37
F/m	1.09-1.24	0.79-1.09	0.79-0.94	M/°	175-180	169-175	165-169
G/m	2.17-2.34	2.0-2.17	1.83-2.0	N/°	175-180	170-175	165-170
H/m	0.642-0.72	0.563-0.642	0.485-0.563	O/°	175-180	170-175	165-170

According to the quantities and levels of design factors and error factors in this study, inner and outer arrays are respectively selected orthogonal array L27（3^{13}） and L4（2^3）. From inner array can obtain 27 designs of the products. Divide the participants into four groups G1, G2, G3, and G4 based on the outer array.

In the process of evaluating feelings in this case, totally 200 questionnaires were sent out, and 180 effective questionnaires were received. Customers' overall preference values (see table 3) are gained by analyzing the resulted data. Because the values of customers' overall preference represents customer preferences, the higher the better, the authors use the highest value of S/N ratio to calculate the S/N ratio of customers' overall preference (see table 3). Different levels of S/N ratio for each design variables can be calculated by customers' overall preference values (see table 4). Table 4 shows that the optimum combination scheme is A2, B3, E1, F1, G3, H2, I3, J1, K3, M1, N2 and O1. The combination car pictures as shown in Figure 4.

Table 3 The sensory quality assessment values and S/N ratios of 27 sedans design schemes

Scheme Number	G1	G2	G3	G4	S/N
1	13.76	14.72	15.34	14.18	23.20
2	12.63	13.34	13.31	13.85	22.45
3	10.79	8.36	10.04	11.35	19.94
4	4.23	4.94	4.28	4.56	13.02
5	5.18	5.60	4.11	3.61	12.90
6	7.94	6.96	7.30	7.32	17.33
7	5.87	5.21	5.57	4.64	14.43
8	15.72	15.24	13.34	15.19	23.39
9	12.35	10.60	12.41	11.48	21.32
10	5.92	4.23	5.04	4.89	13.83
11	10.23	9.52	10.82	11.37	20.35
12	9.45	9.59	10.05	10.81	19.94
13	14.84	13.76	13.55	15.71	23.16
14	5.19	5.62	6.23	5.16	14.81
15	5.92	6.92	5.20	5.69	15.33
16	16.32	11.17	14.96	12.10	22.39
17	5.89	5.06	5.27	5.69	14.72
18	20.48	18.09	17.41	18.63	25.37
19	8.23	5.62	9.58	8.70	17.54
20	5.53	5.89	5.62	5.33	14.93
21	8.20	7.06	8.79	9.00	18.22
22	4.23	4.62	4.31	4.62	12.94
23	18.58	15.30	17.55	17.89	24.71
24	15.08	14.35	13.86	14.40	23.17
25	11.35	11.60	11.54	11.03	21.12
26	11.74	11.05	11.78	11.42	21.20
27	4.04	4.79	4.21	4.97	12.97

Table 4 The S/N ratios of each parameter level of design variables

Level	Level 1	Level 2	Level 3	Level	Level 1	Level 2	Level 3
A/m	18.66	18.88	18.53	I/m	17.21	19	19.87
B/m	18.93	17.49	19.66	J/°	18.86	18.5	18.72
E/m	20.99	16.96	18.13	K/°	17.99	18.3	19.78
F/m	19.35	17.38	19.34	M/°	21.87	19.54	19.78
G/m	17.96	18.83	19.29	N/°	18.52	19.06	18.5
H/m	19.13	19.16	17.79	O/°	19.39	19.06	18.85

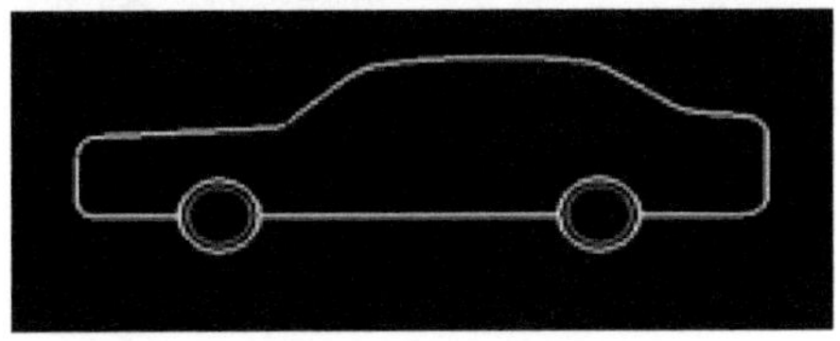

Figure 4 The sidewise two-dimensional contour of the best sedan

（4）Results analysis.

In order to check the effectiveness of the optimum results, a comparative study is made between optimization results and the initial design and calculates out the S/N ratio of every scheme. If the S/N ratio of optimization results is larger than the initial design's, it can confirm the validity of the optimization design.

In determining the optimal combination of parameters in level range three combinations of parameters are selected: one is the optimal combination above, another is close to the upper limit of the range of optimal combination level, and the last is near the lower limit of the range of optimal combination level. Then compare these three combinations with the original design.

Verification process of emotional evaluation and survey data analysis are similar to parameter designs, where sensory quality evaluation values and SN ratios of 4 schemes are obtained (see table 5). And then figure out the feeling discrepancy (FD) of each design, they are 8.88, 6.46 and 5.30 respectively. The value of feeling ambiguity (FA) in each scheme is 0.91, 0.45, and 0.50.

The datum above obtain that sensory quality of the product has improved a lot after optimization. Feeling discrepancy (FD) has reduced 35.92% on average, and each optimum design is decreased by 40.26%, 27.23% and 40.30% respectively. Therefore, every sensory quality of design scheme has been greatly improved. On the FA side, the value of feeling ambiguity (FA) has reduced by an average of 46.52%, and each optimum design is decreased by 50.56%, 43.31% and 45.17% respectively. So evaluators' sensitivity for each optimization design is lower than the original schemes.

Table 5 The sensory quality assessment values and S/N ratios of 4 sedans design schemes

Samples \ Evaluator group	G1	G2	G3	G4	S/N ratio
original scheme	12.34	11.17	11.24	10.01	20.91
optimum scheme 1	18.75	17.83	17.23	17.79	25.05
optimum scheme 2	16.17	15.44	16.18	16.35	24.09
optimum scheme 3	18.66	17.62	18.11	18.30	25.18

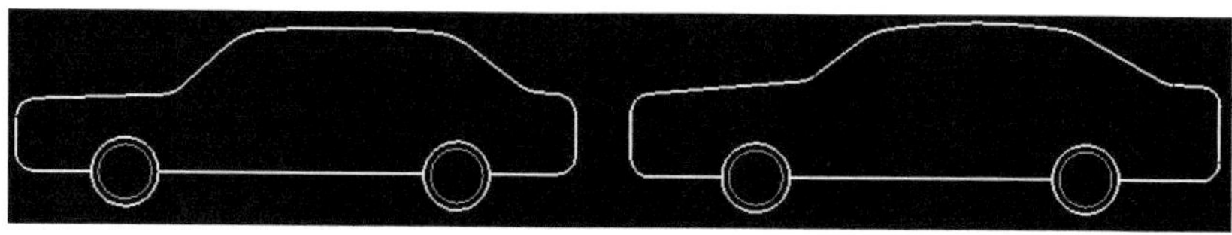

Figure 5 The sidewise two-dimensional contour of two selected sedans

5 CONCLUSIONS

Based on a lot of technical literature, this paper introduces sensory quality evaluation index of products-sensory factor and sensory quality level measure indexes feeling discrepancy (FD) and feeling ambiguity (FA), establishes a product sensory quality evaluation method and proposes an optimization designing method of products sensory quality method based on integrated model of the QFD and Taguchi Method. Finally, the paper makes use of the proposed method to optimize the sensory quality of a sedan. The result indicates that the sensory quality has been greatly improved, which verifies the effectiveness of the optimization design model.

ACKNOWLEDGMENTS

This work was supported by the National Natural Science Foundation of China (No. 70771022 and No. 71171041).

REFERENCES

Eysenck, H. J. 1986. Conceptualizing involvement. Journal of Advertising, Vol.15, 4-14.

Eysenck, H. J. & Eysenck, S.B.G. 1975. Manual of the Eysenck Personality Questionnaire. London: Hodder and Stoughton.

Eysenck, H. J. & Gudjonsson, G. H. 1989. The Causes and Cures of Criminality. New York: Plenum Press.

Guo, F., Liu, G. Y., Chen, C. & Li, S. 2009. Customer-oriented study on design support technology base on perception of car style. Journal of Northeastern University (Natural Science), Vol.30(5), 741-744.

Holbrook, M. B. & Hirschman, E. C. 1982. The experiential aspects of consumption: consumer fantasies, feeling, and fun. Journal of Consumer Research, Vol.12(9), 132–140.

Kolter, P. 1992. Marketing management: analysis planning implementation and control, Englewood: Cliffs.

Lai, H. H., Chang, Y. M., & Chang, H. C. (2005). A robust design approach for enhancing the feeling quality of a product: a car profile case study. International journal of industrial ergonomics, Vol.35, 445-460.

Li, D. Q. 2006. A study on usability design of Kansei quality. Journal of World Standardization & Quality Management, Vol.8, 19-22.

Li, D. Q. & Wei, D. P. 2006. A study on technique for Kansei quality control. Journal of Industrial Engineering and Management, Vol.6, 118-123.

Norretranders, T. 1998. The User Illusion:Cutting consciousness Down to Size. New York: Penguin Press Science.

Su, J. N. & Li, H. Q. 2005. The perceive feature of material in industrial design. Journal of Machine Design and Research, Vol.3, 13-14.

Xu, S. H. & Ye, D. H. 2006. Industrial product design and fashionable color. Journal of Packaging Engineering, Vol.27(2), 205-207.

Zaichkowsky. 1994. The personal involvement inventory: reduction, revision and application to advertising. Journal of Advertising, Vol.23(4), 59-70.

CHAPTER 34

Geometric Dimension Model of Virtual Human for Ergonomic Analysis of Man-machine System

Zhou Qianxiang[1] *Ding Songtao*[2] *Liu Zhongqi*[1] *Zheng Xiaohui*[1, 2]

(1 Key Laboratory of Mechanobiology and Biomechanics, Beihang University, Ministry of Education, Beijing, 100191)
(2 Research Institute of Chemical Defense, Beijing 100191)
Email:zqxg@buaa.edu.cn

Abstract

It is very important to clarify the geometric characteristic of human body segment and constitute analysis model for ergonomic design and the application of ergonomic virtual human. The typical anthropometric data of 1122 Chinese men aged 20~35 years were collected using three-dimensional laser scanner for human body. According to the correlation between different parameters, curve fitting were made between seven trunk parameters and ten body parameters with the SPSS 16.0 software. It can be concluded that hip circumference and shoulder breadth are the most important parameters in the models and the two parameters have high correlation with the others parameters of human body. By comparison with the conventional regressive curves, the present regression equation with the seven trunk parameters is more accurate to forecast the geometric dimensions of head, neck, height and the four limbs with high precision. Therefore, it is greatly valuable for ergonomic design and analysis of man-machine system.

Keywords: Ergonomic virtual human, anthropometry, multiple regressions, ergonomics design, correlation analysis

1 Introduction

In these days, there is a common idea to carry out ergonomic design and evaluation for better safety and health before batch production. Accordingly, anthropometric data acting as the basis for fitting tests are indispensable. Anthropometric data are human physical characteristics, including the basic dimensions (such as height, sitting height, arm length, foot length, etc.), body shape, surface area, volume and weight. These data collected by standard measurement are also important for ergonomics design and analysis software (Jack, Anybody, RAMSIS, SAFEWORKA, etc.).

Ergonomics virtual human is human model constituted for human-machine system, and act as the substitute of real man in virtual man-machine system. The main study focuses on the geometric representation, motion/movement programming, action expression and cognition of virtual human body. The characters of virtual human are as follows[1, 2, 3].

The virtual human emphasize on the geometric and the character of limb movement. Meanwhile, the organs and tissue inside the body are not included when modeling.

The virtual human can represent the characters of the real users.

The oral data for modeling the virtual human are percentage data of samples, which usually be 5%、50%、95%、99%.

At the present time, geometric models are main investigation in ergonomics design and analysis. As we all know, human body is a harmonious integer, and it grows under neural control. Accordingly, there are some relationships between these geometric dimensions. During the Revival of Learning in Italy, Leonardo da Vinci found that the relationship in human body, and pained the famous canvas named "Weiteluweiren". In this canvas, the man expands the limbs to be a cross. Navel can be deemed as the central point. Many dimensions appear to be the golden section scale 0.618, such as height and the distance between navel to the ground, shoulder to the point of finger and elbow to the point of finger, hip to the ground and knee to the ground[4, 5]. The previous study shows that the head and facial dimensions also relate to other body dimensions. The researchers can use these relationships to build human body model, and the result to instruct the ergonomics design and analysis of plane, crew exploration vehicle, costume, furniture and working space. At the present application, height acts as independent variable, and other dimensions act as dependent variable. Because of the distinct errors, adverse effect may be induced. Then, more accurate human body models are necessary [7].

The appearance and parameters of ergonomics human model are characterized in Fig. 1. The body is divided into some units and the model is constituted based on the lengths, circumference of different units and relationships between them.

The emphases of this study lies in the relationships between different dimensions. The anthropometric data were acquired using the standard methods, and model is constituted using the curve fitting of different dimensions.

In past study, human dimensional models were usually characterized using height as independent variable and other parameters as dependant variable to get fitting curve, as shown in Fig. 1 [4]. During application, this simple fitting model induced obvious errors and greatly affects the reliability of accessibility and spatial layout. To resolve this problem, Trotter and Gleser collect a lot of army men samples including while men and black men who died during World War II, and measured the length of humerus, femur, ulna, radio, tibial and fibula. The fitting curves of height and length of bones are obtained and it was found that the lengths of femur and tibial are the optimum parameters to predict the height. Hereby, a reasonable can be obtained using femur and tibial as independent parameters.

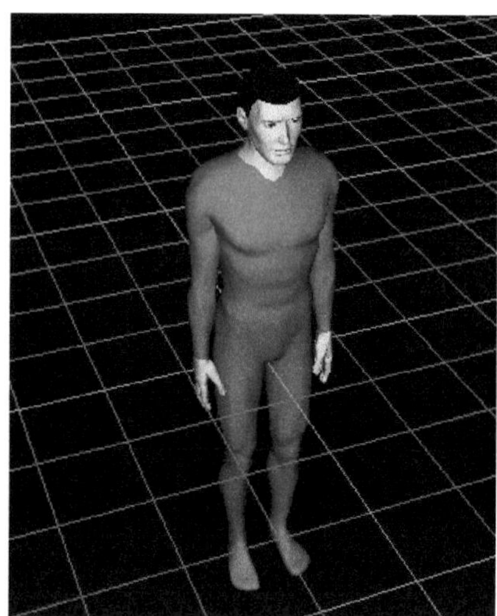

(a) the appearance of virtual human

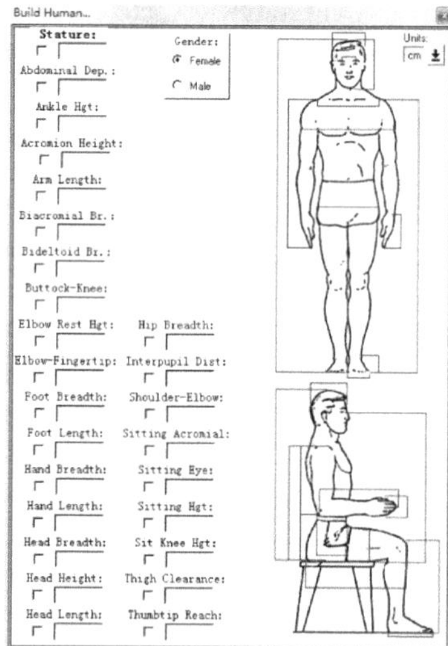

(b) the definition of dimensions of human model

Fig 1 Jack Virtual Human and its parameters

Based on the results in literatures, it can be concluded that trunk is an important part of human body, which includes most of important tissues and organs. Growth and development of trunk affect greatly on enginery of body, and affect the dimensions of trunk and other tagmatas. Accordingly, not only height, but also other parameters should be considered to constitute the dimensional model of human body.

Table 1 Scale of different parameters with height

No.	Items	Man	Woman	No.	Items	Man	Woman
01	Eye height	0.93H	0.93H	09	Foot length	0.15H	0.15H
02	Shoulder height	0.81H	0.81H	10	Arm extending length	1.10H	0.99H
03	Elbow height	0.61H	0.61H	11	Finger lift	1.26H	1.25H
04	Middle finger tip height	0.38H	0.38H	12	Sitting height	0.54H	0.54H
05	Shoulder breadth	0.22H	0.22H	13	Lower extremity length	0.52H	0.52H
06	Upper limb length	0.172H	0.172H	14	thigh length	0.232H	0.232H
07	forearm limb length	0.157H	0.14H	15	shank length	0.247H	0.247H
08	Hand length	0.11H	0.11H				
note:							

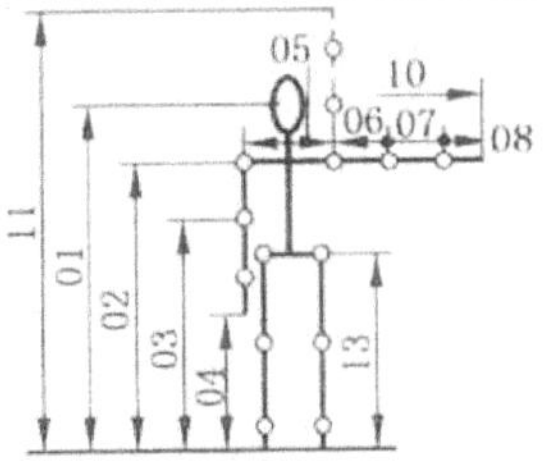

If trunk is deemed as the core of human body, and head, neck, upper limbs and lower limbs as the branches from this core. Using the dimensions of trunk, seven parameters which are waist circumference, chest circumference, crista iliaca breadth, shoulder breadth, 胸背距、hip circumference and chest breadth. And other dimensions as dependant, the regression equation fitting with the seven trunk size items can forecast the geometric size of head, neck, height and the four limbs with high precision.

Many literatures have also put forward this idea. For example, Ozaslan et al. have studied on the relationship of height and the dimensions of lower limbs (greater trochanter height, thigh length, shank length and knee height), and constructed a regression model of multi-variables to forecast height. Some researchers had forecasted height using the dimensions of skull, metacarpus, foot, hand and arm, and the result is acceptable [11].

2 Methods

2.1 The parameters

According to the national standard "Basic human body measurements for technological design" (GB/T5703-1999) and "Human dimensions of Chinese adults", 17 dimensional parameters, including waist circumference, chest circumference, crista iliaca breadth, shoulder breath, 胸背距, hip circumference, chest breadth, shank length, thigh length, head height, head length, upper limbs length, upper arm length, forearm length, height, lower extremity length and neck circumference.

2.2 The subjects

This study involved 1222 young men as subjects. All subjects were born in China mainland. Their ages ranged between 17-34 years, and the summary of native place distribution is shown in Table 2. The classification of native place and age is consistent with the method in GB 10000-88.

Table 2 Native place and age distribution of the subjects

Native place	Northeast-and North area	Southwest Area	Northwest Area	Southeast Area	Central Area	South Area
Number/Percentage	564/46.15%	106/8.67%	300/24.55%	78/6.38%	63/5.16%	111/9.09%
age(year)	17~19	20~24	25~29	30~34	20-25	26-32

2.3 The measurements

All the 17 anthropometry data were acquired from laser scanner VITUS SMART XXL. The measuring speed is quickly and the reliability is high to 1 cm. During measurement, the subjects should only wear special pants and special hat with all hair in hat. The measurement temperature is 20-25℃, and the surrounding noise is 40-50dB.

3 Construction of models

3.1 Correlativity between different dimensions of trunk

As mentioned above, the error of the model constructed based on height is obvious. In present study, a new idea was put forward to construct model using the dimensions of trunk. As we all know, waist circumference, chest circumference, rista iliaca breadth, shoulder breath, 胸背距、hip circumference and chest breadth are the main dimensions of human trunk. The correlativity between these dimensions and dimensions of other parts were calculation using SPSS16.0, and Table 3 shows the results. It can be known that neck circumference, head length, head height, upper limbs length, lower limbs length and dimensions of trunk correlate greatly (P<0.01).

3.2 Regression model of human tagmata dimensions

According to the statistic theory, the main indexes to value regression model are correlation coefficient R and standard error SE. Usually, the more R is, the greater independent variable influent on dependent variable, and the more accurate the regression model is. It is generally agreed that the model is acceptable when R is larger than 0.75. Meanwhile, the model is also acceptable when the standard error SE is small enough.

Table 3 Correlativity coefficient between trunk dimensions and dimensions of other parts

Variable	waist circum-ference	chest circum-ference	rista iliaca breadth	shoulder breath	胸背距	hip circumference	chest breadth
Shank length	0.53	0.55	0.62	0.66	0.58	0.67	0.56
Thigh length	0.48	0.55	0.57	0.69	0.56	0.65	0.58
Head height	0.61	0.61	0.65	0.62	0.63	0.72	0.62
Head length	0.64	0.68	0.67	0.68	0.64	0.75	0.70
Upper limb length	0.56	0.60	0.67	0.74	0.63	0.74	0.64
Upper arm length	0.53	0.58	0.63	0.66	0.60	0.69	0.60
Forearm length	0.43	0.46	0.54	0.68	0.47	0.60	0.53
Height	0.58	0.63	0.69	0.76	0.66	0.78	0.66

Variable	waist circum-ference	chest circum-ference	rista iliaca breadth	shoulder breath	胸背距	hip circumference	chest breadth
Lower extremity length	0.60	0.66	0.70	0.77	0.67	0.77	0.68
Neck circumference	0.89	0.86	0.87	0.69	0.85	0.90	0.84

PS： hypothesis inspection level is p<0.01。

Based on the correlativity of dimensions of human body, new models were calculated and shown in Table 4. The R values of all models are larger than 0.75, which means the reliability of these models is acceptable.

Table 4 Regression models of human body dimensions

Variable	Regression equation	R	SE(mm)
Shank length	102.352-0.17*W1+0.43*W3+0.339*W4+0.205*W5+0.179*W6-0.183*W7	0.78	17.0
Thigh length	152.33-0.133*W1+0.431*W4+0.29*W6	0.84	28.1
Head height	90.716-0.027*W1+0.091*W4+0.141*W6	0.94	8.8
Head length	90.334+0.091*W4+0.074*W6	0.87	6.3
Upper limb length	232.052-0.342*W1+0.705*W3+0.481*W4+0.409*W6	0.88	26.0
Upper arm length	108.086-0.143*W1+0.308*W3+0.179*W4+0.176*W6	0.85	14.3
Forearm length	62.725-0.114*W1+0.215*W3+0.238*W4+0.11*W6	0.91	17.3
height	543.378-0.779*W1+1.306*W3+1.029*W4+1.038*W6	0.88	47.6
Lower extremity length	359.973+0.647*W3+0.297*W6	0.93	25.6
Neck circumference	67.35+0.072*W1+0.227*W5+0.143*W6+0.155*W7	0.85	10.9
Remark	waist circumference: W1; chest circumference: W2; rista iliaca breadth: W3; shoulder breath: W4; 胸背距: W5; hip circumference: W6; chest breadth: W7		

PS： hypothesis inspection level is p<0.01。

4 Discussion

The correlativity between different variables is the base for constructing regression models. It can be concluded that hip circumference and shoulder breadth are the most important regression variables. During these ten regression equations in Table 4, ten equations comprise hip circumference, and eight equations comprise shoulder breadth. In the 100 subjects, the differences between calculated results and measured results are between 0.6～5 cm. During these results, the largest standard error is height 4.76 cm, and the least is head length 0.6 cm. The standard error of head height and neck circumference is about 1 cm, and those of the rest 6 variables are 1.4～3cm. In literature [4] and [12], only thigh length, shank length, forearm length and upper arm length models were studied using height as dependant variable. To compare with the present research, five subjects were chose stochastically to be measured, and error percentage is adopted to value the difference. The results are shown in Table 5 and Table 6. The calculated equation for error percentage is as follows.

$$\text{error percentage} = \frac{\text{calculated value} - \text{measured value}}{\text{measured value}} \times 100\%$$

Table 5 Calculated results of different models

Measured Value (mm)								
Sample No.	Waist circum-ference	Chest circum-ference	Rista iliaca breadth	Shoulder breath	Chest and back margins	Hip circum-ference	Chest breadth	Height
1	718.0	1012.7	265.8	419.3	216.1	929.0	322.0	1698.9
2	721.7	943.8	281.0	429.4	228.6	924.0	322.0	1781.2
3	767.6	973.9	280.0	407.1	221.8	986.0	326.0	1724.7
4	683.9	864.1	274.5	384.9	197.4	861.0	290.0	1627.0
5	698.1	893.2	271.1	400.2	207.7	899.0	295.0	1731.6

Calculated value (mm)						
		Sample 1	Sample 2	Sample 3	Sample 4	Sample 5
Thigh length	MV	498.3	523.4	476.6	469.3	462.2
	CV1	507.0	509.4	511.6	477.0	492.7
	CV2(0.232*H)	394.1	413.2	400.1	377.5	401.7
Shank length	MV	379.1	393.6	375.5	350.2	389.9
	CV1	388.4	399.4	392.6	376.1	385.4

	CV2 (0.247*H)	419.6	440.0	426.0	401.9	427.7
Upper limb length	MV	205.7	231.1	224.0	209.3	220.2
	CV1	240.0	244.7	240.8	230.1	235.6
	CV2 (0.157*H)	266.7	279.6	270.8	255.4	271.9
Upper arm length	MV	332.2	355.7	328.5	317.7	314.1
	CV1	325.8	330.9	331.0	315.3	321.6
	CV2 (0.172*H)	292.2	306.4	296.6	279.8	297.8

注：H: height of sample; MV: Measured value; CV1: Calculated value using the models in present paper; CV2: Calculated value using the models in literatures

Table 6 Difference between measured value and the calculated value using two methods

			Sample1	Sample 2	Sample 3	Sample 4	Sample 5	Mean	Variance
Thigh length	CV1	Percentage (%)	1.74	2.68	7.35	1.63	6.59	4.00	2.47
		Value (mm)	8.66	14.02	35.04	7.65	30.48	19.17	11.39
	CV2 (0.232*H)	Percentage (%)	20.90	21.05	16.04	19.57	13.08	18.13	3.10
		Value (mm)	104.16	110.16	76.47	91.84	60.47	88.62	18.20
Shank length	CV1	Percentage (%)	2.45	1.47	4.55	7.40	1.15	3.40	2.32
		Value (mm)	9.29	5.79	17.07	25.92	4.47	12.51	8.01
	CV2 (0.247*H)	Percentage (%)	10.69	11.78	13.45	14.75	9.70	12.07	1.83
		Value (mm)	40.53	46.36	50.50	51.67	37.81	45.37	5.43
Forearm length	CV1	Percentage (%)	16.68	5.89	7.49	9.94	6.98	9.39	3.88
		Value (mm)	34.30	13.60	16.77	20.79	15.37	20.17	7.46
	CV2 (0.157*H)	Percentage (%)	29.65	20.99	20.89	22.03	22.03	23.12	3.30
		Value (mm)	61.00	48.50	46.80	46.10	51.70	50.82	5.44
Upper arm length	CV1	Percentage (%)	1.92	6.97	0.75	0.77	2.39	2.56	2.30
		Value (mm)	6.36	24.78	2.47	2.43	7.52	8.71	8.29
	CV2 (0.172*H)	Percentage (%)	12.04	13.87	9.70	11.92	5.18	10.54	2.99

			Sample1	Sample 2	Sample 3	Sample 4	Sample 5	Mean	Variance
		Value (mm)	39.99	49.33	31.85	37.86	16.26	35.06	10.95
Height	CV1	Percentage (%)	1.65	1.80	1.67	1.96	1.91	1.80	0.12
		Value (mm)	28.05	32.08	28.77	31.90	33.02	30.76	1.97

It could be concluded that the reliability of the regression prediction is higher than that in literature [4] and [12]. The differences between measured values and calculated values using the present method are lower than 10%. Meanwhile, the differences between measured values and calculated values using the method in literature are close to 23.12%. Ozaslan et al. had forecast height using thigh length, shank length, foot length, foot breadth, knee height and greater trochanter height, and the average difference between calculated value and measured value is 5 ~ 6cm [10]. You et al. had studied the anthropometric data of US. Army men, and constituted graded evaluate model based on the geometric relationship of different sections of human body. This model can calculated 60 anthropometric dimensions and the error covers 0.4-4.8cm. The error of the model in present paper is lower than 2.1 cm. Consequently, the reliability of human body models in this paper is obviously high, and the model is practicable for the ergonomic analysis and evaluation.

5 Conclusion

A novel model is put forward to forecast the anthropometric dimensions of human body using the dimensions of trunk. The dimensions of trunk correlates with that of other part of the body, and hip circumference and shoulder breadth are the most important parameters. The model is constructed using multi-parameters, and the reliability is higher than other method, inspection level is $p<0.01$. The number of selected subjects is large enough to satisfy the requirement of statistics. As a result, the model can be applied to the modeling of Chinese young men and be good for ergonomic design and evaluation.

Furthermore, the subjects in this study came from all over Chinese mainland, and represent the whole characters of Chinese young men. But, the model should be different when applied to people of different race, gender and age. To get higher reliability, this model should be modified for special people group.

Reference

[1] Guo Dandan, Study on Virtual human for man-machine system design, Master paper of Tianjing university, 2004：3-11。

[2] Ke G T., Solving inverse kinematics constraint problems for highly articulated models, University of Waterloo, 2000, CS-2000-19：2-8.

[3] He Yuesheng, Research and Implementation of a Virtual Human Software Platform for

Equipment Maintenance, Master paper of national defense university of technology, 2005：1-2.

[4] Tong Shizhong, Ergonomic design and application manual，2007.7，Press of china standard: 79～82.

[5] Gao Xiulai, anthropotomy，2009.04，Press of Peking university:1-2.

[6] Shu Shuyuan, Zheng Lianbing, Lu Xunhua, Apreliminary study on correlation among somatometry items on human face, Journal of Anatomy, 2001, 24(2): 176～178。

[7] Liu Zhongyu, Zheng Qi, Cao Dongnin, Correlation study of measuring items between head facial part and bodily part in human, Journal of Tianjin normal university-natural science edition, 1999, 19(1): 40～45.

[8] Raxter MH, Auerbach BM, Ruff CB. Revision of the Fully Technique for Estimating Statures. American journal of physical anthropology. 2006, 130:374～384

[9] Burke RM. Can we estimate stature from the scapula? A test considering sex and ancestry. A thesis of the Louisiana State University，2008.

[10] Ozaslan A，Iscan MY, Ozaslan I, et al. Estimation of stature from body parts. Forensic Science International, 2003,3501,1～6.

[11] Shahar S&Bsc NSP. Predictive equations for estimation of stature in Malaysian elderly people. Asia Pacific J Clin Nutr, 2003,12(1):80～84.

[12] Wang Enliang, Industrial engineering manual, 2006.1, Machine press: 273～275.

[13] You H. and Ryu T., Development of a hierarchical estimation method for anthropometric variables. International Journal of Industrial Ergonomics，2005，35：331～343.

CHAPTER 35

Influence of Interface Compatibility on the Spacecraft Tracing Performance in Vibration Environment

Chunhui Wang, Ting Jiang, Peng Teng

Science and Technology on Human Factors Engineering Laboratory
China Astronaut Research and Training Center
Beijing, China
jtingx@yahoo.com.cn

ABSTRACT

In the tracking task in vibration environment, matching extent of Man-Machine interface is one of an important factor of work performance. Study analyses that the important factors of Man-Machine interface design influenced by vibration environment, and build that matching extent evaluation method of Man-Machine interface. Semi-physical simulation system was built, man-in-the-loop typical tracking work was designed, and design indices including visual field, dimension of target, shape of target, and ratio of display-control, were selected. To evaluate the comprehensive level of task performance, some indices including tracking result, tracking precision, and tracking process, were adopted. Based on the index characteristics, multilevel comprehensive evaluation model including expert evaluation model and the entropy analysis was conducted. The result showed that 4 design indices in quiet and vibration environment have statistical difference. As a result, in the ergonomic design and evaluation of Man-Machine Interface, it is important to study the matching method in vibration environment.

Keywords: Man-Machine interface, vibration environment, comprehensive evaluation model

1 INTRODUCTION

With the development of manned space flight, astronauts operating increasingly complex, such as manual control of spacecraft rendezvous and docking, the landing of the lunar module, lunar surface operations, mechanical arm operation. In these tasks, Astronauts need to be completed precise positioning and tracking in the environmental conditions of the vibration. Vibration environment will increase the task difficulty, reduce operating precision, from the perspective of the man-machine interface design, human-computer interface design elements and the match is the important factor affecting the operating performance. Good man-machine interface to help improve the performance of astronauts' precise control of the task. Taking tracking task as a background, this paper analyzed the influence of vibration environment and human-machine interface design elements on the people operating performance, and built the man-machine interface evaluation methods.

2 METHOD

Aimed at the type tracing task, the purpose of this study was to explore: the effect of vibration environment to participants' performance, and the effect of the type design indices extracted from man-machine interface to participants' performance.

2.1 Participants

Sixteen males between the ages of 20 to 35 years participated in the study. All participants have normal visual acuity. We got informed consent from all participants before the experiment.

2.2 Procedure

Participants were required to complete a type tracing task. First, participants can see the tracing target in the simulate scene of semi-physical simulate system, control orientation handle to trace the target, push the locking button when locking the target.

2.3 Measurement Equipment

The function of Semi-physical simulation system was that offering the vibration environment, and display-control loop of tracing task by simulating. This system has three parts:

- Vibration experiment plant: It's used to complete vibration environment. One part was movement plant with six freedoms that can simulate wide range and low frequency vibration like swing and undulation, another part was vertical vibration plant that can simulate high frequency vibration. The functional parameters of movement plant with six freedoms include: the max load absorption was 5t, the max vibrant range was ±50cm, the max velocity range was ±50cm/s, the max acceleration range was ±1.0g. The function parameters of vertical vibration plant include: the max load absorption was 500kg the vibrant frequency range from 2Hz to 80Hz, the vibrant acceleration range was 10g±2g.

Figure.1. Vibration experiment plant

- Simulate Display equipment: it's used to simulate the scene image, tracing target, and assistant information by virtual reality technology. The parameters of design indices can be imputed and modified based on the requirement of experiment.
- Simulate control equipment: the control handle was used to complete the control loop of six orientations.

2.4 Design

Environment condition was the first variable under investigation, and consisted of two levels: quiet environment and vibration environment. Design indices of man-machine interface was the second variable under investigation, Four independent variables were manipulated: visual field, dimension of target, shape of target, and ratio of display-control (tab.1.).

Table 1 Experimental parameter table

Environment condition	Name	Visual field	Shape of target	Dimension of target	Ratio of display-control
Quiet environment	T-1	10°	Triangle	10 mm	1.2
	T-2	10°	Cross	6 mm	1
	T-3	10°	Cross-circle	4 mm	0.8
	T-4	6°	Triangle	6 mm	0.8
	T-5	6°	Cross	4 mm	1.2
	T-6	6°	Cross-circle	10 mm	1
	T-7	4°	Triangle	4 mm	1
	T-8	4°	Cross	10 mm	0.8
	T-9	4°	Cross-circle	6 mm	1.2
Vibration environment	T-1	10°	Triangle	10 mm	1.2
	T-2	10°	Cross	6 mm	1
	T-3	10°	Cross-circle	4 mm	0.8
	T-4	6°	Triangle	6 mm	0.8
	T-5	6°	Cross	4 mm	1.2
	T-6	6°	Cross-circle	10 mm	1
	T-7	4°	Triangle	4 mm	1
	T-8	4°	Cross	10 mm	0.8
	T-9	4°	Cross-circle	6 mm	1.2

Tracing target was moved at invariable velocity that random setting from 0 m/s 、3 m/s、 5 m/s and 7 m/s. Tracer was moved at invariable velocity (4 m/s), come full circle according to the M shape from left to right.

3 EVALUATE INDEX SYSTEM AND SYNTHETIC EVALUATE MODEL

3.1 Evaluate Index System

To evaluate the influence of vibration environment and Man-Machine interface design parameters on tracing task, the performance of subject task must be evaluated. It's difficult to evaluate the performance by one or two Indices, because the tracing process fell into some phases and aspects. It's have to build evaluate index system that have some aspects and some hierarchy, to comprehensively and synthetically evaluate the performance.

Generally, tracing task require subject that the tracing process must be "faster" and tracing accuracy must be "precise". Through the analysis of requests and characteristics of tracing task, the evaluate index system was divided into two levels and three categories by Analytic Hierarchy Process (AHP), as shown in Table. 2. Based on the phases of tracing task, the first level of index system contains Tracing process, Tracing accuracy and Tracing result. And based on analyzing the characteristic of every phase and the consulting result of experts, five indices were confirmed in the second level of index system.

Table.2 Evaluate index system of tracing task performance

Level 1	Level 2	Commentary
Tracing process	Response time	The time from target appearing to subject starting control
	Tracing time	The time from subject starting control to locking target
	Locking time	The time from subject locking target to pushing button
Tracing accuracy	Locking deviation	The deviation between the center of locking maker and the center of target when target locked
Tracing result	Tracing result	Towards a tracing task, the result based on the successfully or unsuccessfully locking target

Looking at table 2 we can immediately see that:

Tracing process was described by time indices, and reflect the "faster" characteristic of tracing task;

Tracing accuracy was described by deviation indices, and reflect the "precise" characteristic of tracing task;

Tracing result was described by successful or unsuccessful result.

3.2 Synthetic Evaluate Model

The purpose of evaluate model was to assign the indices' weight of index system, and calculate the value of synthetic evaluate result. Based on the different characteristic of indices, the different evaluate methods were introduced in synthetic evaluate model.

- The First level of Performance Model

Tracing result was the primary condition of tracing task, so multiplication was applied between tracing result with tracing accuracy and tracing process. The tracing accuracy and tracing process reflect the characteristic of two phases each other, addition was applied between the tracing process and tracing accuracy.

The equation for the first level of performance model is given as follows:

$$S = F_1 * (w_2 * F_2 + w_3 * F_3) \tag{1}$$

Where S is the value of performance, F_1、 F_2 and F_3 are the values of tracing results, tracing accuracy and tracing process, w_2 and w_3 are the weight coefficient of tracing accuracy and the tracing process.

In the paper, the weight coefficient of tracing accuracy and tracing process were analyzed by expert assign method. The values of the coefficients were:

$$w_2 = w_3 = 5.0$$

- The second level of Performance Model

The value of the tracing result was judged by the successfully or unsuccessfully complete the tracing task. The equation for the tracing result of performance model is given as follows:

$$F_1 = \begin{cases} 1 & successful \\ 0 & unsuccessful \end{cases} \tag{2}$$

The Tracing process was described by time indices, and time indices is coordinate so that addition was applied, The equation for the tracing process of performance model is given as follows:

$$F_3 = \sum_{i=1} w_{3i} * x_{3i} \tag{3}$$

Where F_3 is the value of tracing process performance, x_{3i} is the values of the second level indices, w_{3i} is the weight coefficient of second level indices. Correlativity between the tracing accuracy and tracing process was low, so the weight coefficients were analyzed by the entropy analysis method (table.3.).

Table.3 The weight coefficient of Evaluate Performance Model

Level 1	The weight coefficient	Level 2	The weight coefficient
Tracing process	0.5	Response time	0.23
		Tracing time	0.36
		Locking time	0.41
Tracing accuracy	0.5	Locking deviation	-

4 ANALYSIS AND RESULTS

The interviews were analyzed to address following research questions.

4.1 Environment condition

In Figure 2, the Descriptive statistics of performance's values are shown in the context of quiet and vibration environment.

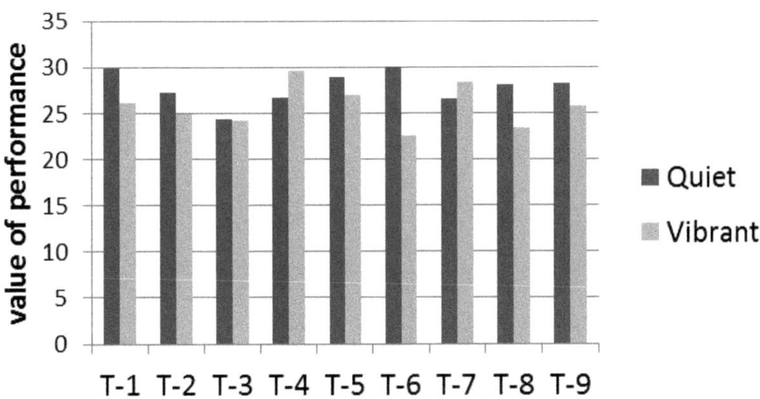

Figure.2. the value of performance of quiet and vibration environment

As a result, there was difference in response time, locking deviation and performance value at the quiet and vibration environment. The main effect of according to Environment condition was significant ($p<0.05$). There was a significant Decrease from quiet environment to vibration environment for performance value (Figure. 2).

4.2 Design indices of man-machine interface

The experiments selected four design indices of man-machine interface, to analyze the operating performance of these design indices. In Figure 3, the Descriptive statistics of performance's values are shown in the context of four design indices.

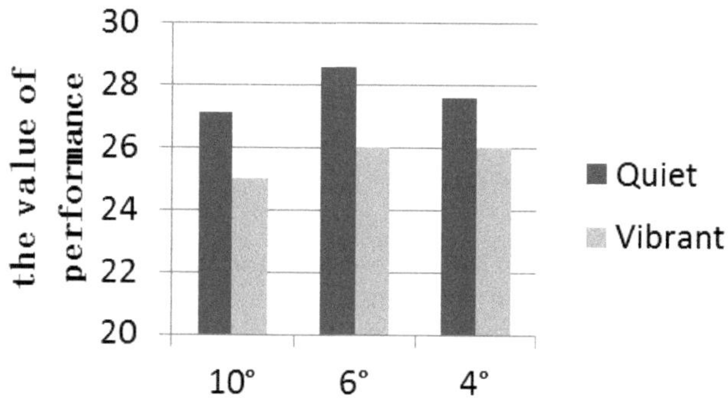

(a) Visual field

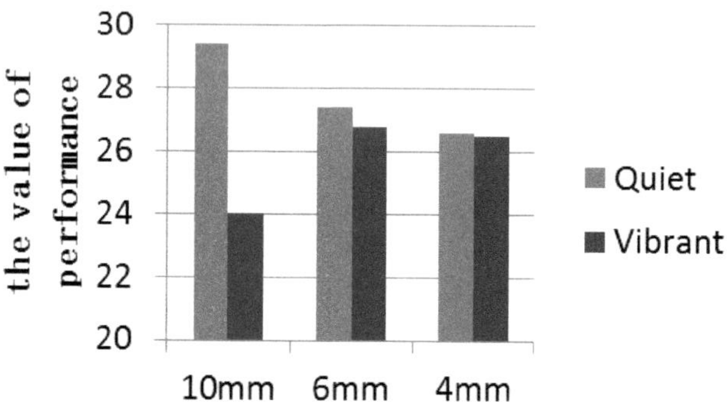

(b) Dimension of target

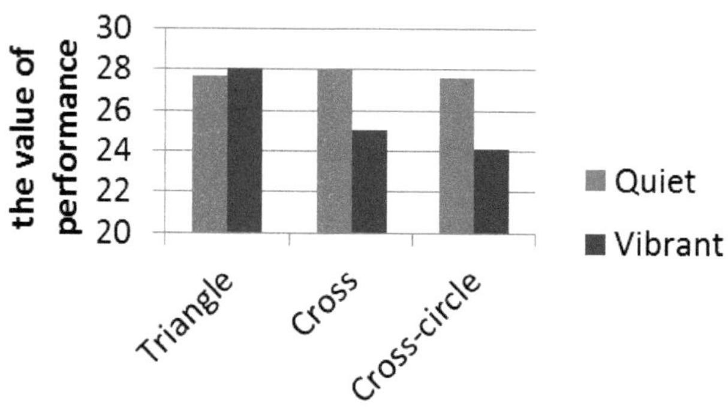

(c) Shape of target

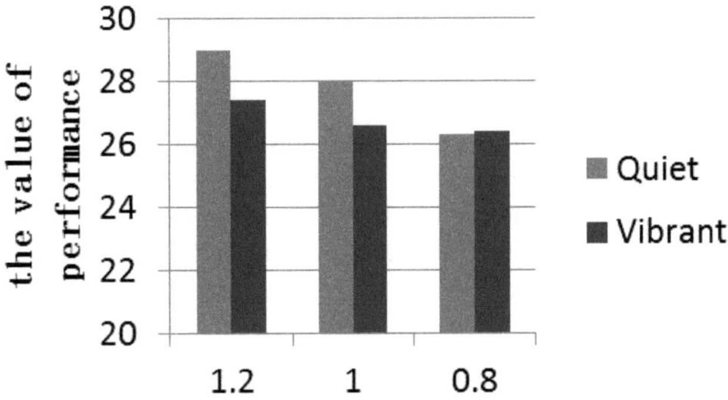

(d) Ratio of display-control

Figure.3. the value of performance of the human-computer interface design indices

- Visual field: As a result, there was difference in tracing time and performance value at the different visual field. The main effect of visual field was significant ($p<0.05$). In the quiet and vibration environment , and the influence of visual field on performance laws are consistent, 6° >4° >10° . This shows that visual field will affect the performance of tracking task, and is the effective factors in human-computer interface design.
- Dimension of target: As a result, there was difference in response time and performance value at the different dimension of target. The main effect of dimension of target was significant ($p<0.05$). In the quiet and vibration environment, and the influence of dimension of target on performance laws are inconsistent. This shows that dimension of target will affect the performance of tracking task, and is the effective factors in human-computer interface design.
- Shape of target: As a result, there was difference in response time and performance value at the different shape of target. The main effect of shape of target was significant ($p<0.05$). In the quiet and vibration environment, and the influence of shape of target on performance laws are inconsistent. This shows that shape of target will affect the performance of tracking task, and is the effective factors in human-computer interface design.
- Ratio of display-control: As a result, there was difference in Locking time, locking deviation and performance value at the different ratio of display-control. The main effect of ratio of display-control was significant ($p<0.05$). In the quiet and vibration environment, and the influence of ratio of display-control on performance laws are consistent, 1.2>1>0.8. This shows that ratio of display-control will affect the performance of tracking task, and is the effective factors in human-computer interface design.

5 CONCLUSIONS

This study has proven that not only vibration environment have impacts on tracing task performances, but also Design indices of man-machine interface do so. In particular, visual field, dimension of target, shape of target, ratio of display-control has significant influences on tracing task performance. Hence, it is necessary to consider the influence of vibration environment and Design indices in design tracing task. Moreover, this study established three stage performance evaluation index system, include tracing result, tracing accuracy and tracing process. Based on indices characteristics, multilevel Comprehensive evaluation model including expert evaluation model and the entropy analysis was built. This model can reflect the operating performance level, and provide a strong technical support on the ergonomic evaluation of the tracking task.

REFERENCES

HE Xiaoqun. 1998. The Methods of Modern Statistics Analysis. *Beijing: Press of Renming University of China.*

Saaty TL. 1994. Fundamentals of Decision Making and Priority Theory with the Analytic Hierarchy Process. *Princeton: RWS Publication.*

Levine P, Pomerol J C. 1986. An interactive program for choosing among multiple attribute alternatives. *European Journal of Operational Research* , 25:272-280.

Coury B G, Terranova M. 1991. Collaborative decision-making in dynamic systems. *In: Proceedings of the Human Factors Society 35th Annual Meeting,* 944-948.

George W R G. 1999. Nonlinear decision weights in choice under uncertainty. *Management Science,* ,45(1):74-85.

Jiang L. 1996. Economic entropy and its application to the structure of the transport system-Quality & quantity. *International Journal of Methodology,* 30(2):161-171.

0 Friends of Florida. Gainesville: University of Florida.

CHAPTER 36

Design and Evaluation of a Standing Platform for Reducing the Physical Strains Faced by Workers on Guyed Telecommunication Towers

Hsieh-Ching Chen[a], *Tan-Long Lin*[b], *Cheng-Lung Lee*[c], *Chih-Yong Chen*[d]

[a]National Taipei University of Technology, Taipei, Taiwan
[b]Chunghwa Telecom Co. Ltd., Taichung, Taiwan
[c]Chaoyang University of Technology, Wufong, Taiwan
[d]Institute of Occupational Safety and Health, Council of Labor Affairs, Taiwan

ABSTRACT

This study designs a novel platform to facilitate work on a guyed telecommunication tower by providing additional foot support to workers. A data logger with six electromyography (EMG) electrodes was used to record the subject exertions of the gastrocnemius, tibialis anterior, and paraspinal muscles during the performance of antenna mounting on an imitated quadratic tower with or without the platform. Six experienced and nine inexperienced subjects were tested. Muscular loads were assessed based on subject ratings of perceived exertion (RPE) and the 50th percentile of the amplitude-probability distribution of the normalized EMG of individual muscles. Subjects using the platform had considerably lower RPE in their lower back ($p<0.001$) and lower limbs ($p<0.001$). When the platform was used, gastrocnemius EMG was reduced by approximately 40% (3 - 4% MVC). The overall and imbalanced muscular loads of bilateral gastrocnemius were also reduced significantly by approximately 40% ($p<0.05$) and 52% ($p<0.01$) while using the platform. Results of this study demonstrate that providing an adequate platform can reduce overall and imbalanced loads in lower limb muscles under conditions where the worksite restricted worker foot placement.

Keywords: Electromyography, physical workload, perceived exertion

1 INTRODUCTION

According to Taiwanese statistics reported by the Directorate General of Telecommunications, Ministry of Transportation and Communications, 21,300 second-generation (2G) base stations and telecommunication towers were constructed in 1994–2000 (Environmental Protection Administration, 2001). According to the number of employees at Chunghwa Telecom Co. Ltd., the largest telecommunication company with a 40% market share in Taiwan, it is estimated at minimum 2000 workers in Taiwan are engaged in installing or repairing antennas. Additionally, over 2,000 towers are estimated to be relocated or newly erected every year. Work on elevated telecommunication towers has become a growth occupation. Consequently, it is rational to expect that the total number of workers involved in tower construction and maintenance is increasing. The safety and performance of workers on communication towers could be affected by environmental factors such as wind, rain, ice, snow, insects, and temperature induced heat stress (ComTrain LLC, 2003; Chad and Brown, 1995). Additionally, Mao *et al.* (2000) reported that high-elevation construction work incurred significantly more reports of dyspnea, low back pain, and unsteadiness while standing than did ground-level work.

Most telecommunication towers in metropolitan areas in Taiwan are guyed towers, unlike the 15 to 50m high self-supporting towers located in suburban districts, because of limitations of space, construction costs, and miscellaneous building permits. These towers are generally positioned on top of tall buildings for better transmission, and are constructed from solid round steel bars, welded together in 4 to 8 m sections. Most of the work performed on these towers is performed in a standing position, with a constrained posture. Based on our worksite observations and informal interviews with 17 experienced tower workers, the main physical workloads of high-elevated workers may include the mechanical work of climbing towers carrying equipment, prolonged muscle strain introduced by extreme and specific operating postures, and muscular loads caused by repetitive movements. Experienced tower workers have stated that continual climbing high towers is physically demanding, can cause soreness and pain in the back and lower extremities when performed for an extended period, and may contribute to workplace accidents, including falls.

Despite the high workload faced by workers on communication towers, no assistance devices were found for alleviating the physical strains resulting from the constrained postures in which they worked. This study designed a novel foldable platform to provide additional foot support for workers performing tasks on guyed towers. The present study evaluated the effectiveness of the platform in reducing muscular loads of the telecommunication tower workers. The aim was to compare worker muscle exertion when installing a cellular antenna on a tower with and without the platform.

2 MATERIALS AND METHODS

2.1 Design of standing platform

A foldable platform with a hollow steel frame and aluminum alloy plate was designed and built (Fig. 1). The platform weighed 3.6kg and can be easily carried and set-up with one hand. The frame was hooked to permit it to be hung on the rungs of a tower. The platform was fitted with a chain and hooks to prevent it from slipping or dropping owing to shear loading or worker carelessness. Since the platform is foldable and the platform width is less than half the rung width, workers can freely step down to a lower rung.

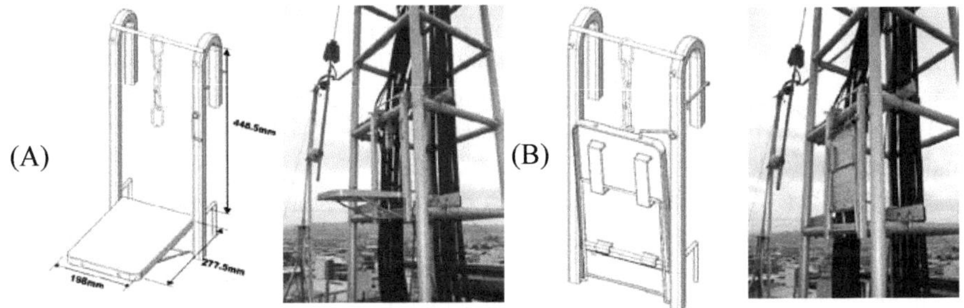

Figure 1. The platform design and installation (A) unfolded (B) folded

2.2 Subject

Fifteen healthy male subjects, divided into two groups of nine inexperienced postgraduate students and six experienced workers of a local telecommunication company, were recruited for this study. Prior to this investigation, no subject reported having a history of cardiovascular and musculoskeletal disorders. The mean age of the inexperienced subjects was 24.6 years (range 23 - 27 years) and the mean age of the experienced workers was 40.3 years (range 30 - 51 years) (Table 1).

Table 1. Overview of subject characteristics (Mean±SD)

Group	N	Age (yr)**	Height (cm)	Weight (kg)
inexperienced	9	24.6±0.4	171.8±2.1	74.4±4.0
experienced	6	40.3±3.8	176.5±2.0	75.0±1.8

** p<0.01 for significant group difference

2.3 Experiment setup and procedure

A 2.4 m-tall 4-sided steel tower with the same conformation as a standard quadratic guyed telecommunication tower was built and erected in a basement parking area of a university building with a 3.6 m height ceiling, a location where no severe environmental factors such as wind, rain, and heat stress would affect subject workload. The tower had 50 cm long rungs, constructed out of solid round steel bars, separated from one another by a distance of 45 cm. A set of rollers and cables was installed on top of the tower for raising a sector type cellular antenna (1.3 m height, 0.185 m width, 0.1 m depth).

The antenna installation task comprised clamping the antenna by bolting it to a steel pole, connecting four waveguide cables to cable connectors, and shielding the cable connectors with three layers of insulation tape. All subjects were asked to complete two antenna installation tasks, with and without the developed platform (WP/WOP), with a 20-minute rest break between each task (Fig. 2). The antenna was installed on the right hand side of the subjects. Installation task without platform (WOP) was randomly assigned to five inexperienced and three experienced subjects as their first task. Immediately following each task, subjective ratings of perceived exertion (RPE) were obtained for their neck, shoulders, upper arms, wrists, upper back, lower back, hip, thighs, lower legs, and feet. The scale for REP was adopted from Borg's CR10 Scale ranging from 0= "nothing at all" to 10= "extremely difficult".

Each subject wore a standard helmet, a pair of hard sole shoes or tennis shoes, and a waist belt, which had a safety cable and a tool bag containing pliers, a knife, bolt nuts, and insulation tapes. The subjects were instructed to climb the tower (1 to 1.5 m above the ground) and perform the installation task according to the standard operating procedure at their own pace and using their own posture. Each task took about 15 to 25 minutes from the time when the antenna was raised to its fixture level. Each subject was video-taped during the task period and a laser pointer with wireless transmitter was used to synchronize the films and a data logger. Two 10-min practice trials, one WOP and one WP, were performed prior to the formal tasks.

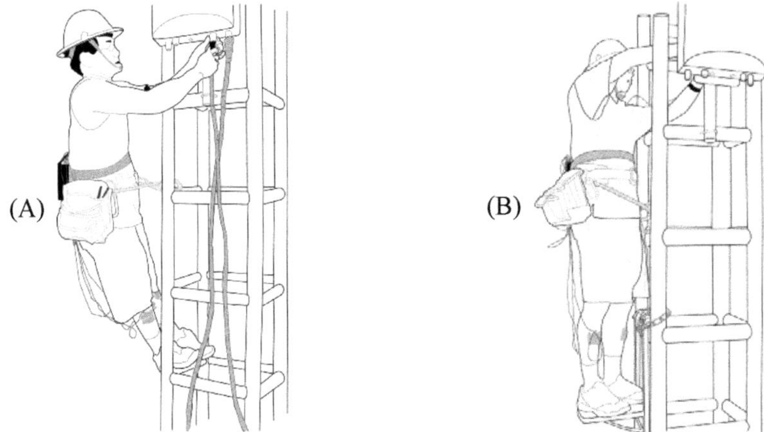

Figure 2. Experimental setup for without platform (WOP) task (A) and with platform (WP) task (B)

2.4 Electromyography recording

The electromyography (EMG) signals were recorded using 20mm intra-distance bipolar surface electrodes (Biometrics, SX230, UK) during the sampling period. The EMG electrodes were placed along the muscle over the left and right gastrocnemius, tibialis anterior, and paraspinal muscles (L5) of individual subjects. After all electrodes were secured on the skin, each subject was instructed to sit down and relax for 5 minutes, and then baseline resting EMG signals were recorded for 20s.

A series of maximal voluntary contractions then were performed to obtain the maximal voluntary capacity of each muscle. All the maximal voluntary contractions were performed in a standing position. Three 3-sec maximal voluntary muscle contractions on each bilateral muscle group were measured with 2-min rest between each attempt. The average maximal voluntary capacity of each muscle was then used to normalize the corresponding EMG recorded during task performance by expressing the EMG magnitude as a percentage of maximal voluntary capacity (% MVC).

The EMG signals were collected for the entire task using a portable data logger (Liu *et al.*, 2006) carried on a belt by the subject. The logger A/D converted EMG signals with a sampling rate of 1000Hz per channel and stored data on a compact flash memory card for further analyses. Recorded EMG data were later downloaded to a personal computer and imported to a signal processing and analysis program, Viewlog, developed by the Institute of Occupational Safety and Health, Taiwan (Chen *et al.*, 2006). Each EMG signal, including EMG signals obtained during maximal voluntary contraction, subtracted individual RMS rest level from the original EMG signal. Each EMG signal was then normalized to %MVC by dividing the magnitude of individual RMS EMG obtained under maximal voluntary contraction. For each task, the RMS values of an EMG signal were calculated for intervals of 0.2 sec to characterize the muscular activity. The amplitude probability distribution function (APDF) developed by Jonsson (1982) was utilized to analyze EMG data for each task. The observed working posture of a tower worker was typically static during antenna installation while standing on a tower rung or platform. Therefore, the median muscular load levels were determined based on the 50th percentile of the APDF for statistical analyses.

2.5 Data analysis

The work periods and trunk angles were determined for all subjects and tasks using video data provided by an investigator. The work period for each task was defined as the time when the subject climbing up the tower until their return to the ground. Trunk posture, namely lateral bending and flexion, was coded as 0, 1, and 2, indicating angles ranging from 0° to 5°, 5° to 10°, and 10° to 15°, respectively. Since subjects' posture remained fairly static during each task, the investigator registered a posture code for each 15sec period and the average code was used for statistical analysis.

SPSS 10.0 for Windows was used for the statistical analysis. Group differences in age, stature, and weight were assessed using the independent t-test. Repeated-measures ANOVA were used to compare differences in the measurements of work periods and the muscular loads by RPE and EMG. For repeated-measures ANOVA, subject group (experienced and inexperienced) denoted the between-subjects factor, and the task (WOP and WP) and side (left and right side of the body) represented the within-subjects factors. If no significant differences were found among groups, Wilcoxon signed-rank test was further used to retest the task effect by taking all subjects as one group. The difference was considered significant at a level of $p<0.05$.

3 RESULTS

No group differences were identified in any analyzed variables tested using repeated-measures ANOVA, except that the experienced group was found to be older ($p<0.01$, Table 1) and had a slightly shorter work period ($p<0.05$, Table 4). The Wilcoxon signed-rank test was thus used to retest the task effect by taking all subjects as one group. Significantly shorter work period was observed in WP task than in WOP task ($p<0.05$, Table 4).

3.1 Perceived exertions

No task differences in perceived exertion were found in the hips, upper back, upper arms, and wrists between tasks with and without the platform. The analytical results demonstrated larger perceived exertion in neck ($p<0.05$), shoulders ($p<0.05$), lower back ($p<0.001$), thighs ($p<0.05$), lower leg ($p<0.001$) and feet ($p<0.001$) in the WOP task than in the WP task (not showed in table). For the lower leg and feet, subjects demonstrated significantly higher perceived exertion on the right side than on the left side ($p<0.05$), as well as in the WOP task compared to the WP task ($p<0.001$). Furthermore, significant 'side by task' interaction was found in both lower leg ($p<0.05$) and foot ($p<0.05$). On average, when the platform was used, subjects reduced their average RPE by 30%, 38%, 25%, 37%, 46%, and 45% in the neck, lower back, shoulders, thighs, lower legs, and feet, respectively. Experimental results demonstrate that the right-left differences in the lower leg and foot were also reduced from an average Borg's scale of 1.4 to 0.8.

3.2 Muscle activity

The results showed that the use of the platform significantly reduced the EMG of both gastrocnemius muscles ($p<0.005$, Table 2). Mean EMG decreased considerably from 7.28% MVC to 4.31% MVC at the left gastrocnemius, and decreased from 9.51% MVC to 5.76% MVC at the right gastrocnemius. Sweat on the back caused short circuiting of the EMG electrodes and resulted in missing data of two subjects for both paraspinal muscles in each group. Notably, the EMG of the left paraspinal muscle significantly exceeded that of the right paraspinal muscle ($p<0.01$, Table 2), and surprisingly was slightly increased by 3.5% MVC when the platform was used, though the increase was not statistically significant. None of the studied factors affected the EMG of the tibialis anterior muscle.

Table 2 Effect of group, task, side on muscle activities (Mean ± SD, unit: %MVC)

Muscle	Tibialis Ant.(n=15)		Gastrocnemius*(n=15)		Paraspinales †(n=13)	
Task	left	right	left	right	left	right
WOP	2.57±1.85	2.77±2.69	7.28±4.50	9.51±5.92	11.54±5.13	5.71±4.32
WP	2.40±1.82	1.82±1.17	4.31±4.00	5.76±4.44	14.81±4.24	5.56±3.34
Wilcoxon p	ns	ns	0.025	0.008	ns	ns

* p<0.005 for significant task difference by repeated-measures ANOVA

† p<0.001 for significant right-left difference by repeated-measures ANOVA

The overall muscular loads were calculated as summation of bilateral muscle EMG across all muscles. The imbalance in muscular loads was calculated as the absolute differences of bilateral muscle EMG. The analytical results revealed the use of the platform only significantly reduced the overall and imbalance muscular loads of the gastrocnemius, by approximately 40% ($p<0.01$) and 52% ($p<0.05$), respectively (Table 3). No significant group and task effects on overall and imbalance muscular loads were noted in tibialis anterior, gastrocnemius, and paraspinal muscles.

Table 3 Effect of task on sum and absolute differences of bilateral muscle activities (Mean±SD, unit: %MVC)

Muscle	Tibialis Ant.(n=15)		Gastrocnemius(n=15)		Paraspinales †(n=13)	
Task	|diff|	sum	|diff| *	sum **	|diff|	sum
WOP	1.42±1.84	5.34±3.98	5.61±3.54	16.79±8.34	5.90±5.07	17.26±7.96
WP	0.95±1.16	4.22±2.71	2.68±2.91	10.07±7.58	9.26±4.23	20.37±6.35
Wilcoxon p	ns	ns	0.025	0.006	ns	ns

* $p<0.05$, ** $p<0.005$ for significant task difference by repeated-measures ANOVA

3.3 Trunk posture

According to the analyzed video data, the experienced group completed the WOP and WP tasks in an average of 22.8 and 18.3 minutes, respectively, which was around 2.8 and 4.5 minutes faster than the time of the inexperienced group ($p<0.05$, Table 4). Subjects had significantly larger lateral bending and slightly larger flexion for WP than for WOP ($p<0.05$, Table 4).

Table 4. Effect of group and task on trunk posture and work period (Mean ± SD)

Task	Group	Lateral bend * (5˚level)	Flexion (5˚level)	Work period † (sec)
WOP	inexp (n=9)	0.33±0.50	0.22±0.44	1538±306
	exp (n=6)	0.67±0.52	0.67±0.52	1370±292
	total (n=15)	0.47±0.52	0.40±0.51	1377±354
WP	inexp (n=9)	0.89±0.60	0.44±0.53	1370±292
	exp (n=6)	1.17±0.75	1.00±0.63	1098±302
	total (n=15)	1.00±0.65	0.67±0.62	1261±317
Wilcoxon p-value		0.011	ns	0.041

* $p<0.01$ for significant task difference by repeated-measures ANOVA

† $p<0.05$ for significant group difference by repeated-measures ANOVA

4 DISCUSSION

Fatigue was reported to markedly decrease motor control performance (Johnston *et al.*, 1998) and increase risk of injury due to loss of balance (Sparto *et al.*, 1997). In the WOP task, the average muscular loads of lower limb muscles ranged from 2.57% MVC to 9.51% MVC, with the right gastrocnemius muscle carrying the greatest load. The average load in the right gastrocnemius muscle exceeded the 8%

MVC limit recommended by Björksten and Jonsson (1977) and is only slightly below the 10% MVC limit recommended by Byström and Fransson-Hall (1994). This phenomenon may indicate that fatigue may become very likely in the right gastrocnemius under prolonged work hours. With the use of the platform, the averaged muscular load in the left and right gastrocnemius was reduced to 4.31% MVC and 5.76% MVC, respectively. These exertion levels only exceeded the 0% MVC limit proposed by Sjøgaard *et al.* (1986) and suggest that subjects can last longer without fatigue under the platform use.

The analytical results presented in this study showed that the use of a platform can increase productivity and reduce the overall and imbalanced loads in lower limb muscles, when worker foot placement was restricted by the worksite. Some previous studies identified the importance of stability when performing manual tasks. Grieve (1979a, b) and Kerk (1992) considered limitations on postural stability in comprehensive models designed to predict the human capabilities of static force exertions. Haslegrave *et al.* (1997) demonstrated that even small workplace constraints on posture and foot placement may significantly affect individual ability to exert force. This finding suggests that a worker would spend more effort than is necessary to accomplish a job task when working under a constrained posture. Tower workers generally support themselves by standing on a tower rung and hooking themselves to the tower structure with one arm, or using a safety belt maneuver by holding themselves with the traction of a hooked safety belt connected to their waist harness. Most tower workers are used to standing on a rung using the middle area of the soles of the feet. In such a supporting position, to stabilize the ankle by providing a counterbalancing moment produced by foot-rung reaction force and the midfoot-ankle level arm, the gastrocnemius muscles must expend more effort than that expended when simply standing on the floor. In this study, the design of the platform focused on providing better foot support for workers performing tasks on a vertical tower. Standing on the platform naturally reduced the overall loading on the gastrocnemius muscles of subjects, particularly for their right feet.

In this study, all subjects adopted the safety belt maneuver for performing tasks with both hands simultaneously, or to reduce muscle strain in their back or limbs for an extended period. Extra support was clearly needed under such conditions. Providing a better foot support helps to facilitate the weight-shifting and posture adjustments of workers. The influence of the platform on subject weight-shifting ability can be demonstrated by the reduction of muscular load imbalance. Moreover, the effect of the platform on individual posture adjustment was further demonstrated by the larger trunk angles adopted by subjects when using the platform. Consequently, the platform may help subjects to reduce imbalances in lower limb muscular load and thus reduce possible hazards resulting from fatigue when working for prolonged periods.

According to RPE results, subjects positively evaluated the use of the platform in reducing their perceived exertion of several body parts, particularly the lower back, lower legs and feet. This finding is consistent with the EMG measurement at the gastrocnemius muscles. The significant task differences of RPE at the lower legs and feet agree with reduced overall muscular load when using the platform. The 'side by task' interaction effect on the RPE at lower legs and feet can also be interpreted by the use of the platform markedly diminishing imbalances in the load borne by the bilateral gastrocnemius. The subjects adopting larger trunk angles in WP may be directly related to the effect of the platform in increasing the flexibility of weight-shifting or the confidence to lean to side more.

The design of the current platform could be further improved to increase its portability by reducing the weight of the platform without sacrificing its structural strength. The same design concept can be applied to various types of vertical ladders, such as those fixed on monopole towers and self-support towers. The platform was designed to be easily manipulated by one hand. However, the extra effort and time spent by the workers in raising/lowering, hooking/unhooking and folding/unfolding the platform was not evaluated in this study. The clothing and shoes workers wore might slightly affect their workloads. A further field study is required to validate the feasibility and efficiency in the real work setting. Results of this study demonstrate that the developed platform is a highly effective means of reducing the musculoskeletal strains while installing the antenna, but is not a substitute for personal protective equipment.

ACKNOWLEDGEMENTS

The authors would like to thank the National Science Council and the Institute of Occupational Safety, Taiwan for financially supporting this research study (NSC 90-2218-E-324-003, IOSH93-H103).

REFERENCES

Björksten, M. and Jonsson, B. 1977. Endurance limit of force in long-term intermittent static contractions. *Scand. J. Work Environ. Health* 3: 23–27.

Byström, S. and Fransson-Hall, C. 1994. Acceptability of intermittent handgrip contractions based on phys response. *Human Factors* 36: 158–171.

Chad, K.E. and Brown, J.M.M. 1995. Climatic stress in the workplace. *Appl. Ergon.* 26: 29–34.

Chen H.C., Chen, C.Y., and Lee, C.L. et al. 2006. Data logging and analysis tools for worksite measurement of physical workload. In. *Proceedings of the 16th World Congress of the IEA*. Maastricht, Netherlands.

ComTrain LLC. 2003. *Tower climbing safety & Rescue*. 3rd ed.

Dewar, M.E. 1977. Body movements in climbing a ladder. *Ergonomics* 20: 67–86.

Environmental Protection Administration 2001. NIR monitoring results for base stations released. *Environmental Policy Monthly*, Taiwan, R.O.C. 4: 20–22.

Grieve, D.W. 1979a. The postural stability diagram (PSD): Personal constraints on the static exertion of force. *Ergonomics* 22: 1155–1164.

Grieve, D.W. 1979b. Environmental constraints on the static exertion of force: PSD analysis in task design. *Ergonomics* 22: 1165–1175.

Haslegrave, C.M., Tracy, M.F., and Corlett, E.N. 1997. Force exertion in awkward working postures - strength capability while twisting or working overhead. *Ergonomics* 40: 1335–1362.

Johnston, R.B. III, Howard, M.E., and Cawley, P.W. et al. 1998. Effect of lower extremity muscular fatigue on motor control performance. *Med. Sci. Sports Exerc.* 30: 1703–1707.

Jonsson, B. 1982. Measurement and evaluation of local muscular strain on the shoulder during constrained work. *J. Human Ergology* 11: 73–88.

Kerk, C.J. 1992. Development and evaluation of a static hand force exertion capability model using strength, stability and coefficient of friction. Doctoral dissertation, University of Michigan, Ann Arbor, MI.

Liu, Y.P., Chen, H.C., and Chen, C.Y. 2006. Portable data logger for worksite measurement of physical workload. *J. Medical and Biological Engineering* 26: 21–28.

Mao I.F., Chen M.L., and Huang J.W. et al. 2000. The subjective symptom of work fatigue, heart rate, blood pressure and life style of the high elevation operators. *J. Occup. Saf. Health* 8: 127–143. (in Chinese)

Perotto, A.O., Morrison, D., and Delagi, E.F. et al. 1994. *Anatomical guide for the electromyographer: the limbs and trunk.* Springfield: Charles C. Thomas.

Sjøgaard, G., Kiens, B., and Jørgensen, K. Et al. 1986. Intramuscular pressure, EMG, and blood flow during low-level prolonged static contraction in man. *Acta Physiologica Scandinavica* 128: 475–484.

Sparto, P.J., Parnianpour, M., and Reinsel, T.E. et al. 1997. The effect of fatigue on multijoint kinematics, coordination, and postural stability during a repetitive lifting test. *J Orthop. Sport Phy. Ther.* 25: 3–12.

CHAPTER 37

An Analysis of Grip Design for Manual Hammer Stapling Tool

Arijit K. Sengupta, Wayne Latta

New Jersey Institute of Technology
Newark, NJ, USA
sengupta@njit.edu

ABSTRACT

Three hammer stapling tools with distinctly different handle designs were evaluated in terms of comfort, safety and hand-arm stress. Sixteen male participants used each tool on two simulated roofs with 4:12 and 6:12 pitches, and stapled roofing underlayment at a frequency of 1 staple per second for two minutes. Tools with smooth, rounded and compressible grips, received significantly better ratings ($p<.05$) in grip comfort and ease of use, than the tool with rectangular grip cross-section employing a hard and serrated grip surface. Tools with grip features that provided protection from unintentional finger pinching received higher safety rating ($p<.05$). The tool with a 10 degree bent handle reduced ($p<.05$) the wrist angle at tool strike. The bent handle tool reduced the wrist flexor muscle activity, but increased the wrist extensor muscle activity. The findings of this study suggest that the hammer stapling tool with smooth and rounded grip cross-section, with a bent handle, improves grip comfort, usability and tool safety, and reduces the risk of repetitive strain injury of the wrist joint.

Keywords: hand tool, grip design, hammer, bent handle

1 INTRODUCTION

Typically a roofer uses a hammer stapling tool (Figure 1) to staple several hundred staples on the paper underlayment to attach it to the plywood roof decking. Since the roofer's hand-arm system experiences repeated impacts from the tool use, the grip design plays an important role in providing of grip comfort and protection from acute trauma and repetitive strain injury. Poor hand tool design is associated with risk of both acute and chronic disorders of hand, wrist and forearm (Aghazadeh and Mital, 1987). Design deficiency of tool or improper selection of tool can generate excessive biomechanical stresses (Chaffin, Anderson and Martin, 1999).

Scientific studies on grip comfort of similar types of tools suggest that, foam rubber grips provides more even distribution of contact pressure than hard unyielding grips (Fellows and Freivalds, 1991), the palmer side of hand is sensitive to serrated grip surface (Fransson and Kilbom, 1991), and grip cross-sections with rounded corners improves grip comfort and functional grip strength compared to grips with less rounded corners (Page and Chaffin, 1999).

The ergonomic principle of "bending to tool, not the wrist" has been studied for hammer. For horizontal and vertical working surfaces, a bent handle hammer reduced the wrist angle at impact (Knowlton and Gilbert, 1983; Schoenmarklin and Marras, 1989), and caused less strength decrement (Knowlton and Gilbert, 1983) compared to a straight handle hammer. A hammer with 10 degree bent handle was preferred than a straight handle hammer by users without any decrement of nailing productivity (Konz and Streets, 1984). Although the action of the hammer stapling tool is similar to an ordinary hammer, the former is associated with an additional risk of inadvertent finger injury by getting pinched against the roof surface. Striking on a slanted roof and guarding against finger pinching might have a different influence on wrist joint than that found in previous studies on hammering task.

The objective of this study is to evaluate the grip design features of hammer stapling tools available in the retail market in terms of grip comfort, safety, usability, wrist angle and muscle activity. Three tool models with distinctly different grip design were selected (Figure 1) for the evaluation. Essentially the three models were comparable in terms of size, weight, magazine capacity and staple size but differed in grip shape, grip material and handle angle.

2 METHOD

2.1 Hand tools

The model HT50 (Figure 1a) and HTX50 (Figure 1b) were manufactured by Arrow Corporation, and the model PC2K (Figure 1c) was manufactured by Bostitch Corporation. Henceforth these tool models will be referred to as Tool#1, Tool#2 and Tool#3, respectively. Tool#1 and Tool#2 had identical length and weight, 28 cm and 0.95 kg, respectively. Tool#3 had slightly longer overall length of 36 cm and weighed 1.0 kg.

Tool#1 grip design was basic, incorporating a straight handle with rectangular grip cross-section, rigid plastic surface with crosswise serrations. The shape of the section along grip axis was uniform, and the serrations were provided to improve gripping friction.

Tool#2 grip had a similar straight handle and rectangular grip cross-section but with more rounded corners. The finger side of grip surface was made of smooth and non-resilient rubber material. The two grip ends had raised sections that acted as shields against unintentional finger pinching during tool strike. The raised sections would also prevent slippage of the tool within the grasp.

The cross-section of Tool#3 grip was oval, and grip surface was smooth and was covered with resilient foam rubber. The thickness along the length of the grip was wedge shaped with a flared section at the end. This shape meant to prevent slippage of the tool along the grip axis within the grasp. This tool had employed a 10 degree upward bend of the handle. The upward bend had provided a clearance from the roof surface, and reduced the risk of finger pinching. Also, the bent handle might possibly promote a more neutral wrist posture during the tool strike.

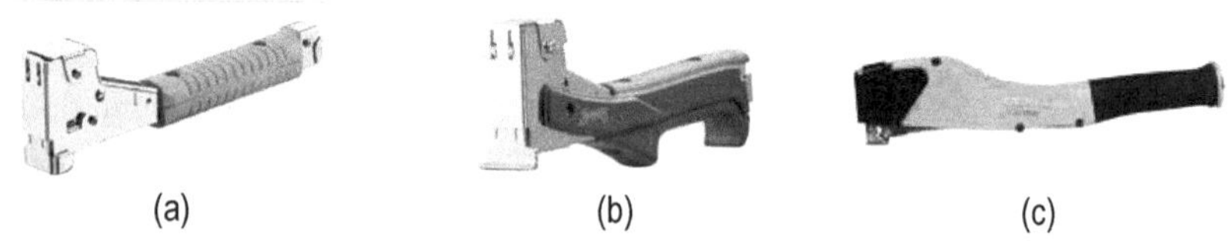

(a) (b) (c)

Figure 1 Hammer stapling tools evaluated in the study

2.2 Participants

Sixteen male university students participated in the study. All participants were in good physical health with no history of musculoskeletal problems and were paid volunteers for this study. Their average (standard deviation) height, weight, and age were 177(8.4) cm, 80(20.0) kg and 22(3.4) years, respectively. The study received approval from the institutional review board.

2.3 Experimental Design

A 6x4-foot wide platform was fabricated with 5/8th inch roofing grade plywood with a pitch of 4 inch rise to 1 foot run (Figure 2). A removable base insert was used to increase the roof pitch to 6-inch rise for 1 foot run. These two pitches are commonly found in residential pitched roofs. The participants stood on the platform facing the roofing underlayment and stapled it onto the plywood.

Each participant completed six separate experimental trials involving the combinations of three different tools and two roof pitches in a randomized order. Participants practiced with the hammer staplers before the experimental session. Five minutes rest break was provided between two experimental trials, to avoid fatigue. Each experimental trial consisted of stapling at a frequency of 1 staple per second for two minutes. The stapling pace was maintained by following an audible metronome. The stapling was done in a pattern following the three rows marked by pre-printed lines on the roofing paper. The pattern consisted of striking on the top, middle and bottom row and repeating this sequence while moving laterally from one side to the other. During stapling operation, the participants were instructed to apply enough force to insert the staples correctly flushed with the paper. Stapling with the hammer stapling tool was a comparatively easy task to accomplish and mistakes were rare.

Figure 2 Test Platform (4:12 Pitch - Left, 6:12 Pitch - Right)

Electromyographic (EMG) activity was monitored for the Flexor Carpi Ulnaris (FCU) and the Extensor Carpi Radialis (ECR) muscles of the forearm, and the Biceps Brachii (BB) muscle of the upper arm (Figure 3). The two forearm muscles, FCU and ECR, have insertion points on the metacarpal bones of wrist joint, and they resist wrist motion from the tool action in the ulnar-radial plane. BB was selected as the main forearm flexor muscle. The skin surface was cleaned and abraded and conductive gel was applied prior to applying the surface electrodes. The surface electrodes (Biometrics Ltd., Model SX 230W) employed a preamplifier (gain 1000), and high pass and low pass filter circuitry to reduce external interference. Two end terminals of an electrogoniometer (Biometrics Ltd. Model SG 110) were affixed to the dorsal skin surface of the forearm and hand by double sided adhesive tape, and the goniometer reading was set to zero while the subject maintained a neutral wrist posture. The EMG and goniometer signals were captured at 1000Hz and were transmitted via a remote patient data acquisition unit attached to the participant's belt to a Biometrics Datalink DLK800 base unit and stored in a personal computer operating Biometrics Datalink software, for further processing.

Prior to the experimental task, EMG for the maximum voluntary contraction (MVC) was recorded. Participants were instructed to hold Tool#3 with the elbow flexed at 90 degree and wrist at the neutral posture so that the tool was in a vertical position. They were then instructed to restrain the tool with their free hand while performing a maximum contraction of their FCU muscle by attempting to rotate the tool away from them (the direction of ulnar deviation of the wrist). They held the maximum contraction for a count of six followed by a rest. The MVC of the ECR muscle was obtained by repeating the same procedure in the opposite direction (the direction of radial deviation of the wrist). The MVC for the BB was measured by having the participant sit in a chair with their elbow flexed 90 degrees. With their fist placed underneath the edge of the desk surface, they performed a maximum contraction of the BB. The MVC efforts were repeated three times for each muscle.

Participants rated their perceived level of discomfort, on a 0-10 scale, in ten key areas of the body, immediately after each trial. The participants also rated each tool in terms of ease of use, grip comfort, and protection from injury potential after each experimental trial, on a 0 -10 scale.

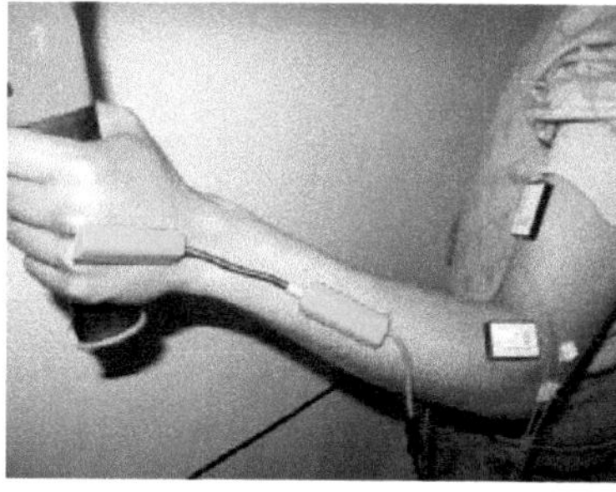

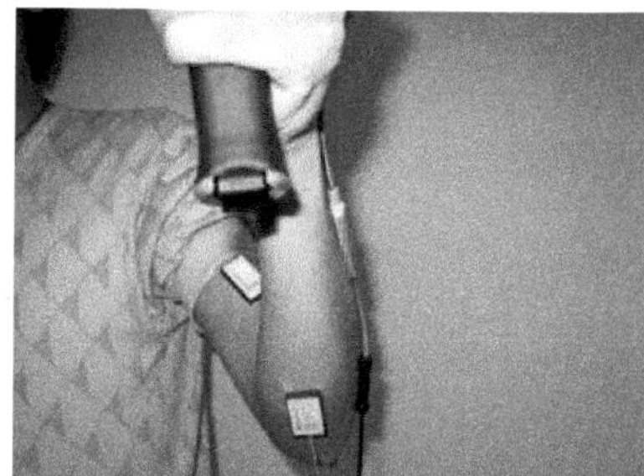

Figure 3 Placement of surface electrodes and goniometer

2.4 Data Analysis

Raw EMG data of task and MVC were first transformed by applying a root-mean-square (RMS) filter with a time constant of 200 millisecond and then were averaged. The maximum of the average RMS of the three MVC data for each muscle was used for normalization. The normalized EMG in percent (%MVC) represented muscle activity.

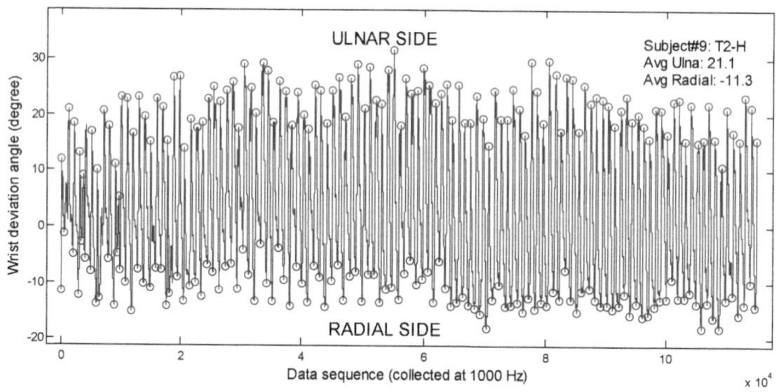

Figure 4 The typical wrist angle variation registered by the goniometer over an experimental trial; the circles at the peaks and valleys represent the wrist angle at impact and windup, respectively.

Figure 4 shows the typical variation of wrist angle in the ulnar/radial plane as registered by the goniometer over an experimental trial. A customized Matlab program pin pointed the impact (ulnar) and windup (radial) wrist angles of each strike. Means of impact wrist angles and windup wrist angles were calculated for each condition, and later used for statistical analysis. The data set from one participant was discarded, because of the detachment of the goniometer during an experimental trial.

All response variables were statistically analyzed using a two-factor (tool and pitch) analysis of variance (ANOVA) model with participant as a blocking factor. Significant differences in the factor level means were determined from Tukey's test of joint confidence interval.

3 RESULTS

No interaction between tool and pitch factors was significant for any of the response variables. Also, the pitch factor was not significant for any response variable, except for the radial deviation angle of the wrist joint.

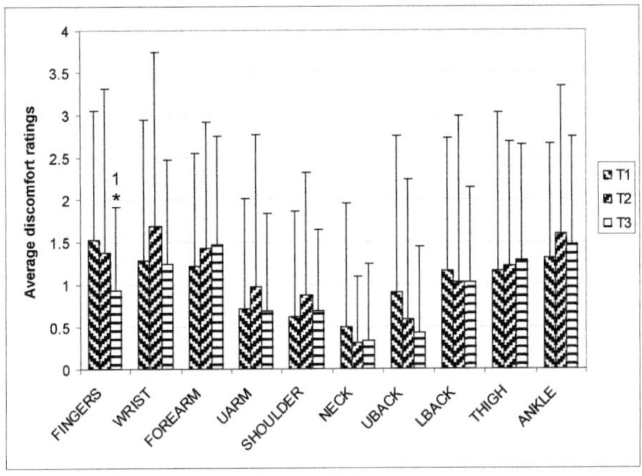

Figure 5 Mean and standard deviation (n=16) of discomfort ratings on a scale 0 to 10; asterisk and the number on the bar represent a significant difference in mean (p<.05) with tool number.

3.1 Discomfort Ratings

The mean and standard deviation of body part discomfort ratings for all participants are illustrated in Figure 5. Although the mean discomfort scores in all body regions were less than 2, the individual ratings varied from 0 to 9. The mean discomfort score for Tool#3 was significantly lower than for Tool#1 in fingers (p=0.03), but no other contrasts of means were statistically significant at α=.05.

3.2 Subjective Perception of Tool Characteristics

The mean and standard deviation of tool characteristics ratings for all participants are illustrated in Figure 6. In terms of ease of use, Tool#3 received significantly better ratings (p=.01) than Tool#2 and Tool#1. In terms of grip comfort, Tool#2 and Tool#3 were not different, but Tool#1 was rated significantly inferior than Tool#2 (p=0.01) and Tool#3 (p=0.00). Similar statistical results were found for perception of protection of injury, ie., no significant difference between Tool#2 and 3, but Tool#1 was rated significantly inferior compared to Tool#2 (p=0.00) and Tool#3 (p=0.00).

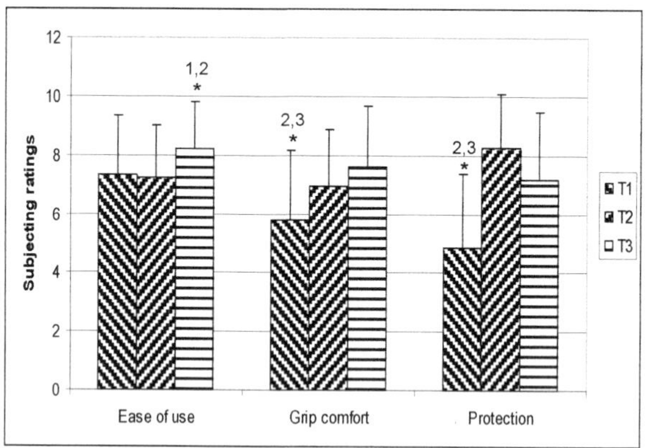

Figure 6 Mean and standard deviation (n=16) of tool characteristics ratings on a scale 0 to 10; asterisk and numbers on a bar represent a significant difference in mean (p<.05) with tool numbers, respectively.

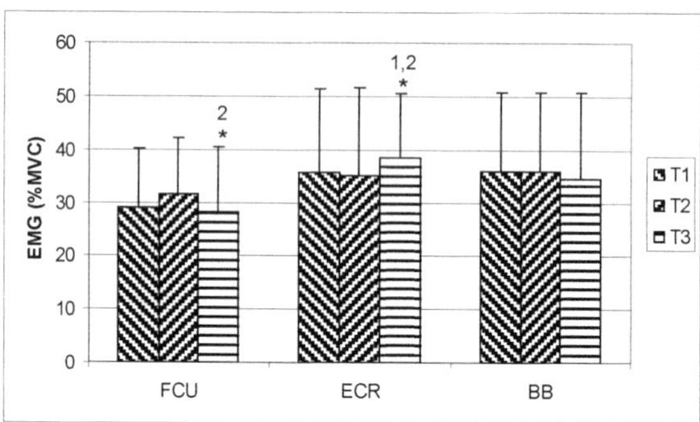

Figure 7 Mean and standard deviation (n=16) of normalized muscle activities; ECR-Extensor Carpi Radialis; FCU-Flexor Carpi Ulnaris; BB-Biceps Brachii; asterisk and the number(s) on a bar represent a significant difference in mean (p<.05) with tool number(s), respectively.

3.3 Muscle Activity (%MVC)

The mean and standard deviation of muscle activity in terms of %MVC are shown in Figure 7. The mean muscle activity of the FCU was lesser for Tool#3 than for Tool#2 (p=.02). The mean muscle activity of the ECR was significantly higher for Tool#3 than for Tool#1 (p=.02) and Tool#2 (p=.00). None of the other contrast of means was significant.

3.4 Wrist Angles at Impact and Windup

The mean and standard deviation of the wrist angle at impact (ulnar deviation) and windup (radial deviation) are plotted against each tool in Figure 8. Tool#3 produced significantly less ulnar deviation as compared to Tool#1 (p=.04) and Tool#2 (p=.00). The radial deviation was not affected by the tool factor.

The wrist angle at windup (radial deviation) was the only variable that showed a significant increase (p=.01) from low pitch to high pitch roof, with means of 17.3 to 19.5 degree, respectively.

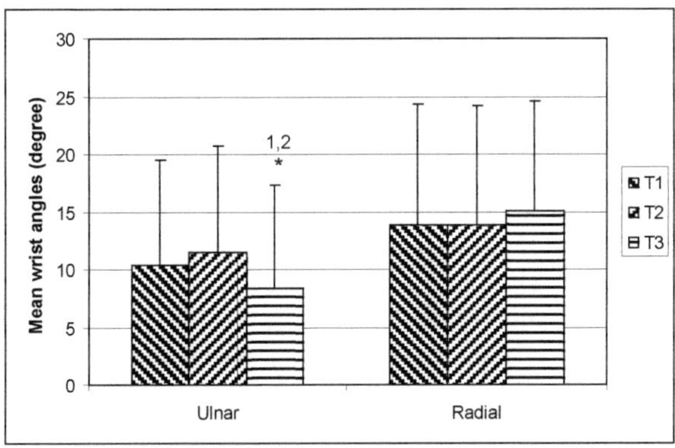

Figure 8 Mean and standard deviation (n=15) of radial and ulnar deviation of wrist; Asterisk and numbers represent a significant difference in mean (p<.05) with tool numbers, respectively.

4 DISCUSSION AND CONCLUSION

Tool#1, with hard serrated plastic grip and less rounded cross-section received significantly inferior mean rating on grip comfort (5.8) as compared to Tool#2 (7.2) and Tool#3 (7.6) that employed smooth, well rounded grip cross-section. This result supports similar findings from the previous studies on grip design (Fransson and Kilbom, 1991; Page and Chaffin, 1999). Tool#1 also produced 60% more mean discomfort at finger region (1.5) as compared to Tool#3 (0.9). Tool#3's oval and compressive foam rubber grip, as opposed to Tool#1's hard serrated plastic grip with less rounded cross-section, should be responsible for such increase in discomfort (Fellows and Freivalds, 1991). However, the mean finger discomfort of Tool#2 (1.4), which also had smooth grip surface, did not reach statistical significance as compared to Tool#1.

The mean rating on protection from injury of Tool#2 (8.2) and Tool#3 (7.2) were significantly higher that Tool#1 (4.9). Safeguards against pinching injury during tool strike is an important tool design aspect (Konz and Johnson, 2008), and Tool#2 and Tool#3 designs incorporated protection strategies from such mishaps. The absence of such a feature in Tool#1 was clearly perceived by the participants as hazardous. In terms of ease of use, Tool#3 was rated (8.3) to be significantly superior to Tool#1 (7.3) and Tool#2 (7.2). In addition to the grip features, the bent handle construction of the tool could also have contributed to this result.

The mean FCU muscle activity from using the bent handle Tool#3 (28% of MVC) was significantly less than that of Tool#2 (32% of MVC). The mean FCU activity of Tool#3 was less than that of Tool#1 (29% of MVC), but the difference did not reach statistical significance. The FCU muscle is the prime mover for resisting the inward rotation of wrist (radial direction) that tends to occur at the time of the tool strike. This reduction in FCU muscle activity, coupled with reduction in the ulnar deviation of wrist (as explained later) from the use of Tool#3, would act synergistically in protecting the soft tissue injury in wrist.

The mean activity of ECU muscle was significantly greater for Tool#3 (38% of MVC) than that of Tool#2 (35% of MVC) and Tool#1 (36% of MVC). Fellows and Freivalds (1991) reported similar increase in EMG of hand flexor muscle from tool grips made from compressible foam rubber. They attributed the increase in EMG from the increased grasping force necessary due to deformation of the foam and a 'loss of control' feeling of the subjects. The compressible foam grip of Tool#3 exhibited a similar trend. Fellows and Freivalds (1991) recommended reduction the thickness of the foam layer on tool grip to reduce the higher grasping force.

The mean ulnar deviation of the bent handle Tool#3 (8.4°) was significantly less than that of Tool#2 (11.4°) and tool#1 (10.4°). Reduction of ulnar deviation is an important factor in reducing the soft tissue injury potential, since the wrist joint is maximally deviated in the ulnar side at the instant of tool strike (Knowlton and Gilbert, 1983; Schenmarklin and Marras, 1989). Most of the muscle tendons develop highest tension at the tool impact, and the wrist joint is most susceptible to soft tissue injury when it is more deviated (Chaffin, Anderson and Martin, 1999). The roof pitch 6:12 produced increased mean radial deviation of wrist (19.5°) than 4:12 pitch (17.4°). The increased radial deviation can be attributed to the higher windup angle of wrist for higher pitched roofs.

In summary, this study concluded that for a hammer stapling tool:

1. Smooth and rounded shaped grip would improve grip comfort than hard, serrated less rounded shaped grip.
2. An upward 10 degree bent handle was proved to be a better approach in protecting inadvertent finger pinching than the raised sections at the grip ends.
3. The bent handle promoted better wrist posture and lesser FCU muscle activity at tool impact, which potentially would reduce the risk of soft tissue injury in the wrist joint.
4. The compressible foam rubber grip, although found preferable in terms of grip comfort and usability, but it was associated with increased ECR muscle activities. Reducing the thickness of the foam layer might

reduce the necessity of higher grasping force and consequently the muscle activity.

The findings of this study are applicable in the design and selection of hammer stapling tools.

REFERENCES

AGHAZADEH, F. & MITAL, A. 1987. Injuries due to handtools: results of a questionnaire. *Applied Ergonomics,* 18, 273-278.

CHAFFIN, D, B., ANDERSSON, G. B. & MARTIN, B. J. 1999. *Occupational Biomechanics 3rd edition.* New York, John Wiley & Sons Inc.

FELLOWS, G. L. & FREIVALDS, A 1991. Ergonomic evaluation of a foam rubber grip for tool handles. *Applied Ergonomics,* 22, 225-230.

FRANSSON, C. & KILBOM, A. 1991. Tools and hand function: the sensitivity of hand to external surface pressure. In. *Designing for everyone,* Taylor & Francis, London, 188-190.

KNOWLTON, R. G. & GILBERT, J. C. 1983. Ulnar deviation and short-term strength reductions as affected by a curve-handle ripping hammer and a conventional claw hammer. *Eronomics*, 26, 173-179.

KONZ, S. & STREETS, B. 1984. Bent hammer handles: Performance and Preference. In. *Proceedings of the Human Factors Society - 28th Annual Meeting,* 438-440.

KONZ, S. & JOHNSON, S. 2008 *Work Design – Occupational Ergonomics.* Scottsdale, AZ, Holcomb Hathaway, Publishers Inc.

PAGE, G. B. & CHAFFIN, D. B. 1999. Functional grip strength: A preliminary study on the comparative effects of fine and gross handle shape attributes. http://www.asbweb.org/conferences/1990s/1999/ACROBAT/030.PDF [Accessed 2/28 2012].

SCHOENMARKLIN, R.W. & MARRAS W. S. 1989. Effects of handle angle and work orientation on hammering: I, wrist motion and hammering performance. *Human Factors*, 31, 397-411.

CHAPTER 38

Space Operation System Based on Human-in-the-Loop

Shengpeng Guo, Dongxu Li, Caizhi Fan

National University of Defense Technology
Changsha, Hunan, China
shengpengg@gmail.com

ABSTRACT

Complicated space missions are becoming more and more difficult to be fulfilled because of insufficient intelligence of space robots and limited extravehicular operating ability of astronauts. In this paper, a concept of space operation system based on human-in-the-loop is presented. This system introduces humans' (i.e. ground operator's) ability of making real-time decision into the space missions, and guides space robots to accomplish tasks with operator's performing behavior. The system measures operational behaviors of operators on the ground, extracts characteristic values, representing a certain operation model and transferred to the remote position. Whole consecutive behavioral commands can be recovered by matching certain characteristic information. Those commands can be used to directly drive the space robot to finish the operation. The simulating results on the ground indicate this method helpful for space robots to improve operating efficiency and to promote the ability of completing the complicated space missions.

Keywords: human-in-the-loop, space robots, ground operator, characteristic values

1 INTRODUCTION

Increasing space tasks such as large complex structure assembly, maintenance and refueling etc. requiring performance on orbit become more difficult to be fulfilled by entirely depending on astronaut Extravehicular Activities (EVA) for the limitations of crew safety, EVA hours and efficiency (Moosavian and Papadopoulos,2007). Nevertheless space robots entrusted with such great responsibilities are also restricted by technical conditions of artificial intelligence. As is known to all, human possesses decision-making ability towards non-programming and unpredictable incidence, which space robots cannot deal with completely autonomously at present. Therefore, in-space robotic operation combined with intelligent decision-making capacity of humans would be an effective approach to overcome the intelligent operation problems in space. Experts are positively doing exploratory researches on this aspect. Various teleoperation methods are developed by which a human operator can perform complex on-orbit tasks in remote environment in cooperation with telerobots (Weisbin and Lavery,1994; Passenberg, Peer et al.,2010).

The application of teleoperation technologies in aerospace has made significant progress. The Shuttle Remote Manipulator System, known as Canadarm, was first launched in 1981 aboard U.S. Columbia (Gibbs and Sachdev,2002). The system is operated by the astronaut in the cabin via joysticks. Interactive teleoperation is introduced when operator on the ground remotely controls robot in the space shuttle in DLR's experiment of ROTEX (Hirzinger, Brunner et al.,1993). After that numbers of advanced experiments are carried out (Yoshida,2009). However, a major problem in practice is that limited by time delays and bandwidth in communication channel the operator receive insufficient feedback information from the remote slave site, which would seriously deteriorate the stability and maneuverability of the teleoperation system (Stoll and Walte,2009). Predictive control as well as a strategy called telepresence which rebuilds virtual model of the remote environment based on pre-experience is introduced to tackle the above problem(Stoll, Wilde et al.,2009). But these strategies are not work so well in unstructured environment.

Primitively on-orbit manipulators are mainly designed to perform typical tasks such as capture and release of target satellites. Precise position control can be achieved by using a joystick- or mouse-type input device. However, the motion degree of freedom (DOF) of the robotic manipulator is increasing in order to complete more complex missions. In 2010 American Robonaut was sent to ISS (International Space Station) who has 42-DOF in total (Ambrose,2000; Diftler, Mehling et al.,2011). Obviously, joysticks and mice are insufficient for controlling such a highly dexterous robot. A new human-machine interface must be designed.

A kind of exoskeleton device is developed to be used as input unit of arm's pose. But this kind of device has complex structure and restricts operator's body type. The end-effector is usually remotely operated by data glove (Lii, Chen et al.,2010), which could measure movement of every joint of each finger. It is an effective device to interact with multi-DOF fingers. However, the fault is that when the operator changed it must be calibrated again. Another fault is that operating force can't be measured. In the field of biomechanics, a new method of identifying movement of fingers by electromyography (EMG) signals is presented. The movement and force can be acquired by analyzing EMG signals because they contain all of the controlling information from nervous system (Rencheng,1996). By now this kind of technology has made great progress in prosthetic controlling, but less in field of aerospace. A big problem for EMG technology encountered in aerospace would be the influence of time delay.

As discussed above, teleoperation technologies have widely application prospect. The common focus is on how to use humans' decision-making ability efficiently to improve performance by research community. Therefor a space operation system based on human-in-the-loop is presented, in which the human intelligence is transmitted from master site to slave site through technologies of behavior sensing, feature extraction and behavior reproducing. The paper is structured as follows: section 2 concerns the system composition and its feature, as well as key technologies. A simulation test bed on the ground for validating key technologies is developed, see section 3. Section 4 shows the status in research and the latest test results. The last section concludes with a summary and an outlook to future work.

2 SYSTEM OVERVIEW

Space operation based on human-in-the-loop is a kind of teleoperation depending on humans' decision making, which means the space robot doing on-orbit missions is under control of the operator on the ground interacting with specific equipment. Here the term of human-in-the-loop includes the control loop between the ground operator and the robot outside the orbital vehicle, and loop between ground operator and the robot inside the orbital vehicle. The difference of this system from an ordinary teleoperation is a novel sensing technology as the input method introduced in the system. As a result, the ground operator is now directly facing operating missions instead of indirect devices such as joysticks and mice etc. In this way the operator could be able to concentrate on how to perform tasks with sufficient flexibility and judgment. The operator's experience, cognition and response are introduced to flexibly control multi-DOF robots dealing with complex tasks and unstructured environment.

2.1 System construction

The space operation system based on human-in-the-loop is designed for controlling slave actuator to express operator's intention. It is composed of three segments, humans, the master site and the slave site, among which the control loop is closed over a communication channel as signals exchanged. The system is capable of reading human's movement when he interacts with the master site, expressing human's intention when the master site interacts with the slave site, and enabling a human to sense remote environment when he interacts with the slave site indirectly.

The term "human" here means the ground operator, who has top decision-making authority in the system. In the course of a mission the ground operator performs manually as if he was being there without thinking about what commands need to be issued to the device. Consequently it is not important for the operator that whether he has received professional training or not, but he must be aware of relevant expertise.

The master site includes human-machine interaction equipment on the ground. It actively senses and recognizes the operator's movement intension, transmits it into the slave site after translating it into machine language. In addition it receives feedback information from the slave site, and rebuilds the remote environment for the operator on the ground.

The slave site usually means a dexterous space robot system interacting with target at remote site. It drives the robot to operate under the guide of the operator's decision, which result in full expression of the operator's intension. Thus it is not necessary for the space robot system to have intelligence. Besides that the slave site feed backward information acquired when interacting with targets.

2.2 Features of the system

The space operation system based on human-in-the-loop expands humans' action capacity by transferring human's intelligence to the remote site. The system has features as follows:

- The operator' body movements are actively sensed and automatically translated into machine language.
- The operator's movements are mapped into space robot's movements at the slave site. This would guide the robot to imitate the operator's behavior.
- The operator interacting directly with the mission target could concentrate on how to perform. This would improve operator's efficiency.
- The system could deal with unknown environment more effectively with the help of operator's expertise, intelligence, and experience.
- The space robot's DOF are controlled by operator's DOF so as to dexterously control complex multi-DOF robots.

2.3 Key technologies of the system

The kernel concept of the system presented in this article is applying humans' intelligent decision and judgment to dexterously control multi-DOF robots at the slave site so as to deal with unknown environment and non-cooperation targets. There exist four key technologies: behavior sensing, feature extracting, behavior reproducing and influence elimination of time delay.

A Behavior sensing technology

Behavior sensing technology is used to read the operator's operating behavior at the master site. The operator doesn't need to tell the robot by joysticks or mice which joint to move, in which direction and how long distance to move them, while facing with mission targets (the target here is reconstructed with feedback information). Instead the operator performs as if he is on the site. Meanwhile his operating intension is understood by the sensors via capturing his body's pose, tracing it continuously and describing it in the math coordinate system.

B Feature extracting technology

This key technology is a guarantee for completely and orderly transmitting the action commands into the slave site through limited bandwidth, including sets of each joint's motion state, position and attitude. Feature extracting means abstracting an ID code from the configuration of operator's behavior as a representation, termed characteristic values. Firstly, sets of the operator's behavior are sorted into different patterns, then from which the characteristic values are extracted. Different configurations have different values. Thus divided into a group of chronological arranged configurations, a sequential action could be transformed into a cluster of characteristic values using the above technology, and vice versa. The characteristic values with no apparent motion information are linked to integrated pose parameters in a model library. Transmitting characteristic values is equated with transmitting motion commands. This kind action-to-characteristic-value mapping diminishes greatly the bandwidth taken up by due-out data.

C Behavior reproducing technology

Behavior reproducing means driving space robot to imitate the operator's behavior, expressing his intention at the remote site. This procedure begins when receiving the characteristic values from the master site. A sequential action is received at the slave site after rearranging the characteristic values chronologically. But it can't be identified by the actuator because that there is no apparent kinematic parameters in characteristic values themselves. However, the operator's behaviors are preliminarily defined in the model library. Each of the models has unique mapping relationship with the characteristic value, therefore the integrated parameters about pose and configuration represented by the characteristic value can be inversed. Acquired by measuring the operator's body DOF these parameters may not well match the space robot's motion DOF. Another mapping method in joint space is needed to convert them into the robot's joint motions.

D Influence elimination of time delay

The time delay significantly decreases the stability and maneuverability of the system, interferes with the operator's judgment and decision. Virtual force feedback and augmented reality are considered to eliminate the negative influence of time delay. The former is to calculate acting force by technology of graphics collision detection when the operator interacts with virtual environment and to feed it back to the operator in real time, so that the time delay is kept out in the bottom of the sys-tem. The latter deals with what the operator feels will be the next step at the slave site; and the executing result would be fed backward to compare with the present virtual scene. Signals from both virtual and real cameras are compared in observer's coordinate system. Prompt adjustment will be made according to the clear-cut errors.

3 EXPERIMENTAL TEST BED ON THE GROUND

An experimental test bed (shown in Figure 1) has been developed on the ground to demonstrate the concept about the discussed system, on which the key technologies are also validated. It is mainly composed of two parts, the master site and the slave site. They are located in two different rooms at intervals, and composed of three main elements, namely, human-machine interaction system, a space robot simulator and a target satellite simulator. Visual feedback and display system are involved. Besides another important element is the wireless communication link between those two parts.

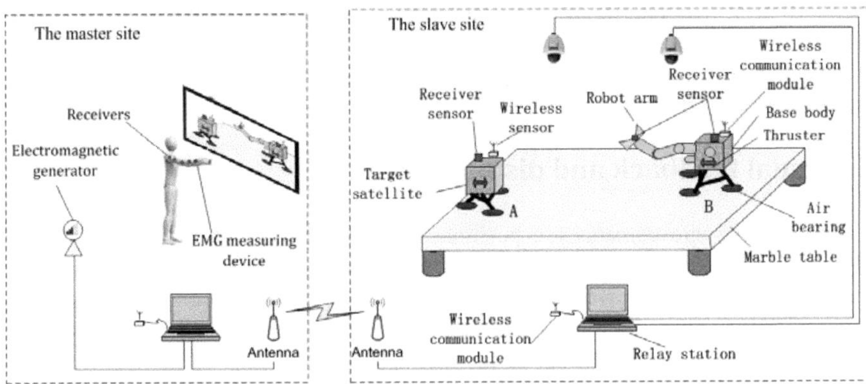

Figure 1 Experimental test bed on the ground

3.1 Human-machine interaction system

The human-machine interaction system actively senses and identifies upper limbs' motion of the operator. Hands and arms are sensed separately. For one thing, an electromagnetic motion measurement system is used to monitor arm's motion. An electromagnetic generator is placed in the room. Three receivers are worn at the shoulder, elbow and wrist. Then build-in software calculates arm's pose in space. For another, the motion of fingers is difficult to measure because that each hand has 21 DOF (Jiashun, Jiamiao et al.,2000) which makes it extremely flexible and changeable in configuration. EMG signal is introduced to monitor fingers' motion. The theoretical basis is that the nervous system uses EMG to control muscles' contraction, which yet is related to the motion of fingers. So once the changing EMG signal is captured by sensors, the contraction of muscles would be detected. Moreover the motions of fingers are sensed. The state of motion of the fingers can be identified through analyzing corresponding EMG signals, because they contain all control messages from nervous system to fingers.

In a word, complete behaviors of the operator's upper limb are acquired after merging the two channels' signals by the human-machine interaction system. Next the characteristic values will be identified from set of behaviors.

3.2 Simulators of space robot and target satellite

At the slave site zero gravity environment is simulated by air bearing system. Two satellite simulators are designed which can float on air bearings above the marble table, as is shown in Figure 1. These air bearings jet downward continuously to counterbalance gravity and support a platform allowing the unit to move and rotate as if it were maneuvering in space. Thrusters are mounted on each simulator to adjust its orbit and attitude. The simulator A simulates target satellite, and the simulator B is used as the base of space robot.

The simulator of space robot in the test bed is composed of the above base and one robot arm mounted on the top surface of the base with a 1-DOF gripper as end-effector. The upper computer on the base directly communicates with the robot arm through Ethernet port, while receiving commands from the relay station through wireless network.

3.3 Visual feedback and display system

At the slave site cameras are positioned around the marble table to provide global vision, which can be used to confirm the simulators' position and keep them from either colliding or moving beyond the table. Another two cameras are mounted on top of space robot's base body, providing local vision when maneuvering. They can pitch and yaw freely because of the active turntable carrier, controlled by the operator. Thus the attitude of cameras is adjusted to accommodate operator's view when necessary.

Display devices at master site receive vision signals from the slave site and create virtual telepresence circumstance by rebuilding models of remote environment. The main device is the three-dimensional (3D) projection system. It can display 2D or 3D signals, or both at the same time. The signals from the slave site are switched by software through which the operator could know what's going on at the other side when operating.

3.4 Communication link

Wireless communication link is created between the master site and the slave site utilizing point-to-point wireless bridge. The available bandwidth provided can ensure efficient transmission of all video, audio and commands. And the time delay in the communication link can be regulated by software to simulate long distance communication between earth and outer space.

4 RESEARCH STATUS AND RESULTS

By far the experimental test bed has been preliminarily completed. Moreover some validation experiments have been conducted as the key technologies below:

- Measuring and identifying the EMG signal. An EMG measuring device has been designed (shown in Figure 2). Being worn on the wrist, the device collects feeble EMG signals when fingers move. After amplifying, filtering and de-noising, the output signal is sent to data processor through wireless network, where features of the signal are extracted. These features will be used to identify behaviors of a hand. By now some simple behaviors have been identified.

- The experiment of simulation of space operation based on human-in-the-loop is carried out on the test bed, as is shown in Figure 2. The operator located at the master site remotely controls the floating simulator of space robot to approach the target. Once getting into the manipulator's workspace the simulator of space robot is switched into mode of holding the relative pose. Under the guidance of the operator, the manipulator slowly moves towards the target. The resulted reaction force is counterbalanced by thrusters on the base body. Wearing behavior sensing device the operator acts like pinching something by a hand. While at the slave site, the gripper of the manipulator successfully catches the target.

Figure 2 Simulation of space operation based on human-in-the-loop.

5 CONCLUSIONS AND FUTURE WORK

In conclusion, a novel concept of space operation system based on human-in-the-loop is presented in the article. The system guided by human's intelligent decision-making ability, utilizes human body's motion DOF instead of conventional input devices to control a dexterous multi-DOF space robot. Accordingly, the efficiency of operation is improved because the operator's concentrations are totally on his tasks. Besides, the problem of robot having insufficient artificial intelligence to deal with unknown environment is solved.

Still key technologies of the system are studied in this article. A novel behavior sensing method depending on non-contact measurement and EMG signal measurement is proposed, which can accurately read the operator's integrated movement by itself; An approach of transmitting movements by characteristic values is presented which promotes the reliability of data transmission but lowers the requirement for bandwidth; A method of driving actuator by database to reproduce operator's behavior is introduced. Moreover, an experimental test bed on the ground is built to demonstrate the concept and validate the key technologies. To date numbers of experiments carried out have got good results.

Admittedly, what has been observed in this article is far from being perfect, and requires further deep efforts. We will continue our work especially in aspects as follows in the future:

- Muscle force identification. The movements of fingers have already been identified from EMG signal, but not the muscle force. If acquired it would contribute to operation accuracy.
- Time delay. Time delay existing in real communication between earth and outer space seriously decreases stability and maneuverability of the system. This problem has been studied separately as a key technology and a corresponding approach to eliminate its impact on the system is presented. But experiments carried on simulating test bed ignored the time delay. In future work, time delay will be integrated into the system and operation under its influence will also be investigated.
- Telepresence. Telepresence circumstance is a guarantee of maneuverability of the system. We have telepresence functions in the system, such as video and audio feedback etc. but not enough. Next, force feedback is going to be added in, and a more transparent telepresence circumstance with immersed feeling will be built as well.

ACKNOWLEDGMENTS

We would like to thank Dr. Y. H. Meng and Dr. H. Chen for their valuable suggestions on the paper.

REFERENCES

Ambrose, R. O. 2000. NASA's Space Humanoid. IEEE INTELLIGENT SYSTEMS: 57-62.

Diftler, M. A., J. S. Mehling, M. E. Abdallah, et al. 2011. Robonaut 2 – The First Humanoid Robot in Space IEEE International Conference on Robotics and Automation. Shanghai,China.

Gibbs, G. and S. Sachdev. 2002. Canada and the International Space Station program: Overview and status. Acta Astronautica 51: 591-600.

Hirzinger, G., B. Brunner, J. Dietrich, et al. 1993. Sensor-Based Space Robotics-ROTEX and Its Telerobotic Features. IEEE TRANSACTIONS ON ROBOTICS AND AUTOMATION 9: 649-663.

Jiashun, W., W. Jiamiao, W. Jun, et al. 2000. Development of A New Teleoperation-Oriented Data Glove. Robot 22: 201-206.

Lii, N. Y., Z. P. Chen, B. Pleintinger, et al. 2010. Toward Understanding the Effects of Visual-Visual- and Force-Feedback on Robotic Hand Grasping Performance for Space Teleoperation. Ieee/Rsj 2010 International Conference on Intelligent Robots and Systems: 3745-3752.

Moosavian, S. A. A. and E. Papadopoulos. 2007. Free-flying robots in space: an overview of dynamics modeling,planning and control. Robotica 25: 537-547.

Passenberg, C.,A. Peer and M. Buss. 2010. A survey of environment-, operator-, and task-adapted controllers for teleoperation systems. Mechatronics 20: 787-801.

Rencheng, W. 1996. Discussion on Various Methods of EMG Processing for the Contron of Prostheses. International Conferece on Biomedical Engineering. Hong Kong.

Stoll, E. and U. Walte. 2009. Ground Verfication of the Feasibility of Telepresent On -Orbit Servicing. Journal of Field Robotics 26: 287-307.

Stoll, E.,M. Wilde and C. Pong. 2009. Using Virtual Reality for Human-Assisted In-Space Robotic Assembly. World Congress on Engineering and Computer Science: 685-690.

Weisbin, C. R. and D. Lavery. 1994. NASA Rover and Telerobotics Technology Program. IEEE Robotics &Automation Magazine 14-20.

Yoshida, K. 2009. Achievements in Space Robotics. IEEE Robotics & Automation Magazine: Magazine: 20-28.

CHAPTER 39

Transformation between Different Local Coordinate Systems of the Scapula

Xu Xu[a*], *Jia-Hua Lin*[a], *Kang Li*[b,c], *and Virak Tan*[c]

[a]Liberty Mutual Research Institute for Safety, 71 Frankland Road, Hopkinton, MA 01748, U.S.A.

[b]Department of Industrial and Systems Engineering, Rutgers, The State University of New Jersey, 96 Frelinghuysen Road, Piscataway, NJ 08854, U.S.A.

[c]Dempartment of Orthopaedics, The University of Medicine and Dentistry of New Jersey- New Jersey Medical School, 90 Bergen Street, Newark, NJ 07101, U.S.A.

ABSTRACT

Different local coordinate systems (LCS) of the scapula have been adopted in previous studies describing the kinematics of the scapula and the positions of muscle attachments. This study provides a transformation between Hogofrs LCS and ISB-recommended LCS. The results make it possible to compare the studies with various LCSs of the scapula.

Keywords: Scapula, local coordinate system, transformation.

INTRODUCTION

International Society of Biomechanics (ISB) recommended a local coordinate system (LCS) for the scapula (Wu et al., 2005). This LCS is based on angulus acromialis (AA), angulus inferior (AI), and trigonum scapulae (TS) with the origin coincident with AA. In recent years, this LCS has been increasingly adopted by other studies of scapula kinematics (van Andel et al., 2009; Picco et al., 2010).

Besides the ISB-recommended LCS of the scapula, there have been various LCSs of scapula describing the scapula orientation (Hogfors et al., 1987; Johnson et al., 1996; Kedgley and Dunning, 2010). In Hogfors et al. (1987), acromiclavicular joint (AC), AI, and angulus superior (AS) are used to define the LCS with the origin located at AC. This referred to as Hogfors LCS. Since the regression-based shoulder rhythm and the coordinates of the shoulder muscle attachments are available under Hogfors LCS (Hogfors et al., 1987; Hogfors et al., 1991; Karlsson and Peterson, 1992), this LCS is later used for building biomechanical models of the shoulder (Karlsson and Peterson, 1992; Dickerson et al., 2007).

The purpose of this work was to build a rotation matrix and a translational vector between the ISB-recommended LCS and Hogfors LCS. With this transformation, the kinematics of the scapula described under the two different LCSs can be compared, and the coordinates of the muscle attachments with respect to one LCS can be transformed to the other LCS.

METHOD

Twelve scapulae (10 left and 2 right) were scanned by CT. All the left scapulae were mirrored to the right scapulae because the ISB-recommended LCS and Hogfors LCS of the scapula were initially proposed for the right scapula. An investigator then digitized the bony landmarks on the scanned scapulae that were required for building the LCSs (Table1).

For each scanned scapula, the ISB-recommended LCS and Hogfors LCS were built (Table 1), and the rotation matrix from Hogfors to the ISB-recommended LCS was then derived. Since taking the average of each element in the rotation matrices among all the scapulae may result in a non-orthogonal matrix, an algorithm proposed in McConville (1980) was used to calculate the average rotation matrix from Hogfors LCS to the ISB-recommended LCS.

The characteristic length of the scapula (l_s) was defined as the distance between AA and AI. The translational vector from AA to AC was measured for each scapula as a proportion of the characteristic length.

Table 1. The definition of the ISB-recommended LCS and Hogfors LCS

	ISB recommended LCS	Hogfors LCS
Origin	AA	AC
X-axis	Perpendicular to the plane formed by AI, AA, and TS, pointing forward	(AI-AC)/|AI-AC|
Y-axis	Perpendicular to X-axis and Z-axis	Perpendicular to X-axis and Z-axis
Z-axis	(AA-TS)/|AA-TS|	Perpendicular to the plane formed by AI, AS, and AC, pointing forward

RESULTS

The average rotation matrix from Hogfors LCS to the ISB-recommended LCS (${}^{ISB}\mathbf{R}^{H}$) was derived as the following:

$$
{}^{ISB}\mathbf{R}^{H} = \begin{bmatrix} -0.141\,(0.050) & 0.160\,(0.065) & 0.977\,(0.007) \\ -0.681\,(0.055) & 0.701\,(0.052) & -0.213\,(0.028) \\ -0.718\,(0.058) & -0.696\,(0.061) & 0.010(0.068) \end{bmatrix}
$$

The numbers in the parenthesis stand for the standard deviations. The translational vector from AA to AC with respect to ISB-recommended LCS (${}^{ISB}(\mathbf{AC}-\mathbf{AA})$) was:

$$
{}^{ISB}(\mathbf{AC}-\mathbf{AA}) = \begin{bmatrix} 0.147(0.052) \\ 0.086(0.029) \\ -0.015(0.033) \end{bmatrix} \cdot l_s
$$

Therefore, the approximate transformation between Hogfors LCS and ISB-recommended LCS can be expressed as:

$$
{}^{ISB}\mathbf{p} = {}^{ISB}\mathbf{R}^{H} \cdot {}^{H}\mathbf{p} + {}^{ISB}(\mathbf{AC}-\mathbf{AA})
$$

where ${}^{ISB}\mathbf{p}$ and ${}^{H}\mathbf{p}$ are the coordinate of point $\mathbf{p}$ under the ISB-recommended LCS and Hogfors LCS, respectively.

DISCUSSION

With different scapula LCSs, the same motion pattern of the scapula would result in different kinematics data, and the coordinates of the same shoulder muscle attachment vary. This study provides an approximate transformation between Hogfors LCS and the ISB-recommended LCS, which makes it possible to compare the coordinates of muscle attachments and kinematics of the scapula from various studies.

A limitation to be addressed is that only one investigator digitized the bony land marks so the bias introduced by the investigator cannot be investigated.

REFERENCES

Dickerson, C.R., Chaffin, D.B., Hughes, R.E., 2007. A mathematical musculoskeletal shoulder model for proactive ergonomic analysis. Computer Methods in Biomechanics and Biomedical Engineering 10.

Hogfors, C., Sigholm, G., Herberts, P., 1987. BIOMECHANICAL MODEL OF THE HUMAN SHOULDER .1. ELEMENTS. J. Biomech. 20, 157-166.

Hogfors, C., Peterson, B., Sigholm, G., Herberts, P., 1991. Biomechanical model of the human shoulder joint .2. the shoulder rhythm. J. Biomech. 24, 699-709.

Johnson, G.R., Spalding, D., Nowitzke, A., Bogduk, N., 1996. Modelling the muscles of the scapula morphometric and coordinate data and functional implications. J. Biomech. 29, 1039-1051.

Karlsson, D., Peterson, B., 1992. Towards a model for force predictions in the human shoulder. J. Biomech. 25, 189-199.

Kedgley, A.E., Dunning, C.E., 2010. An alternative definition of the scapular coordinate system for use with RSA. J. Biomech. 43, 1527-1531.

McConville, J.T., Churchill, T.D., Kaleps, I. Clauser, C.E., Cuzzi, J., 1980. Anthropometric relationships of body and body segment moments of inertia. Air force aerospace medical research laboratory, Wright-Patterson air force base, OH, AFAMRL-TR-80-119.

Picco, B.R., Fischer, S.L., Dickerson, C.R., 2010. Quantifying scapula orientation and its influence on maximal hand force capability and shoulder muscle activity. Clin. Biomech. 25, 29-36.

van Andel, C., van Hutten, K., Eversdijk, M., Veeger, D., Harlaar, J., 2009. Recording scapular motion using an acromion marker cluster. Gait Posture 29, 123-128.

Wu, G., van der Helm, F.C.T., Veeger, H.E.J., Makhsous, M., Van Roy, P., Anglin, C., Nagels, J., Karduna, A.R., McQuade, K., Wang, X.G., Werner, F.W., Buchholz, B., 2005. ISB recommendation on definitions of joint coordinate systems of various joints for the reporting of human joint motion - Part II: shoulder, elbow, wrist and hand. J. Biomech. 38, 981-992.

CHAPTER 40

The System of Coordinating Security for the Construction of Nuclear Power Plant

Jan Donic, Andrea Lezovicova, Karol Habina

BOZPO, s.r.o.
Prievidza, Slovakia
Jan.donic@boz.sk

ABSTRACT

Project completion of units 3 and 4 in Mochovce nuclear power plant known as MO34 in Slovakia is among the most important projects in Europe. Safety, safety project and his staff for all their projects and works are one of the most important priorities and values that the investor prefers. Nowadays, the "safety" of nuclear power inflects in different contexts, the extension or expansion of this project is sharply observed and required not only by the client but also the general public.

The task and the goal of our company that provides outsourcing in the field of safety work is to achieve a “zero accident rate” on this project respectively approach to the limit closer as possible.

The project of the "zero accident" is a huge challenge for all workers share in the management of safety. Achieving this goal is possible only on condition that the safety work will also become part of the mindset of all staff who work on site without distinction of status and job position.

Combining of the legislative requirements of the world's best practice in OSH and applying of supportive measures into practice are necessary measures for the management of safety at work and in meeting the objective of “zero accident”. These measures can be divided in the system on site safety coordination on major areas such as prevention - preventative action, control activities, communication and information exchange.

Prevention is a tool that can be applied in addition to the standard procedures prescribed by the legislation, as well as what can be termed above standard. One of the basic tools of prevention and training system is informing about the risks.

In order to achieve reasonable success in the inspection activities, the site had to be divided into several sections of control depending on the size and scope of the site and determine the system of control activity. One or more safety oversight (OSH technicians) are assigned to each section that coordinate with contractors own performance. Success solution of deficiencies in the performance of control activities is in the range from 80% to 90%, thereby significantly minimizing the risk of accidents.

Prevention and control activity itself, mutual communication and sharing of important information are very important. Regular coordination meetings with suppliers and operations where safety coordinator provides all the important safety information are important tools of the communication. This tool is one of the most important tools for the rapid exchange of information necessary for dragging skills and proven way to cascade to all staff on site.

Keywords: health and safety, coordination, construction site, workers, preventive activity, control activity, documentation

1. INTRODUCTION

At present, Slovakia and other countries in the European Union placed a high emphasis on safety and health at work for all employees. This emphasis stems from the strict rules, taken and identified themselves with the rules adopted in the EU and the need to reduce and minimize the occurrence of occupational accidents.

Project completion of 3 and 4 block NPP called for MO 34 in Slovakia is among the most important projects in Europe. Currently are involved in the construction of more than 3000 employees. One of the most important priorities and values that the developer prefers to all his projects and workplaces is safety, security and the project itself and his employees. Nowadays, the "safety" of nuclear power inflected in different contexts and extension or expansion of this project is sharply observed and required not only the client itself but also the general public.

The role and also purpose of our society, which provides outsourcing of security work has been and still remains secure and to achieve this project or be as close to "0" injuries.

The project "0" accident rate is a huge challenge for all workers, who share in the management of safety of work. Achieving such a goal can only be provided, that job security will also become part of the mindset of all staff, who work on site without distinction of status and work position.

To do that we can fulfill this goal it is necessary to combine the management of occupational safety legal requirements together with the world's best practice in OSH, and apply in practice a number of supporting measures. These measures can be divided in **the system of coordination of security at the site** of the main areas as prevent - preventive action, control activities, communication and information exchange.

2. PREVENTIVE ACTIVITY:

2.1. Training and familiarization of the OSH

Due to the fact, that the project completion 3 and 4 block of NPP is currently one of the largest construction projects, is very important, that will be the mode of entry of persons on site organized and ensuring of quality system both accurate records of workers moving to the site while the accountancy movement and work only those workers, which the input received training on safety and health at work. This system must work seamlessly to provide work for a large number of workers of all requirements, that are placed on a high level of safety.

All workers who want to perform the work on completion of the 3 and 4 block must go through this phase of construction day training and familiarity. The training is divided into several important parts, such as occupational safety, fire protection, radiation protection, understand the hazards and etc. A person who successfully completes the training receives a certificate, which entitles the owner to obtain an entry card into the designated locations place where most of the activities related to the completion.

Basic requirement for successful completion of initial training is control of necessary and required professional documents, which demonstrating professional and medical competence of workers for perform their specific work activity. This is one of the preventive measures, to prevent unauthorized operation execution of works on which are needed for specific certificates, documents, or authorization.

As a developer interested in building awareness of workers does not end with initial training, but instead begins, personnel suppliers during the works will also receive a large amount of additional training. Part of them is a direct response to development of situation on the site and others are designed for specific categories of workers. Introducing the news is established daily pre-job trainings. The content of these trainings is to familiarize personnel everyday in the workplace for new risks, safe work practices, rules and so on. Records of this training must register the head worker. Continuously in these pre-meetings of working as consultants involved in the safety oversight.

Because, the construction is dynamic and constantly changing process, it is necessary, that all workers receive repeated familiarization on safety and health at work. This familiarity, which makes extradition conditional the entry card on workplace is conducted and organized by builder for construction workers in blue collar occupations in the period once a year, for head of workers supply companies is shorter interval to once every 6 months.

Among the preventive measures against damage workers' health due to the late intervention of urgent medical, includes the establishment of a permanent emergency services on site. During the 24 hours 7 days a week is a doctor available on site and manned, who can immediately provide in case of injury. This measure builder clearly show, the cost of human health is priceless and has the highest priority.

2.2. Documentation activity

Each contractor is responsible for carrying out the contract subject and must hold an appropriate valid license as required by law in Slovakia resp. EU, contractual performances can be realized only by professionally qualified personnel with the appropriate certificate. All contractor licenses and certificates about professional competency of his staff, must be in force throughout the term of the contract. The client as customer has the right to check the validity of these documents.

Any change - a new employee, and termination employee work performance, change his the general, mental, health, professional competence (intermediate or full) notify the purchaser in the form of record by the Building / Mounting daily/Diary of services, or special letter.

The contractor must have an efficient system for managing health and safety of the workplace and in case, that absorbed in that it co-operated in working more than one organization, must as part of his supplies to carry out the coordination of security. For this activity must appoint authorized people - Safety Coordinators. This is reflected in preparing plan of OSH. In developing the Plan OHS applies applicable safety and health requirements „Project rules and instruction for processing Integrated Safety Plan for the site / workplace", hereinafter "The rules of the project". OHS plan must a contractor submit to the client for approval no later than 7 working days before the date of receipt of the workplace.

Supplier to develop activities related with the execution of the contract, to be carried out to identify on the site MO34 and assessment of security risks (hereinafter risk register). In developing the risk register using the applicable requirements „Rules project". The Contractor shall submit this information to the client, first time as part of "Plan OHS" before starting work on site, and during the conduct of their regular (at least 1 times per month) update.

Supplier for each part of the fulfillment of the subject of contract also develop „Project Organization assembly - POA", which apply safe working and technological practices. POA contractor must submit to the client for consideration no later than 7 working days before commencement of work under the current schedule. Control application of safe work practices and technology made by a customer (coordinator responsible for client project documentation for this activity).

The Contractor shall provide the basis for creating and updating training materials for the admission of OSH and actively participate in the evaluation of training effectiveness.

As observed for the necessary documentation of OSH, the supplier shall like manner similarly attach also the documentation of fire protection, environmental documentation, great emphasis is placed on keeping site diary, which is conducted in accordance with the Law on Urban Planning and building regulations and internal policies by customer. The supplier is required under the character of the contract performance lead Installation / Construction Diary / Diary Services / Service statement.

3. CONTROL ACTIVITY

Coordination of security activities are carried out on site completion of the 3 and 4 block and related premises construction equipment, and on the premises operated 1. and 2 block of the completion of this completion, outsourcing contract - a specialized company to perform activities of OSH, which is the coordination of OHS of many years of experience in Slovakia and also in abroad. Performance security coordination includes the "Health and Safety at Work (OHS), Fire Protection (FP), Protecting the environment (PE) and Emergency preparedness (EP).

3.1. Preventive control activity

One of the most important activities in the performance of coordination in the context of prevention is regularly inspected by state OHS. Using journal entries („That is not rigt") reports are documenting all the minor flaws, in occupational safety (hereinafter referred to OHS), fire protection (hereinafter referred FP), environment ((hereinafter referred the environment). All these shortcomings were collected monthly to evaluate the level and based on these evaluations can be checked, which controlled area (indicator) is deteriorating, or in which area is progress recorded and improvement.

3.2. Control activities on site

For the purpose of the supervisory and control activities, over effectiveness of management processes, safety and prevention (OHS, FP, environmental) for vendors contracted for the completion of 3. and 4. blocks and regular checks are made on the sites of suppliers focused on the principles and Prevention (OSH, FP, Environmental). Each supplier has its own safety coordinator.

The regulatory regime security coordination is ensured through continuous participation of coordinators and safety representatives of outsourcing compa-nies in OHS on site in the form of continuous operation e.g.. mode (24/7). Coordinating instructions and findings activities are recorded in documents of construction management. Under this system, the building is regularly moni-tored and preventive measures to shorten the time to identify and eliminate non-conformities. Shorter time has a positive impact on the possibility of accidents.

Supplier allows to customer safety coordinator and people responsible for this activity the client full access to their workplace, to OHS inspections of the workplace in accordance with applicable legislation and documentation issued by the sponsor / developer in the area of OHS coordination. The object controls coordinator may be a review of the OSH management contractor, and control of personnel, technical, material and organizational conditions for implementing the subject matter in terms of OHS.

3.3. The imposition of sanctions

In the event of a breach or non-safety of principles and prevention may be awarded penalties for individual contractors under instructions bezpečnostnotechnické performance of construction. Penalties for infringement are as follows:

3.3.1. Cessation of work

In the event of the defect(s) relating to safety and prevention, which can not be removed on the spot and create a risk of health to employees, event of an emergency event (environmental accidents, serious industrial accidents) there is a cessation of work, register it and is also possible to continue by only agreed way to resolve the deficiencies. Stopping work is a very effective sanction and suppliers are very flexible and react quickly and remove the deficiencies, which are a common reason for stopping work.

3.3.2. The system of award yellow cards

In order to increase the level of safety at work of workers of suppliers introduced a system of "yellow cards". These are granted to suppliers workers engaged in construction, installation, management and other activities for completion, who violate the rules and regulations in the field of OHS, FP, and the environment.

3.3.3. Procedure for issuing cards

An employee who violates OHS rules (for example: failure to perform duties of the principal, not wearing PPE, failure to technological procedure, non-compliance manuals, work without a license and authorization to work if this certificate is owned, but but not with himself, work without FP permission, failure to comply with order, incorrect storage of chemicals, improper handling of flammable substances, smoking outside designated smoking places) gets for the infringement "yellow card".

For each award card worker must be duly notified also with explaining the breach and follow procedure which will grant cards dealt:

a.) for any subsequent similar violation will get another "yellow card"
b.) in the event that violates the rules for the third time, gets' red card" and then it will be blocked the entrance to the complex electronic structure
c.) red card worker can get a grant without award yellow cards for gross violation OHS, FP, environment for example: work of height

near the edge of collapse without the individual protection, carry out activities without a license or authorization, unsafe handling of flammable liquids in close flammable objects, smoking in areas with increased fire risk, hazardous and mismanagement of waste, gross breach of duty of the head of worker and etc.

The yellow cards are registered with the safety coordinator and is conducted electronically, but also in writing form.

The yellow card may be granted for completion of the safety coordinator, OSH personnel department builder, coordinators security vendors (for their workers), foreman builder.

3.3.4. The system of positive motivation workers and supply companies

In order to increase the share of workers at a higher level of safety, is system of positive incentives. As the site is a large number of workers, whose behavior is dangerous, there are many workers, whose behavior and habits of the OHS, FP and the environment are more than positive. This system aims to find also these workers, which, when they fulfill the required criteria, initiative approaching to identify risks, proposed action and provide an example for others, can be rewarded. Positive motivation in this case the material or financial donation, or delete the yellow card.

Running this system of positive incentives is also made for the supply company. Those firms that recorded during the evaluation period, the best progress in the field of OHS, FP and the environment are rewarded with a certificate „Safety company of month" (Safety company of month) , which is signed and delivered the project director for construction.

4. COMMUNICATION AND SHARING INFORMATION - COORDINATION MEETINGS

An important part of managing and coordinating the construction of safety is the regular transmission of new information on OHS, FP, environment, all suppliers. For this purpose, OHS coordinator organizes weekly coordination meetings. The agenda of these meetings follows the current problems in the construction period. Participation in coordination meetings is required for safety coordinators and coordinators documentation suppliers. The role of all suppliers, who is to participate in these regular meetings is to push actual information all its employees and workers. Important outcomes of this consultation are also Annex for example most of the photo documentation of deficiencies, current situation of activity plans for individual floors, table activities with indicating places where is blasting operations concurrence of suppliers, identification of hazards on the common workplaces as well as many other current and relevant information.

In order to improve the awareness of top management builder regularly organizes "Safety Day" and "Safety Week“. These actions are aimed at improving levels of OHS but also give space for the supplier, who can present their achievements in OSH, possibly open up possibilities for dialogue between the two parties in an effort to improve.

Projects support in area of the communication is publishing the monthly magazine, which is determined to the workers of the site. In this magazine are provided to the staff all necessary information from the OHS, FP, environment. This monthly magazine, entitled "Together and safely" is also space for a poll, where workers can express their views, observations and suggestions for improvement. The aim of Editorial Board will evaluate proposals and applied in practice all the positive recommendations.

5. CONCLUSION

Completion of construction of the 3 and 4 block is scheduled for years 2013 and 2014. In terms of assessment is therefore possible to evaluate the previous period. This assessment clearly shows, that the high level of safety is ensured by the fulfillment of legislative obligations in area of coordination together with the arrangements set by the builder side but also quality work of experts in the field of OHS from outsourcing company and other professional and managerial staff involved on the construction. Positively reflected in the statistics of occupational accidents.

The fact, that thanks to the developer, workers, who are involved in the management and OHS performance, but also workers of the actual construction the project of construction 3. and 4. block become truly successful project demonstrate our statistics:

number of hours worked: 12 378 547
registered occupational accidents: 8
registered work accidents: 31
serious work accidents:1
fatal work injuries:0

These statistics clearly prove that we are doing well in meeting the objective that all employees return home safe after work and we believe that by a common hard work and we will do well in achieving this goal in the future.

CHAPTER 41

Performance Assessment and Optimization of HSE Management Systems with Human Error and Ambiguity by an Integrated Fuzzy Multivariate Approach in a Large Conventional Power Plant Manufacturer

A. Azadeh 1, Z. Jiryaei 2, A. Hasani Farmand 1

[1] Department of Industrial Engineering, College of Engineering, University of Tehran, Iran

[2] Department of Industrial Engineering, College of Engineering, University of Tafresh, Iran

aazadeh@ut.ac.ir, z.jiryaei@yahoo.com

ABSTRACT

This study presents an integrated approach performance assessment and optimization of integrated health safety environment (HSE) management system based on fuzzy data envelopment analysis (FDEA) considering the possible human error and data vagueness in a conventional power plant manufacturer. In doing so, it corresponds and integrates its registered HSE-MS with OHSAS 18001:2007 and

ISO 14001:2004 to evaluate multiple inputs and outputs of over 35 subsidiary HSE divisions with parallel mission and objectives simultaneously. The HSE divisions of each subsidiary are considered as decision making units (DMUs). Not only doing this refutable method rank their relevant performance efficiencies in certain and uncertain conditions, but also it determines efficient target indices for each DMU, and could assure continuous improvement in the organization. This would help managers to identify the areas of strengths and weaknesses in their HSE management system and set improvement target plan for the related HSE management system. In this model based on Deming's continuous improvement cycle, managers are also able to evaluate the prevailing strengths and weaknesses and target their improvement strategies at the relevant stages of the cycle.

Keywords: Performance Assessment; Data Envelopment Analysis (DEA); Fuzzy Data Envelopment Analysis (FDEA); Decision Making Unit (DMU); Health, Safety, Environment (HSE); Integrated Management System (IMS); Human Error; Ambiguity

1 INTRODUCTION

Integration of HSE Management Systems could only bring about fundamental changes unless the careful selection and constant monitoring of reactive and proactive performance measures enriches its practice. These permit evaluation of implemented HSE Management Systems efficiency in quantifiable terms hence enable the HSE managers to carefully assess and diligently launch performance benchmarking schemes at organizational and divisional levels.

Based on Deming's Continuous Improvement Cycle; the Health, Safety, and Environment Management Systems are the acclaimed systems that not only would significantly minimize the risks to the enterprise human, natural, and capital resources; but also could boost their performances through continuous learning from the past experiences as well as effective benchmarking of their rivals. Indeed, management systems as such have become the main organizational pillars and the key prerequisite for their survival (Zutschi and Sohal, 2003). Given their dynamic corporate culture, some organizations attempt to implement management systems that go beyond the boundaries of the prevailing HSE standards to achieve a better performance and to build a polished public image.

Zutschi and Sohal (2003) highlighted the potential 'tangible and non-quantifiable benefits' from integration. This is also applicable to management systems that register to health, safety, and environmental standards (Basso et al., 2004). Santos-Reyes and Beard (2009) has been developed a Systemic Safety Management System (SSMS) model. The SSMS aims to maintain risk within an acceptable range in the operations of any organization in a coherent way. Graves et al. (2004) reviewed the development of the risk assessment tools contained in the revised guidance. Marhavilas et al. (2011) developed a new hybrid risk assessment process (HRAP) and applied it in the Greek Public Power Corporation (PPC), by using occupational accidents that have been recorded, during the 12-year period of 1993–

2004.Azadeh et al. (2006) described an integrated macroergonomics model for operation and maintenance of power plant.

Data envelopment analysis (DEA) is a good solution to evaluate the performance of systems in a multi input-multi output environment. DEA is one popular optimization method used for measuring the relative efficiency of DMUs. Camanho and Dyson, (2005) developed measures "enable the decision making units' internal inefficiencies to be distinguished from those associated with their group (or program) characteristics". The case study of their works is bank branches. However, they don't consider any approach to deal with DEA sensitivity to noise as a banking system encounter with noisy environment. Wu et al. (2010), used data envelopment analysis (DEA) to construct a scalar measure of efficiency for all police precincts. Mercan et al. (2003) employ DEA to determine fundamental financial ratios using data from commercial banks in Turkey form 1989 to 1999. Tyagi et al. (2009) have study the performance of academic programs in India. They have used DEA with sensitivity analysis in their study. Habibov and Fan (2010) utilized Data Envelopment Analysis (DEA) for a comparison of poverty reduction performances of jurisdictional social welfare programs across Canadian provinces.

This study takes proactive and reactive stance and submits a proposed method for performance assessment and optimization of HSE management systems to further enrich this body of research. It demonstrates how Data Envelopment Analysis (DEA) with the help of fuzzy enables a conventional power plant with uncertain data sets to simultaneously analyze multiple output/input indicators to: a) precisely determine technical efficiencies, b) completely rank the divisional efficiencies, and c) diligently set efficient target indices for its HSE management systems.

2 METHODOLOGY

This study adopts Data Envelopment Analysis (DEA) and fuzzy DEA– a non-parametric method for efficiency evaluation – to assess and optimize the performance of a set of peer HSE divisions as its Decision Making Units (DMUs), which convert multiple inputs to multiple outputs. Given the nature of the DMUs under evaluation – where the change in output is not a function of direct change in input values – an output-oriented DEA model with a Variable Returns to Scale (VRS) frontier type is selected. To choose the appropriate inputs and outputs for effective assessment of the HSE management systems, this study subscribes to its relevant standards contents and clauses respectively. That is, for selection of inputs, the study conforms to the content of OHSAS 18001:2007, which stresses on human resources, specialized skills, organizational infrastructure, technology, and financial recourses. In the same manner, for selection of outputs, it initially identified the implemented – but not necessarily certified – management systems pertinent to HSE and then corresponds and integrates the clauses of registered standards in these divisions.

The adopted VRS output-oriented DEA model can be expressed as follow: (1)

$$\max \phi + \varepsilon(\sum_{i=1}^{m} s_i^- + \sum_{r=1}^{s} s_r^+)$$
$$\sum_{j=1}^{n} \lambda_j x_{ij} + s_i^- = x_{io} \qquad i = 1,2,\ldots,m;$$
$$\sum_{j=1}^{n} \lambda_j y_{ij} - s_r^+ = \phi y_{ro} \qquad r = 1,2,\ldots,s;$$
$$\sum_{j=1}^{n} \lambda_j = 1 \qquad j = 1,2,\ldots,n;$$
$$\lambda_j \geq 0$$
$$s_i^- \geq 0$$
$$s_r^+ \geq 0$$

Considering the set of n observations on the DMUs, each observation for DMU_j $(j = 1, \ldots, n)$, uses m inputs x_{ij} $(i = 1, 2, \ldots, m)$ to produce s outputs y_{rj} $(r = 1, 2, \ldots, s)$. Where DMU0 represents one of the n DMUs under evaluation, ε is a positive non-Archimedean infinitesimal and λ are unknown weights. The optimal value of θ^* represents the technical efficiencies of the DMUs. Therefore, the most technically efficient DMUs are said to have $\theta^*=1$ & $s_i^{-*} = s_r^{+*} = 0$ and the inefficient DMUs will have a $\theta^* < 1$. Methods of DEA are run for each DMU separately.

In cases with multiple efficient DMUs with $\phi^* = 1$, development of methods for complete ranking becomes necessary. For this purpose, this study subscribes to the suggested model by Jahanshahloo et al. (2004), which uses common set of weights (CSW). Following this way, in addition to determining the efficiency and ranking of DMUs, suppose that A = *{j: DMU_j is efficient in the model (1)}*. To rank efficient DMUs, we omit the corresponding constraints of all efficient DMUs (A), and then evaluate efficient DMUs. To more explain for how of determining the efficiency and ranking of DMUs refer to Jahanshahloo et al. (2004) reference.

Also the lack of output indicators can be identified and efficient targets ($\hat{y}_{ro}$) of each DMU can be calculated through the following equation: (2)

$$\hat{y}_{ro} = \phi^* y_{ro} + s_r^{+*} \quad r = 1,2,\ldots,s$$

It is important to note that the competency of the above model can be limited in conditions of uncertainty. A method therefore is needed to add to the potency of this model at this particular situation. Fuzzy DEA is a tool for evaluation of performance under uncertain conditions, which uses the theory of fuzzy sets to demonstrate uncertain data and analyze them more accurately. Saati et al. (2002) proposed a fuzzy version of DEA using triangular fuzzy numbers substituting $\tilde{x}_{ij} = (x_{ij}^m, x_{ij}^l, x_{ij}^u)$ and $\tilde{y}_{ij} = (y_{ij}^m, y_{ij}^l, y_{ij}^u)$ into the model. They offered a new idea by α-cut in fuzzy DEA version converted into certain intervals and choose a point in intervals variables to satisfy the limitations and at the same time optimize the objective function. This fuzzy DEA model can be expressed as follow: (3)

$$\max \phi + \varepsilon(\sum_{i=1}^{m} s_i^- + \sum_{r=1}^{s} s_r^+)$$
$$\sum_{j=1}^{n} \lambda_j (x_{ij}^m, x_{ij}^l, x_{ij}^u) + s_i^- = x_{io} \qquad i = 1,2,\ldots,m;$$
$$\sum_{j=1}^{n} \lambda_j (y_{ij}^m, y_{ij}^l, y_{ij}^u) - s_r^+ = \phi \tilde{y}_{ro} \qquad r = 1,2,\ldots,s;$$
$$\sum_{j=1}^{n} \lambda_j = 1 \quad \lambda_j \geq 0,\ s_i^- \geq 0,\ s_r^+ \geq 0$$

Efficiency of a DMU in Fuzzy DEA is not a crisp number rather it is a fuzzy number. Using α-cuts, also called α-level sets, the inputs and outputs can be represented by different level of confidence intervals. The fuzzy DEA model is, therefore, transformed to a family of crisp DEA models with different α-level sets.

The selected case study for this research is a conventional power plant and its 35 subsidiaries engaged in development and implementation of power plant projects under EPC and IP schemes as well as manufacturing the relative equipments. The HSE divisions of each subsidiary are considered as the model's Decision Making Units. The E&P Forum's HSE-MS, OHSAS 18001:2007, ISO 14001:2004 management systems have been implemented by the group's HSE divisions. For assessing the performance of an integrated HSE management system, a correspondence table has been developed to link the clauses and sub-clauses of each standard.

The number of HSE personnel (x_1), HSE training man-hours for all personnel (x_2), and allocated budget for HSE divisions (x_3) have been selected as three input indicators which correspond to the aforementioned OHSAS 18001:2007. The selective inputs of HSE system that are used in this study have coincidence with the proposed HSE-MS and have been extracted of the HSE-MS, ISO14001 and OHSAS 18001 standards. Clause 4-4-1 from ISO14001:2004 and OHSAS 18001:2007 standard represents that resources, include human resources, specialized skills, organizational infrastructure, technology and financial resources. Also clause 3-3 from HSE–MS standard represents some resources include personnel, equipment, facilities and resources. The most important factor in the selection of input parameters is their coincidence with the HSE–IMS system (Zutschi and Sohal, 2003). All three indicators are proposed as useable resources (inputs) of HSE division to establish and meet the requirements of HSE-MS systems. In addition all three indicators are per capita to considering scale of project or company in consumption measure from resources (inputs). Thus input indicators are normalized due to the volume of activities, size of company or project. The verified quantitative data (for the year of 2009) for each input indicator have been collected from available information in company with the approval of the HSE divisions.

Comprising all the main clauses of the above mentioned management systems, the resulting correspondence table provides the seven qualitative output indicators for assessment purposes and includes: leadership and commitment (y_1); policy and strategic objective (y_2); planning (y_3); organization, resources, and documentation (y_4); evaluation and risk management (y_5); implementation and monitoring (y_6); and, auditing and reviewing (y_7). Output indicators in this study are completely consistent with the clauses of standard; therefore the indicators are system-based. These indicators include two attitudes, passive (event-based) and active (prospective) therefore these indicators can be seen as whole-based indicators. Continuous improvement with the Deming cycle (PDCA) concept is most important characteristics in integration of systems that are considered in output indicators. A specific checklist has been developed for the certified auditors to assess these indicators. The main clauses this checklist are based on HSE-IMS and sub-clauses based on HSE-MS, ISO 14001:2004, OHSAS 18001:2007, HSE workbook and standards indicators that available in this field. For evaluation of each sub clauses some questions are designed that by qualified and trained auditors certified.

3 EMPIRICAL APPLICATION

The results fully conform to the Deming's Continuous Improvement Cycle and are presented in Table 1. In the next stage, inputs and outputs values for all of 35 HSE divisions are collected.

Table 1: Correspondence between OHSAS 18001:2007, ISO 14001:2004, and HSE-MS

Deming' s Cycle	Clauses	HSE-MS	Clauses	ISO 14001:2004	Clauses	OHSAS 18001:2007
Plan	1	Leadership and Commitment	4.1	General requirements	4.1	General requirements
	2	Policy and Strategic Objectives	4.2	Environmental policy	4.2	OH&S policy
	5	Planning	4.3	Planning	4.3	Planning
Do	3	Organization, resources, and documentations	4.4	Implementation and Operation	4.4	Implementation and Operation
	4	Evaluation and Risk Management	4.3.1	Environmental Aspects	4.3.1	Hazard identification, risk assessment
Check	6	Implementation and Monitoring	4.5	Checking	4.5	Checking
Act	7	Auditing and Reviewing	4.5.5	Internal Audit	4.5.5	Internal Audit
			4.6	Management review	4.6	Management review

Considering the set of $n=35$ observations on the DMUs (HSE divisions), each observation for DMU_j $(j=1,\ldots,35)$, uses $m=3$ inputs x_{ij} $(i = 1,2,3)$ to produce $s=7$ outputs y_{rj} $(r = 1,2,\ldots,7)$. The selected DEA model for this case is run and applied for each DMU separately.

After solving the above DEA model by use of Auto Assess software (Azadeh et al., 2007), the results of corresponding technical efficiencies are illustrated in Table 2. As stated before, values of technical efficiencies in 15 DMUs equal to 1. Subsequently ranking of DMUs by use of these values is impossible. Thus, for complete ranking, efficient output targets, by use of common set of weights (CSW) method (Jahanshahloo et al., 2004) are shown in Tables 3, 4 respectively.

Table 2: Technical Efficiency of DMUs with DEA

TE	DMU_j	TE	DMU_j	TE	DMU_j	TE	DMU_j	TE	DMU_j	TE	DMU_j
1	DMU_{31}	0.9	DMU_{25}	0.9	DMU_{19}	1	DMU_{13}	1	DMU_7	1	DMU_1
1	DMU_{32}	0.9	DMU_{26}	0.8	DMU_{20}	1	DMU_{14}	0.8	DMU_8	1	DMU_2
1	DMU_{33}	0.8	DMU_{27}	0.7	DMU_{21}	0.8	DMU_{15}	0.7	DMU_9	1	DMU_3
1	DMU_{34}	1	DMU_{28}	0.9	DMU_{22}	1	DMU_{16}	1	DMU_{10}	1	DMU_4
0.8	DMU_{35}	0.8	DMU_{29}	1	DMU_{23}	0.8	DMU_{17}	1	DMU_{11}	1	DMU_5
		1	DMU_{30}	1	DMU_{24}	1	DMU_{18}	1	DMU_{12}	0.7	DMU_6

Table 3: Complete ranking of all DMUs with DEA

TE	DMU_j	TE	DMU_j	TE	DMU_j	TE	DMU_j	TE	DMU_j
0.96	DMU_{29}	0.852	DMU_{22}	0.95	DMU_{15}	0.915	DMU_8	1.127	DMU_1
1.054	DMU_{30}	0.979	DMU_{23}	1.127	DMU_{16}	0.924	DMU_9	1.007	DMU_2
1.008	DMU_{31}	0.989	DMU_{24}	0.957	DMU_{17}	1.025	DMU_{10}	1.024	DMU_3
0.985	DMU_{32}	0.949	DMU_{25}	1.072	DMU_{18}	0.989	DMU_{11}	1.133	DMU_4
0.998	DMU_{33}	0.971	DMU_{26}	0.938	DMU_{19}	0.991	DMU_{12}	1.007	DMU_5
1.077	DMU_{34}	0.929	DMU_{27}	0.768	DMU_{20}	0.983	DMU_{13}	0.899	DMU_6
0.986	DMU_{35}	1.092	DMU_{28}	0.924	DMU_{21}	1.03	DMU_{14}	0.973	DMU_7

Table 4: Values of efficient output targets

Unit Name	$\hat{y}_1$	$\hat{y}_2$	$\hat{y}_3$	$\hat{y}_4$	$\hat{y}_5$	$\hat{y}_6$	$\hat{y}_7$
DMU1	95	90	90	70	90	100	95
DMU2	$\hat{y}_{1j}$	$\hat{y}_{2j}$	$\hat{y}_{3j}$	$\hat{y}_{4j}$	$\hat{y}_{5j}$	$\hat{y}_{6j}$	$\hat{y}_{7j}$
DMU35	79.73	78.8	74.41	63.93	78.73	89.84	82.45

To reduce the possible errors from the auditors' judgments in correctly assigning scores to the perceived performance of output indicators, this study subscribed to triangular fuzzy numbers and applied fuzzy DEA. To do so, scores 1 and 100 representing the lowest and highest possible values have been assigned to y_{ij}^{l} and y_{ij}^{u} respectively. The applied fuzzy model, as expected, enabled more accurate analysis of uncertain data, which is presented as equation 3. Using α-cuts method, for instance Table 5 shows the results at selected α-levels from the fuzzy DEA model which have been solved (DMU 1 and 35).

Table 5: Fuzzy DEA results - Technical Efficiencies (TE) and ranks at different α-levels

Unit Name	α = 0.1		α = 0.3		α = 0.5		α = 0.7		α = 0.9		α = 1	
	TE	Ran	TE	Ran	TE	Ran	TE	Ran	TE	Ran	TE	Ran
DMU1	12.	11	4.1	9	2.3	9	1.6	9	1.1	9	1.2	2
DMU35	12.	9	4.1	11	2.2	13	1.4	23	0.9	26	1.0	18

4 ANALYSIS AND DISCUSSION

The result from solving the DEA model provides technical efficiencies, ranking, and efficient targets for all DMUs. Table 2 reports 15 DMUs with TE = 1 (as efficient units) and 20 DMUs with TE < 1 (as inefficient units). As discussed earlier, with multiple numbers of efficient DMUs, complete ranking between them becomes necessary. Table 3 illustrates the result from the application of a complete ranking method and reports DMU_4 with TE = 1.133 and DMU_{20} with TE = 0.768 to rank the highest and the lowest respectively. It is important to note that, assigning TE > 1 to DMUs is idiosyncratic to complete ranking models, which rank the most efficient DMU by the highest TE. The technical efficiency of 91% (calculated on average of all DMUs), implies the overall inefficiency of 9% in entire enterprise's HSE management system.

Slacks and efficient targets for each DMU – calculated through equation (2) – are presented in Table 4 and 5. Given the orientation type of the selected DEA model, these targets are specifically calculated for the output indicators. It is important to note that, the output target values ($\hat{y}_i$) for the efficient DMUs (e.g., DMU_1) fully corresponds to their measured output values (y_i), which verifies their location on the efficiency frontier. In the opposite manner, these target values vary from those initially measured for inefficient DMUs (e.g., DMU_{35}) and signifies the need for improvement measures. To locate itself on the efficiency frontier, each inefficient unit can then individually identify its potentials for improvement by calculating the differences between these two values ($\hat{y}_i - y_i$).

In the same manner, the collective potentials of all DMUs can be derived from the difference between the average target and measured values. Putting this into perspective, managers can now identify the areas of strengths and weaknesses in their HSE management system and set improvement target plan for the related HSE management system. As this model fully complies with Continuous Improvement Cycle, managers are also able to analyze the prevailing strengths and weaknesses and target their improvement strategies at the relevant stages of the PDCA cycle.

For instance, as Table 6 presents the average measured values of 50.943 and 54.543 indicates 'evaluation and risk management' (y_5) as well as 'organization, resources, and documentation' (y_4) activities to respectively demonstrate the lowest efficiency levels. These amounts can enhance by potential values of 29.192 and 16.031 to reach the target values of 80.135 and 70.574 for the HSE management system. In the same manner, analysis of these amounts reveals that the HSE management system fails to be efficient at the second stage of PDCA cycle (DO) and alerts additional improvement measures. Other results from Table 8 indicate that the studied HSE management system exhibits highest efficiency in 'auditing and reviewing (y_7) activities (with average measured values of 69.429) at the final stage of PDCA cycle (ACT).

Table 6: Average values of output indicators

PDCA Stages	Output Indicators	Average Measured Values	Average Efficient Target Values	Average potential Improvement Values
Plan	Leadership and Commitment (y_1)	61.057	81.019	19.962
	Policy and strategic objective (y_2)	61.971	82.318	20.347
	Planning (y_3)	62.086	76.433	14.347
Do	Organization, resources, and	54.543	70.574	16.031
	Evaluation and risk management (y_5)	50.943	80.135	29.192
Check	Implementation and monitoring	64.229	81.679	17.45
Act	Auditing and reviewing (y_7)	69.429	84.431	15.003

To justify the fuzzy DEA results, Spearman correlation test has been applied to compare the ranks of crisp DEA and fuzzy DEA. It is obvious that, as α-level increases, the results of DEA and fuzzy DEA get closer, and their correlation increases. Not only a relatively high degree of correlation between the two was reported, but also at $\alpha = 1$ (lowest level of uncertainty) it has a highest degree of correlation ($\rho = 0.95$) which verifies the ranking of fuzzy DEA. In cases that the results are not verified by this test, the fuzzy DEA results are still valid and present the rankings of DMUs in a complete ambiguous environment. The results of Spearman correlation test between DEA and FDEA for different α-values are shown in Figure 1.

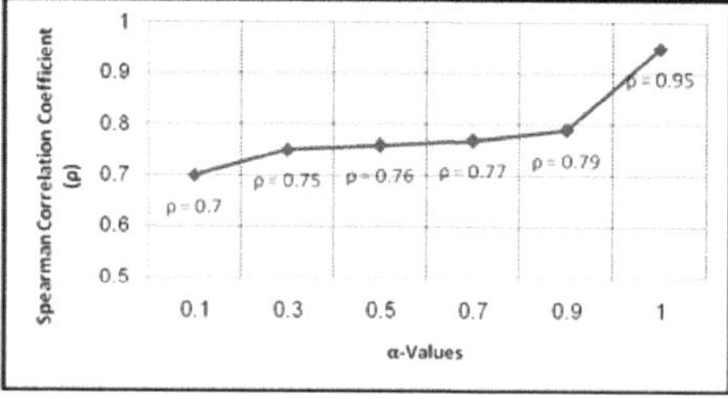

Figure 1: The results of Spearman correlation test between DEA and FDEA for different α-values

5 CONCLUDING REMARKS

HSE Management Systems play an important role in enhancement of safety and human and organizational productivity. This study presented an efficient and integrated approach for performance assessment and optimization of HSE-MS in an actual conventional power plant manufacturer and its 35 subsidiary divisions. The suggested DEA method could evaluate the relative efficiency of HSE management systems through considering multiple input and output indicators simultaneously. In addition to identify strengths and weaknesses of efficient and inefficient units, this method can provide each division and the whole enterprise with efficient targets to reach its potential efficiency frontier. The results show that the second stage of PDCA cycle (Do) and final stage of PDCA cycle (Act) have lowest and highest efficiency respectively in power plant manufacturer. Furthermore, this study subscribed theory of fuzzy sets and adopted fuzzy DEA method to reduce the human error and the uncertainty existing in qualitative indicators. Not only doing this refutable method (integrated DEA and fuzzy DEA) enables the managers to formulate development strategies for each division, but also could assure continuous improvement in their enterprise at its entirety. This study also advocates incorporating different approaches for solving and complete ranking of crisp and fuzzy DEA models. It encourages selection of other valid input and output indicators to evaluate the performance of different HSE management systems. Moreover, development of novel approaches to provide efficient DMUs with improved efficiency targets is highly recommended. Furthermore, utilization of parametric methods, e.g., regression analysis for modeling and analyzing such multivariate systems is also suggested.

REFERENCES

Azadeh, A., Keramati, A., Mohammad Fam, I. and Jamshidnejad, B. (2006). Enhancing the availability and reliability of power plants through macroergonomics approach, Journal of Scientific and Industrial research, 65, 873-878.

Basso, B., Carpegna, C., Dibitonto, C., Gaido, G., Robotto, A. and Zonato, C. (2004). Reviewing the safety management system by incident investigation and performance indicators. Journal of Loss Prevention in the Process Industries, 17 (3), 225–231.

Camanho, A.S. and Dyson, R.G. (2005). Cost efficiency measurement with price uncertainty: a DEA application to bank branch assessments, European Journal of Operational Research, 161(2), 432-446.

Graves, R.J., Way, K., Riley, D., Lawton, C. and Morris, L. (2004). Development of risk filter and risk assessment worksheets for HSE guidance—'Upper Limb Disorders in the Workplace' 2002, Applied Ergonomics, 35 (5), 475-484.

Habibov, N.N. and Fan, L. (2010). Comparing and contrasting poverty reduction performance of social welfare programs across jurisdictions in Canada using Data Envelopment Analysis (DEA): An exploratory study of the era of devolution, Evaluation and Program Planning, 33 (4), 457-467.

International Organization for Standardization (1996). Environmental management systems-Specification with guidance for use, ISO 14001:1996, First Edition.

Jahanshahloo, G.R., Memariani, A., Hosseinzadeh Lotfi, F. and Rezaei, H.Z. (2004). A note on some of DEA models and finding efficiency and complete ranking using common set of weights, Applied Mathematics and Computation, article in press.

Marhavilas, P.K., Koulouriotis, D.E. and Mitrakas, C. (2011). On the development of a new hybrid risk assessment process using occupational accidents' data: Application on the Greek Public Electric Power Provider, Journal of Loss Prevention in the Process Industries, 24 (5), 671-687.

Mercan, M., Reisman, A., Yolalan, R. and Emel, A.B. (2003). The effect of scale and mode of ownership on the financial performance of the Turkish banking sector: results of a DEA-based analysis, Socio-Economic Planning Sciences, 37 (3), 185-202.

Santos-Reyes, J. and Beard, A.N. (2009). A SSMS model with application to the oil and gas industry, Journal of Loss Prevention in the Process Industries, 22 (6), 958-970.

Saati, M.S., Memariani, A. and Jahanshahloo, G.R. (2002). Efficiency Analysis and Ranking Of DMUs with Fuzzy Data, Journal of Fuzzy Optimization and Decision Making, 11 (3), 255-267.

Tyagi, P., Yadav, S.P. and Singh, S.P. (2009). Relative performance of academic departments using DEA with sensitivity analysis, Evaluation and Program Planning, 32 (2), 168-177.

Wu, T.H., Chen, M.Sh. and Yeh, J.Y. (2010). Measuring the performance of police forces in Taiwan using data envelopment analysis, Evaluation and Program Planning, 33 (3), 246-254.

Zutschi, A. and Sohal, A. (2003). Integrated management system: The experience of three Australian organizations, Journal of Manufacturing Technology Management, 16 (2), 211-232.

CHAPTER 42

The Analysis on Force of Door Operation of Applying Universal Design and the Mechanism Design of Measurement Device

Shih-Bin Wang, Kai-Chieh Lin**, Chih-Fu Wu***

*PhD Program of Design Science, Tatung University, Taipei, Taiwan;
Lee-Ming Institute of Technology, New Taipei City, Taiwan
**Tatung University, Taipei, Taiwan
wsb@mail.lit.edu.tw

ABSTRACT

Door operation composed of a series of clever actions is the most common problems confronted in our daily life. However, the lacks of operating force standards and measurement methods make universal design of doors difficult. In this study, a force measurement system for door operation has been developed, which was proved to have the capability of measuring the door operating force.

Thus, experiments of measuring operating force proceeded including hinged doors and sliding doors either with or without door closers. As door closers were applied, the results showed that the operating forces increased greatly, and the fluctuated phenomenon of force signals obviously reduced. Also, both two kinds of measurements composed of the initial force and the maximum force were significant. As opening the door faster, the operating forces rose relatively, and these two measurements became closer even without difference. However, as opening the door slower, a behavior that occurred in the people with smaller body force such as women, children and the elderly, the initial force was smaller obviously than the maximum force. Also, a bigger size of door closer applied for hinged doors would bring about a larger operating force. The results suggest that the use of the door closer makes the operational behavior more complex, also caused a substantial increase of operating forces. Thus, for universal design of a door and a door closer, the consideration of operating force became very crucial.

Key words: door, force measurement system, operating force, universal design

1. INTRODUCTION

The operation of the door has great influence because it happens constantly in our daily life, and thus it is necessary to meet the universal design (UD). UD respects the individual for all kinds of people as the idea of the prerequisite to design. Japanese scholar Mr. Nakagawa Satoshi (2008) referred that due to different ethnicities person, using operation mode of a product is also different. With observing and exploring questions neglected in the life, the implementation of UD can make our life safer and more convenient.

Especially for public occasions with service facilities, the higher, wider and heavier doors are usually adopt, so the door operability must be paid more attention. Not only considering human factor and the demands of convenience, UD also gives overall checks of the disadvantaged and looks over a wider range of users such as the physical disabled, the elder and children. Therefore, how to design a door with versatility so that the elderly and disabled persons can live more conveniently and safely is indeed an important issue.

1.1 Classification of door

According to the manner of operational force, doors can be divided into force normal to door, e.g. push or pull of hinged doors, and force in plane (parallel with the door horizontally or vertically), e.g. sliding or folding doors (Shih-Kai and, Colin G , 2007). Due to the function of use based on type of pressure applying on the door, the door was classified by Thompson (1972) as four types of door: hinged doors, swinging doors, sliding/folding doors and revolving doors.

With the progress of technology, the applications of auxiliary devices such varied kinds of door closers become popular especially in the door of public occasions. The door closer is commonly used as auxiliary device of the door, which is a mechanical device that can close a door at a controlled speed. In general, someone opens a door, after the door is automatically closed. Thus, it's a self-closing device, most commonly used on fire doors and main doors to help prevent spread of the fire and smoke in the case of fire happening.

1.2 Literature reviews

Door exists in the environment around us; human beings use it every day with high frequency. However, studies have shown that doors are still causing many people injured, but analyses due to the viewpoints of human factors are very small. Chang et al. (2007) pointed out that it was more important for the consideration of groups more than 65 years, and the main four factors of door design should be noted as the door location, door material, the type of door and door on sharp part.

In addition, many researchers investigated on the design about the use of visual perception at the doors. Norman (2002) revealed that the design of the door, for example, could give a successfully implication for operators to push or pull without having to rely on the symbols indicate. From the extensive collection of daily life's object with a operation of push or pull, Su, June-Yi (2005) further explored the affordance of push and pull in daily life between the operator themselves and objects to rethink the "push" and "pull operation design of the door.

For the operating force of a door, a lot of specifications have been set in the world. Without a door closer, the maximum force is limited at 22N for internal doors, and the maximum initial force to open the door is up to 132N; with a door closer that a latch force is given, the maximum force is not to exceed 66N (BOCA Code, 1999; the Uniform of Building Code, 1997; 2000 International Building Code Handbook, 2000). Chang (2007) analyzed a series of operational behavior of door; their results showed that people of a shorter stature used to have a larger force to open a bigger door. They suggested that the restoring torque of the door should not be greater than 30 Nm and handle position should be between about 1000 and 1500mm above the floor and about 250 - 350mm from the free edge of the door. Although there are specification limits of door operating force, but no measurement evaluation system are further discussed.

1.3 Purpose of this study

Although operation of doors is a very simple action, but there are also many implicit human factors for designing the doors. For children, part of the upper extremity disabilities or the elderly who cannot operate clever movements such as sliding, rotating actions, the universal design of barrier-free facilities to the operation of the door should give special consideration. Furthermore, although specification for the door operation have been developed, but it is still necessary to establish an appropriate measurement system of door operating force in order to further evaluate the specification.

In this study, a set of operating force measurement system and testing method will be developed, and thus the operating forces in the whole process can be recorded and further evaluated. The analyzed results of the experiments will be exerted as the reference in designing more common and easy operational door.

2. EXPERIMENT

The door operator must apply sufficient force to open the door successfully. People are observed to use any available manners to increase their force production, such as application of their body weight to help their arm force, two hands instead of one hand, feet or other items, etc. Especially, doors of public occasions or fire doors are not only heavy, but also usually installed with door closers so as to avoid the entrance of fire or wind. The door closers usually make people feel more difficult to open the door, and not to mention children, the elderly and physically disabled persons. These seemingly convenient hydraulic devices in the viewpoint of universal design have become obstacles for door operation. Thus, a suitable measuring systems and methods for door operation will be developed, and thus the operating force and its variation of speed can be obtained to further discuss.

2.1 Methods

A door operating force signal processing system was developed to measure the operating forces, which consisted of four parts: (1) force sensor (Load Cell), (2) signal amplifier, (3) analog to digital (A/D) signal capture card, (4) computer (with the software of WaveScan, the analysis software of MS Excel), as shown in Fig. 1.

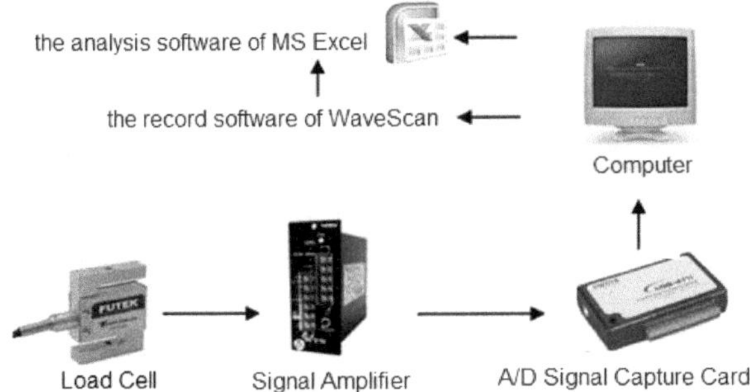

Fig. 1 door operating force signal processing system

A pulling force was exerted on one end of the load cell, and the detected force signals through the signal line were transmitted to the signal amplifier on the other end. Then, by A / D signal capture device, analog signals were transferred into digital signals, which sent to the computer finally. The software of the WavesScan was used to read and record the data, and then the MS Excel software was applied for data processing to obtain the time-varying force diagram.

Two kinds of doors including hinged doors and sliding doors were exerted. The door was pulled by a motor as the operation could be repetitive. The power of the motor was up to 100W, and maximum force output transferred by a gear speed reducer arrived at 120 Kg. Through a reel, the steel rope was droved by the motor at the speed ranging from 8cm/s to 50cm/s, similar to people's behaviors. Also, the motor could adjust the output force according to the load. As the door was heavier, the door was pulled slower, a condition similar to people's behavior.

2.1.1 Operating force measurements of hinged doors

The entire force measuring system of a hinged door was set up as shown in Figure 2. In order to measure the operating force at the position of handle, the door handle was first removed and a hole was generated, and then load cell was fixed into the hole. The pulling force of the door was given at one end of load cell, which was driven by the motor through a reel to pull the steel rope directly. The motor and the reel were placed on a metal base whose height was adjustable. To ensure the pulling force of the steel rope keeping normal to the door, a quarter arc guiding track made by aluminum alloy was designed and mounted on the door at the handle position. The guiding track had a smooth surface, so the friction force could be avoided as possible when the steel rope sliding on it. Thus, the pulling force could maintain the same arm as the door angle changed from zero to 90°.

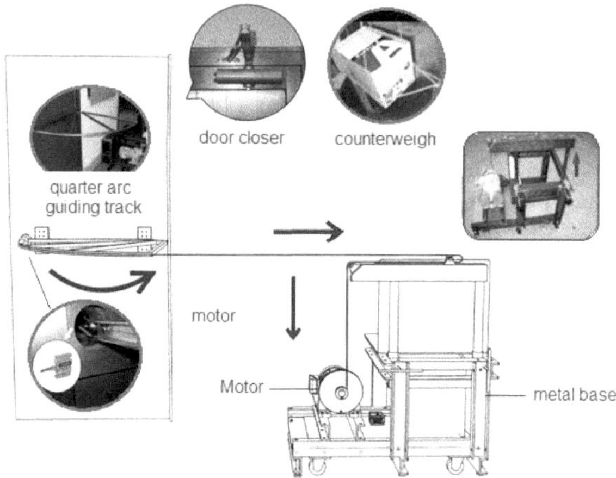

Fig. 2 a force measuring system of a hinged door

Table 1 three sizes of door closer

big size (40-65kg)	medium size (25-45kg)	small size (15-30kg)

In this experiment, two kinds of hinged doors were adopted. One had a door closer which consisted of three sizes: big, medium and small, as listed in Table 1. The other had no door closer with three different weights: 15Kg, 30 Kg and 45Kg, which some packs of papers hanged on the door were used as counterweights to adjust the weight of the door. The rotary speed of the motor was controlled within 400rpm to 2400rpm, and the steel rope speed was within 8.37cm/s to 50.24cm/s.

2.1.2 Operating force measurements of sliding doors

The entire force measurement system of a sliding door was set up as shown in Figure 3. In order to measure the operating force at the position of handle, the load cell was directly fixed at the handle position. The pulling force of the door was given at one end of load cell, which was driven by the motor through a reel to pull the steel rope directly.

In this experiment, two kinds of sliding doors were applied. One had a hydraulic sliding door closer, and the other had no door closer. The rotary speed of the motor to pull the door was controlled within 400rpm to 2400rpm.

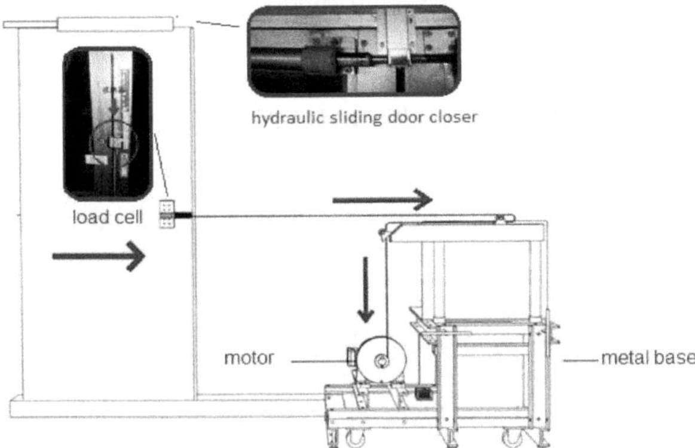

Fig. 3 a force measuring system of a sliding door

2.2 Results

2.2.1 Operating force of hinged doors without a door closer

The hinged door used some objects hanged on the door to adjust the door to have three different weights. The position of the door handle had not changed, so the pulling arm size of the door wasn't unchanged. As the load cell was pulled by the steel rope driven by the motor in according with a reel, the door was directly pulled by the force without any buffer. If the generated torque exceeded the rotation friction torque of the hinge, an inertial force would produce that made the door instantly rotate over the speed of the steel rope. Thus, the steel rope would be relaxed. Until the movement of the steel rope caught up with the movement of the door, the steel rope tightened and the measuring force lifted again. Therefore, a vibrating phenomenon for the measuring force happened. An experimental example was as follows: a hinged door was pulled by the steel rope at the speed of 16.75cm/s, as shown in Fig.4. The maximum force could be obtained at the top value of the first peak as 5N. It was noted that, for all the diagrams of force-time varying curves in this study, the horizontal axis was the unit of time, 0.02s per unit. The vertical axis was the force (N), and due to the strength to pull so the value was negative.

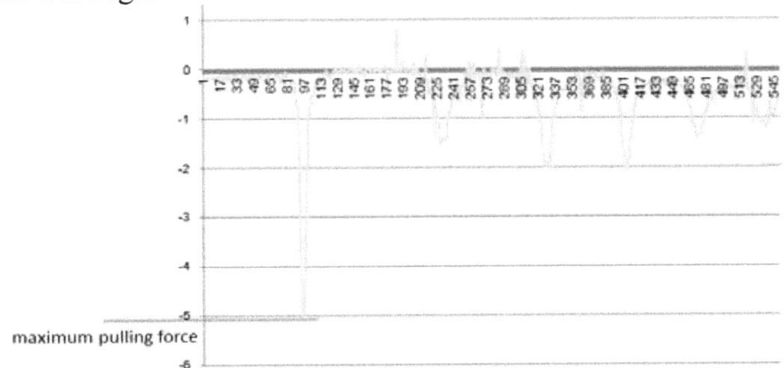

Fig. 4 force measurement for a hinged door weighting 30 Kg without a door closer (steel rope speed at 16.75cm/s)

The maximum force and its operating time for three kinds of weights of the hinged door (15Kg, 30Kg and 45Kg) at varied rotary speeds of motor were recorded, as shown in Table 2. The results revealed that when the door was heavier, the measuring force was greater. Also, a quicker steel rope speed caused a larger operating force in a shorter operating time.

Table 2 force measurements for hinged doors with different weights

No.	steel rope speed (cm/s)	15 Kg		30Kg		45 Kg	
		maximum force(N)	operating time(sec)	maximum force(N)	operating time(sec)	maximum force(N)	operating time(sec)
1	8.37	2.7	11.34	6.7	10.16	9.7	10.44
2	16.75	3	7.56	6	6.12	8	6.4
3	25.12	2.3	7.98	8	4.12	8.6	4.38
4	33.49	2.9	7.6	7.2	3.7	6.6	3.74
5	41.87	4.4	2.74	7.6	3.12	7.1	3.58
6	50.24	4.4	2.66	7.5	3.2	8.2	3.1

2.2.2 Operating force of hinged doors with door closers

In this experiment, the hinged doors were adjusted to three weights of the doors, such as weights of 15Kg, 30Kg and 45Kg, and thus three sizes of door closers, small one, medium one and big one were used in accordance with three weights of doors, respectively. The speeds of the motor were controlled at six speeds. The results were shown in Figure 5, Figure 6 and Figure 7, respectively.

No.	steel rope speed(cm/s)	initial force(N)	maximum force(N)	operating time(sec)
1	8.37	22.7	23.7	9.08
2	16.75	26.9	26.9	4.94
3	25.12	28.4	28.4	3.32
4	33.49	25.9	25.9	2.76
5	41.87	25.7	25.7	2.88
6	50.24	26.8	26.8	2.52

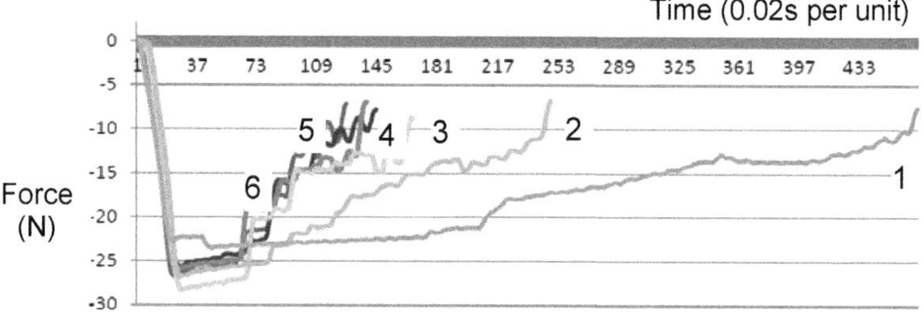

Fig. 5 force measurement for a sliding door weighting 15 kg with a small size of door closer

No.	steel rope speed(cm/s)	initial force(N)	maximum force(N)	operating time(sec)
1	8.37	45.5	45.5	9.2
2	16.75	45.8	45.8	4.68
3	25.12	47.9	47.9	3.2
4	33.49	48.7	48.7	2.56
5	41.87	48.2	48.2	2.96
6	50.24	47.7	47.7	2.8

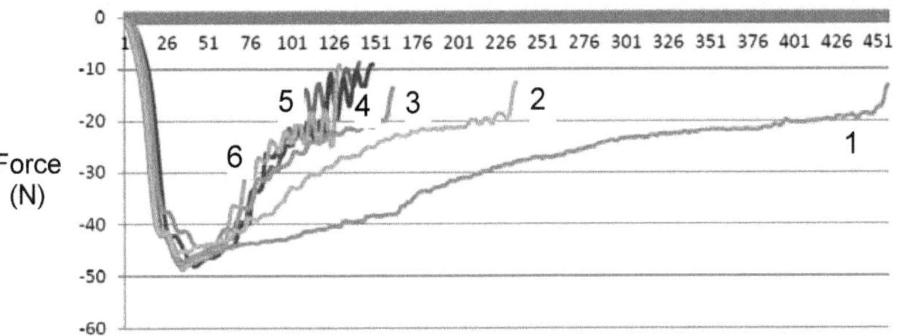

Fig. 6 force measurement for a sliding door weighting 30 kg with a medium size of door closer

No.	steel rope speed(cm/s)	initial force(N)	maximum force(N)	operating time(sec)
1	8.37	44.4	50.9	9.08
2	16.75	49.7	50	4.8
3	25.12	51	51	3.42
4	33.49	51	51	2.72
5	41.87	49.2	54.6	2.3
6	50.24	49.2	53.3	1.7

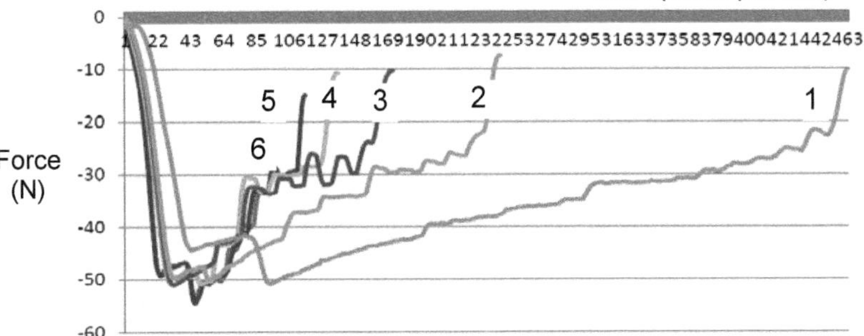

Fig. 7 force measurement for a sliding door weighting 45 kg with a big size of door closer

The results showed a common trend, when the size of the door closer was larger, the operating force to pull the door was greater. Also, a quicker speed of the steel rope, i.e. a faster velocity to open the door, resulted in a larger operating force and a shorter operating time. When the steel rope speeds were lower, two peaks of the operating force occurred. The value of the first peak was the initial force to open the door. With the increasing door angle, the operating force would rise up to the highest peak, and it was the maximum force to open the door. The phenomenon of these two peaks were less obvious when the steel rope became faster and only a peak occurred clearly, where the initial force equaled to the maximum force.

2.2.3 Force measurements of sliding doors

As the sliding door of no hydraulic sliding door closer was applied, the operating force vibrated, a phenomenon as same as those of hinged doors of no door closer. This could be interpreted as the door subjected to a greater pulling force than the sliding friction force, the generated inertia force made the door to move faster than the steel rope, and thus the steel rope relaxed and the operating force reduced suddenly. When the move of the steel rope caught up with the move of the door, the rope tightened again and the operating force suddenly increased, so the vibrating phenomenon of the force signal happened. Most of the maximum force signal occurred at the first peak, where the static friction was overcame, and the values of follow-up peaks due to dynamic friction decreased gradually. Therefore, the most important measurement was the maximum force.

As to hydraulic sliding doors, the hydraulic sliding door closer had a damping effect on absorbing the vibration. Thus, the fluctuation of the force signal was less intense. The first peak could be regarded as the initial force to overcome the static friction force, and the largest one of the following peaks was the maximum force. As shown in Fig. 8, for hydraulic sliding door in the reel speed of 16 rpm, the initial force was of the value at 43.5 N and the maximum force was of the value at 46 N.

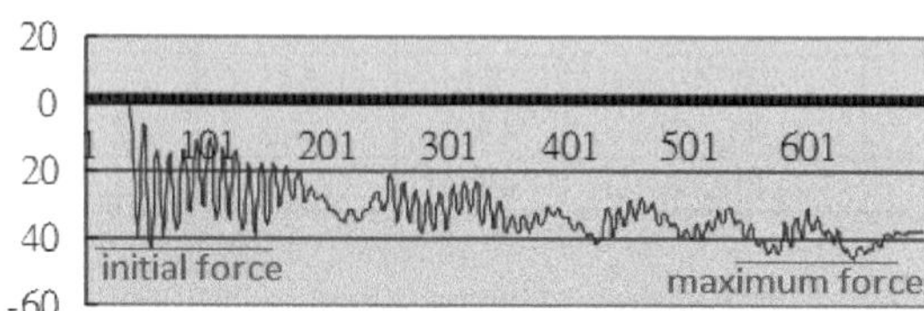

Fig. 8 a force measurement of a sliding door without hydraulic sliding door closer

For sliding doors with and without a door closer, the results of the force measurements were showed in Fig. 9 and Fig. 10, respectively.

The results showed that as pulled faster, the operating force became greater and the operating time became shorter. Thus, to pull door quickly for a same distance, a surge of the force should be required. Also, the maximum forces of the hydraulic sliding door were obviously larger than those of the sliding door without a door closer, their values were all over 30N. As a steel rope speed was lower, the maximum force occurred at the peak of rear section and not the first peak. Obviously, a quite difference between the initial force and the maximum force existed as the door was pulled slower.

No.	steel rope speed(cm/s)	maximum force(N)	operating time(sec)
1	8.37	20.7	6.86
2	16.75	32.1	2.86
3	25.12	25.4	2.4
4	33.49	28	3.6
5	41.87	54	1.12
6	50.24	52	2.4

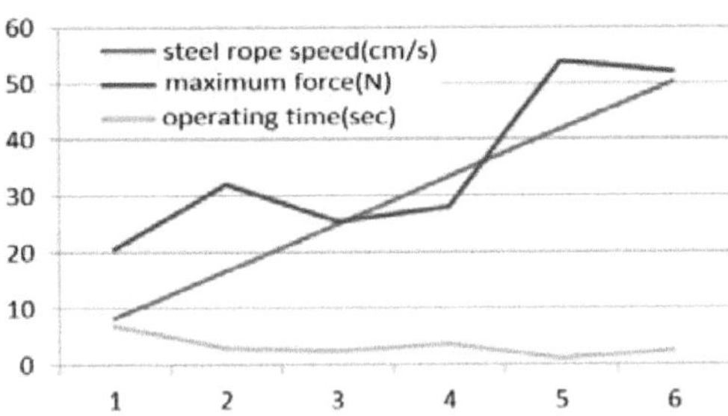

Fig. 9 force measurements for sliding doors without hydraulic sliding door closer

No.	steel rope speed(cm/s)	initial force(N)	maximum force(N)	operating time(sec)
1	8.37	34.8	54	6.32
2	16.75	48.1	48.35	2.76
3	25.12	50.8	52.7	2.4
4	33.49	51.7	51.7	2.58
5	41.87	64.1	64.1	2.14
6	50.24	70.3	70.3	2.28

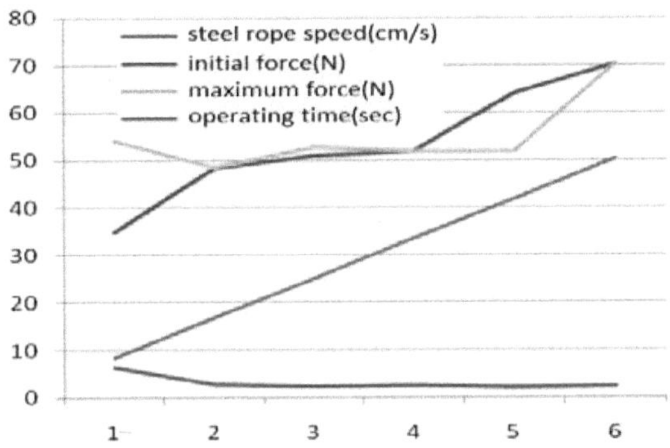

Fig. 10 force measurements for sliding doors with hydraulic sliding door closer

2.3 Discussions

2.3.1 Comparison of the use of hinged doors and sliding doors

To open a sliding door, it had to overcome both the inertia force of the door itself and the sliding friction force. However, as the operating force of the hinged door was normal to the door, the force arm should be especially calculated by the distance from the handle to the door axis. Thus, the operating torque could be obtained by multiplication of the operation force and the arm. To a same operating torque, a larger arm would lower the operating force. Thus, to open a hinged door, it had to overcome both the rotating inertia torque and the torque caused by the friction force of the door hinge. It was easier for the hinged door which rotated around the hinge to overcome the frictional resistance and the door inertia force than the sliding door which moved along rails and by pulleys. The measured results proved that, for the hinged door without closer weighting even to 45Kg, the operating force was as small as 9.7N, a value greatly smaller than 18.6N, the smallest operating forces for all experiments of sliding doors.

2.3.2 Use of door closers

The results showed that the use of the door closer increased the door operating force significantly. For examples, at a steel rope speed of 16.75cm/s, the operating force of the 15Kg door increased from 3N to 26.9N; the operating force of 45Kg door increased from 8N to 50N. Also, a faster steel rope speed, namely, a larger speed to open the door caused a greater operating force. By contrast, the use of the hydraulic sliding door closer enhanced the operating force insignificantly from 32.1N to 48.35N at a steel rope speed of 16.75cm/s.

The results also showed that, without a door closer, whether it was a hinged door or a sliding door, a phenomenon of fluctuated force signal occurred. The maximum force happened in the first peak which could be explained as the motor force was much larger than the door required operating force, once the static friction in the beginning was overcome, the door would be opened smoothly.

However, with a door closer, a phenomenon of fluctuated force signal was not obvious. The experimental results showed that often a smaller peak of the initial force occurred, and then a higher peak of the maximum force arose. This could be explained as the damping effects of the hydraulic door closer reduced the vibrating phenomenon of the force signal. With increasing door angle, the hydraulic damping force became greater, so the maximum force was the required force to overcome the hydraulic damping force. Especially for the sliding door after half of the operating process, the experiments showed that the increasing hydraulic resistance resulted in a longer time to open the door. Therefore, it needs to especially consider the damping effect as applying a door closer to design a universal door.

The results showed that only a value of the maximum force was obvious for doors without a door closer. However, both the initial force and the maximum force could be measured as using a door closer. A smaller difference between these two forces happened as the steel rope speed was higher; whereas an obvious difference existed as the steel rope speed was lower. This illustrated that the difference of these two forces was unobvious and became a same value as opened the door with greater force. For door operators with the smaller body force, such as children, women, the elderly, wheelchair users, it would be more difficult to open the door because a large gap between the initial force and the maximum force. Thus, it should take the operating force into consideration to help the universal design of a door.

3. CONCLUSIONS

In this study, an appropriate measurement systems and methods of operating force have established. Experiments of measuring operating force proceeded including hinged doors and sliding doors, especially considering the use of a door closer. The results revealed the use of door closers would result in a substantial increase in operating force. As a bigger size of door close was applied for hinged doors, the required operating force became greater. Also, the required operating force would relatively increase as the speed of opening the door increased. For both hinged doors and sliding doors, the installation of door closers would obviously reduce fluctuated phenomenon of force signals. Also, an initial force was needed in the beginning to open the door, and then the maximum force was mainly required to overcome the damping resistance of the door closer. The results suggest that the use of the door closer will make the operational behavior more complex, also caused a substantial increase of the door operating forces. Thus, the universal design of the door closer for doors is very important.

REFERENCES

Wikipedia, http://en.wikipedia.org/wiki/Revolving_door, paragraph 7.

Thompson, R.M., 1972. Design of Multi-machine Work Area, Chapter 10. Human Engineering Guide to Equipment Design. American Institute for Research, Washington, DC (pp. 454-455).

BOCA Code, 1999; Uniform Building Code, 1997; 2000 International Building Code Handbook, 2000

Donald A. Norman, 2002, Design of Everyday Things. Basic Book, New York, U.S.

Su, June-Yi, 2005, The Affordance of Push and Pull in Daily Life, thesis ,National Cheng Kung University Department of Industrial Design, Taiwan (R.O.C), pp. 1-2.

Chang, Shih-Kai, Drury, Colin G., 2007, Task demands and human capabilities in door use, Applied Ergonomics 38(3), pp. 325–335.

Architecture and Building Research Institute, 2008, Explanation manual of building barrier-free facilities design specifications, Ministry of the Interior, R.O.C.

ACKNOWLEDGMENTS

The authors would like to acknowledge Architecture and Building Research Institute, Ministry of the Interior, R.O.C.(Taiwan) for their financial support under the project code of ISBN：978-986-03-0751-1.

CHAPTER 43

Ergonomic Design of Classroom Furniture for Elementary Schools

Maria Antónia Gonçalves [1], *Pedro M. Arezes* [2]

[1] School of Managements and Industrial Studies, Porto Polytechnic Institute, 4480-876 Vila do Conde, Portugal

[2] Production and Systems Department, University of Minho, 800-058 Guimarães, Portugal

ABSTRACT

Classroom activities require a high concentration and visual, auditory, motor and cognitive mechanisms are constantly stimulated, which makes educational tasks complex. Students' anthropometric dimensions vary not only between school years in elementary school, but also when comparing individuals of the same age. Thus it is unlikely that furniture with fixed dimensions for all school years in elementary schools will be compatible with the vast majority of students. This work proposes the adoption of adjustable furniture and the use of a (mis)match criteria to address this issue. It is expected that this adjustability and the propose criteria can cope with the differences found in anthropometric dimensions of the students, giving them a better accommodation and ease in their use (usability). The current proposal will, hopefully, increase the comfort and decrease the cases of body aches complaints. For the determination of the adjustability of the school furniture, anthropometric measures of the population of users are necessary and also to find the relationship among them, so that the levels of adjustability can match the anthropometric measures with the dimensions of school furniture.

Keywords: Ergonomics, Elementary, School, Students Furniture

---

[1] Corresponding author: mag@eu.ipp.pt

1 INTRODUCTION

School furniture, as part of a school's physical space, is an essential element with high importance in the school organization. Its characteristics are strongly associated with back pain and neck pain, referred to by schoolchildren (Mandal, 1994; Marshall et al., 1995; Parcells et al., 1999; Knight & Noyes, 1999; Murphy et al., 2007).

The design of school furniture is key for the children to acquiring good postural habits, in the long term (Yeats, 1997) and, according to Hira (1980), the facilitator role of learning, allowing and encouraging good posture when sitting, and must be designed according to the physical structure and biomechanics of the individuals that use it.

Students accommodated in dimensionally suitable furniture, have a better body posture, better calligraphy and faster formation of letters (Parush et al., 1998), which is an evidence of a positive correlation between ergonomic factors, such as the seated posture and positioning and calligraphy performance, in terms of readability and faster. It turns out that there is a significant improvement in the performance of tasks and in sitting posture in children sitting on a chair designed ergonomically, when compared with the demonstrated performance using standard furniture (Knight Noyes, 1999).

The purpose of the seat is to provide a stable support to the body, in a posture that is comfortable for a period of time, psychologically satisfying and appropriate to the task or activity in question. For an individual to feel comfortable in sitting position, we combine three factors: the characteristics of the seat, the user's characteristics and the characteristics of the task. For the seat, height, depth and width are the characteristics that may influence the posture.

But it is not just the seat which has implications on the posture adopted.

Murphy et al. (2007) cites a research done by Salminen et al. (1992) for the height of school desks, in which infers that a low table force children to forward inclination, thus straining the structures of the spine and causing back pain. On the other hand, a work desk too high requires an exaggerated abduction of the upper limb by moving the centre of mass from the side, increasing the load on the column and leading to the emergence of pains in the neck and shoulders.

The ergonomic design of school desks must provide a reduction of medium and lower trunk muscle activity, the maintenance of natural lumbar lordosis and the reduction of the neck's flexion angle (Marshall et al., 1995). The authors concluded that the maintenance of a good postural alignment associated with decreased muscle activity during the school period, could decrease muscle fatigue. This condition would have a positive impact on the learning process and at the same time, would prevent the development of future erratic postural habits, and likelihood of back pain.

School furniture, as part of the school physical space, is an essential and highly important issue within this environment. It will influence the physical and psychological comfort of schoolchildren and consequently a direct impact on their health and learning output.

The objective of this work is focused on setting the benchmark for the design of school furniture for schoolchildren of primary schools, on the basis of data collected from a sample taken from the population of catwalk of establishments, aged between 6 and 10 years.

The definition of the main ergonomic aspects to consider, as well as the definition of the reference values for these parameters, will thus promote the necessary adaptations to minimize the physical consequences and, eventually, cognitive in the main users of these spaces.

The main aim of this study was to set the benchmarks for the design of school furniture for primary schools, taking into account the characterization and analysis of a sample selected in advance.

As a result of this project, a specific methodology for furniture adjustment was developed, based on the children's stature.

2 METHOD

This work included three stages.

The first stage was a collection of anthropometric data of a elementary school, since there is no information, or is not available or published, necessary for completion of the work. It was developed an anthropometric chair (Figure 1) for this. With the anthropometric data, anthropometric tables were built for children from 6 to 10 years.

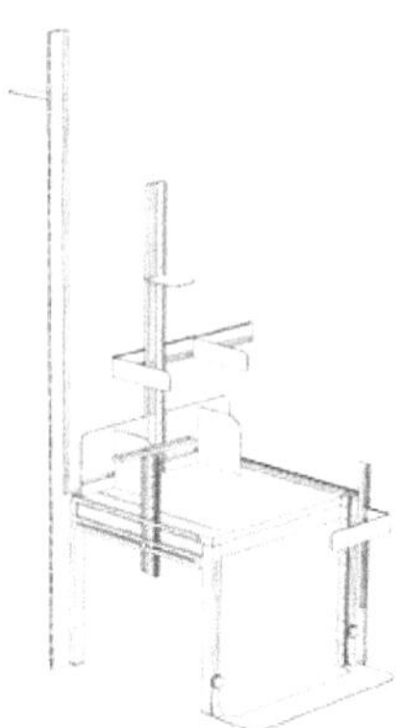

Figure 1 – Anthropometric chair

The analysis of the anthropometric data that was been collected, as well as its practical applicability, were the scope of the second stage of this work. The possible relationships between different anthropometric measures and their percentiles were analyzed and a set of parameters were established for the adjustment levels of the school furniture, so that it is compatible with the anthropometric measures of its users. These data were complemented with the criteria of match/mismatch between the dimensions of the furniture and the anthropometric dimensions used in others studies in this area.

In possession of relevant data, the dimensional parameters for the design of ergonomic furniture for elementary schools were defined and a methodological guide for teachers was prepared, aiming at the proper use of the ergonomic furniture. It will be expected that, with this guide, teachers will be able to obtain the desired results for the students/furniture interface in the elementary schools.

3 RESULTS AND DISCUSSION

3.1 Collection of anthropometric data

Thirteen static anthropometric measures were identified, three of them with the subject standing and the remaining ten with the subject sitting.

The thirteen selected measures, described in table 1, were basically the measures deemed relevant in other studies applied to school furniture (Garcia-Acosta & Lange-Morales, 2007; Molenbroek et al., 2003, Panagiotopoulou et al., 2004, Milanese & Grimmer, 2004; Parcells et al., 1999)

Table 1- Relevant anthropometric dimensions

Anthropometric Dimensions		
Standing	Sitting	
	Vertical measures	Horizontal measures
Stature	Sitting height	Buttock-popliteal length
Eye height	Eye height (sitting)	Buttock-knee
Shoulder height	Popliteal height	Shoulder width
	Seat-shoulder distance	Hip Width
	Seat-elbow distance	
	Thigh Thickness	

The sample consists of 432 volunteers students (216 male, 216 female) of 9 schools, belonging to the elementary schools, aged between 6 and 10 years (average 8.5 ± 1.2). The sample size (432 children) has an associated confidence level of 87.8%.

Anthropometric measurements were taken with the subject seated in an upright position and relaxed in the anthropometric chair, with the legs bent at an angle of 90°, and the feet supported on an adjustable feet support. Exception of this scenario were the measures of stature, eye height standing and shoulder height standing, which was carried out with the individual standing in an upright and relaxed position using the tape measure bound to anthropometric chair (figure 2).

The sampling method of determination followed the ISO 7250: 1996 norm.

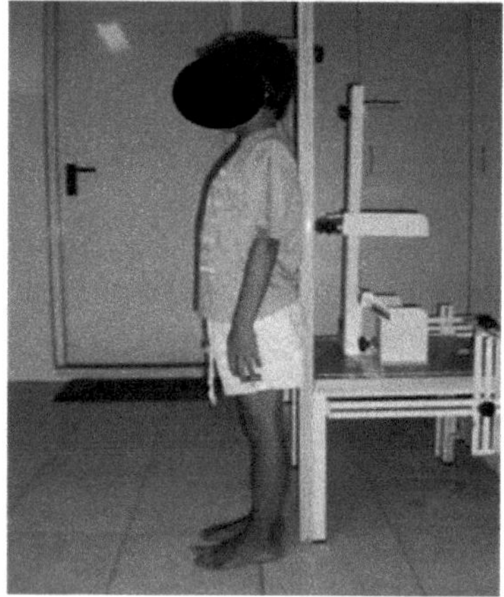
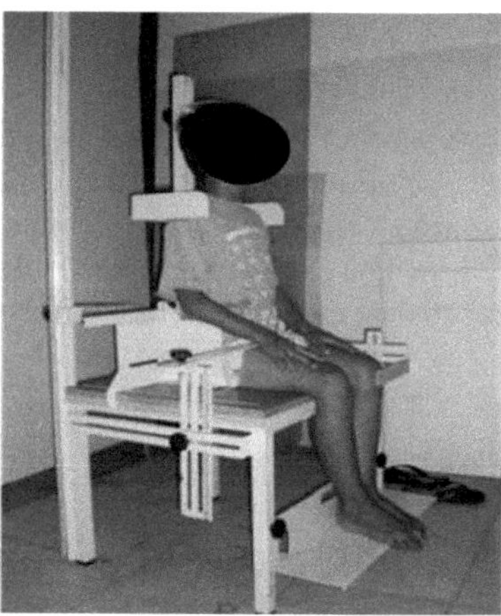

Figure 2 – Collection of Anthropometric Measures

3.2 Correlation between anthropometric measures

The use of percentiles is very useful in applied anthropometry and in particular in design projects. However, when using the percentiles it is necessary to take into consideration two important issues:

(1) the percentiles are specific of the characterized population and;

(2) the described dimension (Pheasant & Haslegrave, 2006).

The second issue is corroborated by research done by Vasu and Mital (1999), in which they concluded that it is wrong to assume the fixity of percentiles, meaning that an individual with a anthropometric dimension at a given percentile, will have the remaining dimensions in the same percentile.

In an attempt to prove this, a study was carried out with two dimensions that have a strong correlation (Pearson's Correlation = 0.897; $p<0$ .01): the popliteal height and stature, of male children. The results prove that an individual with the stature of a specific percentile does not necessarily have the remaining dimensions in the same percentile.

This observation leads to assume that it is necessary to find another relationship between anthropometric measures, other than correspondence of percentiles to allow for the establishment of a compromise between these, which will lead to the establishment of the dimensions required for the school furniture.

For Roebuck et al. (1975), some body segments can be expressed as a ratio of stature.

To analyze the correlation between the dimensions involved, it is noted that the stature is the one that presents the most significant correlation coefficient ($p<0$ .01) with the majority of anthropometric measures, which is the main indicator for the prediction of others anthropometric measures.

Knowing that these dimensions follow a normal distribution, regression analysis was made, for the relationship between the two variables in order to predict anthropometric measures required from stature:

Popliteal height (PH) = -5.89 + 0.302 * Stature (S)

Buttock-popliteal length (BPL) = -2.919 + 0.313 * Stature (S)

Seat-elbow distance (SED) = 2.659 + 0.105 * Stature (S)

Seat-shoulder distance (SSD) = 2.670 + 0.318 Stature (S)

3.3 Anthropometric dimensions relevant for the sizing of school furniture

Taking into account others studies (Garcia-Acosta & Lange-Morales, 2007; Gouvali & Boudolos, 2006; Panagiotopoulou et al., 2004; Molenbroek et al., 2003; Miller, 2000; Parcells et al., 1999; Knight & Noyes, 1999), were established a parallel between the anthropometric measures and the furniture dimensions (table 2).

Table 2 - Association between anthropometric and furniture dimensions

Anthropometric dimensions	Furniture dimensions
Popliteal height (PH)	Seat height (SH)
Hip Width (HW)	Seat width (SW)
Buttock-popliteal length (BPL)	Seat depth (SD)
Seat-shoulder distance (SSD)	Backrest height (BH)
Seat-elbow distance (SED)	Seat-desk distance (SDD)
Thigh Thickness (TT)	Clearance between seat and table (CST)

The match/mismatch between anthropometric dimensions and furniture dimensions are shown in table 3.

Table 3 – Match/Mismatch Equations

Match/Mismatch Equations	Source Authors
88% PH ≤ SH ≤ 95% PH	Parcells et al. (1999)
HW < SW	Evans et al. (1988); Helander (1997); Mondelo et al. (2000); Pheasant & Haslegreve (2006)
80% BPL ≤ SD ≤ 95% BPL	Parcells et al. (1999)
60% SSD ≤ BH ≤ 80% SSD	Gouvali & Boudolos (2006)
SED – 5cm ≤ SDD ≤ SED	Poulakakis & Marmaras (1998); Pheasant & Haslegrave (2006)
TT + 2cm ≤ CST	Parcells et al. (1999)

Based on this information, sizing values were established for school furniture, with four levels of adjustability, for four ranges of stature values.

The match of the furniture dimensions established has been tested using the equations of table 3 and the anthropometric data of the sample (table 4).

Table 4 – Match of the furniture dimensions

Furniture vs anthropometric dimensions	Match
PH *vs* SH	> 60%[1]
HW *vs* SW	> 94%
BPL *vs* SD	> 84%
SSD *vs* BH	> 96%
SED *vs* SDD	> 73%
TT *vs* CST	> 99%

[1] When looking at the percentage of children whose popliteal height is lower than the seat height (ergonomic criteria cited for a wide range of authors, as Dul Weerdmeester, 1998; Helander, 1997; Molenbroek & Ramaeekers, 1996; Oxford, 1969), the scenario is far more favorable, with more than 93% of match.

Thus, depending on the stature, the school furniture should be adjusted according to the illustration of the methodological guide that will be made available to teachers, for the proper use of furniture (figure 3).

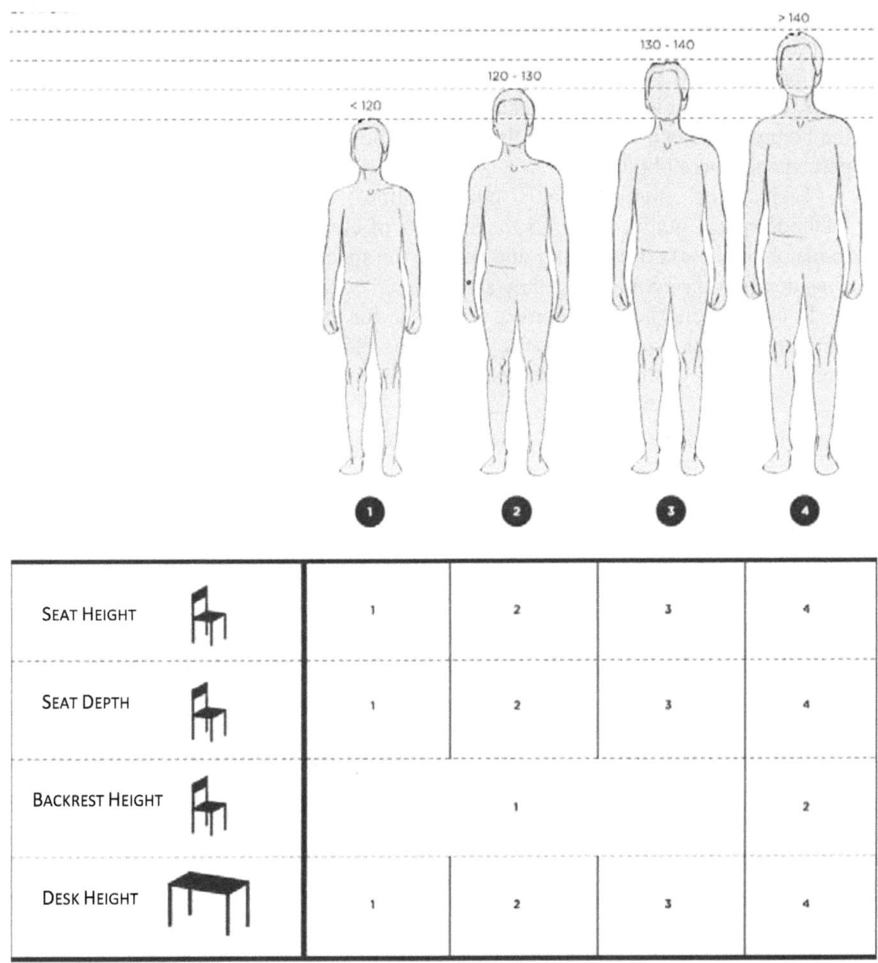

Figure 3 – Methodological guide for furniture adjustment

4 CONCLUSIONS

The design of school furniture in Portugal do not seems to be supported by any scientific or practical explanation, and is, in most cases, in contradiction with those that are the scientific evidences relevant to the design of this type of equipment.

Stature proved to be a good predictor for determining other anthropometric measures, and there is a significant correlation ($p<0.01$) with the anthropometric measures considered in this study. Thus, using regression equations, with the stature as independent variable, values of desired anthropometric measure were determined (dependent variable). In accordance with the principles of ergonomic suitability proposed in literature, compatibility equations were used between the dimensions of the furniture and the dimensions, the dimensional values studied for tables and seats, for different levels of adjustability. When confronted the values obtained from the compatibility equation with the sample, some interesting compatibility (match) percentages were obtained.

Taking the previous into consideration, it was possible to define a methodological guide, intended for teachers of elementary schools, allowing them, to match the height of the chair and desk to the specific child stature, and taking into consideration the activity(ies) they are doing.

In conclusion, the conditions are in place for the development of a typology of furniture, which accompanies the growth of children, without compromising the smooth running of their apprenticeship, comfort and health.

REFERENCES

Dul J., Weerdmeester, B. (1998). Ergonomics for Beginners – A Reference Guide. Taylor & Fancis, London.

Evans, W.A., Courtney, A.J., Fok, K.F. (1988). The design of school furniture of Hong Kong school children: an anthropometric case study. Applied Ergonomics, 19 (2), 122–134.

Garcia-Acosta, G.; Lange-Morales, K. (2007). Definition of sizes for the design of school furniture for Bogotá schools based on anthropometric criteria. Ergonomics, 50 (10), 1626-1642.

Gouvali, M.K., Boudolos, K. (2006), Match between school furniture dimensions and children's anthropometry. Applied Ergonomics, 37, 765-773.

Helander, M. (1997). A Guide to the Ergonomics of Manufacturing. Taylor & Fancis, London, 210 pgs.

Hira, D.S. (1980). An ergonomic appraisal of educational desks. Ergonomics, 23 (3), 213-221.

ISO 7250: 1996 – Medidas básicas do corpo humano para o design tecnológico

Knight, G., Noyes, J. (1999). Children's behaviour and the design of school furniture. Ergonomics, 42, 747-760.

Mandal, A.C. (1994). The prevention of back pain in school children, In: Lueder, R., Noro, K. (Eds.), The Ergonomics of seating. Taylor & Fancis, London, 269-277.

Marschall, M., Harrington, A.C., Steele, J.R. (1995). Effect of work station design on sitting posture in young children. Ergonomics, 38 (9), 1932-4190.

Milanese, S., Grimmer, K. (2004). School furniture and the user population: an anthropometric perspective. Ergonomics 47, 416-426.

Miller, H. (2000). Workplace research. Consultado em Setembro de 2010, disponível em http://hermanmiller.com/research/.

Molenbroek, J.F.M., Kroon-Ramaekers, Y.M.T. (1996). Anthropometric design of a size system for school furniture, In: Robertson, S.A., (Ed.), Proceedings of the Annual Conference of the Ergonomic Society: Contemporary Ergonomics. Taylor & Fancis, London, 130-135.

Molenbroek, J.F.M., Kroon-Ramaekers, Y.M.T, Snijders, C.J. (2003). Revision of the design of a standard for the dimensions of school furniture. Ergonomics, 46 (7), 681-694.

Mondelo, P., Gregori E., Barrau, P. (2000). Ergonomía: Fundamentos, 3rd Edition, Alfaomega Grupo Editor – UPC, México, 186 pgs.

Murphy, S., Buckle, P., Stubbs, D. (2007) A cross-sectional study of self-reported back and neck pain among English schoolchildren and associated physical and psychological risk factors, Applied Ergonomics 38, 797-804.

Oxford, H.W. (1969), Anthropometric data for educational chairs. Ergonomics 12 (2), 140-161

Panagiotopoulou, G., Christoulas K., Papanckolaou, A., Mandroukas, K. (2004). Classroom furniture dimensions and anthropometric measures in primary school. Applied Ergonomics, 35, 121-128.

Parcells, C., Stommel, M., Hubbard, R. (1999). Mismatch of Classroom Furniture and Student Body Dimensions: Empirical Findings and Health Implications. Journal of Adolescent Health, 24: 265–273.

Parush, S., Levanon-Erez, N., Weintraub, N. (1998). Ergonomic factors influencing handwriting performance, Work: A Journal of Assessment, Prevention, and Rehabilitation 11, 295-305.

Pheasant, S., Haslegrave, C. (2006). Bodyspace: anthropometry, ergonomics and the design of work, 3rd Edition, CRC Press, 332 pgs.

Poulakakis, G., Marmaras, N. (1998). A model for the ergonomic design of office, In: Scott, P.A., Bridger,R.S., Charteris, J.(Eds.), Proceedings of the Ergonomics Conference in Cape Town: Global Ergonomics. Elsevier Ltd., 500–504.

Roebuck, J., Kroemer, K.H.E., Thomsonm W.G. (1975). Engineering Anthropometry Methods, John Wiley and Sons, New York, 459 pgs.

Vasu, M., Mital, A. (1999). Evaluation of the validity of anthropometric design assumptions International Journal of Industrial Ergonomics, 26 (1), 19-37.

Yeats, B. (1997). Factors who may influence the postural health of schoolchildren. Work, 9 (1), 45-55.

CHAPTER 44

Materialization of the Storage Halls Lighting

Ivana Tureková, Miroslav Rusko, Jozef Harangozó

Institute of Safety and Environmental Engineering
Faculty of Materials Science and Technology
Slovak University of Technology in Bratislava
Botanická 49, 917 24 Trnava
Slovak Republic

ivana.turekova@stuba.sk
miroslav.rusko@stuba.sk
jozef.harangozo@stuba.sk

ABSTRACT

The article deals with materialization of artificial lighting at chosen workstation of industrial logistical hall. Light is an important factor of the working environment, its importance and influence on health are undeniable. The lighting intensity is dependent on the sort of working activity done in the relationship with the sight tasks demandingness. These are the main criteria of the work-lighting demands. The evaluation of artificial lighting described in this article was implemented by screening measurement and its results were compared with the standard. Next attitudinal evaluation of illumination was accomplished by a form of questionnaire. After both the materialization and the employees responses evaluation there are offered some proposals for practical solutions in the article conclusions. Software support for illumination proposals was used in the proposals.

Keywords: illumination, workplace, health and safety

1 INTRODUCTION

„The brain has found the means how to look at an outer world. The eye is a part of the brain that touches the light."

Richard P. Feynman

While investigating we are using the sight as one of the most important senses. Daylight is an ideal element ensuring optimal illumination of internal building areas and industrial objects. In spite of considerable progress in the area of illumination techniques it is impossible to create ideal conditions by artificial illumination (Kaňka, 2010).

2 WORK ILLUMINATION

Research works have proved that light effect plays an important role for human organism and its psyche (Figure 1). Light influences biochemical processes of human organism (Plch, 1999).

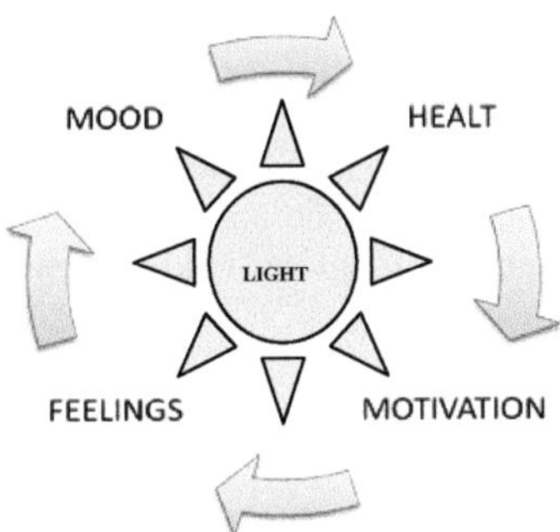

Figure 1 Influence of light on a man [2]

European Unions´ creed for development and support of light technics prefers as primary human needs at lighting proposals (Plch, 1999).

Implementation of work-tasks in the Slovak Republic are qualified by the relevant objects´ fulfilment of conditions given in the Government of the SR Decree No. 391/2006 Coll. On minimal safety and health demands in workplace and No. 541/2007 Coll. that specifies notions connected to the illumination and its materialization with referring to standards. The Standard STN EN 12665 describes terms from this field and recommends criteria of illumination demands determination. Standard STN EN 12464-1 deals with workplaces illumination. The standard prescribes the light flow measurement procedure with a sphere integrator (Hnilica, 2010).

The light can be described also by these quantities and their units:

- **illumination** (lighting intensity) ⇨ (Figure 2), unit: lux (lx),
- **luminosity of source** in given direction: candela (cd),
- **luminous flow of the light source**: glowing flow evaluated by normal human eye; unit: lumen (lm),
- **light performance** ⇨ value given at light sources; unit: $lm.W^{-1}$ (Balog, Tureková and Turňová, 2006).

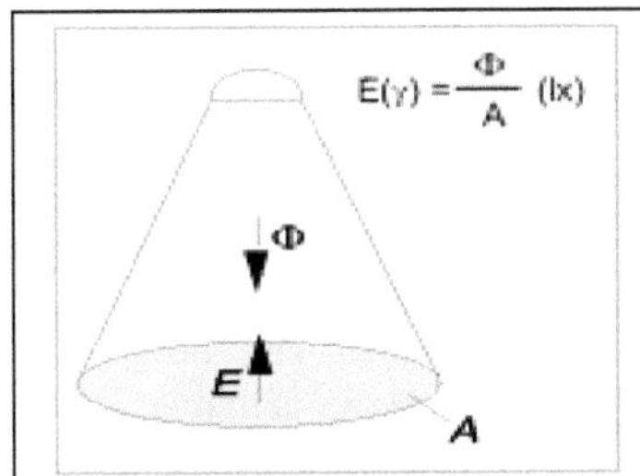

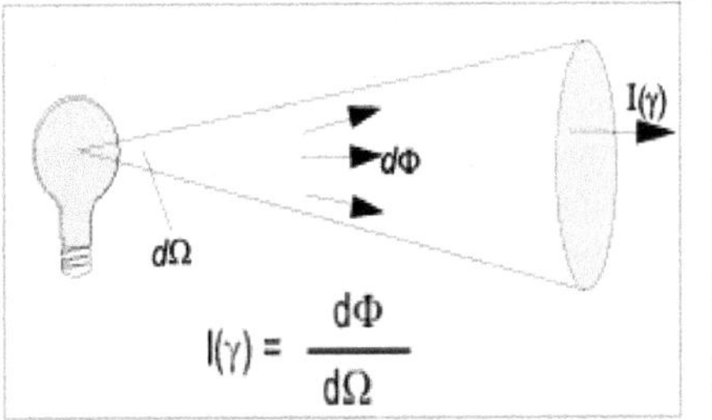

Figure 2 Illumination and luminosity of source in given direction (AMI, 2011)

The sight comfort expresses pleasant psycho-physiologic condition that is needed for work and relax. The sight is accomodated to various brightnesses and contrasts of the light – there is adaptation process (Balog, Tureková and Turňová, 2006).

The visual perception quality depends on the watched object during working. Conditions of perception are determined by the size, brightness, contrast of brightness of the object, its surrounding and the time of perception. The time needed for watching given detail depends on the contrast and other quantities, and ranges between 0,075 - 0,3 s. This time is shortened by rising of the lighting intensity (Sokanský, 2003). Internal workstation areas have to be illuminated by daily or artificial light for ensuring safe workstation and performance demanded. We can distinguish following work environment illumination:

- day-light illumination,
- artificial illumination,
- associated illumination.

Day-light illumination depends on outer daylight changes, facility construction, arrangement of lighting openings, shadows and reflections of surrounding barriers. The day-light coequality is its basic demand given (Balog, Tureková and Turňová, 2006). It is assigned in industrial halls by some upper or side windows or by their combination (Monzer, 2011).

In case of business where technology does not allow the day-light, the demands of illumination are risen. It is related also to the workstastions where elderly workers operate, from the decline of accommodation reason. Choice of the illuminating system parameters depends on the sight tasks performed in the workstation (Fišerová, 2011). For raising of illumination that is prescribed by the Standard, we can take into consideration following conditions that significantly influence illumination demands: lasting staying of persons, sight activity at work is determining, errors can be costly liquidated, accuracy and productivity are very important, sight abilities of workers are lowered, sight tasks are or very small or less contrast, task is performed at unusually long time (STN EN 12464-1, 2005).

Minimal safety and healthy demands on workstation according to the Government of the SR Decree No. 391/2006 Coll. prescribe periods of the light-fittings and windows cleaning (OSRAM SK, 2010).

3 METHODS OF THE ILLUMINATION MEASUREMENT IN THE HALL OF LOGISTIC CENTRE

Measurement methods watch various aspects that can complement or overlap at measurement of specific photometric quantity (OSRAM SK, 2010). Not only measurement methods but measurement procedures have to be prescribed for ensuring unified measurements (ASB SK, 2010).

Production hall workstation of the logistic centre measured can be characterised from the viewpoint of employees stay longitude as an activity with long-lasting stay of employees. In the logistic centre there are overstocked and assembled materials. In the Admission zone run these three basic operations: **discharging ⇨ checking ⇨ shipping.**

The following scheme in Figure 3 describes processing together with safety rules. This procedure accompanies every input material flow. Working team at the workstation performs its activities in the following constitution:

- monitor of working group – work-place is the whole Admission zone without any fixed work-place of working task (2 employees),
- high-lift truck driver - work-place is the whole Admission zone without any fixed work-place of working task (16 employees),
- controller – with fixed work-place of working task (6 employees).

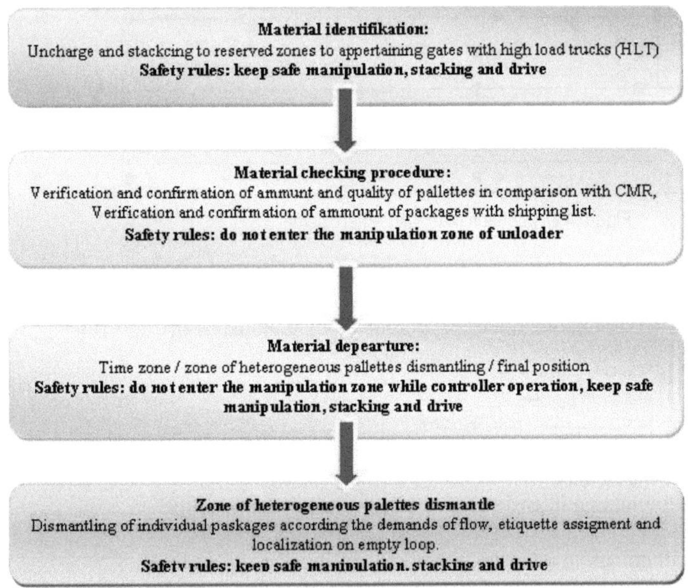

Figure 3 Scheme of working process procedure (Trebatická, 2011)

Among the most important tasks at the admission of material from the viewpoint of illumination demands is checking.

4 MEASUREMENT RESULTS

The assessment of illumination appropriatedness at workstation was focused at:

- evaluation of overall artificial illumination,
- evaluation of illumination of the places for working tasks,
- attitudinal evaluation by employees.

While determination of checking measuring points there was respected the following (Figure 4):

- to alternate the position of lines identified with the position of lighting devices and the lines that lay between the axes of lighting fittings,
- points were on quadratic or rectangle base in functionally determined part in regular distribution and measurement places are repeated on equal base in central part,

- position of points started 2,30 m from the wall and distance of points was ca 4 – 6 m for large areas,
- at the determination of the points there was respected the arrangement of the lighting fittings.
- measurement points for total illumination **P, T, Q, R**; **1, 2, 3, 4** – places for sight tasks, ⊗ - discharge lamp, fluorescent lamp, concrete column

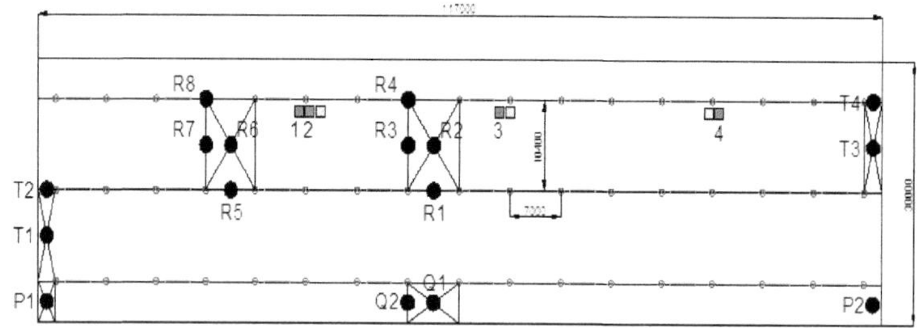

Figure 4 Distribution of measurement points

By the measurement there was investigated:

$\bar{E}_m$ - average kept illuminating intensity (illumination), counted as an arithmetic mean of all accomplished measurements

$$\bar{E}_m = \frac{\sum E_n}{n} \quad \text{(lx)},$$

E_n – measured values in lux,
n – number of measurements,
E_{min} - minimal illuminating intensity, represents the lowest value of the measured ones,
E_{max} – maximal illuminating intensity, represents the highest value of the measured ones,
r – equability of the illuminating, is given by the relation of to average illuminating intensity

$$r = \frac{E_m}{\bar{E}_m},$$

U – measurement uncertainty, represents the parameter that determinates the interval of values around the measurement result, is set - binded to the value measured.

Callibrated luxmeter LX-1108, that measures with the accuracy of ± 3 %, was used for the measurements. The values measured and measurement accuracy for total illumination are in the Table 1.

Table 1 Measurement results for total illumination

$\bar{E}_m$ (lx)	E_{max} (lx)	E_{min} (lx)	r	U (%)	$\bar{E}_m$ - U (lx)	$\bar{E}_m$ + U (lx)
126,49	177,00	85,90	0,68	3	122,69	130,28

Sites of sight task are work-tables serving mostly to administrative works. Measured values are in Table 2.

Table 2 Measurement results for illumination of the place of sight task - work places 1 - 4

Place of sight task	$\bar{E}_m$ (lx)	E_{max} (lx)	E_{min} (lx)	r	U (%)	$\bar{E}_m$ - U (lx)	$\bar{E}_m$ + U (lx)
R6 1 2	171,82	174,30	168,90	0,98	3	166,67	176,97
	148,90	156,40	141,00	0,95	3	144,43	153,37
R2 3	130,08	132,40	126,70	0,97	3	126,18	133,98
4	106,38	121,90	93,60	0,88	3	103,19	109,57

While comparing the measured values we proceeded according to the Standard STN EN 12 464 – 1 Light and Illumination, Working Places Illumination, Part 1: Internal working places, and according to the Decree MoH SR No. 541/2007 Coll. On Details and Demands for Illumination at Work.

For illumination evaluated workstation there are demanded values of total average kept illumination on comparing level of working places, or functionally determined part for total illumination at sustained stay of persons according to the Table 3.

Table 3 Illumination demands according legislation and technical regulations of the SR

Types of room, task, activity	Values
Storehouses and reservoirs - at sustain stay	**$\bar{E}$m = 200 lx***
workstations without daylight	**$\bar{E}$m = 1500 lx****
workstations without daylight - if of reserve measures are proofed	**$\bar{E}$m = 500 lx, ***
place of sight task – illumination of task – illumination of close surrounding of the task	****** **300 lx** **500 lx**

* Decree of MoH SR No.541/2007 Coll., ** STN EN 12464 - 1

While comparing the measured values it can be said that the total illumination as well as both places of sight task are not up to the standard of illumination, they are not consistent and even incorrect designed for activity that is applied in this area.

The reason of this status was the fact that the original workstation was declared as the Zone of unloading and by redefinition of the activity the workstation was not adapted to the demands of the sight activities done in this zone.

For the abatement of unfavourable impact on health, safety and quality of work there were proposed some technical and organizational measures as follow: to decrease of existing light fittings by 1 m and to raise the flurescent laps performance, to concentrate light fittings just over the Zones of control, to reassess maintenance plan, regular cleaning of light fittings twice a year, organization of work, more frequent pauses, time for eye exercises and regeneration, medical care check-up – eye examination once a year, edification of employees on risks.

Among the technical measures at the places of administrative performance there was proposed completion of local illumination. Such a solving asks for investitions because at the places given an electric fitting lacks. It is appropriate to deal with version of completion of dormer windows or change of light system for betterment of workstation visual comfort.

2 SUBJECTIVE QUESTIONNAIRE EVALUATIONS

A questionnaire was used for the attitudinal evaluation of visual welfare.The questionnaire was a tool for supplementary method for quick ascertainment of opinions, attitudes as well as needs of respondents. It was divided in more parts that deal with sight welfare, quality and performance, illumination and total maintenance of workstation. Employees declared their feelings and observations of given working environment by answer yeas or no. and The questionnaire issues are in Table 4 and the results are in Figure 5.

Table 4 Questionaire of employees (answers: yes - no) (Trebatická, 2011)

Occurrence of the eyestrain symptoms	
1.	Do you feel dry eyes?
2.	Do you feel burning eyes?
3.	Do you watch reddened eyes?
4.	4. Do your eyes tear?
5.	5. Do you feel a pressure in eyes?
6.	Do your troubles outlast after work?
7.	Do you have headache belonging to sight tiredness?
8.	Do you feel discomfort from bad illumination?
9.	Do you feel worsening of your eyes since last doctor inspection?
10.	Do you absolve the eye-doctors´ inspection when some difficulties?
Quality and efficiency	
11.	Do you feel that illumination influences your work quality?
12.	Do you feel that illumination influences your work efficiency?
llumination	
13.	Workstation illimination is rather infavourable?
14.	Would you like raising total illumination?
15.	Would you like mor light at som places?
16.	Does it occur at your work to dazzlement?
Maintenance of windows, lamps, walls	
17.	Are the windows dirty?
18.	Are the lamps dirty?
19.	Are the walls dirty?
20.	Are the lamps out of operation (2 times in month)?

The most critical points referred to the complex evaluation of working environment by all employees were identified by regressive arrangement of individual answers “yes” (Figure 6) an especially for the post of controller (Figure 6b).

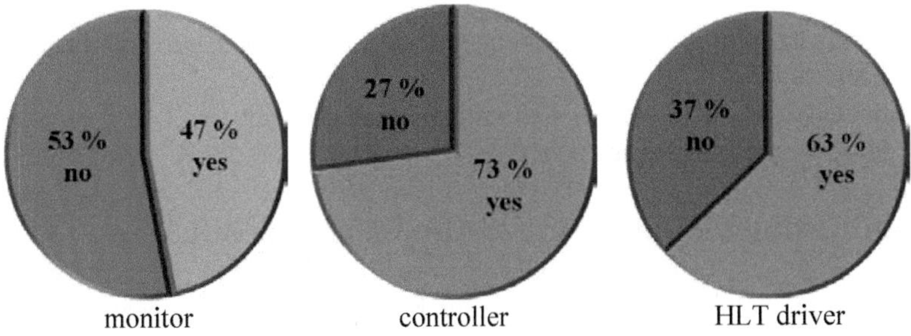

Figure 5 Percentual distribution of individual posts answers

The most critical points referred to the complex evaluation of working environment by all employees were identified by regressive arrangement of individual answers "yes" (Figure 6) an especially for the post of controller (Figure 6b). High demand for raising illumination was declared by 88 % of employees (question No. 15). 83 % of employees declared in the questionnaire feeling of their eyes burning what can be evaluated as some marks of eyestrain. The respondents sustained that the illumination is a risk factor not only for health byt importandly influences quality of the given workstation service.

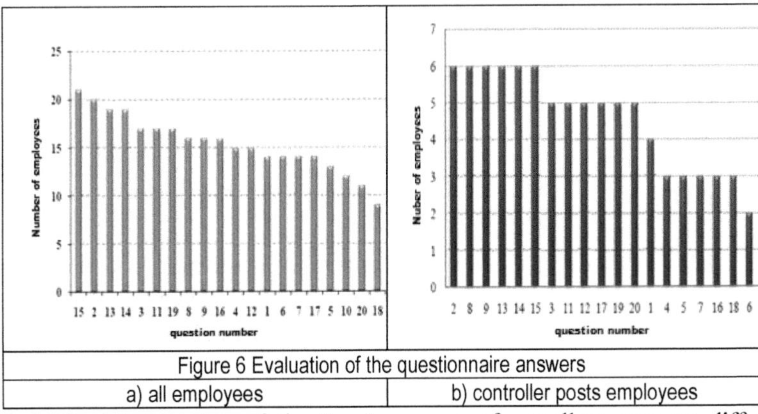

Figure 6 Evaluation of the questionnaire answers

a) all employees	b) controller posts employees

The interpretation of the "yes" answers of cotrollers says on different predication of this group of respondents. All controllers confirmed symptoms of eyestrain from work that manifested:

- by eyes burning,
- discomfort at workstation,
- sight ingraxescence,
- unsufficient illumination.

For decreasing unfavourable impact on health, safety and quality of work it would be appropriate to deal with some possibilities that firstly only decrease harmful influence on working team but can be easily implemented in praxis. Among technical and organizational measures belong:

- to decrease of existing light fittings by 1 m and to raise the flurescent laps performance,

- to concentrate light fittings just over the Zones of control,
- to reassess maintenance plan – regular cleaning of light fittings twice a year,
- organization of work – more frequent pauses, time for eye exercises and regeneration,
- medical care check-up – eye examination once a year,
- edification of employees on risks.

Illumination visualization in the Program WILS

For work objectification of lighting was used WILS 6.3 program. Modern money-saving solution with using Modus Maxis light fitting appeared as more preferable for logistic centre purposes. Compared to discharge lamps it gives cost savings for electricity according to the Table 5.

Table 5 Using different lamps - Halogene (HL) and fluorescent (FL) (Astra, 2007)

Indicator	Lighting system with HL	Lighting system with FL	Savings
System without lamp	112 pieces	70 pieces	38 %
Input of lamp	450 W	525 W	-
Total input of system	50,4 kW	36,75 kW	27 %
Electric energy of 3200 hrs [kWh]	161280	117600	43680

The savings consist also from number of light fittings and investment return comparing to more expensive fluorescent lamps is approximately 18 months. The producer declares a possibility of energy saving rauising by illumination regulation that can be aimed by the discharge lamps.

DISCUSSION

The goal of this experiment was illumination assessment at workstation chosen by technical measurement supported with attitudinal evaluation by workstation employees. Topical goal of this evaluation combination was to show an importance of working conditions assessment from viewpoint of employees supported by measurement and modelling.

Mesaurement with luxmeter showed that total illuminations of the hall and at individual working places do not comply with the values of legal regulations. The demands of the Decree of MoH SR No. 541/2007 Coll. for Workstation Illumination without Daylight are 1500 lx, or after providing substitute measures according to the Appendix No. 4 are 500 lx. This reality was confirmed by attitudinal evaluation of illumination by questionnaire. Results of the attitudinal evaluation declared employees discontent, significant symptoms of sight discomfort and deterioration at more than 64 % of employees.

Visual assessment of illumination solution possibilities was done with Wils Program. Technical measures represent a priority from the reason of highest effect and purpose of group prevention:

1 total exchange of light fittings and light sources,
2 illumination system simulation resulted in more saving variant,
3 dormer windows for betterment of visual comfort in work station,

4 lowering of contemporary light fittings and raising of light performances,
5 supplementary illumination of places for sight tasks,
6 illumination maintenance plan.

CONCLUSION

Utilisation of day-light is a first rate rule for workstation visual comfort ensurance. When the demands of the Standards are not complied there is necessary to reassess the workstation categorization. In case of the logistic hall there was shown inappropriate illumination of work places not only by objective measurement but attitudinal evaluation as well. It was confirmed that at every change of work activity there has to be assessed also illumination conditions relevant to work activities changed.

The assessment results confirm an importance of annual risk assessment at workstations. Some continual activities changes, accommodation of workstations to new projects and production demands can totally change originally projected workstations and proposed infrastructure associated with them. It is important not to lower the illumination system properties due lowering the input investment. Total environment of the workstation influences importandly on psychic comfort at the workstation and on the employee motivation what was shown by the questionnaire survey.

REFERENCES

KAŇKA, J., Day lighting rooms. Light 2010/1, s. 30 - 34.

PLCH, J., Lighting in practice. IN-EL, Prague1999, 207 s., ISBN 80-86230-09-0.

HNILICA, R., Effect of light on safety. FET in Zvolen, [cited 2010- 01-11 09:33 SEČ]. Available online at: < http://www.bozpinfo.cz/knihovnabozp/citarna/tema_tydne/vplyv08.html>

BALOG, K., TUREKOVÁ, I., TURŇOVÁ, Z., Engineering work environment. Bratislava: STU, 2006, s. 30-39.

AMI Nové Zámky, Center lighting technology [cited 2011-10-01 19:06 SEČ]. Available online at: < http://ami.proxia.sk/>

Sokanský, K, et al. Intelligent lighting systems interior lighting. ČEA, VŠB-TU Ostrava, FEI. 2003.

MONZER, L., Artificial lighting in the living quarters second part - selection intensity lighting, [cited 2011-05-01 20:05 SEČ]. Available online at: < http://www.earchitekt.cz/index.php?KatId=131&PId=1846>

FIŠEROVÁ, S. Applied ergonomics methodology for current engineering practice. Communications - Scientific Letters of the University of Zilina, 2011, roč. 13, č. 2/2011, s. 123-128. ISSN 1335-4205.

STN EN 12464-1: 2005. Light and lighting. Lighting jobs. Part 1: Inside jobs.Electric interior lighting.

OSRAM SK, Light Design, [cited 2010-06-12 10:05 SEČ]. Available online at: <http://www.osram.sk/osram_sk/Sveteln_design/index.html>

ASB SK, Measuring Method 1 part: measurement of light and lighting - Basic terms, [cited 2010-06-12 10:05 SEČ]. Available online at: < www.asb.sk/>

TREBATICKÁ, Martina: Objectivization lighting at selected department. [Thesis] .MTF STU in Trnava; Trnava: MtF STU, 2011.

ASTRA, Wils 6.3 Calculation of artificial lighting [online]. Zlín: Astra 92 a.s., 2007. [cited 2011-31-03 20:07 SEČ], Available online at: <http://www.astrasw.cz/>

CHAPTER 45

Nail Clipper Design for Elderly Population

Hsin-Chieh Wu[a], *Yu-Cheng Lin*[b,*], *Cheng-Heng Hou*[a]

[a]Chaoyang University of Technology, Taichung, Taiwan
[b]Overseas Chinese University, Taichung, Taiwan
* Corresponding author; Email: charlesyclin@gmail.com

ABSTRACT

The elder people generally have the problems of physical-function degeneration. These problems mostly are failing to successfully use the tools or equipments that are originally designed for healthy and young people. This research focused on evaluating the problems and requirements with respect to nail clipper usage by elder people. This research tries to find out the problems of elder people in cutting nails by questionnaire survey. Based on the ergonomics design principals, we redesigned the traditional nail clipper for satisfying the requirements of the elder people. A new handgrip and a footrest were incorporated with this newly-designed nail clipper. An experiment was conducted to compare the newly-designed nail clipper with the traditional one. The task completion time and total number of cutting were collected from 24 elder people during the experiment. These participants also filled an questionnaire after each trial to evaluate the usability of the tested nail clipper. The experimental results indicated that the newly-designed nail clipper was better than the traditional one in the rating of holding-comfort and perceived effort. However, no significant differences in the task completion time and total number of cutting were found between these two nail clippers. In summary, this newly-designed nail clipper has similar performance to the traditional one, but it has less perceived effort and more comfortable feeling when cutting the nails.

Keywords: Aging, nail clipper, usability, hand tool, ergonomic design

1 INTRODUCTION

Nail is a part of the skin, and as individuals grow older, their skin conditions also change. Some major signs of aging include hardening, dryness, and thickening of nails (Tsai, 2002). Therefore, as opposed to young people, it is more difficult for the elderly to clip their nails. The pinch distance may influence the nail clipping pinch strength. Imrhan and Rahman (1995) indicated that the pinch distance has a significant effect on pinch strength when the distance within 2-9.2 cm. In addition, the study also found that, if the pinch distance is more than 9.2 cm, some individuals may fail to complete the pinching motion because their fingers are not long enough. This finding is consistent with the suggestion for the largest hand-tool handle to be the width of 8.8 cm, as proposed by Dababneh, et al. (2004). In 1977, Greenderg and Chaffin (1997) found that when pinch distance is 2.3-3.5 in, most individuals can perform pinching motion of the largest strength. Further, the material of the handle is also associated with the comfort level for pinching a nail clipper. Therefore, how to redesign the handle of a nail clipper is the main issue of this study.

The purpose of this study is to re-design a nail clipper according to the needs of the elderly. The knowledge and technology of ergononic engineering are applied to amend the inappropriate postures required for nail clipping. In addition, it is also expected that the amended nail clipper can enable the elderly to clip their nails more easily.

2 QUESTIONNAIRE SURVEY

A questionnaire survey was conducted to understand the inconvenience and potential problems experienced by individuals during the use of nail clippers. 121 people filled the questionnaire for investigating the usage of nail clippers. 119 valid respondents were returned. The subjects consisted of 52 male and 67 female. The age of the subjects ranged between 15-90 years old, with the average age of 44. The age groups of the subjects were divided into three groups, the young (under the age of 29), the middle-aged (30-39 years old), and the elderly (over the age of 60). A descriptive statistical analysis was conducted on the three age groups and three common types of commercial available nail clippers, as shown in Table 1.

In general, most people (81.7%) used the “two-point pinch” to clip the fingernails of both the right hand and the left. As for toenail clipping, the “sole-pronate” posture was most frequently used (68.8%).

According to the results obtained from the assessment on subjective discomfort of the usage of nail clippers, no matter which type of nail clipper was used during fingernail clipping, the major parts experiencing discomfort were the thumb and the forefinger. In terms of toenail clipping, the major parts experiencing discomfort by most of the people were neck and lower back (lumbar vertebra).

Table 1 The usage of three types of nail clippers

Types	Usage	Usage ratio
Scissor-form nail clipper	Frequently used	23%
	Had used	41%
	Never used	40%
General nail clipper	Frequently used	85%
	Had used	12%
	Never used	3%
Round-handle nail clipper	Frequently used	18%
	Had used	52%
	Never used	40%

3 REDISIGN OF THE CONVENTIONAL NAIL CLIPPER

In regard to design concept, larger contact surface for fingers was adopted, and soft materials were added to the handle, which reduce the discomfort caused by fingers' local force application on a hard handle. The amendment made in the newly-developed nail clipper is a change of the angle of the knife edge, which could reduce the bending of wrist. The knife edge directly rotate to 114°, as shown in Figure 1.

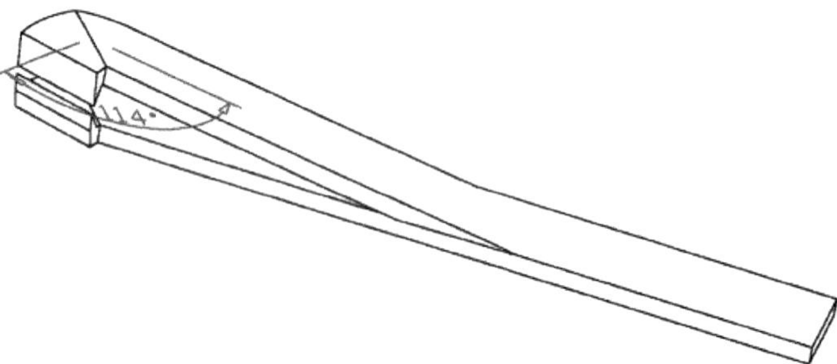

Figure 1 The angle of the knife edge for the newly-developed nail clipper

It was mentioned that the best pinch distance falls between 2 and 8 cm, with the best at 4.4 cm. Appropriate width not only enables users to pinch more freely, but also reduces the discomfort caused by an inappropriate width of the handle. Therefore, 6.8 cm pinch distance was adopted in this design. Further, to reduce the pressure experienced by users between their fingers and contact surface of the nail clippers is also very important. In addition, a slight addition of friction can reduce slipping caused by handle materials. Therefore, not only an outer plastic layer was attached to the newly developed nail clipper to increase friction, but a foam layer was further attached to the original steel handle of the nail clipper to disperse the pressure between the direct contact of fingers/palm and the handle. As opposed to the direct use of steel as a contact surface of conventional nail clippers, the friction and softness was added to the newly developed clipper, as shown in Figure 2. The handle shape was further amended from that of the general conventional nail clippers to that similar to the handle shape of pliers, which makes it more convenient for users to adopt the grasping posture for nail clipping.

Figure 2 The newly-developed nail clipper

During the redesign process, a pedal plate was simultaneously designed in order to reduce users' lumbar angle during the toenail clipping. In addition, JACK software was used to simulate the most appropriate bending angles for ankles and sole of foot. According to the simulation results, the lateral and forward angles of the pedal should be set as 35° and 10°, respectively. The design chart of the pedal plate is shown in Figure 3. In order to fit various body heights, the adjustable pedal height was designed in the range of 358 and 413 mm, which was based on the knee height of Taiwanese adults aged 45-65.

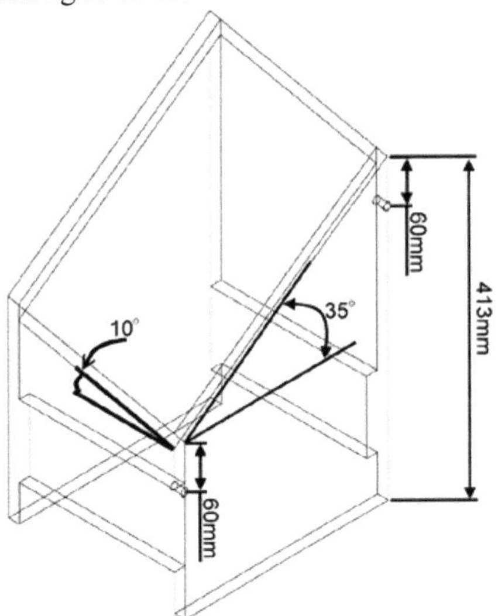

Figure 3 Proposed auxiliary pedal plate for clipping toenails

4 EXPERIMENTAL TESTS

4.1 Methods

This study enrolled 24 volunteer subjects. The dominant hand of all the subjects was the right hand. The average age of male subjects was 71.1 ± 6.7, while that of females was 72.3 ± 8.6. In the experiment, a half of the subjects were requested to use their right hand to hold the newly developed nail clipper to clip their left hand nails at first, and then to use their left hand to hold the conventional nail clipper to clip their right hand nails. The other subjects were contrary. After the nails were completely clipped, the subject was requested to use their dominant hand to clip their toenails. Each nail clipper was adopted to clip either left or right toenails. The experiment was conducted under the premise that all the hand nails and toenails have to be completely clipped at one time. During the experiment, the dependent variables collected were as follows, the number of clipping, completion time, and ratings of usability (5-point scale; 1 = not agree, 5 = strongly agree).

4.2 Results

The t-test results of the comparison of number of clipping and completion time between the conventional nail clipper and the newly developed clipper are shown in Table 2. The results indicated that, neither the number of clipping or task completion time had significant influence.

A nonparametric two-related-samples test (Wilcoxon test) was performed on the ratings in usability to compare these two tested nail clippers. The results are shown in Table 3.

Table 2 T-test results for comparing the two tested nail clippers

Dependent variables	P value
When clipping hand nails	
Number of clipping	0.930
Task completion time	0.400
When clipping toenails	
Number of clipping	0.851
Task completion time	0.219

Table 3 Nonparametric analysis on the ratings in usability

	Conventional nail clippers (Mean ± SD)	Newly developed nail clippers (Mean ± SD)	P value
Ease of clipping	3.7 ± 0.7	4.4 ± 0.7	0.002*
Ease of slipping	2.5 ± 0.9	2.2 ± 0.8	0.194
Discomfort after clipping	2.4 ± 0.8	2.8 ± 0.9	0.219
I have confidence to use this nail clipper	3.5 ± 0.7	3.5 ± 0.7	0.670
The overall feeling for the use of pedal plate and nail clipper is good	No pedal plate	4.0 ± 1.0	NA

**P<0.05*

5 DISCUSSION

The average ratings in "The overall feeling for the use of pedal plate and nail clipper is good" was 4.0, suggesting that, after the use of pedal plate, the subjects agreed with "a good feeling of use". The ratings in "Is it easy to pinch the nail clipper?" for the conventional nail clippers was 3.7, while that for the newly developed clipper was 4.4. In addition, significant P=0.002 was obtained from the nonparametric two-related-samples test (Wilcoxon test), indicating that, the subjects suggested it is easier to pinch the newly developed nail clippers.

The average ratings in the discomfort with the tested nail clippers were little (less than 3). It could not be concluded that either of the two types of nail clippers could contribute to a sense of discomfort. The reason might be that, in the experiment, the period of nail clipping test may be too short, which resulted in little feeling of discomfort.

6 CONCLUSIONS

The major findings obtained from the experiments on a comparison of newly-developed nail clippers and conventional clippers are as follows: (1) the newly-developed nail clipper makes no difference for left or right hand use; (2) there is no significant difference in the performance between the newly-developed nail clipper and conventional clippers, suggesting that the performance of the newly-developed nail clipper is similar to that of the conventional clipper; (3) as opposed to the conventional nail clippers, there is a significant change in the pinch posture for the newly-developed nail clipper, and users felt it was significantly easier to perform the posture; (4) for the newly developed nail clipper, there is a need to provide a detailed explanation regarding usage prior to use in order to enable the users to understand and become more comfortable with the strength-saving method. Because most people are used to the two-point pinch, the grasping-styled nail clipper may be less frequently seen.

ACKNOWLEDGMENTS

The authors would like to acknowledge the National Science Council of the Republic of China for financially supporting this research.

REFERENCES

Dababneh, A., Lowe, B., Krieg, E., Kong, Y.K. & Waters, T. 2004. A Checklist for the Ergonomic Evaluation of Nonpowered Hand Tools. *Journal of Occupational and Environmental Hygiene* 1:D135–D145.

Greenderg, L. & Chaffin, D. 1997. *Workers and Their Tools*, Mi. : Pendell.

Imrhan, N., & Rahman, R. (995. The effects of pinch width on pinch strengths of adult males. *International Journal of Industrial Ergonomics*, 16, 123-134.

Tsai, C.F. 2002. Ageing of nail. *Health World* 193:85-90. (in Chinese).

CHAPTER 46

An Adjustable Assistive Chair Design for Older People

Chiwu Huang

Department of Industrial Design/ Graduate Institute of Innovation and Design
National Taipei University of Technology
1 Sec. 3, Chung-Hsiao East Road,
Taipei 10608
Taiwan
E-mail: chiwu@ntut.edu.tw

ABSTRACT

In a previous study, an assistive chair design was proposed to assist older people sit-down and rise-up from the chair easily. The user's weight was supported partially by a gas spring mechanism underneath the chair. As a result, despite a few of heavier and lighter users, most of the users felt easier in both rising-up and sitting-down. This study proposes a new assistive chair design in which the gas spring mechanism can be adjusted in different angles to fit different weights of people. An evaluation was carried out. 30 older people aged between 62 and 84 years (mean=70, std. dev. =6.3, 15 females and 15 males, 14 were healthy and 16 had minor arthritic rheumatics or waist pain) were recruited to evaluate the design. Speed for sitting-down and rising-up were measured. Video was taken to observe changes of respondents' trunk-angle during use. The user's satisfaction was also measured in a ten-point scale. The result shows the design works better than the previous one which the gas spring is fixed. The lager the angle of gas spring the larger the force provided for helping the subject rising up and sitting down to the chair. The heavier people need a larger angle. Users rising up from the chair need a larger angle than sitting down.

Keywords: older people, sit-to-stand, stand-to-sit, assistive chair

1 INTRODUCTION

Sit-to-stand and stand-to-sit maneuvers have been considered to be one of the most challenging movements older people encounter in their daily life (Manchoundia et al. 2006). Many fall accidents occur during these movements. Lehtola et al. (2006) reported, after a two-year observation of 555 elders (the average age is 85), that 12% of falls occurred during the daily routines of standing up and sitting down. It has been observed that elderly subjects experienced difficulties moving in and out of a chair (Kao and Huang, 2008), and therefore it is crucial to design a chair that can assist older people to safely stand up and sit down from a chair.

Huang (2010) proposed a chair design that uses a gas spring installed underneath the seat to support part of the user's weight and to reduce the pressure on his/her lower limbs. The design was evaluated by 22 older subjects. It found the subjects using the chair had a smaller leaning angle than using a regular chair. This finding indicates that older people might be benefitted from using the design. 80% of subjects claimed that the design is easy to use. However, one thinner subject was unable to sit down due to his light weight while other two heavier subjects felt no difference when rise up from the chair. Based on the findings of the above study, it can be concluded that there may be a correlation between the weight of the subjects and the force of the gas spring. The heavier the weight, the more assistive force is needed. Moreover, it was also found rising up from chair need more assistive force than sitting down. The design can be improved further.

Base on the mechanism design of a seat assist product on the market, UpLift Seat Assist (North Coast Medical, Inc.), an adjustable assistive chair design is proposed (figure 1). The gas spring can be adjusted to various angles to cope with user's different weight and both sit-down and rise-up situation (figure 2). The angle can be set in seven settings, i.e. 5 to 35 degree with every 5 degree ascending. To test the usability of the chair an evaluation experiment was carried out.

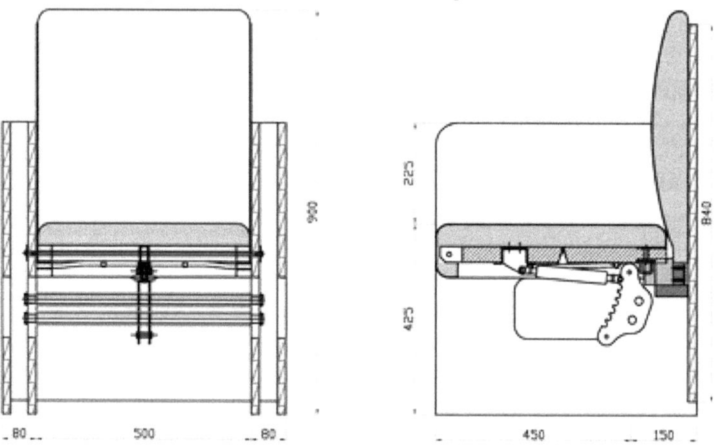

Figure 1 Front view and side view of the adjustable assistive chair

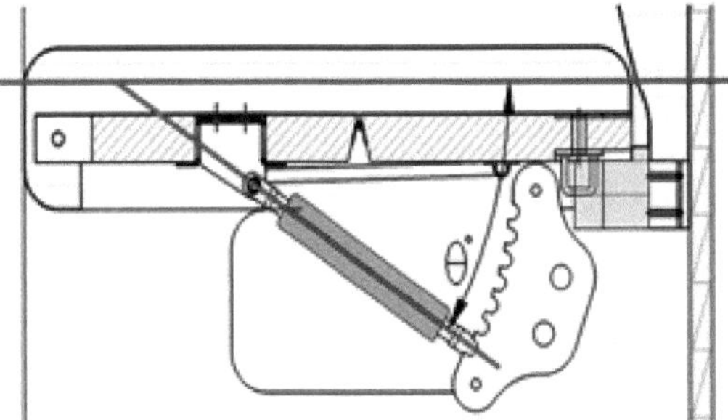

Figure 2 Adjustable angle of gas spring

2 METHODS

To test the usability of the proposed design an experiment was carried out. A prototype (Figure 3) was made. The stratified sampling method according to the normal distribution of Taiwanese stature and weight (Wang et al. 2002) was adopted to render the sampling process. 30 subjects, fifteen males and fifteen females, aged between 62 and 84 years old (mean =70) were recruited to test the chair. The subjects have to be able to handle their daily living independently.

Figure 3 Prototype of an adjustable assistive chair

The experiment was carried out in following steps:

1. Measuring the subject's weight and height
2. subject was initiated the test on the 65kg gas spring and set at 5 degree in angle. If he/she cannot sit down, a lighter, i.e. 50kg, gas spring was used instead. If she/he sits down too easy, a heavier, i.e. 80kg, gas spring was used.

3. Once gas spring power was decided the subject tests the chair from 5 to 35 degree upward until he/she cannot sit down.
4. Every angle was tested three times, with 1 minute break in between.
5. An evaluation was taken against every angle setting.
6. Repeat step 2 to 5 for rise-up test from 35 to 5 degree downward until the subject cannot rise up by the pushing of the gas spring.

Outcome measures and data analysis are as following: still images of standing-up and sitting-down movements were extracted from the videos for analysis. AutoCAD 2006 was used to measure the leaning-forward angles on the pictures. The leaning angle was defined as the angle between the trunk and vertical plane (Figure 4). SPSS 12 was used to analyze the data. The independent variables include: subjects' age, stature, and weight; dependent variables include: subjective preference, time required, and the leaning angle during standing up and sitting down.

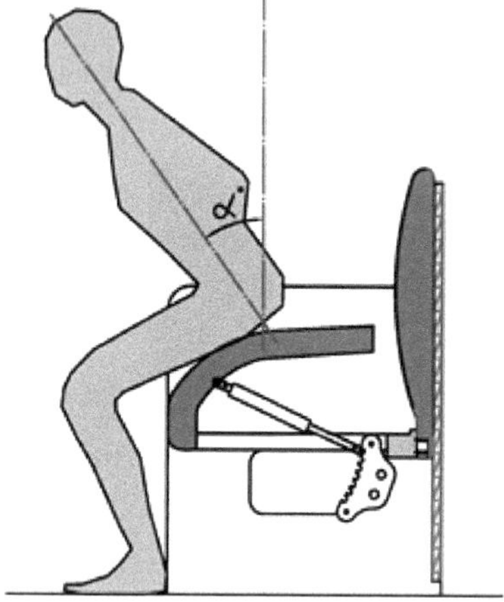

Figure 4 Trunk leaning angle

3 RESULTS AND DISCUSSION

3.1 gas spring selected according to subject's weight

The result found that the force of gas spring was selected in accordance with subjects' weight. The heavier the weight the larger force of gas spring was selected. Table 1 shows 9 (30%) subjects weighted in 45-64 kg chose gas spring in 500N. 12 subjects weighted in 55-69 kg chose 650N gas spring. The rest 9 subjects (30%) weighted over 70 kg chose 800N gas spring. The preferred gas spring force is illustrated in shading area in Table 1.

Table 1 The gas spring force selected by subjects

Body Weight (kg)	Gas spring force (N)			subtotal
	500	650	800	
40-44	0	0	0	0
45-49	2	0	0	2
50-54	5	0	0	5
55-59	1	3	0	4
60-64	1	2	0	3
65-69	0	7	2	9
70-74	0	0	3	3
75-80	0	0	2	2
81-84	0	0	1	1
85-90	0	0	1	1
subtotal	9	12	9	30
Percentage (%)	30.0	40.0	30.0	100.0

3.2 Gas spring angle for sit-down

Table 2 shows that 13 subjects (43.3%), 9 males, 4 females prefer gas spring was set at 20 degree while 10 subjects (33.3%), 3 males and 7 females prefer 15 degree. 15 and 20 degree were thought as the most appropriate angle for gas spring in sit-down.

Table 2 The gas spring's angle selected for sitting down

ΘDN (degree)	male		female		Total (n=30)	
	n	%	n	%	n	%
5	0	0.0	1	3.3	1	3.3
10	2	6.7	2	6.7	4	13.3
15	3	10.0	7	23.3	10	33.3
20	9	30.0	4	13.3	13	43.3
25	1	3.3	1	3.3	2	6.7
30	0	0.0	0	0.0	0	0.0
35	0	0.0	0	0.0	0	0.0
total	15	50.0	15	50.0	30	100.0

3.3 Gas spring angle for rise-up

Table 3 shows 23 subjects (76.7%), 12 males and 11 females prefer gas spring was set at 35 degree while 7 subjects (23.3%), 3 males and 4 females prefer 30 degree when rise up from chair. The result shows subjects prefer a larger angle when rise-up from chair for it release larger force.

Table 3 The gas spring's angle selected for rising-up correspondent to gender

ΘUP (degree)	male		female		Total (n=30)	
	n	%	n	%	n	%
35	12	40.0	11	36.7	23	76.7
30	3	10.0	4	13.3	7	23.3
total	15	50.0	15	50.0	30	100.0

3.4 Trunk leaning angles and subjective satisfaction

Trunk leaning angles were measured. The result shows in Table 4. The average leaning angle is 20 degree with 5.59 degree standard deviation. By comparing to the former chair design (mean=33), the angle is 13 degree smaller. This means the new design needs less effort to use. However, when ask the subjects if they are satisfied with new design, 24 out of 30 (80%) thought the chair is helpful for siting and rising (Table 5). 4 subjects have no comments and 2 subjects thought the chair is unhelpful. The satisfaction rate is almost as same as the former design.

Table 4 Mean trunk leaning angle (N=30)

	Minimum degree	Maximum degree	mean	Std. dev.
Leaning angle	5	28	20	5.59

Table 5 Subjective satisfaction

	n	%
helpful	24	80.0
No comment	4	13.3
unhelpful	2	6.7
total	30	100.0

4 CONCLUSION

This study proposed an adjustable gas spring chair design. The design is evaluated by 30 older people. The result shows the design works better than the previous one which the gas spring is fixed. The lager the angle of gas spring the larger the force provided for helping the subject rising up and sitting down to the chair. The heavier people need a larger angle. Rising up from the chair the user needs a larger angle than sitting down. Findings are described as follows.

1. The output power of gas spring lifter should be selected according to the user's weight. The lighter the user the lesser power of gas spring should be applied. In the evaluation, it was found that the users whose weight under 55 kg, 55-70 and above 70 kg, were applied to 500, 650 and 800 N gas spring respectively.

2. Sitting-down needs less supportive power than rising-up. This can be achieved by adjusting the angle of the gas spring. In the proposed design, 15 degree is for sitting-down whereas 35 degree is for rising-up.

3. 24 out of 30 (80%) respondents thought that the design can make sitting-down and rising-up more easily than ordinary chairs.

4. The average trunk angle for rising-up is 20 degree which is significantly less than the previous design (33 degree).

ACKNOWLEDGEMENT

This study is sponsored by the National Science Council of Taiwan (NSC100-2221-E-027-069)

REFERENCES

Huang, C. 2010. The Evaluation of an Assistive Chair Design for the Elderly. Proceedings, 3rd International Conference on Applied Human Factors and Ergonomics (AHFE), Miami, Florida USA, 17-20 July

Kao, L. and C. Huang. 2008. A study on the rising and sitting chair design for the elderly (in Chinese). Proceedings, Technology and Education 2008, Mingchi University of Technology, Taipei, Taiwan.

Lehtola, S., P. Koistinen, and H. Luukinen. 2006. Falls and injurious falls late in home-dwelling life. Archives of Gerontology and Geriatrics, 42, 217-224.

Manckoundia, P., F. Mourey, P. Pfitzenmeyer and C. Papaxanthis. 2006. Comparison of motor strategies in sit-to-stand and back-to-sit motions between healthy and Alzheimer's disease elderly subjects. Neuroscience, 137, 385-392.

North Coast Medical, Inc. UpLift Seat Assist. Accessed Sept. 2010 http://upliftproducts.com/seat-assist/product-specifications.html

Wang, M. J., E. M. Wang and Y. Lin. 2002. Anthropometric Data Book of the Chinese people in Taiwan. The Ergonomics Society of Taiwan.

CHAPTER 47

A Comparative study of Static Balance Before and After a Simulated Firefighting Drill in Career and Volunteer Italian Firefighters

Massimiliano Pau, Giulia Cadoni, Bruno Leban and Andrea Loi

Department of Mechanical, Chemical and Materials Engineering
University of Cagliari, ITALY
pau@dimeca.unica.it

ABSTRACT

In many countries, the firefighting service relies on the support of a relevant amount of volunteer personnel. Although volunteer firefighters (FFs) undergo basic training similar to that administered to regular career units, they often perform a lower number of operation shifts and training updates. However, both volunteer and career FFs are required to perform on the same standard level, and no differences are allowed as regards technical and psychophysical skills necessary to perform efficiently. Among the most important physical features, balance plays a crucial role in firefighting and search and rescue operations, as often such activities are performed under unstable conditions, with a reduced base of support and in dark or smoky environments. On the basis of the aforementioned considerations, this study aims to investigate the static balance capabilities of two groups (n=10 each) of career and volunteer Italian FFs. Balance was assessed by measuring postural sway parameters calculated on the basis of center-of-pressure (COP) time series acquired in two conditions: rest (basal value) and after a fatigue protocol which included several activities commonly performed during firefighting and rescue services. The results show that the basal sway features of volunteer and career FFs are basically similar, despite the significant different in age (which should penalize the older individuals). When fatigued, all participants showed significantly poorer balance, with volunteer FFs generally characterized by larger increases in most sway parameters with respect to their career colleagues. Although further investigations on larger samples are required, the results obtained here appear to suggest that the reduced activity of volunteer FFs may represent a co-factor in increasing the risk of slips and falls during operations.

Keywords: firefighters, balance, postural sway

1 INTRODUCTION

In many western countries, a significant part of firefighting personnel is composed of volunteers. In the United States and Sweden, for example, they represent the majority (800,000 units vs. 300,000 career FFs in the U.S., Centers for Disease Control and Prevention CDC, 2006, 10,700 vs. 5,100 in Sweden, Swedish Civil Contingencies Agency, 2008) and serve mainly small communities. In the UK and Italy, volunteer FFs (also indicated as "retained" FFs) represent a percentage ranging from 20 to 30% (Department for Communities and Local Government, UK, 2011, Corpo Nazionale dei Vigili del Fuoco, Italia, 2011). According to Italian laws and regulations, volunteer and career FFs must undergo the same basic training, receive the same salary and perform the same type of duties. Nevertheless, volunteer FFs are limited to a maximum of 160 days of service per year. This fact unavoidably reduces the number of shifts (and thus field experience) and "on fire-station" training updates, which are regularly administered to career personnel. The way such differences reflect on the effectiveness and quality of the performance is still partly unexplored and, above all, if volunteer FFs are exposed to a greater risk of accidents and injuries due to their special condition. However, statistics show that volunteer FFs are more likely to receive injuries at the fireground and during training than all firefighters combined (National Fire Protection Association, USA, 2012).

Among the psychophysical skills required to safely and optimally perform firefighting and rescue operations, balance certainly plays a crucial role, especially if one considers that FFs must often operate on slippery surfaces, to move on a reduced base of support (e.g. ladders, roofs) and in dark environments. Moreover, balance can be impaired by a number of factors which include primarily age (i.e. postural control declines with aging, Colledge et al., 1994, Du Pasquier et al., 2003, Abrahamová and Hlavačka, 2008) physical activity (Jakobsen et al., 2010), anxiety (Ohno, 2004), difficulty in the cognitive task (Pellecchia, 2003) and fatigue (Paillard, 2012).

Thus, it appears important to understand if, from the point of view of basic balance skills, career and volunteer FFs exhibit different performances due to possible anthropometric and training differences. and, in particular, if fatigue originated by physical activities necessary to perform operational duties differently affects the two categories. Balance is usually assessed on the basis of objective measurements of postural sway, that is, the small oscillations performed by the body to ensure the maintaining of the upright standing position against gravity and other external perturbations. This is achieved in practice by analyzing the movement, in time, of the center of pressure (COP), which identifies the position of the resultant of the forces exchanged between body and ground through the plantar region, and whose trajectory is related to that of the body center of mass.

A number of previous studies specifically performed on FFs analyzed time series of COP to assess postural sway under a range of different conditions: Sobeih et al. (2006) investigated the effect of turnout gear and long work shift on FFs postural stability, finding that prolonged shifts may contribute to increasing the risk of slips and falls. Punakallio et al. (2003, 2004, 2005) performed several extensive experimental campaigns aimed at investigating the existence of a relationship between work ability and risk of slips and falls in FFs and balance, considering variables such as age, muscular capacities and presence of protective equipment. While there is a certain agreement in literature on the importance of balance as a predictor of functional performance and as a co-factor able to influence the risk of harmful events, other issues appear still not fully clear such as, for example, the effect of protective equipment also as a function of subjects' ages.

The issues related to the different amount of experience and training/service frequency appears relatively less explored, and this justifies the need for specific studies as the one proposed here. Given that differences between career and volunteer FFs are likely to exist in terms of age, field experience, hours of training updates and number of performed fire and rescue services per year, the purpose of the present research is to assess whether such differences also reflect on static balance capabilities of personnel. In particular, the study proposes to answer four basic questions: 1) are basic balance skills similar in career and volunteer FFs? 2) to what extent does the presence of protective equipment modify physiologic postural sway? 3) what is the role of fatigue (originated by a simulated firefighting and search and rescue activity) in altering postural control features? and 4) does fatigue differently impair balance in career and volunteer FFs?

2 METHODS

2.1 Participants

Twenty career and volunteer FFs currently in service at the Fire Department of the city of Oristano (Sardinia, Italy) were recruited for the study on a voluntary basis. Their anthropometric and service features are reported in Table 1: as visible from the data, the volunteer FFs were significantly younger than career ones. Moreover, according to the official time-sheet supplied by the Department

Command, the two categories of FFs performed a significantly different number of working days in 2010. Test procedures and purposes of the study were explained to all participants, and a signed informed consent form was obtained from them. Participants were also asked to fill in a questionnaire containing basic information about the physical activity independently performed during a typical week, as well as the existence of either acute or chronic musculoskeletal diseases. The whole study was carried out in compliance with the ethical principles for research involving human subjects expressed in the Declaration of Helsinki, and was approved by the Departmental Review Board.

Table 1 Anthropometric and service data of the firefighters participating in the study

	Career FFs (n=10)	Volunteer FFs (n=10)	p-value
Age (years)	46.4 (5.3)	29.2 (6.2)	**<0.001***
Height (cm)	171.4 (3.7)	172.2 (4.0)	0.456
Weight (Kg)	77.0 (7.4)	78.4 (11.4)	0.963
BMI (kg m^{-2})	26.3 (3.0)	26.8 (3.7)	0.819
Working days in 2010	265.6 (0.8)	75.3 (47.6)	**<0.001**

* denotes a significant difference between Career and Volunteer FFs

2.2 Data acquisition and post-processing

Postural sway parameters were calculated on the basis of COP time series acquired by means of a pressure platform Footscan® 0.5 system (RS Scan International, Belgium) composed of a pressure-sensitive plate (4096 7.62 x 5.08 mm sensing elements based on piezoresistive technology arranged in a 64 x 64 matrix) and a USB interface box connected to a personal computer. As the plate management software allows the collection of a fixed number of 1000 events (frames) regardless of trial length, the acquisition frequency was automatically set to 33 Hz. The raw COP time series were then post-processed with a custom-developed Matlab® routine to calculate the following sway parameters: sway area (SA, 95% confidence ellipse), COP path length (COP PL, the overall distance travelled by the COP during the trial), COP maximum displacement (MDISP., i.e. the difference between the maximum and minimum value of the selected coordinate recorded during the trial) in the mediolateral (ML) and anteroposterior (AP) directions, and COP velocity (V_{COP} calculated as the average of 1000 values calculated for each temporal event into which the trial was subdivided) in AP and ML.

The firefighters underwent three series of experiments during their regular shifts directly at the fire station in a dedicated quiet room. Two trials were to assess the baseline reference sway values, and in this case the firefighters were tested with and without the additional fire protective equipment (FPE) that is, Nomex jacket, brush pants, gloves and helmet (Figure 1). The participants were asked to stand barefoot as motionless as possible for 30 s on the pressure plate with the feet placed on a sheet of paper with two footprints oriented at approximately 30°, keeping a stable and relaxed position with arms freely positioned along their sides and their gaze fixed on a target image.

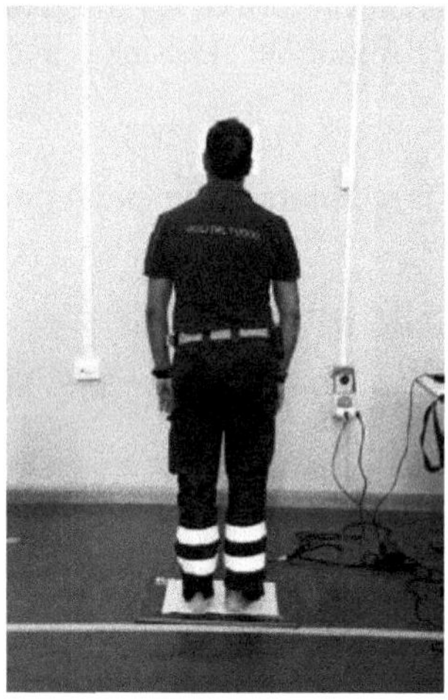

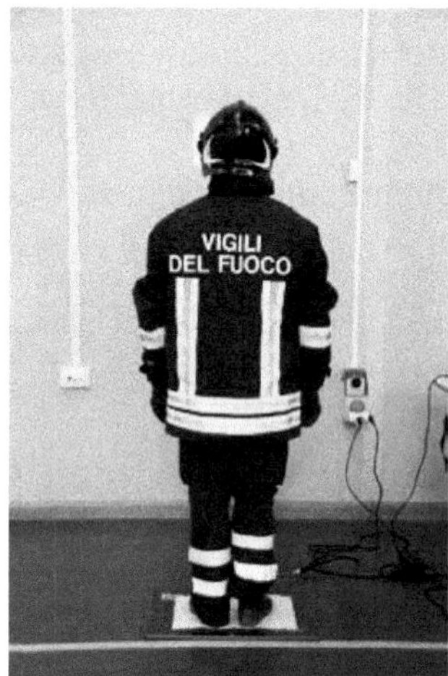

Figure 1 Position of the firefighters on the pressure platform for postural sway measurements without and with the additional fire protective equipment

The tests were then repeated after a fatigue protocol that included a number of activities commonly performed in actual firefighting and rescue operations as follows (Figure 2):

1. Complete unrolling of a 70 mm diameter, 20 m long fire hose
2. Connection of the fire hose terminal to the fire engine
3. Wearing of the Self Contained Breathing Apparatus (SCBA)
4. Ascending two floors of the fire training tower
5. Raising the fire hose from the ground level through a fire training tower window, using a rope
6. Lowering the fire hose
7. Returning to ground level and removal of the SCBA
8. Simulating the rescue of an unconscious survivor by transporting a 25 Kg dummy for 40 meters at the maximum possible speed.

For completion the whole sequence of operations required a time of approximately 3 minutes. The statistical *t*-test revealed no significant differences in performance time between volunteer and career FFs (182.2 (44.9) and 163.0 (15.7) seconds respectively, p=0.218).

Figure 3 Firefighters during the execution of the fatigue protocol.

Apart from step 8, in which firefighters were expressly required to run as fast as possible, for all the preceding steps the pace was self-selected according to the specific skills and energy management strategy of each participant, also considering that haste may adversely affect search and rescue activities that need lucidity to be efficiently performed.

The statistical significance of the possible differences in postural sway parameters introduced by fatigue was assessed using the three-way analysis of variance (ANOVA), where independent variables were firefighter status (volunteer/career) and body status (rest/fatigue) and dependent variables SA, COP PL, MDISP AP and ML, V_{COP} AP and ML, by setting the level of significance at $p<0.05$. When necessary, a post-hoc Holm-Sidak test for pairwise comparison was carried out to assess intra- and inter-group differences. Data were checked for normality (using the Shapiro-Wilk test) and equal variance before any ANOVA test.

3 RESULTS

A preliminary screening (carried out by means of a *t*-test) was performed to verify whether the presence of the FPE induced significant changes in sway reference values or not, as in literature contrasting results have been proposed as regards this issue. The results of the statistical analysis revealed that wearing FPE induced significant increases in SA (p=0.032) and COP MDISP in the ML direction (p=0.033), thus it was decided to select, as sway baseline values, those obtained from trials in which the FPE was worn.

The fatigue originated by the simulated operation protocol significantly affected (in terms of the main effect found by the ANOVA test) all sway parameters, although with different modalities. In the case of SA (see Figure 4), while similar basal values were found for both career and volunteer FFs, the post-hoc analysis revealed that the increased post-fatigue was significant only in the case of volunteer FFs (p=0.041). On the contrary, the COP PL increased significantly (and similarly even from a quantitative point of view) in both categories of FFs (p=0.014 and p=0.048 respectively for volunteer and career FFs).

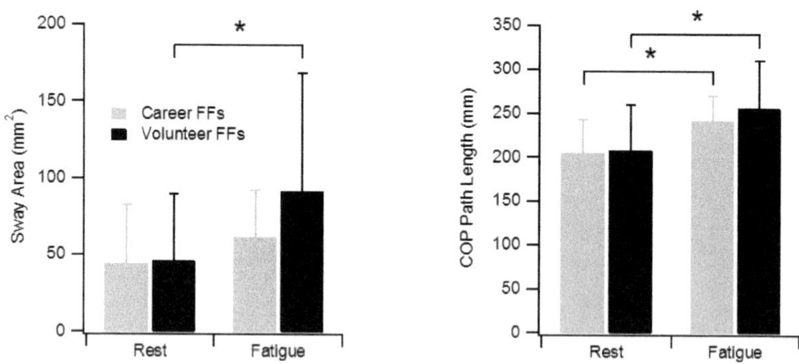

Figure 3 Sway area (left) and COP Path Length (right) before and after the fatigue test (*p<0.05)

As regards the COP displacements (Figure 4) the main effect of fatigue was found for both directions even though the post-hoc analysis revealed that in the ML direction a significant increase in displacement values were detected only for volunteer FFs (p=0.026).

In the case of the AP direction, post-hoc tests produced no significant results, probably due to the limited size of the samples, but it is noteworthy that the p value of career FFs was borderline (p=0.062). The COP velocities (Figure 5), increased significantly in both groups in the AP direction (p=0.001 volunteers, p=0.012 career), while in the ML direction, again no significant results came from the post-hoc test, although for the volunteer group the p value was 0.076.

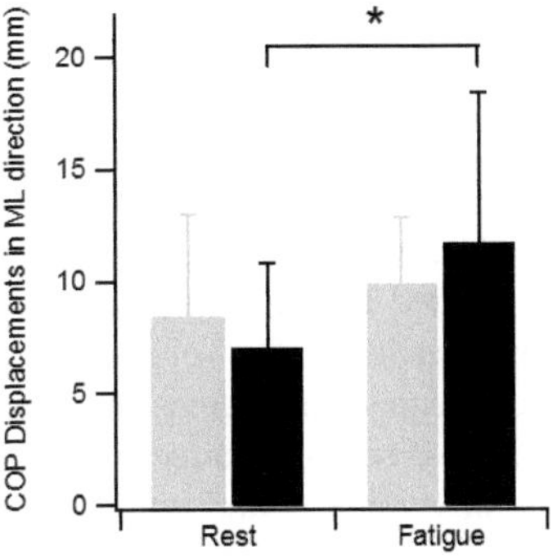

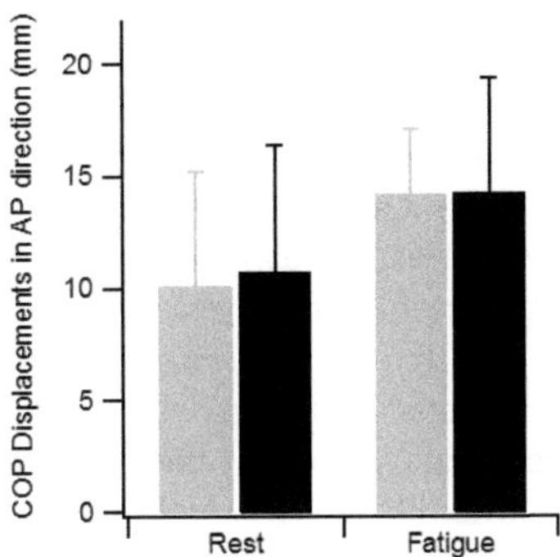

Figure 4 COP displacements in ML (left) and AP (right) directions before and after the fatigue test

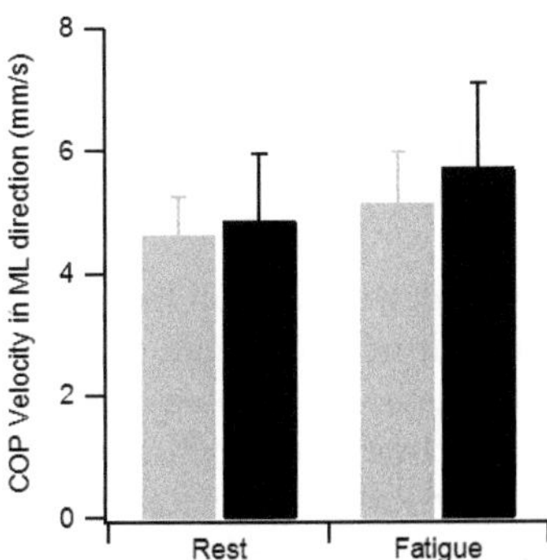

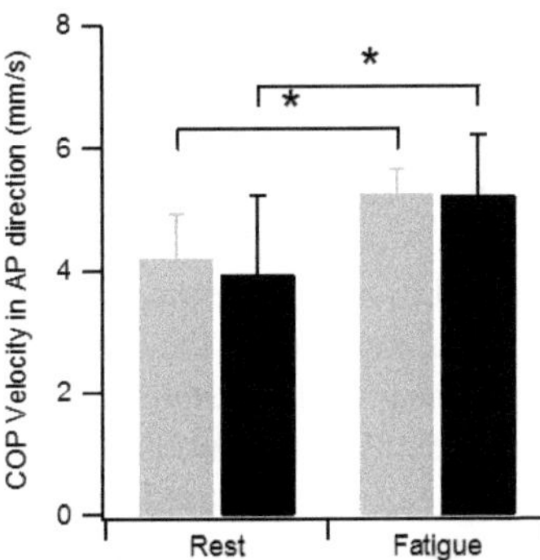

Figure 5 COP velocities in ML (left) and AP (right) directions before and after the fatigue test

The results of the questionnaire on physical activity performed outside the work period showed that career and volunteer FFs were active in a similar manner, as 76 (career) to 90% (volunteer) of them regularly performed some kind of activity and 40% did that twice per week. The majority of participants (40% for both groups) dedicated 2-4 hours per week to exercise, while a similar percentage (30-34%) increased this value up to 4-6 hours.

4 DISCUSSION

It was somewhat surprising to observe that the reference values of the postural sway parameters analyzed (i.e. in rest conditions while wearing FPE) were similar in both groups of FFs, despite the significant difference in age. There is agreement in literature on the fact that some degree of balance impairment occurs with age: Colledge et al. (1994) found a linear increase in sway path with increasing age,

Abrahamová and Hlavačka (2008) observed a nonlinear growth trend of COP displacements and velocity as the age of individuals increases and Du Pasquier et al. (2003) reported a linear increase in COP velocity in AP directions with age. On the contrary, our results showed no significant increase in sway parameters in the group of older subjects. Given that physical activity (as self-reported in the questionnaires) was similar, at least in quantity, for the two groups tested here, we may hypothesize that the absence of differences in sway amplitude is related to the constant and superior rescue activity performed by the career FFs with respect to the volunteers. After the simulated drill protocol, both groups exhibited a generalized increase in all sway parameters, although the volunteers performed worse than career FFs, especially in terms of SA (which doubled after fatigue in volunteers and was only 38% higher in career FFs), while in the other parameters the differences were quite small (on the order of 5%). The adverse effects of whole-body and localized fatigue are well known in literature (Paillard, 2012) but few studies have investigated the effect of age on fatigue-induced balance impairment. It has been observed that the disturbance effect of postural control is similar in young and older subjects and is sometimes more evident in young ones. This is partly due to different postural strategies (i.e. younger individuals preferentially adopt an ankle strategy, older ones mainly use a hip strategy) and compensatory mechanisms (more efficient in the older) but also because older subjects are characterized by a higher percentage of type I muscular fibers which ensure a better resistance to fatigue in general (Paillard, 2012).

Of course, this study is limited from several point of view: first of all the size of the sample is quite small and thus some of the effects originated by fatigue, which at the moment appear as a not statistically significant trend, should be more carefully evaluated. Secondly, the simulated fire and rescue activity, although realistic, probably does not fully reproduce the maximum physical effort necessary to perform an actual emergency service, especially in the case of large fire accidents, where heat, brightness and smoke play a crucial role in altering the FFs body response.

Moreover, although participants were stimulated to give their best performance when performing the fatigue protocol, the experiments were characterized by a very low level of stress and anxiety which are instead certainly present during a real emergency. As postural sway is known to be affected by anxiety, the results obtained here might be underestimated to some extent.

5 CONCLUSIONS

The disparities existing between career and volunteer FFs, in terms of age, field experience and training schedule, must be carefully investigated in order to predict possible consequences on service performance and health of personnel, also considering that volunteer FFs appear to be more exposed to injury while on duty. The present study analyzed static balance features before and after a fatigue protocol of two groups of FFs who were in service for a significantly different number of days in 2010. The results are encouraging since the career FFs, probably due to their habit of daily facing situations that require optimal balance skills, exhibit basal values of postural sway similar to those measured in their younger colleagues. On the other hand, fatigue appears to have a stronger impact on volunteers in terms of balance impairment. Further studies are needed to clarify this issue but, in any case, it appears necessary to plan proper training programs (including specific balance training) to be regularly administered with similar scheduling to both categories of FFs.

REFERENCES

Abrahamová, D. and Hlavačka, F. 2008. Age-related changes of human balance during quiet stance. *Physiological Research* 57:957-964.

Center for Disease Control and Prevention (CDC) *"Fatalities Among Volunteer and Career Firefighters - USA, 1994—2004"*, Accessed February 24 2012, from http://www.cdc.gov/mmwr/preview/mmwrhtml/mm5516a3.htm

Colledge, N.R., Cantley, P., Peaston, I. et al. 1994. Ageing and balance: the measurement of spontaneous sway by posturography. *Gerontology* 40: 273-278.

Corpo Nazionale Vigili del Fuoco (Italia), *"Il Corpo Nazionale dei Vigili del Fuoco"*, Accessed February 24 2012, from http://www.vigilfuoco.it/

Department for Communities and Local Government (UK). 2011. "*Fire and Rescue Service Operational Statistics Bulletin for England 2010 -1", 7*

Du Pasquier, R.A., Blanc, Y., Sinnreich, M. et al. 2003. The effect of aging on postural stability. *Neurophysiologie Clinique* 33:213-218.

Jakobsen, M. D., Sundstrup, E., Krustrup, P., et al. 2011. The effect of recreational soccer training and running on postural balance in untrained men. *European Journal of Applied Physiology* 111 (3), 521–530.

National Fire Protection Association (USA). 2012. *An Analysis of Volunteer Firefighter Injuries, 2008-2010,* 1

Ohno, H., Wada, M., Saitoh, J. et al. 2004. The effect of anxiety on postural control in humans depends on visual information processing *Neuroscience Letters* 364, 37-39

Paillard, T. 2012. Effects of general and local fatigue on postural control: a review. *Neuroscience and Biobehavioral Reviews* 36, 162-176.

Pellecchia, G.L. 2003. Postural sway increases with attentional demands of concurrent cognitive task. *Gait and Posture* 18, 29-34.

Punakallio, A., Lusa, S. and Luukkonen R. 2004. Functional, postural and perceived balance for predicting the work ability of firefighters. *International Archives of Occupational Environmental Health* 77, 482-490.

Punakallio, A., Hirvonen, M. and Grönqvist, R. 2005. Slip and fall risk among firefighters in relation to balance, muscular capacities and age. *Safety Science* 43, 455-468.

Punakallio, A., Lusa, S. and Luukkonen R. 2003. Protective equipment affects balance abilities differently in younger and older firefighters. *Aviation, Space and Environmental Medicine* 74, 1151-1156.

Sobeih, T.M., Davis, K.G., Succop, P.A. et al., 2006. Postural Balance Changes in On-Duty Firefighters: Effect of Gear and Long Work Shifts *Journal of Occupational Environmental Medicine* 48, 68-75

Swedish Civil Contingencies Agency. 2008. "*The Swedish Rescue Services in Figures",* 12 Accessed February 24 2012, from http://www.msb.se/RibData/Filer/pdf/25586.pdf

Section IV

Ergonomic Methods and Task Analysis

CHAPTER 48

Prediction of Drowsiness Using Multivariate Analysis of Biological Information and Driving Performance

Atsuo Murata, Yutaka Ohkubo, Takehito Hayami, and Makoto Moriwaka

Graduate School of Natural Science and Technology, Okayama University
Okayama, Japan
murata@iims.sys.okayama-u.ac.jp

ABSTRACT

The aim of this study was to predict drowsy states by applying multivariate analysis such as discrimination analysis and logistic regression model to biological information and establish a method to properly warn drivers of drowsy state. EEG, heart rate variability, EOG, and tracking error were used as evaluation measures of drowsiness. The drowsy states were predicted by applying discrimination analysis and logistic regression to these evaluation measures. The percentage correct prediction for discrimination analysis and logistic regression were 85% and 93%, respectively. The polynominal logistic regression model was found to lead to higher prediction accuracy. The biological data might be used for the long-term prediction of drowsiness, while the performance data such as tracking error can be used only for the short-term prediction.

Keywords: multivariate analysis, discrimination analysis, polynominal logistic regression, biological information, drowsiness, prediction technique

1 INTRODUCTION

Many studies used psychophysiological measures such as blink, EEG, saccade, and heart rate to assess fatigue or drowsiness. No measures alone can be used reliably to assess drowsiness, because each has advantages and disadvantages. The results of these studies must be integrated and effectively applied to the prevention of drowsy driving. To prevent drivers from driving under drowsy state and causing a disastrous traffic accident, not the gross tendency of reduced arousal level but the more accurate identification of timing when the drowsy state occurs is necessary. It is not until such accurate measures to predict the timing of occurrence of drowsy driving is established that we apply this to the development of ITS (Intelligent Transportation System) which can surely and reliably avoid unsafe and unintentional driving under drowsy and low arousal state.

Brookhuis et al., 1993 imposed participants on an on-road experiment to assess driver status using biological information such as EEG and ECG. They found that changes in EEG and ECG reflected changes in driver status. Kecklund et al., 1993 measured EEG continuously during a night or evening drive for eighteen truck drivers. During a night drive, a significant intra-individual correlation was observed between subjective sleepiness and the EEG alpha burst activity. The subjective sleepiness and the EEG alpha burst activity were significantly correlated with total work hours. As a result of a regression analysis, total work hours and total break time predicted about 66% of the variance of EEG alpha burst activity during the end of drive. Galley, 1993 made an attempt to overcome a few disadvantages of EOG in the measurement of gaze behavior by using on-line computer identification of saccades and additional keyboard masking of relevant gazes by the experimenter. As EOG, especially saccades and blinks, is regarded as one of useful measures to evaluate drivers' drowsiness, such an improvement might be useful to easily and conveniently detect the low arousal state of drivers. Wright et al., 2001 investigated sleepiness in aircrew during long-haul flights, and showed that EEG and EOG are potentially promising measures on which to base an alarm system. Skipper et al., 1986 carried out an experiment to detect drowsiness of driver using discrimination analysis, and showed that the false alarm or miss would occur in such an attempt.

Murata and Hiramatsu, 2008 succeeded in detecting the increase of RRV3 when the arousal level is low. Murata and Nishijima, 2008 also detected the decrease of EEG-MPF for the drowsy state. However, it is not possible to predict the drowsiness on the basis of the time series of EEG-MPF or RRV3. Although detecting the arousal level of a driver automatically by ITS and warn drivers of the drowsy state is an ultimate goal in such studies, it is impossible to develop such a system unless such studies (Murata and Hiramatsu, 2008 and Murata and Nishijima, 2008) are further enhanced and the prediction method on the basis of some useful methodology is established. Few studies made an attempt to predict the arousal level systematically on the basis of physiological measures.

Murata and Hiramatsu, 2009, using Bayesian probability, made an attempt to predict the timing when drivers actually felt drowsy. In this study, the likelihood and the conditional probability must be calculated on the basis of the time series of biological information such as RRV3, EEG-MPF and EEG-α/β. In order to use this method for predicting the timing of low arousal level, the basic data for each participant must be collected before the prediction is carried out. Therefore, easier prediction of drowsiness is desirable. Promising candidates for such a prediction method include multivariate analysis methods such as discrimination analysis and logistic regression model.

This study made an attempt to predict the drowsy state by applying multivariate analysis (polynominal logistic regression model and discrimination analysis) to the biological information and the performance data (tracking error in driving simulator task). The prediction accuracy was derived for both discrimination analysis and polymoninal logistic regression model and was compared between two methods. Although the prediction of drowsiness was basically executed using 1-min data 1-min before the prediction, the effects of the location of 1-min data used for the prediction on the prediction accuracy have also been examined.

2 METHOD

2.1 Participants

Five male graduate or undergraduates (from 21 to 26 years old) participated in the experiment. They were all healthy and had no orthopedic or neurological diseases.

2.2 Apparatus

EEG (*EEG-MPF*), heart rate variability (*RRV3*), EOG, tracking error and subjective rating of drowsiness were measured while performing a simulated driving task for one hour. The relation between these measurements and drowsiness was analyzed. The psychological rating of drowsiness was also checked every 1 min in order to use this as a baseline of change of drowsiness with time. The psychological rating included the following three categories: 1: arousal, 2: a little drowsy, 3: very drowsy.
Electroencephalography (EEG), Electrocardiography (ECG), and Electorooculography (EOG) activities were acquired with the following measurement equipment. An A/D instrument PowerLab8/30 and bio-amplifier ML132 were used. Surface EEG was recorded using A/D instrument silver/silver chloride surface electrodes (MLAWBT9), and sampled with a sampling frequency of 1kHz. According to international 10-20 standard, EEGs were led from O1 and O2. The participants sat on an automobile seat, and were required to carry out a simulated driving task.

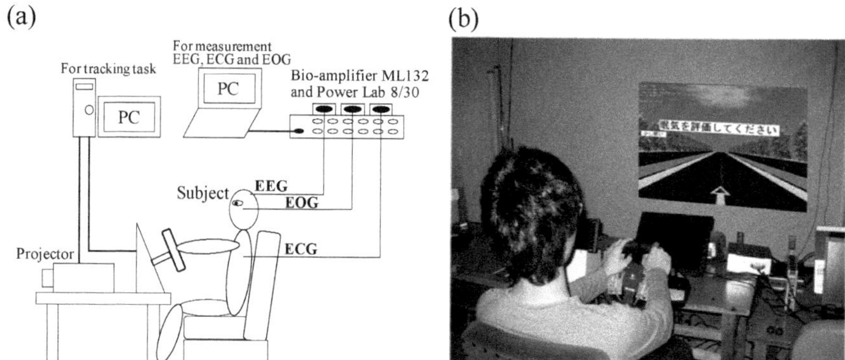

Figure 1 Outline of experimental system (a) and sketch of experiment (b).

FFT was carried out every 1024 data (1.024s). Before the EEG data were entered into FFT program, the data were passed through a cosine taper window. Based on this, the mean power frequency was calculated. This was plotted as a X-bar control chart Using a X-bar control chart, the judgment of drowsiness of participants was carried out. The ECG was led from V5 using BiolaoDL-2000(S&ME). On the basis of ECG waveform, R-R intervals (inter-beat intervals) were obtained. Heart rate variability (HRV) measure *RRV3* was derived as follows. The moving average per ten inter-beta intervals was calculated. Variance of past three inter-beat intervals was calculated as *RRV3*, which is regarded to represent the functions of parasympathetic nervous systems. The drowsy state leads to the dominance of parasympathetic nervous system. When the parasympathetic system is dominant, it is known that *RRV3* gets larger. ECG was used to count the number of blinks per one minute. The outline of experimental system and the sketch of experiment are shown in Figure 1(a) and (b), respectively. An example of EEG, EOG, and ECG measurements is demonstrated in Figure 2.

2.3 Design and procedure

The experiment was continued for one hour. *EEG-MPF*, *RRV3*, the number of blinks, and the mean tracking error were obtained every one minute. Applying these measures to the logistic regression model and the discrimination analysis, the drowsy state was judged and compared with the psychological evaluation every one minute. A total of 60 predictions were carried out for each participant. If the prediction corresponded with the psychological evaluation (rating), the prediction was regarded as correct. For every participant, the percentage correct prediction was calculated for both logistic regression model and discrimination analysis.

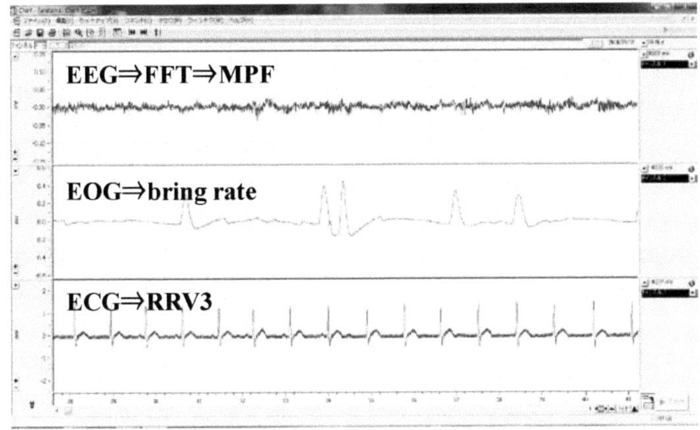

Figure 2 Example of EEG, EOG, and ECG measurements.

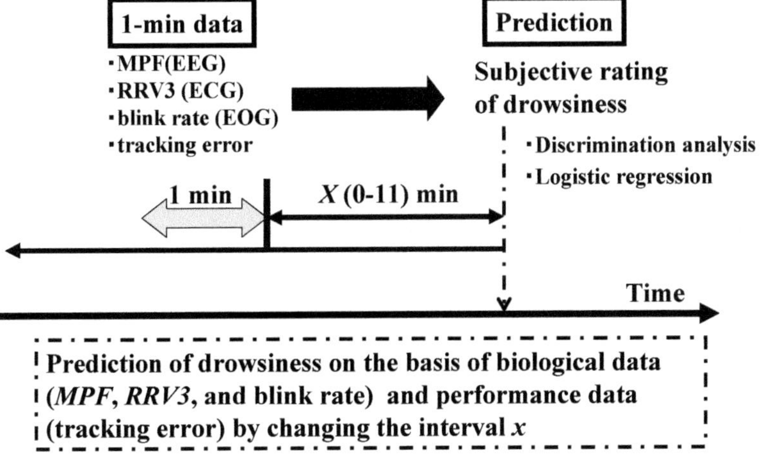

Figure 3 Procedure for predicting drowsiness.

2.4 Prediction method of drowsiness

In Figure 3, the procedure for predicting drowsiness using a discrimination analysis or a logistic regression model is summarized. Prediction of drowsiness is performed on the basis of biological data (MPF, RRV3, and blink rate) and performance data (tracking error). The information every 1 min is used to predict drowsiness. The predicted result is compared with that of subjective rating of drowsiness in order to obtain the prediction accuracy.

In the discrimination analysis, the following discrimination function (Eq.(1)) was used to predict drowsiness on the basis of biological information and tracking data. The parameters a, b, c, d, and e were obtained using a statistical package SPSS. Here, f, MPF, $RRV3$, BLK and TE represent predicted arousal level (drowsiness), mean power frequency, RRV3, and blink rate, and tracking error, respectively.

$$f = a \times MPF + b \times RRV3 + c \times BLK + d \times TE + e \qquad (1)$$

In the logistic regression model, the arousal level (drowsiness) y is predicted using the following equation. The parameters a, b, c, d, and e are obtained for each prediction trial.

$$y = \frac{1}{1 + e^{-(a \times MPF + b \times RRV3 + c \times BLK + d \times TE + e)}} \qquad (2)$$

Different from the discrimination analysis which enables us to discriminate more than two categories, the logistic regression is able to discriminate only two categories.

The prediction accuracy was obtained for both discrimination analysis and logistic regression. Using 1-min data 1-min before the prediction, the prediction of drowsiness was executed.

3 RESULTS

The aim of this study was to predict drowsy states by applying multivariate analysis such as discrimination analysis and logistic regression model to biological information and driving performance. EEG, heart rate variability, EOG, and tracking error were used as evaluation measures of drowsiness. The drowsy states were predicted by applying discrimination analysis and logistic regression to these evaluation measures.

The arousal level was predicted using four measurements (*EEG-MPF*, *RRV3*, number of blinks per one minute, and mean tracking error) as independent variables and the arousal level as a dependent variable in both logistic regression analysis and discrimination analysis. The results of prediction are shown in Figure 4 and Figure 5.

The percentage correct discrimination for the discrimination analysis and the logistic regression were calculated as 85% and 93%, respectively (Figure 4). The logistic regression model was found to lead to higher prediction accuracy.

It must be noted that three states cannot be discriminated by the logistic regression model due to its mathematical property. The percentage correct discrimination (prediction accuracy) among three categories (1: arousal, 2: a little drowsy, 3: very drowsy) by the discrimination analysis was the lowest. The percentage correct discrimination between 1: arousal and 2: a little drowsy, and between 1: arousal and 3: drowsy was higher than that between 2: a little bit drowsy and 3: drowsy. The percentage correct discrimination (prediction accuracy) between 1: arousal and 3: drowsy was the highest for both logistic regression and discrimination analysis. This means that the drowsy state can be discriminated from the arousal state with high accuracy. Both discrimination analysis and logistic regression might be a promising means to predict the arousal level.

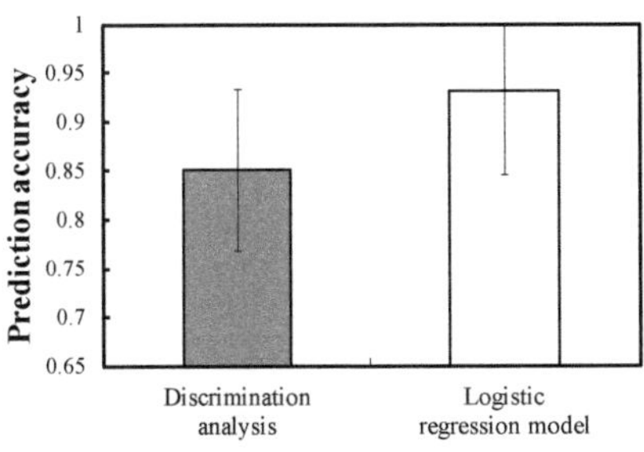

Figure 4 Comparison of prediction accuracy (percentage correct prediction) between discrimination analysis and logistic regression model.

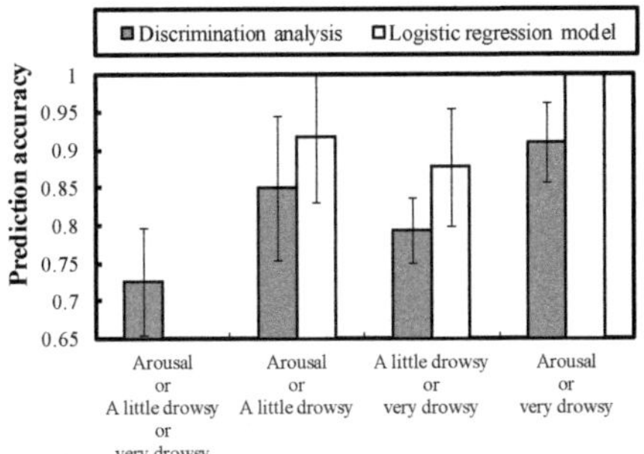

Figure 5 Results of prediction (Percentage correct prediction).

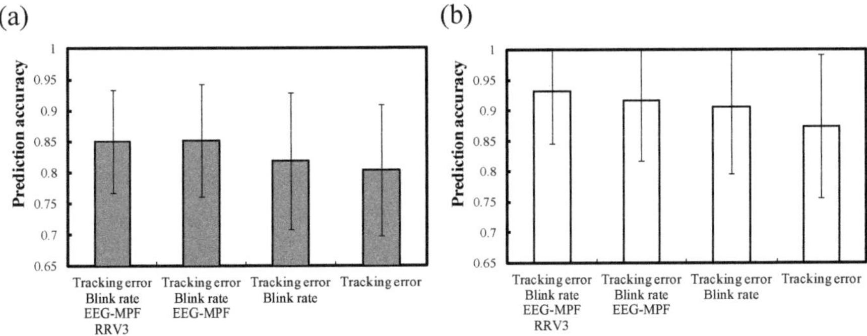

Figure 6 Prediction accuracy (percentage correct prediction) when four, three, two, or one data were used ((a) discrimination analysis, (b) logistic regression model).

In Figure 6, the percentage correct prediction (prediction accuracy) is plotted as a function of the number of evaluation measures. Here, the percentage correct prediction was obtained when four data (*EEG-MPF*, *RRV3*, blink rate, and tracking error), three data (EEG-MPF, blink rate, and tracking error), two data (blink rate and tracking error), and only one data (tracking error) were used. The more data were used for the prediction, the higher the prediction accuracy was. The prediction accuracy was higher for the logistic regression model than for the discrimination analysis.

For both prediction methods, the prediction accuracy when only one measure was used was not so low as compared with that when four measures were used. In order to attain higher prediction accuracy, more data should be made use of in the prediction.

In Figure 7, the mean standardized canonical discrimination function coefficients are compared among variables (*EEG-MPF*, *RRV3*, blink rate, and

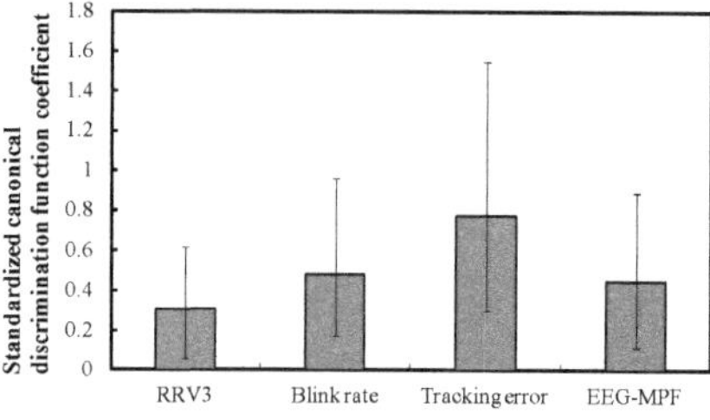

Figure 7 Mean standardized canonical discrimination function coefficient compared among variables used in the discrimination analysis.

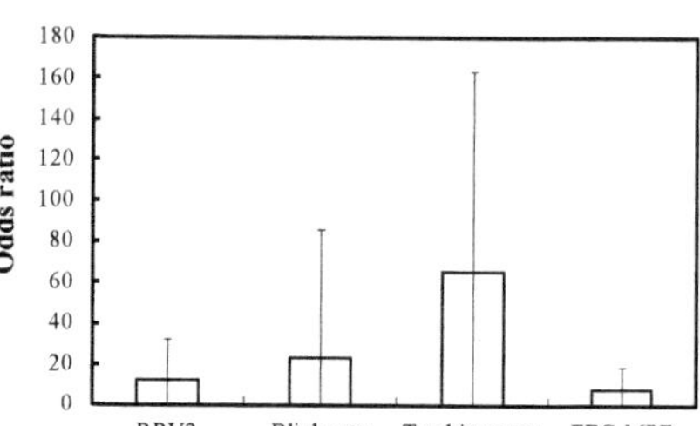

Figure 8 Mean odds ratio compared among variables used in the logistic regression model..

tracking error) used in the discrimination analysis. It is clear that the tracking error generally contributes to account for the subjective drowsiness. Figure 8 shows the mean odds ratio compared among variables (*EEG-MPF*, *RRV3*, blink rate, and tracking error) used in the logistic regression model. This also indicates that the tracking error affects the subjective drowsiness more remarkably than other evaluation measures (*EEG-MPF*, *RRV3*, blink rate). These results show that the tracking error is a very critical factor in explaining and predicting drowsiness of drivers. However, it must be noted that the tracking error alone cannot predict the drowsy state with high reliability as shown in Figure 6(a) and (b). Combining as many measures as possible is sure to lead to high reliability of drowsiness prediction.

4 DISCUSSION

In this study, an attempt was made to predict drowsiness by multivariate analysis (logistic regression model and discrimination analysis). As a result, the logistic regression model and the discrimination analysis led to prediction accuracy of 93% and 85%, respectively. The logistic regression discriminated the arousal and the very drowsy states with the accuracy of 100%. In future work, the prediction of drowsiness ahead by more than five minutes should be carried out to establish a more elaborated prediction technique of drowsiness.
Subjective rating of drowsiness is variable among individuals. In short, the criterion of subjective drowsiness cannot be objectively and definitely determined. Future work should propose a method that takes individual differences into account. In such a way, the prediction accuracy might be further enhanced.

The results above are based on the data one minute before the prediction (X was set to 1). The prediction of drowsiness should also be executed using 1-min data X-min before the prediction. It is important to investigate the effects of X values on the prediction accuracy. Therefore, a variety of X values more than 1

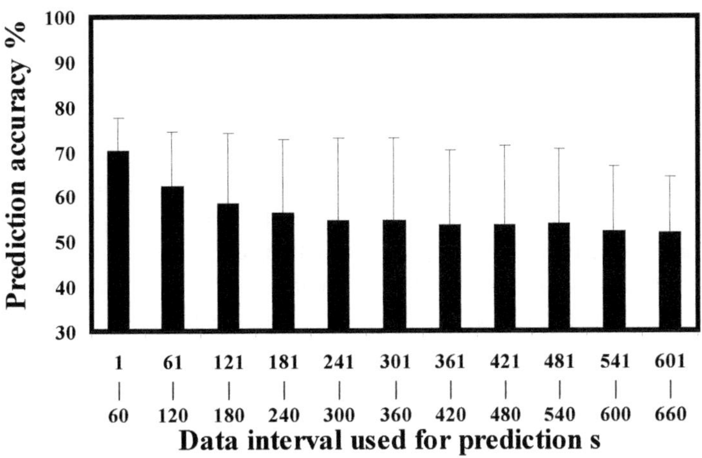

Figure 9 Prediction accuracy of multinominal logistic regression with a tracking error used as a dependent variable as a function of data interval.

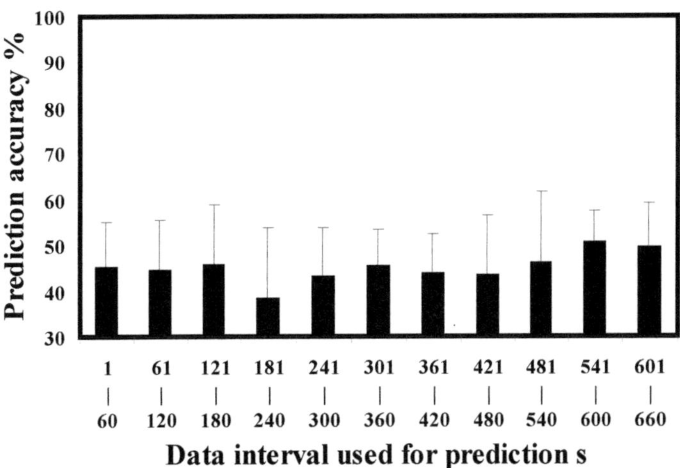

Figure 10 Prediction accuracy of multinominal logistic regression with *EEG-MPF* used as a dependent variable as a function of data interval.

was used, and the prediction accuracy was compared among X values. The detailed explanation is summarized in Figure 3. Multinomial logistic regression analysis was carried out with a tracking error or *EEG-MPF* used as a dependent variable in order to examine the prediction accuracy as a function of X. Eleven kinds of X (1-60s, 61-120s, 121-180s, 181-240s, 241-300s, 301-360s, 361-420s, 421-480s, 481-540s, 541-600s, and 601-660s) were used in the analysis. Only the nominal logistic regression was used in the analysis.

In Figure 9, the prediction accuracy for the case with a tracking error used as a dependent variable is plotted as a function of *X*. The prediction accuracy was about 70 % when *X* was 1. Going back in time far away from the prediction point in time, the prediction accuracy decreased monotonously. The result indicates that only the tracking data immediately followed by the prediction point in time can be used effectively for the prediction of drowsiness. The change of tracking data must be reflected immediately before the prediction. If the data is far away from the prediction time in time, the contribution of the tracking data to the prediction accuracy is weakened.

Figure 10 shows that the prediction accuracy for the case with *EEG-MPF* used as a dependent variable is not so remarkably affected by *X*. This might indicate that the biological data such as *EEG-MPF* can be used even if the measurement point in time is far away from the prediction point in time. The biological data might be used for the long-term prediction of drowsiness. The analysis of the effects of *X* on the prediction accuracy revealed that the mechanism how the drowsiness is induced with time and appears differed between performance data such as tracking error and the biological data such as *EEG-MPF*.

Future research should explore more systematically the effects of *X* on the prediction accuracy by constructing a more general prediction model. The optimal *X* used for the prediction of drowsiness should be clarified separately for performance and biological data.

REFERENCES

Brookhuis, K.A. and Waard, D. 1993. The use of psychophysiology to assess driver status, *Ergonomics* 36: 1099-1110.

Galley, N. 1993. The evaluation of the electrooculogram as a psychophysiological measuring instrument in the driver study of driver behavior, *Ergonomics* 36: 1063-1070.

Kecklund, G. and Akersted, T. 1993. Sleepiness in long distance truck driving: An ambulatory EEG study of night driving, *Ergonomics* 36: 1007-1017.

Murata,A. and Hiramatsu,Y. 2008a. Evaluation of Drowsiness by HRV Measures -Basic Study for Drowsy Driver Detection-, *Proc. of IWCIA2008*: 99-102.

Murata,A. and Hiramatsu,Y. 2009. Evaluation of Drowsiness by HRV Measures -Proposal of prediction method of low arousal level-, *Proc. of IWCIA2009*: 348-353.

Murata,A. and Nishijima,K. 2008b. Evaluation of Drowsiness by EEG analysis -Basic Study on ITS Development for the Prevention of Drowsy Driving-, *Proc. of IWCIA2008*: 95-98.

Skipper, J.H. and Wierwillie, W. 1986. Drowsy driver detection using discrimination analysis, *Human Factors* 28: 527-540.

Wright, N. and McGown, A. 2001. Vigilance on the civil flight deck: incidence of sleepiness and sleep during long-haul flights and associated changes in physiological parameters, *Ergonomics* 44: 82- 106.

CHAPTER 49

Development of a Measurement Method for Individual Hazard Perception

Ayako Hirose, Daisuke Takeda*

Human Factors Research Center
Central Research Institute of Electric Power Industry
Tokyo
*a-hirose@criepi.denken.or.jp

ABSTRACT

The decrease of hazard perception among workers, especially young workers, has become a problem at hazardous work sites in Japan. To improve hazard perception among these workers, it is necessary to visualize and measure individual hazard perception. Thus, the purpose of this study is to develop a Hazard Detection and Imagination Test (HDIT) designed to measure individual core hazard perception. In the context of applying HDIT to young workers, we consider whether remedial education in hazard perception according to trainees' level of hazard perception is needed. We divided participants into three groups by their level of hazard perception, and examined the features of young workers who had low hazard perceptions. The results reveal that HDIT can successfully measure individual core hazard perception. We also identified certain features of hazard recognition in young workers with low hazard perception. We suggest that it is necessary to understand level of hazard perception and to provide education accordingly.

Keywords: Hazard Perception, Occupational Accidents, Equipment-related Accidents, Young Workers

1 INTRODUCTION

A decrease in workers' hazard perception, especially among young workers, has become a problem at hazardous work sites in Japan, as the mass retirement of the postwar baby-boom generation results in huge losses of valuable skills and knowledge for work safety. Another reason is that a decrease in recent years in occupational accidents due to improvement in facilities has reduced the chance for workers to learn from workplace accidents. Not only in Japan but also in many other countries, decrease in young workers' hazard perception has been pointed out, and an increasing number of young workers are involved in accidents in the workplace (European Agency for Safety and Health at Work, 2009). To improve this situation, increased safety education and safety activities need to be carried out.

Although the level of hazard perception is different for each individual, most safety education and training is carried out in a group setting. Therefore, there is a concern that these activities will be ineffective if their level is inconsistent with trainees' level of hazard perception. To carry out safety education and training at an appropriate level for the individual, it is necessary to measure individual hazard perception. However, previous studies on hazard perception are restricted to the area of transportation (e.g. Scialfa et al., 2011), and may not be generally applicable. Unfortunately, the diversity of possible hazards and consequences among workplaces makes the design of a general hazard perception measure difficult.

In general, hazard perception is affected by experience and knowledge about the work. Thus, most previous studies have been comparisons of novices with experts (Fin & Bragg, 1986; Crick & Mckenna, 1991), but experience does not necessarily line up completely with accident-proneness. We suggest that each person has a core level of hazard perception that is not dependent on the task-specific knowledge or experience, and therefore that a general hazard perception measure is possible.

The present study gives the results of some experiments conducted among office workers toward the development of such a measure (Hirose, et al., 2009; Hirose, et al., 2010). Participants were asked to watch a film and identify as many hazards as possible along with the expected consequences. A correlation was found between ability to identify hazard in the film about daily life and the film about overhauling a valve, even among participants without experience or expertise in this work. This result supports the existence of individual core hazard perception.

At the same time, the following results were also obtained: 1) it is possible to measure individual core hazard perception by how many hazards can be identified and how accurately people can anticipate the bad consequences of a hazard; 2) it is necessary to measure whether people can identify both immediate and latent hazards; and 3) after classifying hazards by content, it is possible to measure individual bias in hazard identification by assessing what types of hazards people identify well.

On the basis of the above, the purpose of this study is to develop a Hazard Detection and Imagination Test (HDIT) designed to measure individual core hazard perception, and applying HDIT to young workers, to consider whether the improvement of hazard perception by targeted education according to trainees' level of hazard perception is needed.

2 DEVELPOMENT OF THE HDIT

2.1 Outline

The HDIT is composed of a film divided into 10 scenes (one of the scenes is for trial) and an answer sheet. In HDIT, participants are asked to watch a film with various hazards depicted in it, write down all the hazards they identify on the answer sheet, and answer some supplementary questions.

2.2 Film and inserted hazards

The film depicts an office worker's day at work. Total footage is about 3.5 minutes (each scene is about 20 seconds).

Using previous data for 86 workers (75 male, mean age 38 years, mean work experience 17 years), 31 points were identified as hazardous. These 31 "essential points" comprised by the more than 20% of respondents identified. All of these points (e.g., walking on wet tiles, luggage blocking passage) are commonly encountered in daily life. Table 1 shows specific content of the footage and the number of hazardous points in each scene.

Table 1 Content, footage, and number of essential points of each scene
Takeda is the main character of this film and Hanawa is his colleague

Scene		Content	Footage (s)	The number of hazards
Trial		Takeda pass a cutter to Hanawa	10	(1)
1		Takeda comes to his office and enters the office building in a downpour	25	3
2		Takeda walks along a hallway	21	4
3		Hanawa conveys some packages on a cart	24	5
4		Hanawa drops off the packages from a cart	21	4
5		Hanawa opens the package with a cutter	20	3
6		Takeda pours a cup of coffee from a coffee machine	7	3
7		Takeda brings the cup to his desk, and power up a PC	23	5
8		Takeda fields a phone call	53	1
9		Hanawa drops off a baggege from a height. Takeda descends stairs to go home	20	3

2.3 Procedures

For each scene, participants followed four steps: 1) count the hazards in one's head while watching the scene, 2) write down the number of identified hazards on the answer sheet, 3) circle the identified hazards using the photographs in the sheet, and write why they are hazardous, and 4) following 2.4, give the subjective degree of risk and potential damage for each identified hazard. Figure 1 shows a sample answer sheet.

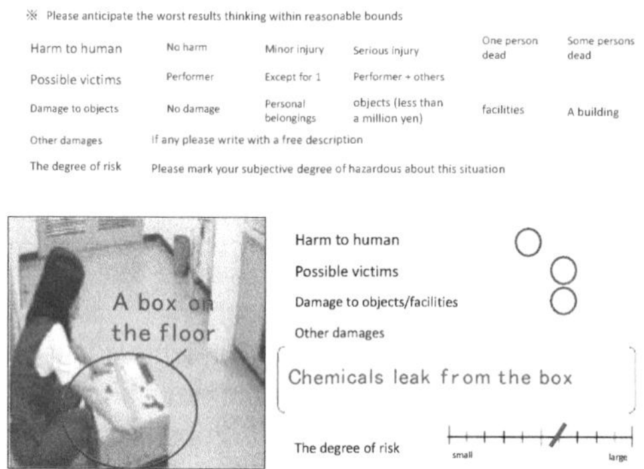

Figure 1 Answer sheet and example of an answer

2.4 Measurements

Participants replied to the following six questions about the hazards they identified.

(1) Hazards: Participants circled the identified hazards in each scene.
(2) Potential extent of harm to human: Participants chose from the following 5 categories; 1) no harm, 2) minor injury, 3) severe injury, 4) one person dead, and 5) more than one person dead.
(3) Possible victims: Choices were 1) the "performer," the person acting unsafely; and 2) others except for the performer, 3) performer + others.
(4) Extent of damage to objects or facilities: Choices were: 1) no damage, 2) personal belongings, 3) objects (less than 1 million yen), 4) facilities, and 5) a building.
(5) Other damage: Answers were given by free description.
(6) Subjective degree of risk: See Figure1.

2.5 Indices

To assess individual hazard perception, three indices were estimated: Detection (D) score, Imagination (I) score, and total (T) score. D score indicated how many hazards the participants could identify, I score how accurately they could anticipate the extent of harm or damage, and T score their level of core hazard perception. Table 2 shows examples of correct answers. Calculations were made as follows:

(1) D score: Number of essential points identified divided by total number of essential points (31).
(2) I score: Divided into three subscores; serious harm to humans (I-h score), possible victims (I-v score), and serious damage to objects or facilities (I-o score). For 14 of 31 essential points, it was anticipated that more than one person would die; each of these was worth two points for a correct answer and one for a nearly correct answer. I-h score was calculated by dividing total points by 28 (14 x 2).

    Twenty-one of 31 essential points assumed that more than one person was harmed. I-v score was calculated by dividing total points by 42 (21 x 2).

    For three essential points, there was serious damage assumed to facilities; I-o score was calculated in the same way as I-h score.
(3) T score: First, sub-T scores for each essential point were calculated. Each sub-T score was composed of presence or absence of hazard identification and accuracy of a correct answer. The total of 31 sub-T scores was the T score.
(4) Total number of hazards: The total number of hazards identified by a participant (not only essential points) was calculated.
(5) Hazard category scores: As described in the introduction, it is possible to measure individual bias in hazard identification by assessing how many categories and what types of hazards people identify well. The detailed method grouping and scoring hazards is described later.

Table 2 Example of correct answers

Hazards (example)			
Harm to human	⑤ more than one person dead	⑤ more than one person dead	② minor injury
Possible victims	③ performer+others	③ performer+others	① the performer who act against safety
Damage to objects	⑤ a building	② personal belongings	③ objects (less than a million yen)

3 TRIAL FOR YOUNG WORKERS

3.1 Participants

The participants in this study were 384 new employees who worked for the same utility company. After excluding flawed data, the final number of participants was 333, with a mean age of 20.7 (range 18 to 27). There were 49 females. The response rate was 86.7%.

3.2 Procedures

The HDIT was carried out on April 4, 2011, before the safety training for new recruits was executed at full scale, to measure their initial core hazard perceptions. The HDIT was given to 100 people at a time in a large hall, projected on a screen.

After instruction on the purpose of the HDIT and answer methods, a trial was conducted using a trial scene. The experimenters and training checked participants' answers and gave guidance where they were incomplete. After the trial, the scenes were presented in the order in Table 1. The film was stopped every scene and participants given 2.5 minutes to answer. The process took about 30 minutes in total.

3.3 Supervisors' subjective evaluation

The supervisors were required to evaluate each participant's hazard perception subjectively on a five-point scale from 1 (very low) to 5 (very high). The evaluation was done at the end of the training period when the supervisors had become familiar with the participants. Ninety-five participants were excluded because their training period was too short for the supervisors to become familiar with them. Evaluations were used to validate the HDIT.

3.4 For analysis

Each index described in 2.5 was calculated. Hazard category scores were calculated using the following procedures. At first, all 195 hazards obtained were gathered into 18 midsized categories according to similarity of contents using the KJ method. Then, these categories were gathered into five large categories by the same method. The number of hazards in each category was divided by the total number of hazards; and thus 18 midsized category scores and five large category scores were obtained. How broadly the participants viewed hazards was also assessed based on the number of categories.

3.5 Results

(1) Confirmation of validity of HDIT

Although HDIT validity had already been confirmed using supervisor evaluations in the previous study (Hirose et al., 2008; Hirose et al., 2011), this result was for a few participants only. Therefore, the validity of HDIT would be confirmed again.

The supervisors' evaluation described above was used for the confirmation of validity. The average score was about 2.99 out of 5 (SD = .628). The majority of evaluations scored 3 (hazard perception is normal) and no participant scored 5 (hazard perception is very high). Thus, participants were classified into three groups as follows; 1) high hazard perception (4 or more; 42 participants), 2) medium hazard perception (3 to 3.99; 142 participants), and 3) low hazard perception (less than 3).

To investigate the differences between these three groups, a one-way ANOVA was conducted for T score. There was a significant difference in groups (F (2, 233) = 12.166, p < .001). Multiple comparison (Tukey's HSD method) revealed significant differences among groups (see Figure 2). It was suggested that the higher the evaluation, the higher the T score, and the validity of HDIT was confirmed.

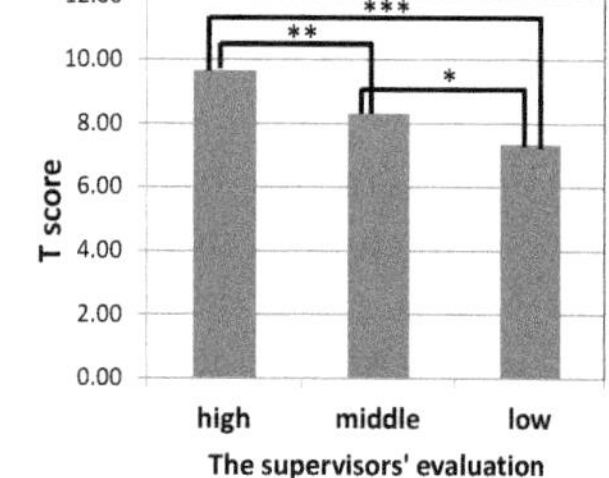

Figure 2 T scores according to supervisor-evaluation group

(2) Comparison between young workers

We wished to examine the features of young workers who had low hazard perception to clarify the necessity of educating trainees by hazard perception. For this purpose, 333 participants were placed in three groups by T score.

After we calculated T score (average and SD), 55 participants with more than one standard deviation above the mean were classified as the "high" hazard perception group, 62 participants who had more than one SD less than the mean were classified as the "low" hazard perception group, and the other 216 participants were classified as the "middle" hazard perception group. These groups were compared using one-way ANOVA and Tukey's HSD method.

The "high" group identified 24.5 total hazards, the "middle" group 19.3, and the "low" group 14.7. The ANOVA revealed significant differences between groups ($F(2, 330) = 110.241$, $p < .001$). Multiple comparisons showed significant differences between all groups ($p < .001$).

D score and I scores also showed significant differences between all groups. Each index showed a higher score for the higher groups.

Supervisors' evaluations averaged "high" group 3.36, "middle" 2.98 and "low" 2.67. One-way ANOVA revealed significant differences between groups ($F(2, 233)=14.036$, $p < .001$). Multiple comparison showed significant differences between all groups ($p < .001$).

To investigate the significant bias in hazard identification by group, a chi-squared test was carried out. If the chi-squared test was significant, residual analysis was carried out to investigate where the bias existed. Significant biases were seen in 22 of 31 essential points, mostly between "high" and "low" groups. The "high" group identified each hazard more effectively than the other groups, and the "low" group less. The "high" group also identified more hazards that were not included in the essential points than the other groups, and the "low" group fewer.

There were significant differences for sub-T scores in 28 of 31 essential points between groups. Multiple comparison revealed that the "high" group scored higher for all of these points than the "low" group. Moreover, significant differences were seen in 21 points between the "high" and "middle" groups and in 19 points between the "middle" and "low" groups. It is of course to be expected that sub-T scores would reflect the significant differences in hazard identification described above. However, significant differences were also seen in sub-T score at points where 50% or more of the participants as a whole were able to identify a hazard and no significant differences had been shown between groups in identification. Extracting these points, the average scores of each group and the results of the analysis are shown in Table 3. There are significant differences between groups on some of these points, with higher scores in the "high" group, suggesting that this group could not only identify many hazards but also anticipate the results of a hazard.

On average, the "high" group identified 12.4 categories, the "middle" group 10.8, and the "low" group 9.1. One-way ANOVA revealed significant differences between groups ($F(2, 330) = 75.101$, $p < .001$), as did multiple comparison ($p < .001$). There were significant differences in "bias of attention" ($F(2, 330) = 8.865$, $p < .001$) and "work environment" ($F(2, 330) = 11.436$, $p < .001$). Multiple comparison revealed that the "low" group had a higher score than the other groups in the former, and lower than the others in the latter ($p < .001$). For the medium hazard category, "collision with someone who comes out from the cover, side, and the door", "lack of a necessary act", and "shortage of 4S" were significantly higher

Table 3 Sub-T score according to level of hazard perception

Scene	hazards	Ratio of identification in all participants	Level of Hazard Perception High	Middle	Low	F	p	Multiple Comarison H-L	H-M	M-L
1	sweeping rain drops on an umbrella on the passage	0.847	0.664	0.523	0.374	17.801	***	***	**	***
3	opening a door behind	0.874	0.648	0.509	0.347	24.269	***	***	***	***
3	building packages up to head height and conveing	0.814	0.603	0.529	0.355	13.470	***	***		***
4	dropping a package to the floor loughly	0.910	0.661	0.620	0.518	5.372	**	**		*
5	putting a cutter on the floor without putting back the blade	0.574	0.502	0.374	0.239	8.626	***	**	*	*
7	putting a cup of coffee on the edge of a desk	0.940	0.918	0.781	0.611	20.139	***	***	**	***
7	a small bump on the passage	0.517	0.560	0.345	0.168	18.401	***	***	***	**
8	not noting down a client's phone number	0.550	0.607	0.474	0.261	10.096	***	***		**
9	descending stairs using a cell phone	0.964	0.748	0.611	0.495	24.492	***	***	***	***
9	dropping off a baggege getting on a pipe chair	0.862	0.800	0.565	0.339	46.914	***	***	***	***

※ ***:p<.001, **: p<.01, *:p<.05

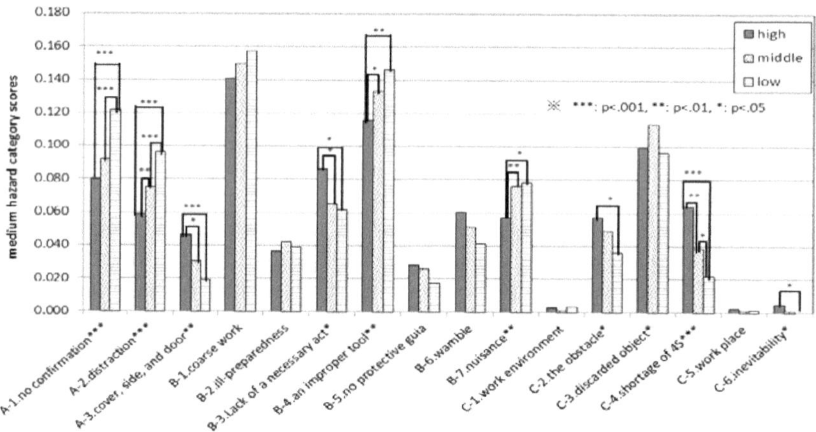

Figure 3 Medium-hazard category scores according to level of hazard perception

in the "high" group than the others. Also, "existence of an obstacle" and "inevitability" were significantly higher in the "high" group than the "low" group. On the other hand, "no confirmation" and "distraction" were significantly higher in "low" group than the others. Also, "use of an improper tool" and "nuisance" were significantly higher in "low" group than "high" group (Figure 3).

4 DISCUSSION

After dividing new employees into three groups by T score, the following features were suggested of young workers with low hazard perception: 1) all indices which measured hazard perception were low, 2) they couldn't identify some hazards which 85% or more could identify in the other groups, 3) their guess at the harm or damage brought by the hazard was too low even when they could identify it, 4) hazards concerning bias of attention were identified often, and those concerning work environment rarely, 5) they had biases regarding hazard identification, 6) supervisors' subjective evaluation of them was low.

This suggests that HDIT could be used when superiors would like to know newcomers' hazard perception or measure the effect of safety activities or trainings.

In addition, new employees with low hazard perception disproportionately identified hazards related to the behavior of the protagonist in the film, like bias of attention. In contrast, the "high" group identified various hazards, paying attention to the situation and environment. However, it is necessary to consider participants' age in this regard. The mean age was 19.3 years old in the "low" group, 20.7 years old in the "middle" group, and 22.4 years old in the "high" group. In various foreign countries, there are special regulations on workers 18 or younger (European Agency for Safety and Health at Work, 2009). Although HDIT measures individual core hazard perception, which is not dependent on task-specific knowledge and experiences, it can be supposed that teenagers' core hazard perception is still immature and that they may be less able to identify hazards in daily situations. (However, this does not mean that there will not be older workers with low as well as high hazard perception.) In a previous study (Hirose et al., 2008), HDIT could distinguish workers by the number of hazards identified and the accuracy of the bad results they anticipated. To wrap up, it can be supposed that core hazard perception will increase until a certain age through the experiences of childhood and youth, and will then fix. There is room for future study of the relationship between age and core hazard perception, and also the process of formation of core hazard perception.

5 CONCLUSIONS

HDIT successfully illuminated individual differences in core hazard perception, and showed that they existed even across novice employees. It is to be hoped that educational methods for improvement of core hazard perception will be developed in the future, and it will be useful to consider whether it is possible to conduct this education efficiently by changing methods according to level of hazard perception.

REFERENCES

Charles T.Scialfa, et al., 2011. A hazard perception test for novice drivers, *Accident Analysis and Prevention*, 43(1), 204-208

Crick, J & McKenna, F. P., 1991. Hazard perception: Can it be trained?, *Behavioural Research in Road Safety*, 2, 100-107

European Agency for Safety and Health at Work, 2009. Preventing risks to young workers: policy, programmes and workplace practices

Finn, P. & Bragg, B. W. E, 1986. Perception of the risk of an accident by young and older drivers, *Accident Analysis and Prevention*, 18(4), 289-298

Hirose, A. et al., 2009. Study on the Possibility of Measurement of Individual Risk Perception, *CRIEPI Report*, Y08017 (in Japanese)

Hirose, A. et al., 2010. A Study on Measurement of Individual Hazard Perception (Part 2): Development of Rating Scales Based on the Content, Severity and latency of Hazards, *CRIEPI Report*, Y09011 (in Japanese)

Hirose, A. et al., 2011. A Study on Measurement of Individual Hazard Perception (Part 3): - Development of Measurement Method of Individual Hazard Perception -, *CRIEPI Report*, Y10012 (in Japanese)

CHAPTER 50

The Evaluation of Heat Transfer Using the Foot Manikin

Uwe Reischl[*], *Ivana Salopek Čubrić*[**], *Zenun Skenderi*[**], *Budimir Mijović*[**]
[*]Boise State University
Boise, USA
[**]University of Zagreb
Zagreb, Croatia
ureischl@boisestate.edu

ABSTRACT

Heat transfer characteristics were determined for five types of men's shoes using a thermal foot manikin system capable of measuring heat resistance levels of shoes. The manikin included 13 separate thermal segments that provided an opportunity to differentiate thermal characteristics within selected regions of the shoe. Measurements included the whole foot (i.e. all 13 segments), but focus was placed on segments that were known to be most relevant to conductive heat loss. It was found that total heat resistance for a man's winter boot was 1.8 times higher than the heat resistance of a man's dress shoe. The measurements revealed a significant difference between shoes in the lower foot segments. The dress shoe exhibited resistance of 0.13 m^2KW^{-1}, while the resistance of the winter boot was 0.17 m^2KW^{-1}. This information is helpful in providing feedback to footwear manufacturers who can design new products that offer better thermal comfort and can provide improved shoe safety under extreme temperature conditions.

Keywords: thermal foot manikin, heat transfer, footwear

1 COMFORT

The feet are important components of human body heat exchange. The feet are usually the only parts of the body that come in direct contact with the ground. In cold environments, this contact creates the major pathway for conductive heat loss from the body. Therefore, footwear in cold environments must provide high thermal insulation in order to offer reasonable comfort and provide safety for the wearer. In general, if a person feels cold, the person perceives the cold through the feet. This is the result of vasoconstriction. It is believed that symptoms of cold feet are representative of whole body cold discomfort (Kuklane, 1999).

Various methods can be used for determining the thermal insulation characteristics of footwear. The most effective method is o use a thermal manikina (Figure 1). Manikins have many advantages over the use of human subjects. Manikin measurements are accurate and are repeatable when carried out in temperature controlled environments. Using manikins eliminates subjective factors that are always present when using human subjects. Additionally, when experiments on footwear are conducted under very cold conditions, test subjects may be exposed to significant health risks. Experiments carried out on foot manikins do not present such risks. Furthermore, if an environmental chamber is used to conduct manikin tests, a wide range of conditions can simulated within a relatively short period of time because thermal equilibrium can be achieved more quickly using a manikin than with a human subject (Kuklane, 1999).

Thermal protection and comfort are important performance requirements for footwear when used in extreme temperature environments. Inadequate protection may cause irreversible injuries. To determine the thermal performance of specific footwear, the thermal and evaporative resistance characteristics should be evaluated. The most effective method for making such evaluations is the use of thermal manikins.

Thermal manikins have a number of advantages over the use of human subjects including the following:

- Measurements are highly accurate and repeatable
- Subjective factors are not present
- Testing under extreme conditions represents no health risks
- Manikins can simulate relevant physiological conditions
- Thermal and evaporative resistance of each manikin segment is based on physical measurements.

Thermal and evaporative resistance data are not only relevant to footwear that is used in occupational settings but are also relevant to footwear designed for everyday use. Additionally, assessing the thermal performance of footwear can provide manufacturers of footwear to optimize comfort and safety in their products (UCS, 2011).

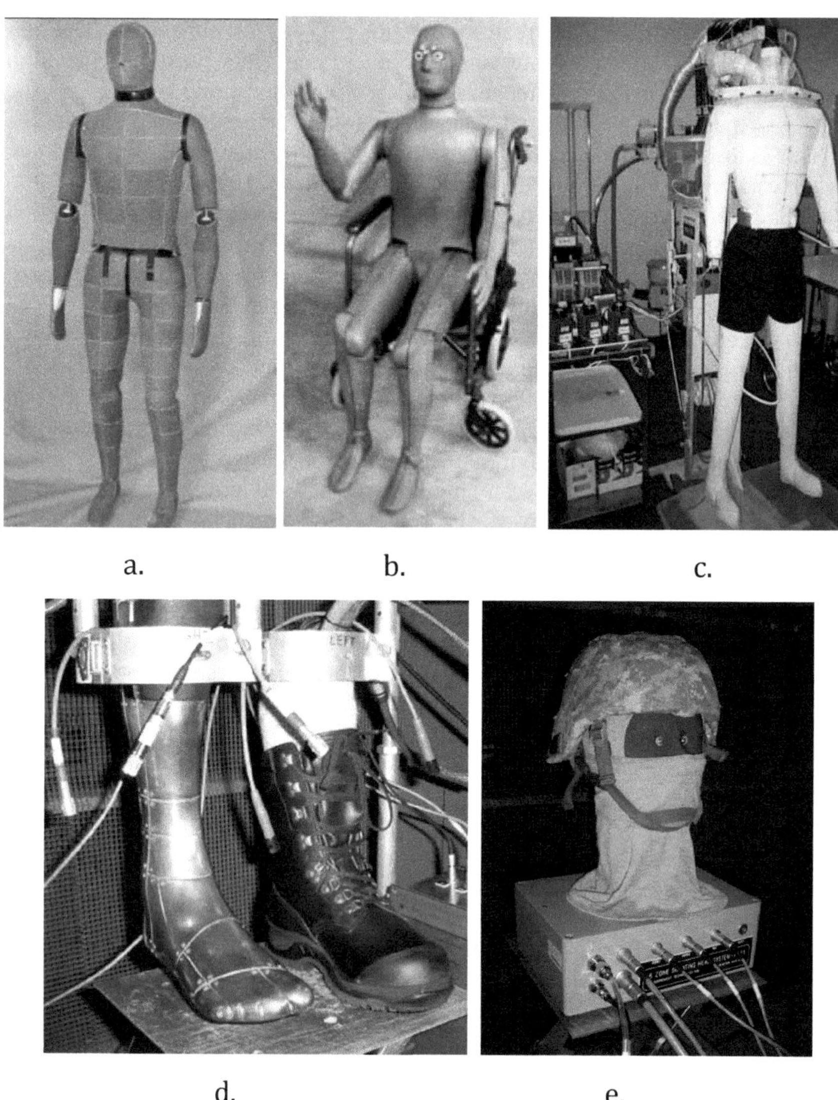

Figure 1 The thermal manikins: a. ADAM (From: Measurement Technology Northwest: Advanced Thermal Manikin ADAM 2008), b. Nemo (From: Measurement Technology Northwest: Nemo – submersible thermal manikin, 2008), c. Thermal manikin (From: Reischl, U., B. Mijović, Z. Skenderi, I. and Salopek Čubrić. 2010. Heat Transfer Dynamics in Clothing Exposed to Infrared Radiation. In. *Advances in Ergonomics Modeling and Usability Evaluation,* eds. Khalid, H., Hedge, A. and Ahram T. Z. Boca Raton: Taylor & Francis Group, USA.) d. Thermal foot (From: Terry Rice: USARIEM Uses Thermal Manikin to Test Warfighter Clothing and Individual Equipment (CIE)), e. Thermal head (From: Terry Rice: USARIEM Uses Thermal Manikin to Test Warfighter Clothing and Individual Equipment (CIE)).

2 METHODOLOGY

Heat exchange characteristics were determined for three different types of men's shoes including the following (Figure 2):
Shoe Design 1. Low cut business shoes
- Sample F1 (upper part: buffalo hide)
- Sample F2 (upper part: cow hide)

Shoe Design 2. High cut business shoes
- Sample F3 (upper part: cow hide + synthetic material)
- Sample F4 (upper part: cow hide)

Shoe Design 3. Boot
- Sample F5 (upper part: cow hide)

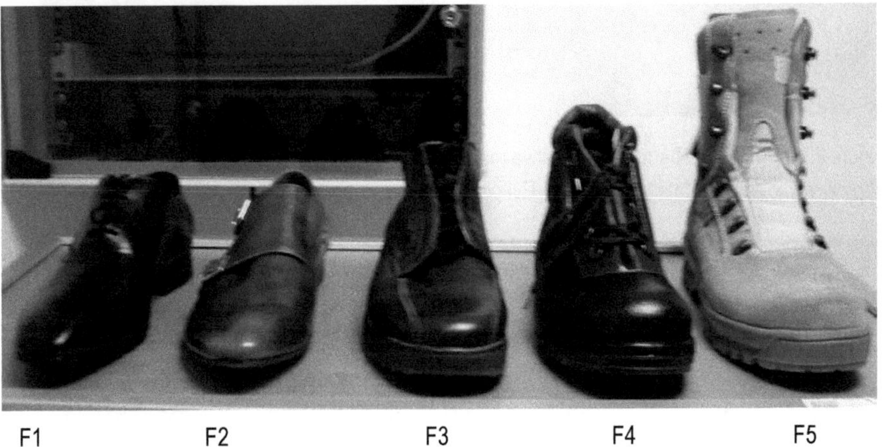

Figure 2 Measured footwear

The heat resistance characteristics of the above listed footwear were determined using the UCS Male Foot System. This technology consisted of the following components:

1. Foot model (Figure 3 and 6)
   - Laminated with 13 highly conductive surface thermal-segments made of silver
   - Individually controlled heaters and temperature sensors
   - Sweat "glands" distributed uniformly over the manikin surface
2. Manikin control unit (Figure 4)
   - Control unit to manipulate the temperature of the Manikin
3. Control software (Figure 5)
   - Provides setup of programmable parameters for both manikin and peripheral devices
   - Provides graphical variable displays

- Performs calculations to provide integral statistics with user-defined time intervals, as well as the export of data to independent data formats for custom post-processing (UCS 2011).

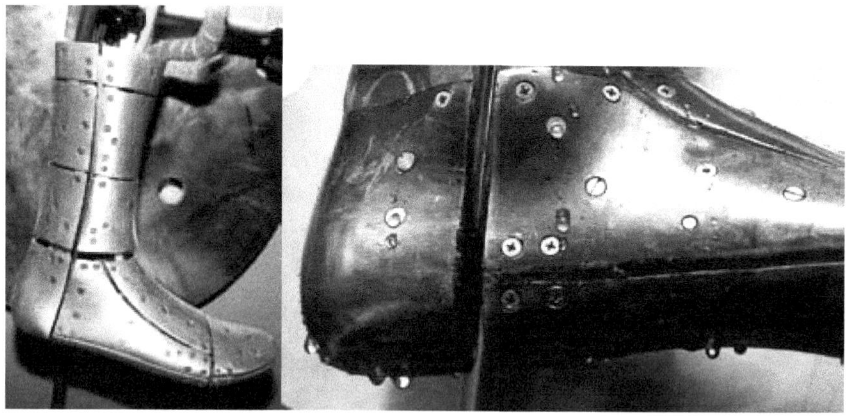

Figure 3 (From: UCS: Thermal-evaporative manikin system, available at http://www.ucstech.eu/images/Thermal_Evaporative_Manikins.pdf)

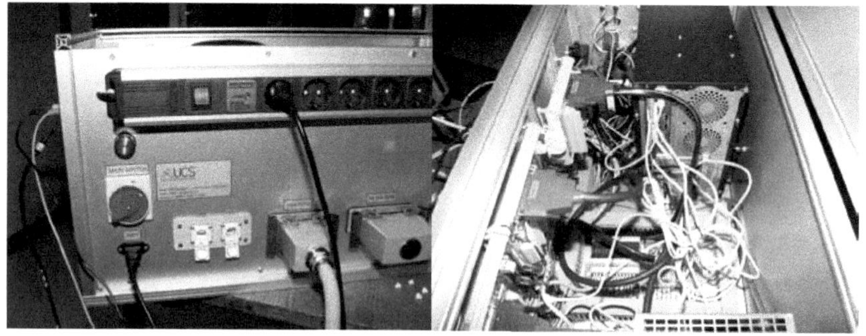

Figure 4 The control unit of the foot manikin system. (From Aleš Jurca, Universal Customization Systems)

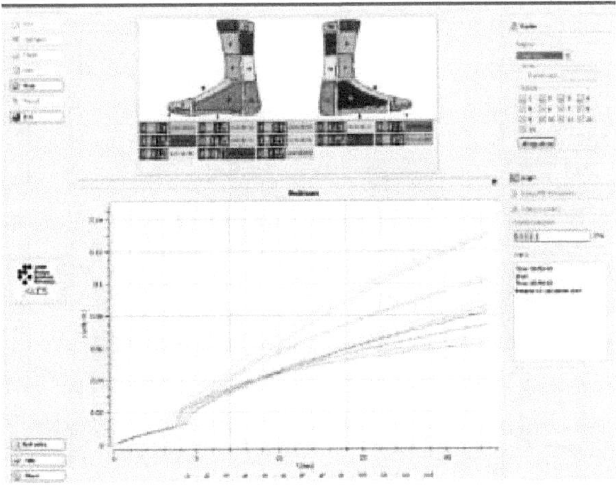

Figure 5 Software screenshot (From: UCS: Thermal-evaporative manikin system, available at http://www.ucstech.eu/images/Thermal_Evaporative_Manikins.pdf)

Used manikin consists of 13 highly thermal-conductive surface segments made of silver (Figure 6). In order to compare the heat resistances of chosen footwear, the results have also been observed for the zones covered by footwear in all the cases – i.e. segments 1, 2 and 3.

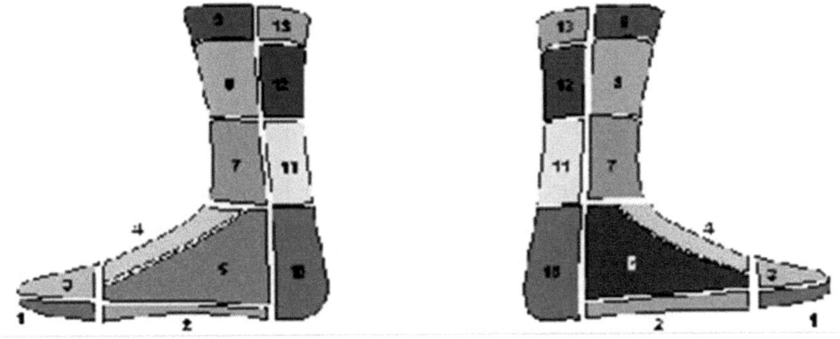

Figure 6 Foot model segments (From: UCS: Thermal-evaporative manikin system, available at http://www.ucstech.eu/images/Thermal_Evaporative_Manikins.pdf)

3 RESULTS AND DISCUSSION

The results presented here are based on data collected on 5 different types of footwear. This includes heat exchange for all segments of the foot manikin (Figure 7), segments covered by footwear only (Figure 8), and heat exchange values for a single segment only (Figure 9). The footwear types included men's formal dress shoes as well as winter boots.

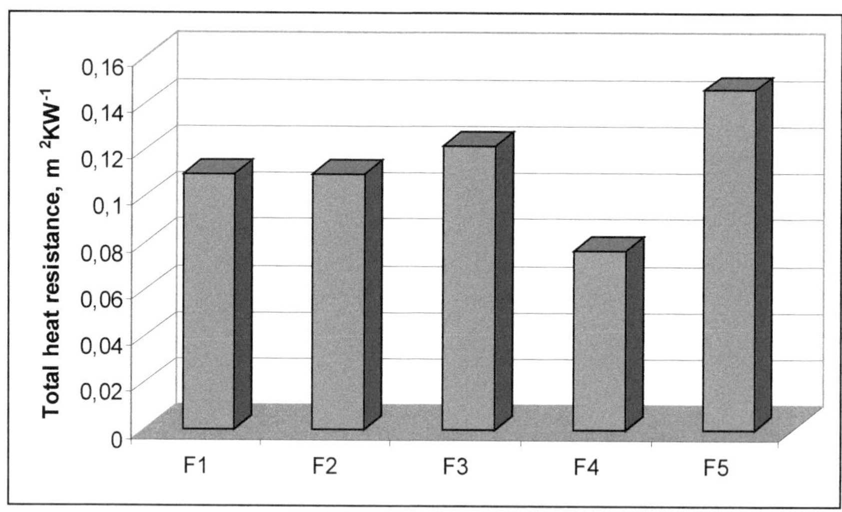

Figure 7 Total heat resistance

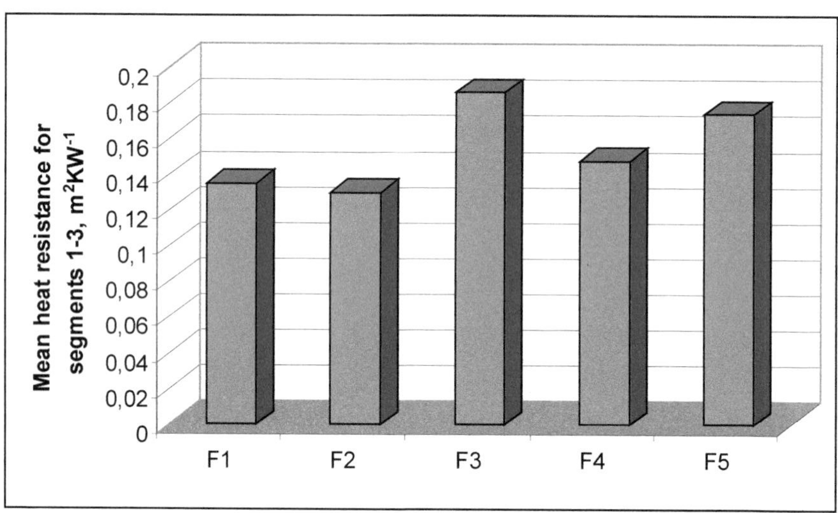

Figure 8 Heat resistance for segments 1-3

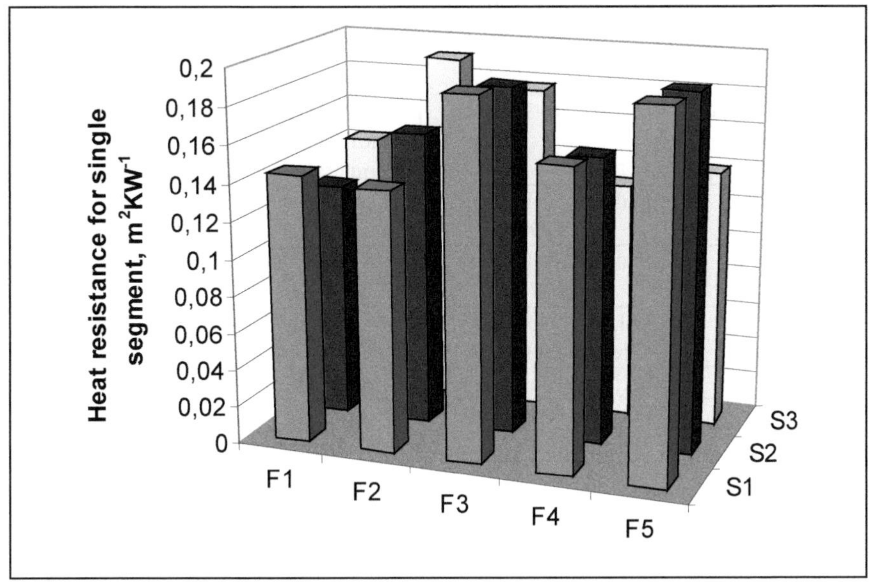

Figure 9 Heat resistance for a single segment

Although the data collected always included 13 thermal segments, focus was placed on the segments most relevant to conductive heat loss from the foot. It could be seen that the total system heat resistance exhibited by the winter boot was 1.8 times higher than the heat resistance exhibited by the dress shoe. These measurements revealed significant differences in the lower foot segments. For the dress shoe, these values were 0.13 m^2KW^{-1}, while the resistance level for the winter boot was 0.17 m^2KW^{-1}.

As shown in Figure 7, dress shoes F1 and F2, were similar in design and exhibited similar heat resistance characteristics (total heat resistance of 0.11 m^2KW^{-1}). Differences between the average values are less than 3%. However, there is a significant difference when compared to shoe F3 which exhibited a higher value than shoes F1 and F2. Although one might expect increasing resistance values with an increasing shoe height, the measured value is actually lower for shoe F4. This difference is specifically due to the design of shoe F4. This shoe is wider than the shoe F3 which allows greater heat dissipation. Boot (F5) has the highest total heat resistance. However, reviewing the values for manikin segments 1-3 only, it can be seen that shoe F5 exhibits a similar heat resistance characteristic than shoe F3.

Figure 10 provides a comparative overview of our results in relationship to the results presented in recent literature. When our results are viewed in context of data presented in recent literature, it can be noted that the values observed for our boot is in the range of values presented in studies IV and V. In these investigations, the test conditions were similar to the conditions used in our experiments.

Table 1 The overview of results from literature and results presented in the paper

Investigation number	Investigators (year)	Environmental conditions/ambient temperature (°C)	Motion
I	Kuklane & Holmer (1999)	Cold/ +3	Static
II	Kuklane & Holmer (1997)	Cold/ +4	Dynamic
III	Kuklane, Holmer & Giesbrecht (1999)	Extreme/-10	Dynamic
IV	Mekjavić et al. (2005)	Moderate/+15	Static
V	Mekjavić et al. (2005)	Moderate/+15	Dynamic
VI	Reischl et al. (2012)	Moderate/ +17	Static

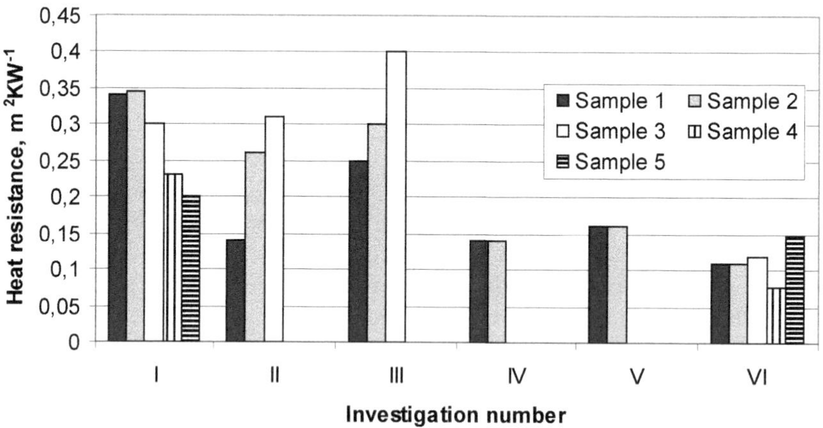

Figure 10 The comparison of obtained results with ones from the literature

If the results of measured heat resistance are set in the context of other types of footwear shown in the literature, it can be noted that the resistance of the tested boots is in the range of values shown for the boots in the investigations IV and V. In these investigations, the test conditions were similar to the conditions established in this experiment.

4 CONCLUSIONS

Heat resistance values obtained for selected shoes provided practical information about the thermal properties of typical men's footwear. These data can now be used to estimate the potential comfort offered by the shoes when worn under various environmental conditions. The goal of this research is to provide manufacturers with information that can contribute to the design of new footwear which will achieve thermal comfort and safety.

ACKNOWLEDGMENTS

The paper is a part of research within scientific projects Multifunctional technical, nonwoven and knitted fabrics, composites and yarns (number 117-0000000-2984) and Ergonomic design of the worker-furniture-environment system (number 117-0680720-3051) conducted with the support of the Ministry of Science, Education and Sports of the Republic of Croatia.

REFERENCES

Measurement Technology Northwest: Advanced Thermal Manikin ADAM, Accessed November 3, 2009, http://www.mtnw-usa.com/thermalsystems/fullbody_manikins.php.

Measurement Technology Northwest: NEMO - submersible thermal manikin, Accessed November 3, 2011, http://www.mtnw-usa.com/thermalsystems/fullbody_manikins.php.

Reischl, U., B. Mijović, Z. Skenderi, I. and Salopek Čubrić. 2010. Heat Transfer Dynamics in Clothing Exposed to Infrared Radiation. In. *Advances in Ergonomics Modeling and Usability Evaluation,* eds. Khalid, H., Hedge, A. and Ahram T. Z. Boca Raton: Taylor & Francis Group, USA.

Terry Rice. USARIEM Uses Thermal Manikin to Test Warfighter Clothing and Individual Equipment (CIE). Accessed November 3, 2011
http://www.usaasc.info/alt_online/article.cfm?iID=0912&aid=02.

UCS: Thermal-evaporative manikin system, available at http://www.ucstech.eu/images/Thermal_Evaporative_Manikins.pdf.

Kuklane, K. and I. Holmér 1997. Reduction of footwear insulation due to walking and sweating: a preliminary study. *Problems with cold work.* Arbete och Hälsa, National Institute for Working Life, Stockholm, Sweden

Kuklane K. and I. Holmer. 1997. Effect of sweating on insulation of footwear. *International Journal of Occupational Safety and Ergonomics*. 123-136.

Kuklane K. Footwear for cold environments – thermal properties, performance and testing, Doctoral thesis, 1999.

Kuklane, K., I. Holmér, and G. Giesbrecht. 1999. One week sweating simulation test with a thermal foot model. *The Proceedings of the Third International Meeting on Thermal Manikin Testing*, National Institute for Working Life, Stockholm, Sweden.

Mekjavić, I. B. et al. 2005. Static and Dynamic Evaluation of Biophysical Properties of Footwear: The Jozef Stefan Institute Sweating Thermal Foot Manikin System, *Proceedings RTO-MP-HFM-126*, 6-1- 6-8, Neuilly-sur-Seine, France.

CHAPTER 51

A Bayesian Based Model Application to the Prediction of Task Accuracy – Case Study on Call Center's Agents Dialog

Eko Wahyu Tyas Darmaningrat[1], *Shu-Chiang Lin*[2]

Department of Industrial Management
National Taiwan University of Science and Technology, Taiwan, R.O.C
M9901804@mail.ntust.edu.tw, slin@mail.ntust.edu.tw

ABSTRACT

This paper investigates whether an application of a hybrid Bayesian based semi-automated task analysis model is able to learn and predict subtask categories from the narrative telephone conversations between agent and customer. A total of 126 customer calls were collected from a call center and were transcribed into word document. These data were then classified into 72 subtask categories by a human expert and were verified by 22 other persons. The data was used to train a Bayesian based model to predict the subtask categories. Hit rate, false alarm rate, and sensitivity value are used to test if the Bayesian based machine learning tool is able to learn or predict subtask categories. The preliminary results show that the Bayesian based model has overall hit rate of 57.21%, false alarm rate of 0.64%, and sensitivity value d' of 2.67. These results indicate that the model is able to learn subtask categories from the agent/customer narrative telephone conversations and to predict them as well. Further investigations in this study will focus on the comparison of prediction accuracy among thirteen various word combinations, ranging from single word, two-word combination, three-word combination, four-word combination, single-pair combination, single-three combination, single-four combination, pair-three combination, pair-four combination, three-four combination, single-three-four combination, pair-three-four combination, to single-pair-three-four combination. The aim is to examine which combination(s) might fit best for different datasets that include training set, testing set, and a combination of both. The preliminary analyses show that the single-pair-three-four words combination has the highest hit rate of 54.60% compared to other combinations, while the four words combination has the least hit rate of 26.34% among others. A further investigation is needed to check whether the single-pair-three-four word predictors have the highest accuracy to predict the subtask categories.

Keywords: Bayesian model, task analysis, hit rate, false alarm, call center agent

1. INTRODUCTION

Task analysis is an essential part of ergonomic design process. Task analysis involves documenting tasks and task sequences, time and information requirements, and the skills, knowledge, and abilities required to perform the tasks. Training and other means of improving performance will often be specified to help ensure that tasks go with capabilities of people that will be performing them. Throughout task analysis, various activities may also be necessary to help determine how well the tasks will be performed under expected environmental conditions (Lehto and Buck, 2008). Although there are a wide range of existing techniques to human factors specialists performing task analysis, identifying and decomposing user's task into small task components remains difficult, impractically time-consuming, and costly process that involves extensive manual effort (Lin and Lehto, 2009).

Among a range of Bayesian model applications, two Bayesian inferences are often proposed: classic (naive) Bayesian model and fuzzy Bayesian model. If multiple evidences are acquired, classic Bayesian model tries to incorporate these evidences and conceives that they are conditionally independent given that the hypothesis is true. In contrast to Naive Bayesian, Fuzzy Bayesian model avoids building strong independence assumptions of the evidences (Lin and Lehto, 2009). The result of previous study by Lin and Lehto (2009) informed that the tool developed is able to learn and predict subtask categories from telephone conversation between customers and human agents. This result is a starting point that enables future studies of Bayesian theories applications in task analysis. With the profusion of results obtained in this work, finding the most accurate result becomes a crucial part. For that reason, in this study, advanced analyses based on the previous results will be carried out to compare the prediction accuracy among all the combination words obtained from fuzzy Bayes methods.

2. LITERATURE REVIEW

2.1. Task Analysis

2.1.1. Definition of Task Analysis

Numerous books and papers have been discussed task analysis and its applications in various fields. Annett et al. (1971) stated that the process of analysing a task is the process of diagnosing the plan which is needed to achieve a stated goal. According to Swezey and Pearlstein (2001) task analysis is a technique that determines the inputs, tools, and skills or knowledge necessary for successful task performance. Task analysis can involve describing jobs in terms of individual physical actions (e.g., button pushing) or describing acts in terms of their higher level or functional foundations (such as goals, domain concepts, etc.) (Hoffman and Militello, 2008). Task analysis depicts both our current understanding of factors affecting human performance and the information needs of system designers. Modern technology has dramatically changed the nature of human work as well as concepts and techniques of analysis in recent years and they are continuously changing to meet new requirements (Annett and Stanton, 2000a).Various studies indicated that the use of task analysis has been broadened from task specialist such as ergonomists, task designers, and task analysts, to task-related workers such as operators, managers, supervisors, and incumbents. Each individual has played a part in efficiently and effectively integrating the human element into system design and operations (Lin and Lehto, 2009).

2.1.2. Hierarchical Task Analysis

The underlying technique of Hierarchical Task Analysis (HTA), hierarchical decomposition, analyzes and represents the behavioural aspects of complex tasks such as planning, diagnosis and decision making (Annett and Stanton, 2000b). HTA breaks tasks into subtasks and operations or actions. It entails identifying tasks, categorizing them, identifying the subtasks, and checking the overall accuracy of the model (Crystal and Ellington, 2004). According to Crystal and Ellington (2004), HTA is useful for interface designers because it provides a model for task execution, enabling designers to envision the goals, tasks, subtasks, operations, and plans essential to users' activities. HTA is useful for decomposing complex tasks, but has a narrow view of the task, and normally used in conjunction with other task analysis methods to increase its effectiveness. Stanton and Young (1999) mentioned some advantages of HTA such as its easiness to be implemented once the initial concepts have been understood. Besides, its rapid execution provides user satisfaction as good progress is made in little time. However, HTA also has several drawbacks such as it provides more descriptive information than analytical information; there is little in the HTA which can be used to directly provide design solutions, such information is necessarily inferred; HTA does not handle cognitive components of tasks (e.g. decision making), only observable elements.

2.1.3. Cognitive Task Analysis

Compared to HTA, Cognitive Task Analysis (CTA) presents quite different challenges to the analysts. It requires deep relationship with a particular knowledge domain, working closely with subject-matter experts to elicit their knowledge about various tasks (Chipman et al., 2000). CTA represents an attempt to capture task expertise. Since expertise is often tacit, it can be much more difficult to analyze than the explicit actions typically considered by HTA. In fact, CTA requires "making explicit the implicit knowledge and cognitive-processing requirements of jobs" (Dubois and Shalin, 2000). In addition, CTA has increased understanding of many important cognitive aspects of modern task environments. However, it is unclear how effective CTA techniques are in representing these aspects in a systematic and useful way (Shepherd, 2001). Another significant problem with many CTA techniques is that studying high-level cognitive functions in a real task situation is very difficult. Studies may take months or years and rely on the dedicated efforts of senior researchers. Consequently, some practitioners are developing simpler approaches that can form a "practitioner's toolkit" (Militello and Hutton, 2001). When a user interface needs to be designed for a task that depends on information integration, such as navigation, few theoretical tools are available. Combination of CTA and HTA forms the basis of interface definition (Westrenen, 2010).

2.2. Statistical Methods

Among a number of statistical approaches, applications of Bayesian model are proven to be successful in various areas. Zhu and Lehto (1999) put forward Bayesian inference rule as a good statistical model to fit the dependent relationship between index terms and words in the text. Lehto and Sorock (1996) appraise Bayesian inference as a machine learning technique and conclude that it successfully acknowledged and predicted motor vehicle accident categories and subcategories from accident narrative data. Chatterjee (1998) reveals that the fuzzy Bayes model was superior in performance to that of the keyword model by classifying a significant amount of free-text accident narratives that the keyword model had failed to classify. Lehto, et al. (2009) compare two Bayesian methods (Fuzzy and Naive) for classifying injury narratives in large administrative databases into event cause groups. Overall, Naive Bayes provided slightly more accurate predictions than Fuzzy Bayes. Lin and Lehto (2009) applied naïve (classic) bayes, fuzzy bayes, and hybrid bayes model to construct a semi automated task analysis tool to predict subtask categories. The authors utilized recorded dialog of customers and agents at call center as source data to train and test their proposed tool.

3. MODEL DEVELOPMENT

This study utilizes recorded dialog at call center as source data which collected from the observed call center knowledge agent's troubleshooting process. A customer called the call center to report problems relating to his/her printer and is dispatched to a knowledge agent. The knowledge agent is responsible for defining, identifying, and/or judging the problems as well as for troubleshooting, diagnosing, and/or analyzing the problems and constructing a plan to explain, interpret, and/or adjust the instructions to guide the clients through resolution remotely in a timely fashion. The knowledge agents often utilize several types of resources to help the customers discover the problems because of different complexity levels of printer problems. Cases in point are browsing various different on-line troubleshooting tools and/or expert systems, referring to paper based service manuals and documents, performing physical product testing, communicating and consulting with higher level technicians, supervisors, or other knowledge agents, etc. Thus, from a human expert point of view, the troubleshooting job involves not only physical activities but also psychological activities such as decision making. Figure 1 illustrates model of Bayesian based semi-automated task analysis tool derived from call center knowledge agent's troubleshooting process proposed by Lin and Lehto (2009). Arrows in the picture illustrate data flow. The experimental process is accomplished in four phases as shown in Figure 2.

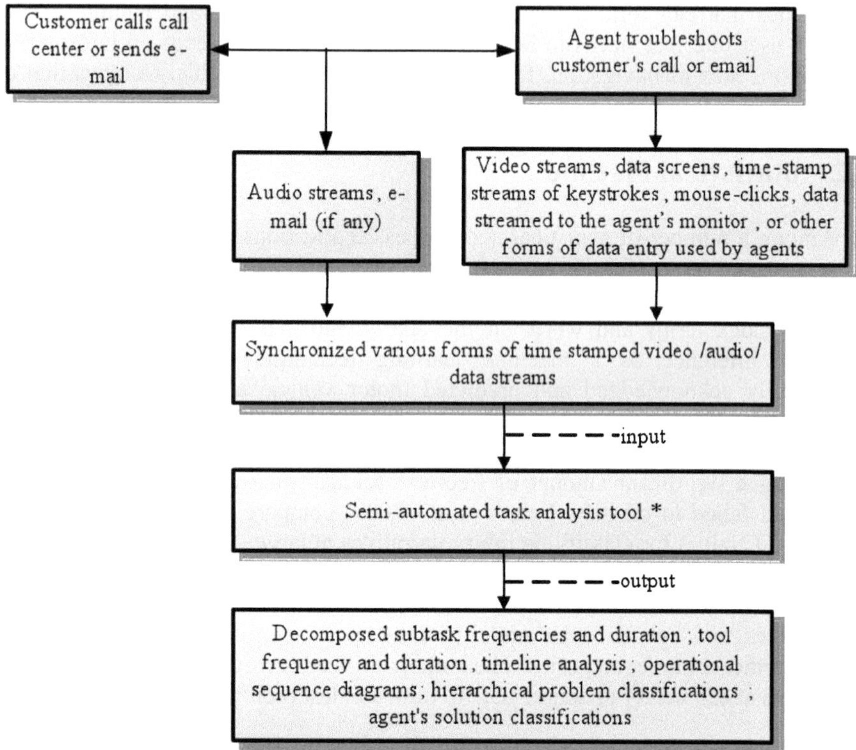

Figure 1 Model of Bayesian Based Semi-automated Task Analysis Tool Derived from Call Center Knowledge Agent's Troubleshooting Process

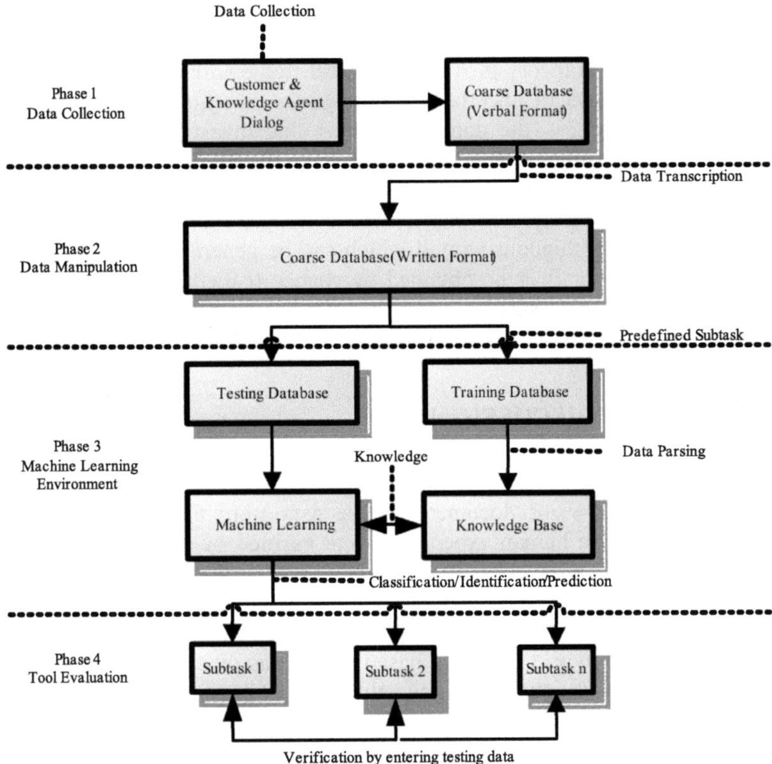

Figure 2 Four Phases of Bayesian Based Semi-Automated Task Analysis Tool Development

3.1. Solution Methodology

3.1.1. Hit Rate and False Alarm Rate

The simplest definition of hit rate and false alarm rate is that if the machine learning tool identified or predicted a subtask category that matches the subtask category we assigned to the same word vectors, then this was a "hit" for the predicted category; if the tool identified or predicted a subtask category that does not match the subtask category we assigned, then this was a "false alarm" for the predicted category and a miss for the actual category (Lin and Lehto, 2009). Generally, the outcomes of hit rate and false alarm rate from an experiment are displayed as follows:

		Pre-assigned	
		True	False
Predicted by Machine Learning	True	Hit	False Alarm
	False	Miss	Correct Rejection

3.1.2. Signal Detection Theory (SDT)

Signal Detection Theory (SDT) was fully developed on the theoretical side by 1960s. Signal-detection theory provides a general framework to describe and study decisions that are made in uncertain or ambiguous situations. It is most widely applied in psychophysics—the domain of study that investigates the relationship between a physical stimulus and its subjective or psychological effect—but the theory has implications about how any type of decision under uncertainty is made (Wickens, 2002). SDT is used to analyze data coming from experiments where the task is to categorize ambiguous stimuli which can be generated either by a known process (called the signal) or be obtained by chance (called the noise in the SDT framework). Particularly, SDT is used to analyze experiments where a binary answer (e.g., 'Yes' or 'No') needs to be provided (Abdi, 2009).

4. RESULT AND DISCUSSION

In this study, a total of 126 customer calls were collected from a call center and were transcribed into word document. These data were then classified into 72 subtask categories by a human expert and were verified by 22 other persons. The data were used to train a Bayesian based model to predict the subtask categories. Hit rate, false alarm rate, and sensitivity value are used to test if the Bayesian based machine learning tool is able to learn or predict subtask categories. The preliminary results show that the hybrid Bayesian based model has overall hit rate of 57.21%, false alarm rate of 0.64%, and sensitivity value d' of 2.67.

The Bayesian based semi-automated task analysis tool generated a total of thirteen prediction results recorded in thirteen prediction tables respectively. These tables predicted the subtask categories for various word combinations ranging from single word to single-pair-three-four word combination. Textminer program was used to generate master word list containing more than 165,000 instances of words used in 5184 dialogs between agent-customer. After excluding repetitive words, a total of around 5400 single words were collected in the master word list. Once data parsing completed, the Textminer learning tool then used the revised master wordlist (single-word) to identify combinations of words appearing in the narratives that could be candidates for subtask category predictors.

From 5184 records of agent-customer dialog, we obtained 2058 records of single word which prediction strength equal to or more than 50%. After eliminating the repetitive words, a total of 290 single word predictors were collected. Table 1 shows partial listing of those single word predictors.

The single-word frequency list was then used to create a list of two-word combinations. As for single words, the two words had to appear together at least 4 times in the dataset to be included in the pair-word frequency list. From 5184 records of agent-customer dialog, we obtained 2416 records of two-word combination which prediction strength equal to or more than 50%.

Table 1 Partial Listing of Single Word Predictors

Cat	SWStrength	SWPredictor	Cat	SWStrength	SWPredictor
111	1	HPCustomerCare	621	1	Dell
	1	LaserJetsupport		0.85714287	feeling
211	1	Stephanie	622	0.5	Mark
	0.92307693	firstname	624		
	0.76190478	lastname	625	0.80000001	southernCalifornia
	0.54471546	Please		0.75	TechSupport
212	1	Yin	626	0.80000001	Holdonaminute
	0.66666669	Burnt		0.5714286	Mind
215	0.80000001	records		0.56060606	Holdonasecond
	0.54098362	phonenumber	627	0.625	Alight

After taking out the repetitive words, a total of 979 two-word predictors were obtained. Table 2 shows partial listing of two-word predictors.

Table 2 Partial Listing of Two-Word Predictors

Cat	PWStrength	PWPredictor	Cat	PWStrength	PWPredictor
111	1	Brian&hello	626	0.666666687	Holdonasecond&I'll
	1	calling&HPCustomerCare		0.875	Holdonasecond&right
	1	calling&Michelle		0.692307711	Holdonasecond&Sure
	0.833333313	calling&Thank		1	Holdonasecond&things
	1	Chris&HPCustomerCare		0.625	I'm&putyouONHOLD
	1	for&Michael		0.5	me&putyouONHOLD
	0.975609779	hello&Hi		1	Okay&putyouonholdforasecond
	1	hello&how		1	problem&putyouonholdforasecond
	1	hello&LaserJetSupport	627	0.555555582	back&Holdon
	1	hello&Thanks		0.5	hang&on

By repeating the steps above, another twelve tables of word combinations predictors' strength and accuracy were obtained. Table 3 summarizes these word combinations predictors, as shown below.

Table 3 Summary of Word Combination Predictors

Predictor	Total records (strength ≥ 50%)	Total without repetitive words	Empty Categories	% Hit Rate
single word	2058	290	18	49.56
two-word	2416	979	21	49.77
three-word	1791	929	25	36.97
four-word	1321	828	30	26.34
single-pair	2694	972	17	**54.15**
single-three	2520	932	18	**51.53**
single-four	2330	857	16	49.29
pair-three	2509	1136	18	**50.68**
three-four	1844	1008	24	37.89
single-three-four	2541	1013	16	**51.82**
pair-three-four	2515	1177	19	**50.70**
single-pair-three-four	2748	1144	16	**54.60**

The preliminary analyses summarized above in Table 3 reveal that the single-pair-three-four words combination has the highest hit rate of 54.60% compared to other combinations, while the four words combination has the least hit rate of 26.34% among others. A further investigation is needed to check whether the single-pair-three-four word predictors have the highest accuracy to predict the subtask categories.

5. CONCLUSIONS

Bayesian based machine learning methods combined with task analysis methods can be used to help practitioners analyze their tasks. Preliminary results indicate this approach successfully learned how to predict subtasks from the telephone conversations between customers and call center agents. Further investigations reveal that on the comparison of prediction accuracy among thirteen word combinations, single-pair-three-four words combination has the highest hit rate compared to other combinations, while the four words combination has the least hit rate among others, an indication that the single-pair-three-four words combination predict better than other word combinations in this research. An advanced study on which combination(s) might fit best for different datasets will be examined next.

ACKNOWLEDGMENTS

I would like to thank Dr. Shu-Chiang Lin, my master thesis advisor, for his guidance, wisdom, knowledge, and support during my study at NTUST and for making the completion of this paper possible.

REFERENCES

Abdi, Hervé. (2009). Signal Detection Theory. In McGaw, B., Peterson, P.L., Baker, E. (Eds.): Encyclopedia of Education (3rd Ed). New York: Elsevier.

Annett, J., Duncan, K.D., Stammers, R.B. and Gray, M.J. (1971). *Task Analysis*. London: HMSO.

Annett, J. and Stanton, N.A. (2000a). *Research and Developments in Task Analysis*. In Annett and Stanton. (2000): Chapter 1 (pp. 1-8).

Annett, J. and Stanton, N.A. (2000b). *Task Analysis*. London: Taylor and Francis.

Chatterjee, S. (1998). *A Connectionist Approach for Classifying Accident Narratives*. Unpublished Ph.D. Dissertation. Purdue University, West Lafayette, IN. Retrieved October 19, 2011, from http://search.proquest.com/docview/304450940

Chipman, S.F., Schraagen, J.M., and Shalin, V.L. (2000). *Cognitive task analysis*. Mahwah, NJ: Lawrence Erlbaum. Retrieved September 27, 2011, from http://www.questia.com/PM.qst?a=o&d=78569828

Crystal, A. and Ellington, B. (2004). Task Analysis and Human-Computer Interaction: Approaches, Techniques, and Levels of Analysis. *Proceedings of the Tenth Americas Conference on Information Systems, New York.*

Dubois, D. and Shalin, V. (2000). *Describing Job Expertise using Cognitively Oriented Task Analyses (COTA).* In Chipman, et al. (2000), Part II (page 41-56).

Hoffman, R.R and Militello, L.G. (2008). *Perspective on Cognitive Task Analysis: Historical Origins and Modern Communities of Practice.* New York: Taylor and Francis.

Lehto, M.R., and Buck, J.R. (2008) Introduction to Human Factors and Ergonomics for Engineer. New York: Lawrence Erlbaum Associates.

Lehto, M.R., Maruci-Wellman, H., and Corns, H., (2009). Bayesian Methods: A Useful Tool for Classifying Injury Narratives into Cause Groups. *Injury Prevention Module 2009;000:0–7. doi:10.1136/ip.2008.021337.*

Lehto, M.R., and Sorock, G.S. (1996). Machine learning of motor vehicle accident categories from narrative data. *Meth. Inform. Med.*, 35 (4/5), 309-316.

Lin, S. and Lehto, M.R. (2009). *A Bayesian Based Machine Learning Application to Task Analysis, Encyclopedia of Data Warehousing and Mining*, Classification B, 1-7, Wang, John (Ed., 2nd Edition).

Militello, L. and Hutton, R. (2000). *Applied Cognitive Task Analysis (ACTA): A Practitioner's Toolkit for Understanding Cognitive Task Demands.* In Annett and Stanton (2000): page 90-113.

Shepherd, A. (2001). *Hierarchical Task Analysis*. New York: Taylor & Francis.

Stanton, N.A. and Young, M.S. (1999). *A Guide to Methodology in Ergonomics: Designing for Human Use*. London: Taylor and Francis.

Swezey, R.W. and Pearlstein, R.B. (2001). Selection, Training, and Development of Personnel. In Salvendy, G. (Eds). *Handbook of Industrial Engineering: Technology and Operations Management (3rd Ed)*. Chapter 35 (pp. 920-947). New York: John Wiley & Sons.

Westrenen, F. (2010). *Cognitive Work Analysis and The Design of User Interfaces*. Cogn Tech Work (2011) 13:31–42. Springer.

Wickens, T.D., (2002). *Elementary Signal Detection Theory.* New York: Oxford University Press.

Zhu, W. and Lehto, M.R. (1999). Decision support for indexing and retrieval of information in hypertext system. *International Journal of Human Computer Interaction*, 11, 349-371.

CHAPTER 52

Study on the Control Rules of X-axis Relative Speed of Chase Spacecraft during the Manual Control Rendezvous and Docking

Tian Zhiqiang, Jiang Ting, Wang Chunhui, Guohua Jiang

Science and Technology on Human Factors Engineering Laboratory
Astronaut Research and Training Center of China
Beijing, China
Tianzhiqiang2000@163.com

ABSTRACT

To study the relationship between X-axis relative speed of chase spacecraft and docking time, fuel wastage, spacecraft stability and docking precision during the manual rendezvous and docking (MRVD), two different difficulty level experiments (large or small initial attitude biases) were designed on a MRVD simulation experiment system and twenty male subjects aged 22-40 participated in the experiments. During the experiments of manual control, the data of relative position and attitude changing between the chase spacecraft and the target spacecraft were recorded. At the end of docking, docking time, result and fuel wastage also were recorded. Experiment results show that the X-axis relative speed within different subjects has large diversity during the docking and its stability has high correlation with docking time ($r = -.664$, $P<.01$) and fuel wastage ($r = -.726$, $P<.01$). At the end of the docking, X-axis relative speed does not influence the docking precision. From initialization state to the end, the X-axis relative speed of two different level tasks has the same changing trend and the optimum control ranges between two different difficulty level tasks have not notable diversity ($P>.05$). From the results, researchers have the conclusions that the stability of the

X-axis relative speed influences the manipulation performance of MRVD. There are suitable control ranges of the X axial relative speed at different phases during the MRVD, the optimum control speed is 0.3m/s ~0.4m/s at tracking control stage, 0.2m/s ~0.3m/s at accurate control stage, 0.15m/s ~0.2m/s at docking moment.

Keywords: MRVD, X-axis Relative Speed, Control Rules, Chase Spacecraft

1 INTRODUCTION

Rendezvous and docking (RVD) is an essential technology for manned space missions (Zhou J.P, 2011 and Lin L.X, 2007 and Zimpfer D, 2005). It is a prerequisite of space station assemblage and cargo supply in orbit. The active one of two spacecrafts who is undergoing rendezvous and docking is called chase spacecraft and the passive one is called target spacecraft. In case of malfunction of automatic control system, as an important backup for automatic rendezvous and docking, the manual rendezvous and docking (MRVD) will enable astronauts to utilize their flexibility in the complex control system. During the MRVD, astronaut observes the outline of the target spacecraft and docking cross drone image, to estimate the relationship of position and attitude of two spacecrafts, when the X-axis speed of chase spacecraft is bigger than that of the target spacecraft, two spacecrafts gradually move closer. Astronaut adjusts the chase spacecraft's position and attitude by translation and attitude handles to maintain the two spacecrafts' relative position and attitude to meet the docking access conditions until the successful docking. Compared with the dynamics of the automatic control and sensor technology researches (Wigbert F, 2003 and Kelsey J.M, 2006 and Wu H.X, 2003), less ergonomics documents could be searched about the MRVD. Research (Zhou Q.X, 2006) showed that astronauts' role should be integrated into the RVD control, it is suitable for them to finish the yaw, pitch and roll control in order to assure the man-machine system high performance. Other research (Zhang Y.J, 2008) argued that docking time is proportional with the control complexity of MRVD, and another research (Hu H.X, 2006) showed that automatic control of relative attitude is in favor of manual control of relative translation of the chase spacecraft in the MRVD mission.

For two spacecrafts flying at high-speed in low orbit, the relative speed of X-axis should satisfy the requirements of astronaut's control ability, and ensure to complete the docking before time limitation and consumption of fuel. It is necessary to study the control rules of relative speed of X-axis. This research, based on the MRVD simulation experiment system, designed two experiments of different difficulty levels and analyzed relationships between performance indexes such as docking time, fuel wastage, docking accuracy and relative speed stability of X-axis in MRVD process, to seek the appropriate control rules of relative speed of X-axis of chase spacecraft.

2 METHODS

2.1 Experiment System

MRVD simulation experiment system can realize the simulation of 100m-0m rendezvous and docking process, the monitor show images of target spacecraft and docking cross drone as well as information of relative attitude and position of the two spacecrafts in the manual rendezvous and docking process. The experimental system partially simulates dynamics control model of chase spacecraft and could set experimental conditions, initial conditions of manual control and parameters of spacecrafts as per purpose of the experiments. Figure1 is the hardware system sketch map of the simulation system and the experimental software interface.

The back-end database of the experimental system can accurately capture, record, display and store related information during the experiments, including the profiles of subjects, the experimental parameters, the initial and final parameters of the two spacecrafts, the operational data and result data of the MRDV experiments, related post-processing analysis can be carried out as well.

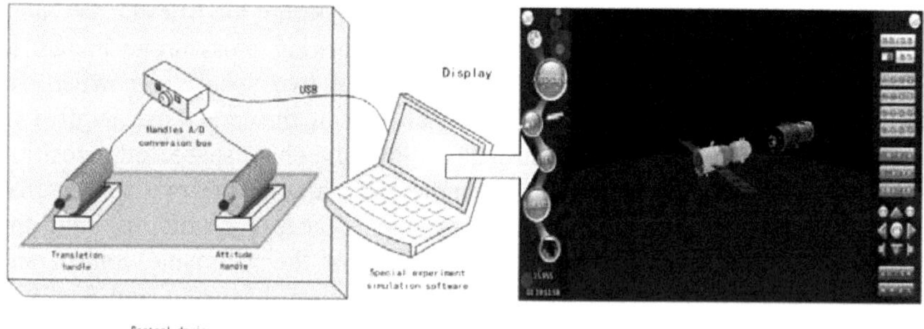

Figure1 Hardware system sketch map and software interface of the experiment system

2.1 Subjects

20 male subjects participated in the experiments, 24-42 years old, right-handed. All the subjects learned knowledge about basic principles of MRVD, and learned how to judge relative position and attitude between two spacecrafts and how to control translation and attitude by handles.

2.2 Protocol

(1) Experimental objective

The experiments verify the size of the initial attitude biases how to influence the MRVD control rules of relative X-axis speed.

(2) Experimental level

The experimental set up two different difficulty level trials include: large initial attitude deviation trial (LADT) and a small attitude deviation trial (SADT). Docking experiments started from the point of 100m, initial velocity of 6 freedom degrees of are zero, the initial position of Y and Z axes are 10m, roll, pitch and yaw initial attitude of large biases are 20 °, attitude of small biases is: roll 5 °, pitch 6 ° and yaw 6 °. Each trial includes eight initial states. Because of difference of coordinate system symbol existed in chase spacecraft relative position and attitude parameters, every initial state will have different initial image on the monitor of subjects observed.

(3) Controlled factors

The monitor displays the images of the target spacecraft's outline, the docking mechanism and docking cross drone only, does not display the information of the value of the relative position and attitude of the two spacecraft. Subjects judged related information of the chase spacecraft according to the size of the target spacecraft and the image of position and attitude. The simulative light environment of space during two spacecrafts docking process is solar area.

2.3 Experimental procedure

The experiments were performed in the RVD ergonomics lab of Astronaut Research and Training Center of China. The experimental procedure included the following steps:(1)Main trailer introduced the guideline of experiments, the subjects confirmed the basic requirements of the docking and filled in the confirmation paper with the necessary personal information.(2) Subjects only granted access to start experiment once passed skill assessment with three time's consecutive successful test.(3) Main trailer setup initial parameters and confirmed firstly, after that subjects clicked the "Start" button, back-end database will automatically record the 6 freedom degrees parameter data created during operation. According to the judgment of docking success or failure, the final result was recorded at the docking moment. (4) The analyzed experimental data of subjects will be exported from database by *. txt format.

2.4 Statistical analysis

SPSS 15.0 for Windows is used to process the experiment data. Analyzed data includes the biases and speed of the docking moment, fuel wastage, docking time, the mean X-axis velocity, the maximum X-axis velocity, and speed stability of the X-axis.

3 RESULTS

3.1 Subjects individual diversity

The experimental data was processed using the Kolmogorov-Smirnov method to test if the sample distribution was normal .All the collected data passed the test.

Seen from descriptive statistic data of experiments, subjects' individual deviation exist during the MRVD operation, 6 freedom degrees curves are different within subjects.

3.2 Tendency comparison of Vx of docking process under SADT and LADT

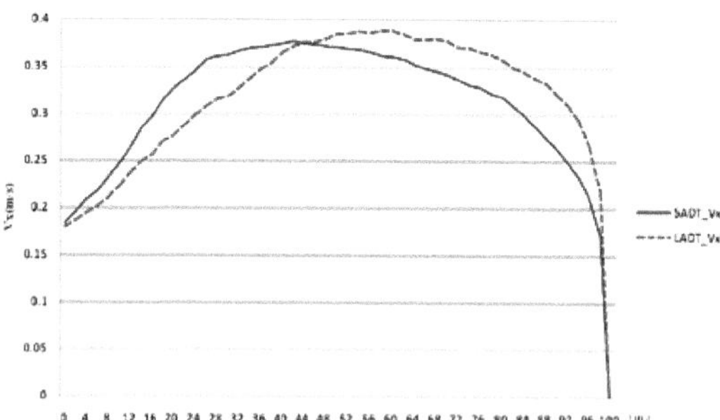

Figure2 Tendency comparison of Vx of docking process under SADT and LADT

Figure 2 illustrates 20 subjects' mean data of X-axis speed variation share the same tendency under SADT and LADT. The data show the former reaches summit of 0.375m/s at around 40m while the later reaches summit of 0.387m/s at around 60m. Vx equals 0.328m/s at 20m for SADT while 0.276m/s 20m for LADT. The study (Jiang T, 2011) showed that MRVD process could be divided into two parts, tracking control stage and accurate control stage. Tracking control stage is about from the 100m distance to 20m, Owing to the far distance of target, detailed features can't be observed, astronaut mainly track the target and adjust chasing spacecraft's attitudes by observing the space shape features. Accurate control stage is form 20m to 0m, astronaut mainly precise control the attitude and translation of vehicle to meet the docking access. The experimental data were compared from tracking control stage and accurate control stage.

3.3 Relationships of Vx and other performance indexes under SADT and LADT

Under two different difficulty experiments, the relationships are weak between X-axis speed and translation and attitude biases at the docking moment. But the relationships between X-axis speed and fuel wastage and docking time have statistics meaning, and there are high correlations between X-axis speed stability and fuel wastage ($r = -.726$, $P<.01$)and docking time($r = -.664$, $P<.01$).

Table 1 Relationships of X-axis speed and fuel wastage and docking time under SADT and LADT(Pearson correlation, Two-tailed)

Phases	Variable	SADT		LADT	
		Fuel wastage	Docking time	Fuel wastage	Docking time
Tracking control stage	Max of Vx	.160*	-.437**	.129	-.461**
	Mean of Vx	-.306**	/	-.371**	/
	Stability of Vx	-.471**	-.418**	-.538**	-.437**
Accurate control stage	Max of Vx	.664**	.021	.175*	-.272**
	Mean of Vx	-.058	/	-.484**	/
	Stability of Vx	-.566**	-.552**	-.726**	-.664**

3.4 Comparisons of performance indexes under SADT and LADT

Under SADT and LADT, performance indexes data have a sample T-test which created by the fuel wastage, the docking time and a variety of Vx value. All the analysis data meet the requirements of the homogeneity of variance.

Table 2 Comparisons of performance indexes of tracking control stage (100m-20m) under SADT and LADT

Performance indexes	Experiment	Mean	St	T- value
Fuel wastage	SADT	11.2	5.01	.445
	LADT	11.0	4.67	
Docking time	SADT	321.6	106.1	1.07
	LADT	309.3	100.6	
Max of Vx	SADT	.443	.17	-.352
	LADT	.449	.14	
Mean of Vx	SADT	.333	.05	-1.00
	LADT	.348	.05	
Stability of Vx	SADT	.437	.18	.951
	LADT	.417	.20	

For 100m-20m tracking control stage, the two experiments do not show clear deviation at the two process indicators: fuel wastage and docking time, and three X-axis velocity indicators: Vx value of docking moment, maximum of Vx, mean of Vx and stability of Vx. Seen from the overall analysis, the size of initial attitude biases has small impact to the performance of tracking control stage of MRVD.

Table3 Comparisons of performance indexes of accurate control stage (20m-0m) under SADT and LADT

Performance indexes	Experiment	Mean	St	T- value
Vx of docking moment	SADT	.183	.05	.637
	LADT	.180	.03	
Fuel wastage	SADT	2.92	2.53	-1.51
	LADT	3.36	2.8	
Docking time	SADT	107.1	54.8	-1.99
	LADT	119.9	59.8	
Max of Vx	SADT	.351	0.15	3.66**
	LADT	.298	0.08	
Mean of Vx	SADT	.252	.05	2.86**
	LADT	.226	.03	
Stability of Vx	SADT	.470	.19	-.040
	LADT	.476	.21	

For 20m-0m accurate control stage, the two experiments do not show clear deviation at the four docking accuracy indicators: roll, yaw and pitch biases of docking moment, translation biases of docking moment. But the data of two experiments show that obvious deviation between maximum of Vx ($P<.01$) and mean of Vx ($P<.01$), while the others indicate the same with tracking control stage analysis.

4 DISCUSSION

Results from the statements of subjects and other reports (Yang J ,2010 and Jiang Z.C, 2007)showed when subjects began manipulation, they judged position and attitude biases according to the location and outline of the target spacecraft on the display and ensured the control's priority of freedom degree, and controlled the target spacecraft to the center of the screen gradually. Upon completion of this process, with the shorten of distance between the two spacecrafts, the subjects observed the cross drone image, focused on controlling those freedom degrees which large biases still existed and ensured the X-axis speed, attitude and position biases access the docking requirements before the last 20m (figure 3).

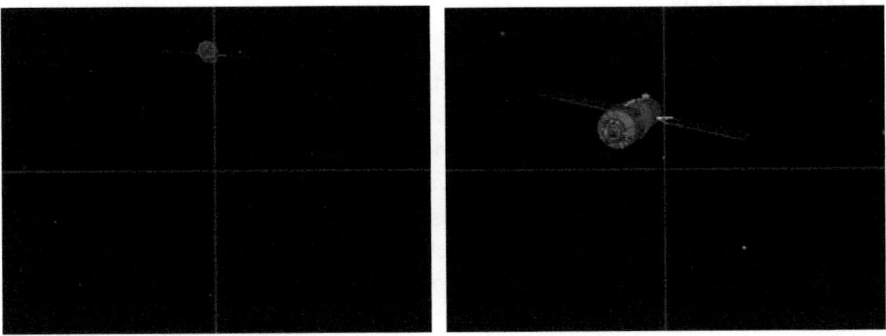

Figure 3 Simulation of the long-range and the medium-range target spacecraft images

It can be seen from figure 1 that X-axis speed control process is the same basically under SADT and LADT. Chase spacecraft accelerates from hover state, X-axis speed improves rapidly and closes to 0.3m/s in the distance of 20m. At the distance of 35m or so before docking, X-axis speed slows down, subjects depend on the size of the biases at that time to select the appropriate final docking speed in the accurate control stage.

T-test results show that the size of the initial attitude biases has a weak influence to fuel wastage and docking time in the MRVD, and the correlation coefficient between Vx values and docking accuracy of docking moment is small. The subjects reflected that too fast docking X-axis speed will cause the target spacecraft zoomed in quickly, which will bring visual strong concussion feeling at the end of docking. X-axis speed, including maximum, average and stability, affect the docking operation time and fuel wastage. The correlation coefficients between X-axis speed stability and docking time and fuel wastage is very high. All the data indicate it is very important that a stable docking X-axis speed is very important in the accurate control stage.

Subjects need complete adjustment of the relative attitude and position of the two spacecrafts in the distance of 20m before docking. The relative attitude initial biases, no matter 20 degrees or 5 degrees, are adjusted to less than 2 degrees before 20m, but the LADT three axes attitude mean biases are little larger than the SADT. This reflects T-test significant differences of Vx maximum and average value at the accurate control stage between the SADT and LADT. At this point, it shows that 20 degrees initial attitude biases impact the docking speed at accurate control stage in 100m distance.

20 subjects' data of X-axis speed in the 100m control process meet the normal distribution. Theoretical calculations shows the amount of data includes in 68.26% total area of normal distribution between ± 1 standard deviation. In other words, according to population distribution infers that the people of middle ability to control the various stages of the docking operation should select a suitable X-axis speed control range, such as the tracking control stage at 0.3m/s ~ 0.4m/s, the accurate control stage 0.2 m/s ~ 0.3m/s, and the moment of docking 0.15m/s ~ 0.2m/s, will help improve the performance of the docking operation.

5 CONCLUSION

The MRVD is a refined tracking control task and skill requirements are quite high. Based on the simulation experiment system, subjects can complete the docking operation within 100 meters under specified conditions after training. But the subjects show individual diversity in various performance indicators. The influences of the size of initial attitude biases between docking time and fuel wastage are not obvious, but it affects the docking speed in accurate control stage. The control stability of X-axis speed with other operating performances of the MRVD are closely related. According to normal distribution theoretical calculations, the control range of X-axis speed in different stage of docking has a suitable range.

ACKNOWLEDGEMENTS

This study is supported by The National Basic Research Program of China (No.2011CB711000). The authors would like to acknowledge Huang W.F and Jiang G.H gave us some suggestions during the course of experiments.

REFERENCES

Zhou J.P.2011. Rendezvous and Docking Technology of Manned Space Flight. *Manned Spance* 2:1-7. (in Chinese)

Lin L.X. 2007. Development of Space Rendezvous and Docking Technology in Past 40 Years. *Spacecraft Engineering* 16(4):70-77 .(in Chinese)

Zimpfer D, Kachmar P, Tuohy S. 2005. Autonomous Rendezvous, Capture and In -Space Assembly: Past, Present and Future. AIAA 2005 -2523

Wigbert F. 2003. Automated rendezvous and docking of spacecraft [Dissertation]. *Cambridge University*, Cambridge, UK

Kelsey J M, Byrne J, Cosgrove M, et al. 2006. Vision-based relative pose estimation for autonomous rendezvous and docking. *Proceedings of the 2006 IEEE Aerospace Conference*: 1-20.

Zhou Q.X, Qu Z.S,and Wang C.H, et al. 2006. Study on performance of integration control by man and machine in stage of final approaching for spaceship rendezvous and docking. *Journal of System Simulation* 18(5): 1379-1383. (in Chinese)

Wu H.X, Hu H.X, Xie Y.C, et al. 2003. Several questions on autonomous rendezvous docking. *Journal of Astronautics* 2(24):132-137. (in Chinese)

Zhang Y.J, Xu Y.Z. and Li Z.Z. et al. 2008. Influence of Monitoring Method and Control Complexity on Operator Performance in Manually Controlled Spacecraft Rendezvous and Docking. *Tsinghua Science and Technology* 13(5):619-624.

Hu H.X, Xie Y.C. 2006. Control method research of space manually controlled rendezvous and docking. *Chinese Space Science and Technology* 5: 10-16.(in Chinese)

Jiang T, Wang C.H, and Tian Z.Q. 2011. Study on Synthetic Evaluation of Human Performance in Manually Controlled Spacecraft Rendezvous and Docking Tasks. *14th International Conference on Human-Computer Ineraction* LNCS 6777:387–393.

Yang J, Jiang G.H, CHAO J.G.2010. A Cross Drone Image:Based Manual Control Rendezvous and Docking Method. *Journal of Astronautics* 31(5):1398-1404 .(in Chinese)

Jiang Z.C, Zhou J.P ,Wang Y.F, et al. 2007. Manual Control Rendezvous and Docking Simulation Based on Cross Drone. *Journal of National University of Defense Technology* 5(29):100-103 (in Chinese)

CHAPTER 53

Ergonomic Work Analysis Applied to Maintenance Activities at a Pilot Plant of Oil and Gas Industry

ZAMBERLAN, M.C.; GUIMARÃES, C.P.; CID,G.L.; PARANHOS, A.G.; OLIVEIRA,J.L.

Instituto Nacional de Tecnologia - INT/ MCT
cristina.zamberlan@int.gov.br

ABSTRACT

The aim of this paper is to present an ergonomic study applied to maintenance activities at a pilot plant on oil and gas industry. The Ergonomic Work Analysis was the main methodological approach. The study concerns the analysis of different maintenance activities previous selected by workers and managers. The Ergonomic Work Analysis methodology involved three stages: reference situations diagnosis and recommendations; establishing an ergonomic design concept; evaluation of the new work condition and/or improvement of the maintenance tools. The reference situation diagnosis were conducted using a questionnaire, previous tested by the ergonomic group and also video capture. The data analysis was conducted on the Ergonomics Laboratory of INT. The results of the study showed that even with plant unity specificity, some ergonomic problems were common, such as lack of space at the unity to work; repetitive, static and forceful activities in awkward postures as squatting, trunk and neck forward bending, shoulder flexion and abduction over 90 degrees, wrist flexion and extension combined with abduction, manual material handling activities, as lifting, carrying, pushing and pulling heavy loads. The discomfort analysis pointed out the lower back and shoulders areas with higher pain and frequency scores. Some environment conditions were also pointed out as common problems in special reference to heat, lighting and vibration. Some safety and risk problems as using ladders with high inclination or navy ladders to achieve the work location, carrying loads and tools can predispose to slipping and falls. The recommendations based on the diagnosis were presented to managers at periodic meetings where the ergonomic researchers, designers and managers have been discussing issues to define and establish ergonomic design concepts to be applied and improve the maintenance work conditions, that can involve proposed part of the unity plant redesign or /and new ergonomic tools design.

Keywords: ergonomic work analysis, maintenance activities, pilot plant

1 CONTEXT

The aim of this paper is to present an ergonomic study applied to maintenance activities at a pilot plant on oil and gas industry. The Ergonomic Work Analysis was the main methodological approach. The Ergonomic Work Analysis (EWA) is a methodology which, as the result of studying behaviors in the work situation, provides an understanding of how the operator builds the problem, indicates any obstacles in the path of this activity, and enables the obstacles to be removed through ergonomic action (Wisner, 1995). The central point of this methodology is the analysis of real work activities performed by workers. Based on that, some ergonomic risk interactions have to be included in the study:

- Risk aspects inherent to the worker - involve physical, psychological and non-work-related activities that may present unique risk factors;
- Risk aspects inherent to the job - concern work procedures, equipment, workstation design that may introduce risk factors;
- Risk aspects inherent to the environment - concern physical and psychosocial "climate" that may introduce risk factors.

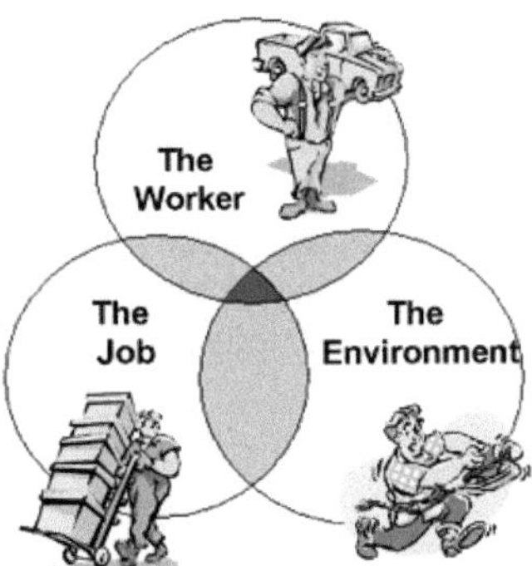

Figure 1 - Ergonomic risk aspects interaction. From Introduction to Ergonomics. How to identify, control, and reduce musculoskeletal disorders in your workplace! OR-OSHA. 101. Revised 01/97 by the Public Education Section Department of Business and Consumer Business, Oregon OSHA, (www.orosha.org/pdf/workshops/201w.pdf) 1997.

At a pilot plant of oil and gas industry, the maintenance activities can be characterized as a work involving varied tasks. This kind of work involving varied tasks is distinguished by a wide variety of tasks that are part of the operator's expertise and know-how and also involves a set of tasks that each of which underlies a large number of operations that are not always organized in a specific work cycle. These tasks can be performed at locations that vary considerably from one to another. For example, a mechanic is required to perform different tasks, such as maintaining and repairing equipments throughout different units of production at the pilot plant.

Despite the great variability in work places and job tasks, the human body serves as a constant. In order to make design decisions take into account human factors, it is important to understand how the body responds to and moves about in its environment. Work in the oil industry involves diverse activities including work in rigs, workshops, and offices. Heat stress as a potential safety and health hazard has been recognized in the literature and guidelines for exposure have been formulated (Hancock and Vasmatzidis, 1998).

2 METHODS

The study concerns the analysis of different maintenance activities previous selected by workers and managers. The maintenance activities were classified as insulation, instrumentation, electric and mechanics.

The Ergonomic Work Analysis methodology involved three stages: reference situation diagnosis and recommendations; establishing an ergonomic design concept; evaluation of the new work condition and/or improvement of the maintenance tools.

The reference situation diagnosis were conducted using a questionnaire, previous tested by the ergonomic group and also video capture (figure 2). The study of body areas discomfort was made using the discomfort / pain diagram (Corlett and Bishop, 1976). The data analysis was conducted at Ergonomic Laboratory of National Institute of Technology.

The ergonomic research group went to pilot plant three times a week for six months at scheduled time. First the supervisors were interviewed and after that the group followed the maintenance group to the unit of the plant where the maintenance activities happen. The maintenance groups were also interviewed and the activities were videotaped, the ergonomic group made all activities registration and took measurements of the workplace, tools and equipments.

Figure 2 - Maintenance activity registration

3 RESULTS

When comparing body discomfort analysis between the maintenance work activities and others work activities in different areas of the pilot plant, classified as transact area, assistance area and laboratory area, the low back pain, neck and upper arm shown the highest scores (figure 3).

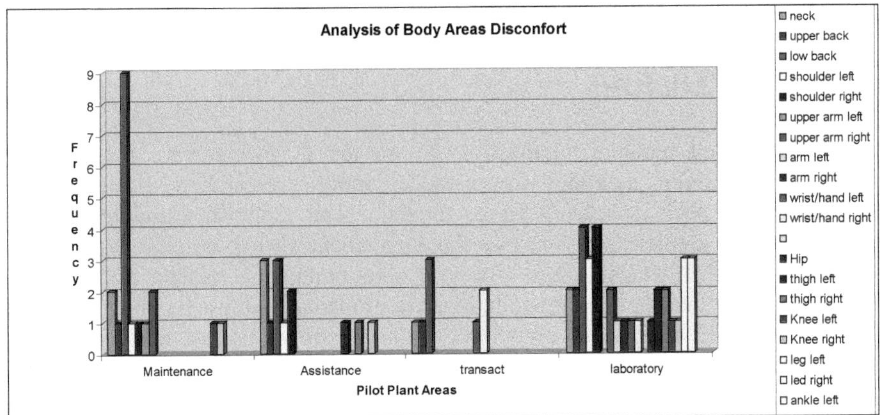

Figure 3 - Body Area Disconfort comparison between different areas at pilot plant

The results of the study also showed that even with plant unity specificity some ergonomic problems were common at maintenance activities, such as lack of space at the unity to work; repetitive, static and forceful activities in awkward postures as squatting, trunk and neck forward bending, shoulder flexion and abduction over 90 degrees, wrist flexion and extension combined with abduction, manual material handling activities as lifting, carrying, pushing and pulling heavy loads. Some environment conditions were also pointed out as common problems in special reference to heat, lighting and vibration. (figure 4).

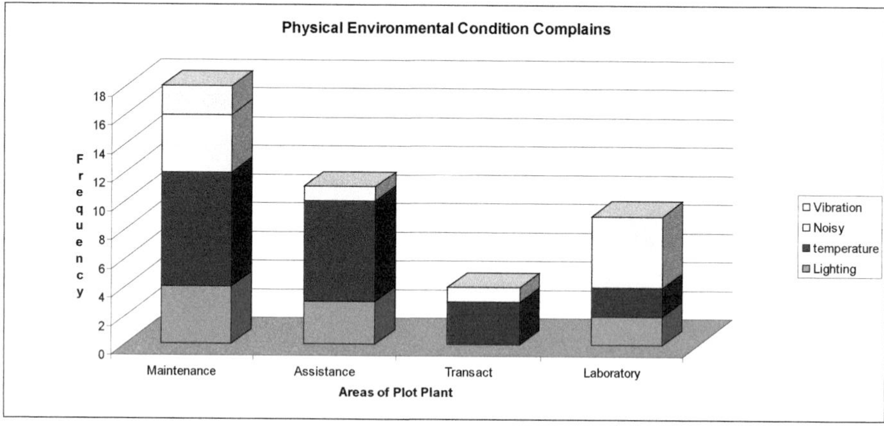

Figure 4 - Physical Environmental Condition Complains at Different areas of Pilot Plant

Some safety and risk problems were also frequent as using ladders with high inclination or navy ladders to achieve the work location with carrying loads and tools that can predispose to slipping and falls.

4 CONCLUSIONS

The results and recommendations based on the diagnosis were presented to managers at periodic meetings where the ergonomic researchers, designers and managers have been discussing issues to define and establish ergonomic design concepts to be applied and improve the maintenance work conditions, that can involve unity plant redesign and / or new ergonomic tools design. Some of the strategies to implant the recommendations were: managers must be knowledgeable and aware of the benefits of ergonomics, and the prevention of injuries through ergonomics implementation, remodeling tools and improvements to maintenance workstations in order to minimize the manual material handling activities; employees need to be trained systematically in ergonomics in order to improve ergonomic conditions and safety; the work and workplace design should be carried out using ergonomic guidelines; acts and recommendations should consider user population in special emphasis to maintenance workers. The environment must also be given adequate consideration, mainly the temperature (heat) and vibration of some equipments that can remodeled or changed to new ergonomics ones.

ACKNOWLEDGMENTS

The authors would like to acknowledge FINEP and CNPq – Brazilian government sponsors agencies.

REFERENCES

Corlett, E. N., Bishop, R.P. 1976 .A technique for assessing postural discomfort. Ergonomics, v. 19, p.175-182.

Hancock, P. A., & Vasmatzidis, I. 1998. Human occupational and performance limits under stress: The thermal environment as a prototype example. Ergonomics, 41, 1169–1191.

OR-OSHA 101 1997 How to identify, control, and reduce musculoskeletal disorders in your workplace! 101. Revised 01/97 by the Public Education Section Department of Business and Consumer Business, Oregon OSHA, www.orosha.org/pdf/workshops/201w.pdf

Shikdar, A. A and Sawaqed N. M 2004. Ergonomics, and occupational health and safety in the oil industry: a managers' response. Computers & Industrial Engineering 47 (2004) 223–232

Shikdar, A. A 2004 Identification of Ergonomic Issues That Affect Workers in Oilrigs in Desert Environments International Journal of Occupational Safety and Ergonomics (JOSE) 2004, Vol. 10, No. 2, 169–177

CHAPTER 54

Fuzzy Bayesian Based Classification of Call Center Agent Activity

Shu-Chiang Lin,[1] Mark R. Lehto[2] and Sugani Leman[3]

[1]Department of Industrial Management,
National Taiwan University of Science and Technology, Taiwan, R.O.C
[2]School of Industrial Engineering, Purdue University, West Lafayette, IN, USA
[3]FedEx Corporation, Nashville, TN, USA
slin@mail.ntust.edu.tw, lehto@purdue.edu, Sugani.leman@fedex.com

ABSTRACT

This paper presents an application of a Fuzzy Bayesian based methodology for task analysis (BMTA) to identify the particular tasks performed by experienced call center agents when responding to calls. A total of 126 customer calls were analyzed pertaining to 55 printer models and 70 software and hardware issues. To develop the model, each exchange of information between customer and agent was first classified into one of 72 task categories by a human expert. This data was then used to train the model to predict which subtask was being performed for a particular exchange of information. Model testing results showed an overall hit rate was above 50%, false alarm rate was less than 1%, and sensitivity value d' was above 2.5, supporting the conclusion that Bayesian methods can serve as a practical tool for identifying some of the tasks performed by a call center agent based on what is said during the conversation with the customer. The findings support the conclusion that Bayesian methods are capable of learning how to classify tasks performed in naturalistic settings based on the results of classifications made by a human expert performing task analysis. This result is important, because identifying the subtasks performed within a job is a difficult, time-consuming process. The most basic application of this Bayesian model would also be used to do off-line analysis of human-computer interaction historical data such as email files, blogs, texting, e-learning, transcribed calls and voice mails. The flexibility and mobility of BMTA approach goes in line with the current explosive growing number of mobile computing applications and speech recognition devices.

Keywords: Bayesian inference, knowledge acquisition, machine learning, task analysis

1. INTRODUCTION

Task analysis is one of the oldest and most widely used techniques for improving jobs, methods, interfaces, or human machine systems in general. Task analysis involves the systematic description of a user's task in terms of subtasks or task elements that are often further analyzed or evaluated using criteria, such as time requirements, mental or physical demands, or information flow. This basic idea dates back to the early 1900s when Taylor and Gilbreth introduced techniques to identify and measure the elements of manual tasks. In subsequent years, many task analysis techniques have been developed and applied, including methods of CTA (cognitive task analysis) for analyzing activities such as information processing, multimedia learning, supervisory control, decision making, and problem solving and planning (Lehto et al, 1991). Although many task analysis techniques have been developed, identifying the subtasks performed within a job remains a difficult, time-consuming, and expensive process (Lin, 2006). Several task modeling methods have been developed to help analysts perform task analysis more efficiently and effectively.

Specific examples include the MHP, GOMS, NGOMS, TAG, SHAPA, TKS, GTA, TOOD, TIC, and other task modeling languages such as ACT-R, AMBR, CORE, CTT, Diane+, EPCI, HCIPA, IMPRINT, ISOLDE, KLM, LOTOS, MAD, SOAR, USN. Much less sophisticated task descriptions can, however, be very useful. At the most basic level, the objective of the initial task analysis may be to develop taxonomy of commonly performed tasks or subtasks, the performance of which is then tracked over time to guide staffing decisions, measure productivity, identify improvement opportunities, collaborative cell production system, developed e-learning system and other purposes (Tan et al., 2008, Yusoff and Yin, 2010). This is often done periodically by an analyst who observes how often and well certain subtasks are performed. A variety of indirect measurement methods can also be used, including audio, video, photographic methods, voice transcripts, computer keystrokes, mouse-clicks, or other computer interfaced instrumentation.

Collecting such data over time, will eventually result in a database containing historical information about the performed tasks (Lehto and Buck, 2008). This information may be the same data the human analyst uses, and often is in the form of transcribed speech (verbalizations of what is being done). This leads to the interesting question of whether it might be possible to develop automated tools capable of learning from previously classified cases, to accurately identify what is being done by people in naturalistic work settings.

2. A BAYESIAN METHODOLOGY FOR TASK ANALYSIS

Previous work has shown that machine learning methods are capable of learning by example from human performance data to duplicate human performance on classification tasks (Lehto et al., 1992; Jagielska et al., 1999; Bishop, 2007). Bayesian methods are another commonly applied machine learning method that might be used to learn from previously classified cases, to accurately identify what is being done by people in naturalistic work settings based on information in verbal transcripts and other information recorded at the time the task was performed. From a pragmatic perspective, simple Bayesian models are attractive, given research showing that they perform well compared to competing approaches, such as neural networks, that require much more computational effort (Lehto and Sorock, 1996). Several studies have shown that simple Bayesian models can learn to perform similar classification tasks with accuracy comparable to that of the human (Lehto et al., 2009; Noorinaeini and Lehto, 2007; Choe et al., 2011).

The implementation of a method capable of learning from historical data how to accurately identify what is being done by people in naturalistic work settings would require, 1) a knowledge base containing the results of previous task analyses, including both the assigned tasks, and information relevant to the particular assignments, 2) a machine learning algorithm to learn the basis for past assignments, and 3) a predictive model used to automatically classify new tasks using task related information. These elements and how they are related are shown in Figure 1.

A database containing the results of previous analysis is perhaps the most critical requirement for the implementation of this approach. The contents of database would include classified cases consisting of subtasks tagged with evidence defined by the analysts. As indicated in the figure, if such data does not yet exist, it will be necessary for the analyst to go through a process of identifying subtask categories and tagging them with evidence. If the evidence is text based, it will consist of keywords or phrases parsed from transcripts or other sources. For example, the evidence associated with a particular subtask might be a set of keywords summarizing what the worker said at the time they were doing the subtask. In the case of a call center, this data would be what the agent and customer said during a particular part of the call, such as trouble-shooting, that corresponds to a subtask of interest. The subtask might also be tagged with other evidence, such as keystrokes or mouse clicks.

The machine learning algorithm is the second important element. The function of the machine learning algorithm is to learn the relationship between the evidence and subtasks, and then use this knowledge to make predictions. As mentioned earlier, many different machine algorithms could be implemented, but simple Bayesian models have the advantage of both simplicity and proven performance. Among such models, Fuzzy Bayes is a particularly simple model that bases its prediction on the single piece of evidence that has the strongest statistical relationship to a predicted category. Despite its simplicity, several studies have shown that this model performs surprisingly well on text classification problems, often giving results very comparable to human coders. Other simple Bayesian models such as Naïve Bayes have also been shown to perform well, in some cases showing modest improvement over Fuzzy Bayes (Lehto et al, 2009).

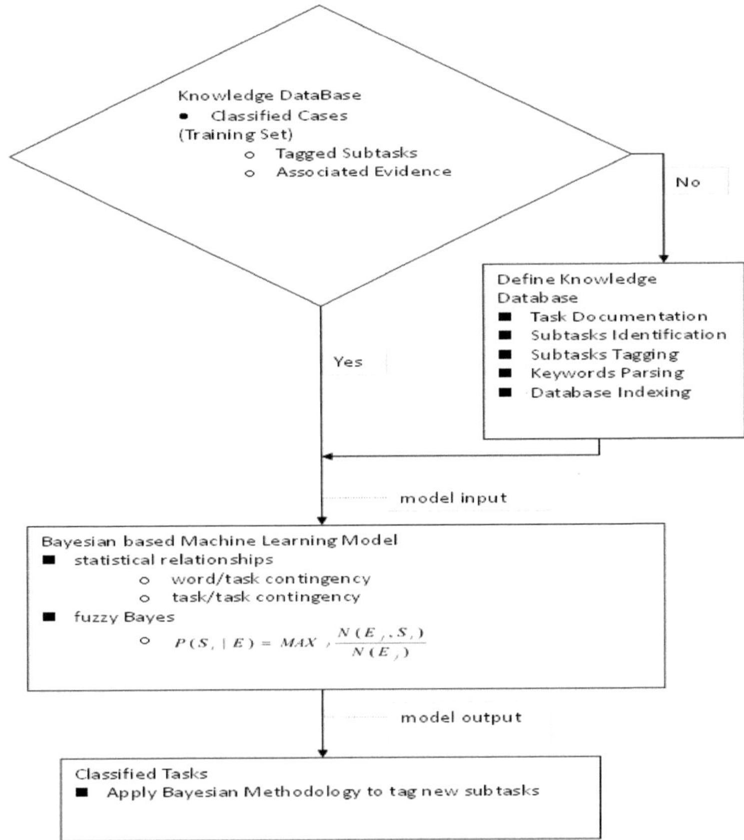

Figure 1 A Bayesian Methodology for Task Analysis (BMTA)

For this particular application, the Fuzzy model describes the relationship between the probability a particular subtask is being performed and the tagged evidence using the expression:

$$P(S_i \mid E) = MAX_j \frac{N(E_j, S_i)}{N(E_j)} \tag{1}$$

where $P(S_i|E)$ is the posterior probability that subtask S_i is being performed given the evidence E is present, $N(E_j, S_i)$ is the number of times subtask S_i is historically tagged with evidence E_j, and $N(E_j)$ is the number of times evidence E_i is historically present for all subtasks. The machine learning part of the process corresponds to determining the values of $N(E_j,S_i)$ and $N(E_j)$ based on the historical data in the database. The Fuzzy Bayes prediction for each subtask is then determined using the MAX operator as shown in equation (1). The use of the MAX operator results in basing the prediction on the strongest single piece of evidence, and is appropriate assuming the different sources of evidence regarding a subtask are highly correlated (Lehto et al., 2009).

3. MODEL DEVELOPMENT

3.1. Call Center Case Study

To test our proposed Bayesian methodology for task analysis (BMTA), we conducted a case study at a large call center located in the Western United States that was operated for Hewlett Packard Company. The study involved 24 experienced call center agents, responding to a total of 126 customer calls pertaining to 55 printer models and 70 software and hardware issues. All of the agents were 25 years of age or older, and had a high school or above education with background in software and hardware. Each agent had received 4 to 6 weeks of full-time training and passed an in-house exam to be certified as a knowledge agent. Each agent also had at least 4 months work experience involving at least 1500 troubleshooting calls. In this particular call center, each agent typically received between 15 to 30 calls a day, and. the duration of a call varied from 10 minutes to 90 minutes depending on the complexity of the problems.

A field study was conducted in which agents were observed over two different weekly intervals. During the first week of on-site observation, the activity of 8 agents taking a total of 45 calls was videotaped, resulting in approximately 22 hours of audio-video data. During the second week, another 16 agents taking a total of 105 calls were observed, resulting in an additional 40 hours of audio-video data. This resulted in a total of 62 hours of audio-video data documenting agent activity for 150 calls. After collecting the data, the audio-video data was analyzed to document agent activity during a call. This process involved determining what subtask was being done at any moment in time, and then tagging the particular subtask with the words spoken while it was being done. The first step in the process was to transcribe the audio portion of the data into text. The next step was to identify the subtasks performed by the agents within the calls. A taxonomy consisting of 72 agent subtask categories was then developed. Several of which are shown in Table 1. The following step in the process was to identify which subtasks were performed in each call, and then tag each of the assigned subtasks with the words spoken by the agent and customer at the time the particular subtask was being performed.

Table 1 Examples of agent subtasks

Subtask category	Subtask Description
252	Verify model/serial number of product
257	Verify whether the product is supported by the agent, or is qualified for services
321	Provide service options for unqualified product
333	Ask to describe specific symptoms of the problems
351	Instruct to print configuration pages
374	Instruct to check if fusing units are functioning normally
375	Instruct to check if transfer kit is functioning normally
383	Instruct to check slots, ports, cables, cards setup or are functioning normally
411	Send out toner cartridge replacement
412	Send out drum kit replacement
413	Send out transfer kit replacement
521	Provide reference number for future contact

3.2. Model Development

After completing the tagging process, the next step was to develop a simple Fuzzy Bayes model for estimating the probability the human analyst would have assigned a particular subtask, based on the statistical relationship between terms in each exchange of information and the manually assigned subtasks. To illustrate how the Fuzzy model approach as described in section 2 might be applied, consider a troubleshooting process where an agent performs subtask Ai, and suppose we consider the words used by the agent and the customer as the sources of evidence, and then describe what the agent said by a word vector

$WA_i = (WA_{i1}, WA_{i2}, \ldots, WA_{iq})$, of length q,

where $WA_{i1}, WA_{i2}, \ldots, WA_{iq}$ are the q words in the dialog.

Similarly, we describe what the customer said in responds to the agent's question by a second word vector

$WC_i = (WC_{i1}, WC_{i2}, \ldots, WC_{iq})$, of length q,

where $WC_{i1}, WC_{i2}, \ldots, WC_{iq}$ are the q words in the dialog.

We also consider that A_i is potentially relevant to WA_i, WC_{i-1}, and WC_i for i greater than 1. Following this approach we can then derive equation (2) from equation (1) to calculate the posterior probability of subtask A_i:

$$P(A_i|WA_i,WC_i,WC_{i-1})=MAX[N(WA_i , A_i)/N(WA_i), N(WC_i, A_i)/N(WC_i), N(WC_{i-1}, A_i)/N(WC_{i-1})]$$
$$=MAX [MAX_j [N(WA_{ij}, A_i)/N(WA_{ij})], MAX_j [N(WC_{ij}, A_i)/N(WC_{ij})], MAX_j [N(WC_{(i-1)j}, A_i)/N(WC_{(i-1)j})]] \quad \text{for } j=1,2,\ldots q \qquad (2)$$

This was done using the Textminer program developed by one of the authors (ML), which is implemented in Visual Basic and provides an interface for analyzing datasets readable by Microsoft Access.

After loading the tagged data set of subtasks into the Textminer program, several preprocessing steps were taken to clean up the dataset. The first step in this process was to extract the individual words used in each exchanges of information. The extracted words were then cleaned up by spell-checking, and removing punctuation marks and non-alphabetical characters. A morph table was then created to transform certain words and word sequences into morphs listed on a morph list. During this process, reference was made to the service manuals to help decide whether particular words should be morphed. Examples include terms like “domain administrator”, “error log”, “event log”, “fuser kit”, “image drum”, “jet admin”, “jet direct card”, “transfer drum” which were merged to create terms such as “domain’administrator” “error’log”, “event’log”, “fuser’kit”, “image’drum”, “jet’admin”, “jet’direct’card”, “transfer’drum” respectively. By substituting the original words with morphs of the same meaning it was possible to reduce the file size, shorten the computation process, increase the number of times a meaningful word is found in the narratives, and possibly allow stronger predictions based on these words.

After assigning the morphs and dropping rarely occurring terms, single word predictors were used to generate two-word combinations, three-word combinations, and four-word combinations as possible predictors by the Textminer program. The next step in the process was to split the data into a training set and prediction set. Two thirds of the information exchanges between agent and customer were randomly assigned to the training set, and the remainder to the prediction set. This process of learning and testing was repeated 10 times using different randomly assigned training and prediction sets to allow for cross validation of the accuracy of the results.

4. RESULT AND DISCUSSION

The initial analysis of model performance focused on examining model hit rate as a function of cross validation run. The first step in this process was to compare model performance for the 10 different randomly selected training and prediction sets, used in the cross validation runs. Table 2 shows the corresponding number of hits, when performance is averaged over all categories, for each of the 10 cross validation runs. The predicted results are further broken down into those obtained for the training and testing sets, as indicated in columns three and four of the table.

Table 2 Number of hits in prediction set for each of the 10 randomized cross-validation runs

Cross-Validation Runs	All data (5183 narratives)	Training set (3472 narratives)	Testing set (1711 narratives)
Run1	2931	2212	719
Run2	2784	2085	699
Run3	2791	2102	689
Run4	2809	2090	719
Run5	2798	2039	758
Run6	2830	2130	701
Run7	2859	2159	701
Run8	2758	2068	689
Run9	2891	2170	720
Run10	2867	2146	722
Mean	2831.8	2120.2	711.8
Standard Deviation	54.15	52.60	20.68
Average Hit Rate	0.5464	0.6106	0.4160

The results in Table 2 show some variation in the number of hits among the different runs, but overall the results are remarkably consistent. ANOVA analysis was conducted to test whether the differences between the ten runs were statistically significant. The results of this analysis are given in Table 3, and confirm that the differences among the ten runs were not statistically significant ($p = 0.24$, $F = 1.28$). The latter statistical result reveals that the overall model predictions were quite robust for our dataset, and supports the conclusion that use of a single randomly generated training set would have given similar results to that observed here for the 10-fold cross validation.

Table 3 ANOVA table to test effect of subtask and cross validation run on the hit rate

Source of Variation	Sum of Squares	DF	Mean Squares	F	P-value	F crit- α=0.05
Subtasks (S)	60.3043	70	0.86149	131.0104	0.000000	1.3161
Runs (R)	0.0759	9	0.008435	1.2827	0.2428	1.8947
Interaction (S*R)	4.1427	630	0.006576			
Total	64.5229	709				

To further investigate the model performance, we examined the false alarm rate and sensitivity value d'. This was done for both overall performance of the model, and for the individual subtasks. The first step in the process was to calculate overall false alarm rates and hit rates. The overall hit rate was above 50%. Also, the overall false alarm rate was estimated be 0.65%. These two numbers were then used to estimate an overall sensitivity value d' of 2.65 by adding the standard Z scores respectively associated with the false alarm and hit rates.

5. CONCLUSIONS

The results of this study supported our hypothesis that a simple Bayesian model would be able to learn and make robust predictions of subtask categories based on the telephone conversation between the customers and the human agents. Somewhat surprisingly, the most difficult part of the project was obtaining the initial set of classified tasks. The positive side of this issue is that, assuming such data is available, the approach is quite easily implemented.

According to Bureau of Labor Statistics, customer service representatives held about 2.3 million jobs in 2008 in US, ranking among the largest occupations. Most customer service representatives do their work by telephone in call centers. In applications for call center business alone, this method could practically and repeatedly be applied in many different ways for a wide variety of purposes, such as work measurement, activity analysis, operator training and feedback, performance assessment, methods improvement, and enhanced service quality. In the most basic "text mining" type of application, the Bayesian model would be used to do off-line analysis of historical data such as email files, blogs, texting, e-learning, or transcribed calls or voice mails. For example, after training the model using a set of earlier transcribed calls, the model might be used to systematically track over time what percentage of operator activity is spent on particular types of tasks, such as trouble-shooting, for particular products. Automatically tracking such information would be extremely beneficial.

In the current application, the Bayesian model predicted the subtask based on particular words listed in a transcription of the call. This research revealed that BMTA serves as a practical and easily applied tool for identifying some of the tasks performed by a call center agent based on what is said during the conversation with the customer. BMTA could immediately be implemented by the task analysis with a laptop. The flexibility and mobility of this task analysis approach goes in line with the current explosive growing number of mobile computing application and automated voice recognition of keywords that have achieved some degrees of success (Helmi and Helmi, 2008; Shneiderman et al., 2009; Rahbar and Broumandnia, 2010). The recognition technology is already in use in iPhone applications such as Google Voice Search that allows users to speak search terms and select among alternatives or correct partial mistakes with onscreen keyboard instead of typing all the terms.

An important caveat is that the results presented here should be viewed not as an end product in themselves, but rather as a starting point and resource that will encourage future studies of the application of Bayesian and other machine learning methods to the very important topic of task analysis. The abundance of results obtained in this work, suggest many more advanced forms of analysis, as well as refined data collection methods, and statistical approaches. Some of the many interesting topics for further research include:

1. examining whether model performance is improved when the subtask categories are merged and classified into bigger clusters of coarse categories that contain similar meanings or job contents;
2. further comparison of the prediction accuracy of different fuzzy Bayes methods including a systematic evaluation of how much performance is improved by including sequences and word combinations as predictors;
3. further investigation of misclassified categories to explore the reason for discrepancies between the human analyst and model predictions;

A second major issue, was that the diversified and sparse nature of the data used in this study created numerous difficulties when assigning subtask categories that adequately matched each dialog between the customer and the agent. Many people contributed a significant amount of time and effort at different stages of this process. Based this experience we can make several suggestions for increasing the efficiency and effectiveness of data collection and analysis:

1. Focus the data collection as much as possible - A quick way to reduce the data collection time while obtaining quality data is to focus the data collection on a specific topic of interest. As mentioned earlier, the 104 phone calls we transcribed in our study addressed over 55 printer models, 75 companies, 110 customers, 35 printer error codes, and over 70 problems or situations ranging from software, operating systems, drivers, networking, print quality, and hardware. Narrowing down the data collection effort, say to a specific print quality issue or a specific error code, would have greatly simplified the process of assigning tasks and assigning morphs.
2. Focus on conversations that apply to the primary problem of interest, and consider removing sections that appear very unusual or irrelevant to the task. Doing so, can make human classification easier and the model's learning and predicting processes more efficient.
3. Introduce statistical tests when necessary: With more clear targets and fewer variables with fewer levels, some advanced analysis can be easily conducted based on stratified data such as subtask categories, agent groups, error codes, etc. Statistical tests such as an Analysis of Variance (ANOVA) can be used to refine the measure of the model performance based on these stratified data.

REFERENCES

Bishop, C. M. (2007). Pattern Recognition and Machine Learning, Springer.

Choe, P., Lehto M. R., Shin G.C., and Choi, K. Y. (2011, accepted). Semi-Automated Identification and Classification of Customer Complaints. *Human Factors and Ergonomics in Manufacturing & Service Industries*

Helmi, N. and Helmi, B. H., (2008). Speech Recognition with Fuzzy Neural Network for Discrete Words. *Fourth International Conference on Natural Computation*. Vol. 7, pp. 265 – 269.

Jagielska, I., Matthews, C., and Whitford, T. (1999). Investigation into the application of neural networks, fuzzy logic, genetic algorithms, and rough sets to automated knowledge acquisition for classification problem. *Neurocomputing*, 24, 37-54.

Lehto, M.R., Boose, J., Sharit, J., and Salvendy, G. (1992). Knowledge Acquisition. In Salvendy, G. (Eds.), *Handbook of Industrial Engineering* (2nd Ed.), Ch. 58 (pp.1495-1545). New York: John Wiley & Sons.

Lehto, M.R. and Buck, J.R. (2008). *An Introduction to Human Factors and Ergonomics for Engineers*, Lawrence Erlbaum Associates: Taylor and Francis Group, NY, 838 pages..

Lehto, M.R., Sharit, J., and Salvendy, G. (1991). The application of cognitive simulation techniques to work measurement and methods analysis of production control tasks. *International Journal of Production Research,* 29 (8), 1565-1586.

Lehto, M.R. and Sorock, G.S. (1996). Machine learning of motor vehicle accident categories from narrative data. *Meth. Inform. Med.*, 35 (4/5), 309-316.

Lehto, M. R., Wellman, H.M., and Corns, H. (2009) Bayesian Methods; A Useful Tool For Classifying Injury Narratives Into Cause Groups, *Injury Prevention*, Vol. 15, No. 4, pp. 259-265.

Lin, S. (2006). A Fuzzy Bayesian Model Based Semi-Automated Task Analysis. Unpublished Ph.D. Dissertation. Purdue University, West Lafayette, IN.

Noorinaeini, A. and Lehto, M.R. (2007). Hybrid Singular Value Decomposition; a Model of Human Text Classification, *Human Interface, Part I, HCII 2007*, pp. 517 – 525. M.J. Smith, G. Salvendy (Eds.).

Rahbar, K., and Broumandnia, A., (2010). Independent-speaker isolated word speech recognition based on mean-shift framing using hybrid HMM/SVM classifier. *18th Iranian Conference on Electrical Engineering (ICEE).*156 – 161.

Shneiderman, B., Plaisant, C., Cohen, M., and Jacobs, S., (2009). *Designing the User Interface: Strategies for Effective Human Computer Interaction 5th edition.* Chapter 8 (349-359). Pearson.

Tan, J.T.C, Duan, F., Zhang, Y., and Arai, T. (2008). Extending Task Analysis in HTA to Model Man-Machine Collaboration in Cell Production. Proceedings of the 2008 IEEE International Conference on Robotics and Biomimetics. Bangkok, Thailand, 542-547.

Yusoff, N.M. and Yin, W.C. (2010). Multimedia Learning System (MMLS): Valuing the Significance of Cognitive Task Analysis Technique and User Interface Design. International Symposium in Information Technology (ITSim) 2010, 1-6.

CHAPTER 55

Testing of Chosen Personal Protective Equipment (PPE) as per Standard STN EN ISO 15025:2003

Linda Makovická Osvaldová 1, Anton Osvald 2, Peter Makovický3

University of Zilina
Zilina , Slovak republic
linda.osvaldova@fsi.uniza.sk
anton.osvald@fsi.uniza.sk
p.makovicky@kia.sk

ABSTRACT

Improving standards of Occupational Health and Safety is not only beneficial for whole society but also for an individual. Using of equipment for personal protection of health at work is increasing. This causes more demand for personal protective equipment. Increased demand could cause also decrease in quality of the personal protective equipment. This raises a question: Are we ready to accept low quality personal protective equipment at the expense of health damage? Therefore the quality of the personal protective equipment, materials used for its production became an important factor. Our article is also dealing with tests on chosen personal protective equipment - protective garments according to the standard STN EN ISO 15025:2003 and quality assessment of the chosen personal protective equipment.

Keywords: Safety and health protection, personal protective equipment, clothing-testing material

1 PERSONAL PROTECTIVE EQUIPMENT

Personal protective equipment is equipment that an employee wears, holds and uses in any way at work, including its accessories, when it is aimed at personal safety and health of an employee. A personal protective equipment can also be other personal protective equipment , when they were defined by the employer according to § 4 and are in agreement with the government regulation No 395/2006 about minimum requirements for the provision and the use of personal protective equipment (Government regulation No. 395/2006, Government regulation No 124/2006)

Problematic regarding testing protective clothing (not only for workers but also in rescue part –fireman's) was and still is very actual as you can see in other article focus on this problematic for example: Fire-fighter Thermal Exposure Workshop: Protective Clothing, Tactics, and Fire Service PPE Training Procedures (JASON, N, 1997, LAWSON, R, 1997) or article Heat transmission of protective clothing is also focus on evaluated characteristic of protective clothing. (HIRSCHMAN, J. E, 1992)

1.2 TESTING MATERIAL

As testing material we chose personal protective equipment and protective clothing which is freely available on the market. For testing we used material from which the personal protective equipment is produced – protective clothing. The chosen textile fabrics were of various material compositions, different value of parameters for basic weight of the textile fabrics and different final coating of the textile fabric. Concrete types of the textile fabrics:

A. Textile fabric green 1 TERA

Material composition: cotton 35% + polyester 65 %

Basic weight of the textile fabric: 241,93 g / m^2

Final coating of the textile fabric: without final coating

B. Textile fabric grey 2 TINA

Material composition: cotton 100 %

Basic weight of the textile fabric: 257,03 g / m^2

Final coating of the textile fabric: without final coating

C. Textile fabric blue 3 TONA

Material composition: cotton 100 %

Basic weight of the textile fabric: 267,54 g / m^2

Final coating of the textile fabric: without final coating

D. Textile fabric blue 4 PROBAN

Material composition: 100 % cotton with fireproof surface coating PROBAN

Basic weight of the textile fabric: 296,49 g / m^2

Final coating of the textile fabric: fireproof coating PROBAN

E. Female boiler suit jacket of blue colour, material composition cotton 100%

Basic weight of the textile fabric: 240,00 g / m^2

Final coating of the textile fabric: without final coating

F. Male boiler suit jacket for welders in grey-red colour, material composition cotton 100% with fireproof surface coating PROBAN.

Basic weight of the textile fabric: 390,00 g / m^2

Final coating of the textile fabric: fireproof surface coating PROBAN.

Materials from which the protective clothing is made and ready made protective clothing used in the experiment were bought in retail chains and we exposed them to testing according to specification of the chosen technical standards. As the main testing method we chose the method for testing of selected protective clothing in accordance with the STN EN ISO 15025 Method of testing for limited flame spread.

2. EXPERIMENT

The testing of personal protective equipment – protective clothing took place in the accredited laboratory of flammability evaluation, accredited according to the STN EN ISO / IEC 17025. According to the procedure STN EN ISO 15025 it is needed that the tested samples used in the testing based on this method have the following parameters (STN EN ISO 15025: 2003). The parameters of the tested samples: 200 mm 2 mm x 160 mm 2 mm. Number of tested samples: 6 pieces from each type of textile fabric If not determined otherwise the tested samples must be conditioned at least 24 hours in an environment with temperature 20 °C 2 °C and relative air humidity 65 % 5 %. If it is not possible to test them straight after conditioning, the tested samples are put after conditioning into a sealed-up container. Each sample must be tested within 2 minutes from taking out of the conditioned environment or from the sealed-up container.

2.1 TESTING EQUIPMENT

The testing equipment for testing of personal protective equipment – protective clothing according to the method of testing for limited flame spread in accordance with the STN EN ISO 15025 is a vertical equipment for flammability measuring SDL M 233 B Shirley Flammability Tester in which the tested samples of given parameters are exposed to the effect of a defined small flame. This equipment is situated in an enclosed fume chamber with sliding front wall. (STN EN ISO 15025: 2003)

2.2 EXPERIMENT IN ACCORDANCE WITH THE STN EN ISO 15025

Significance of the test lies in the fact, that the defined flame of given burner is let to work on the surface for 10s (test procedure A) or on the lower edge (test procedure B) of the sample textile fabric that is situated in a vertical position.

We recorded:

1. time of flame burning
2. time of smoulder

In the experiment we used both procedures methods of testing in accordance with the STN EN ISO 15025 (A, B). Through the method of testing in accordance with the STN EN ISO 15025 from all types of the tested textile fabrics (TERA, TINA, TONA, PROBAN, boiler suit jacket without final coating, boiler suit jacket with final coating) by procedure A as well as by procedure B we found out that the textile fabrics of material composition: cotton 100% with surface coating PROBAN reached minimum time levels of spontaneous flame burning. The values ranged from 0.5s to 0.8s. From the point of view of user's safety these textile fabrics are the most suitable for production of protective clothing when we take into account the spread of flame burning on the surface of the protective clothing. The measured values for the parameter time of spontaneous flame burning of the textile fabric PROBAN fulfilled conditions for minimum basic safety requirements for protective clothing that are aimed at the user's body protection by welding and other processes with comparable threat in accordance with the STN EN ISO 11611.

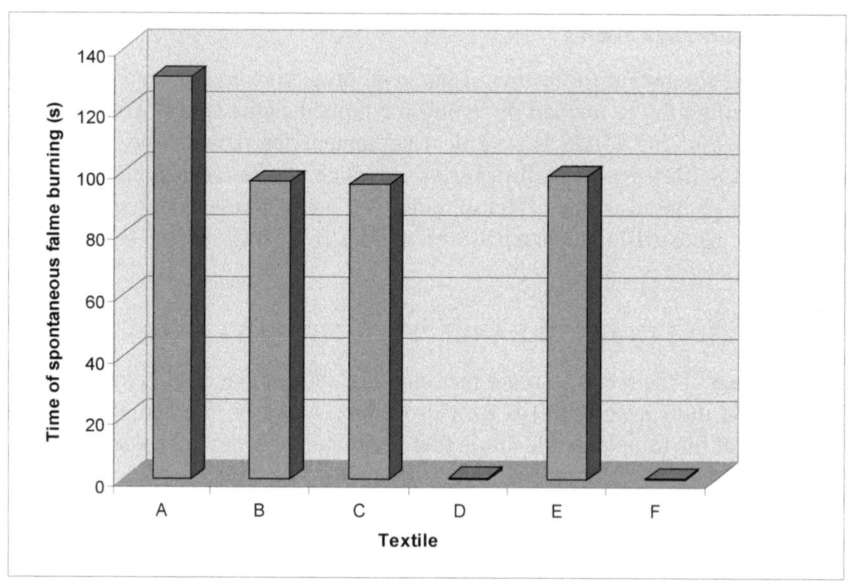

Figure 1Chart of spontaneous flame burning of the textile fabric. Procedure A

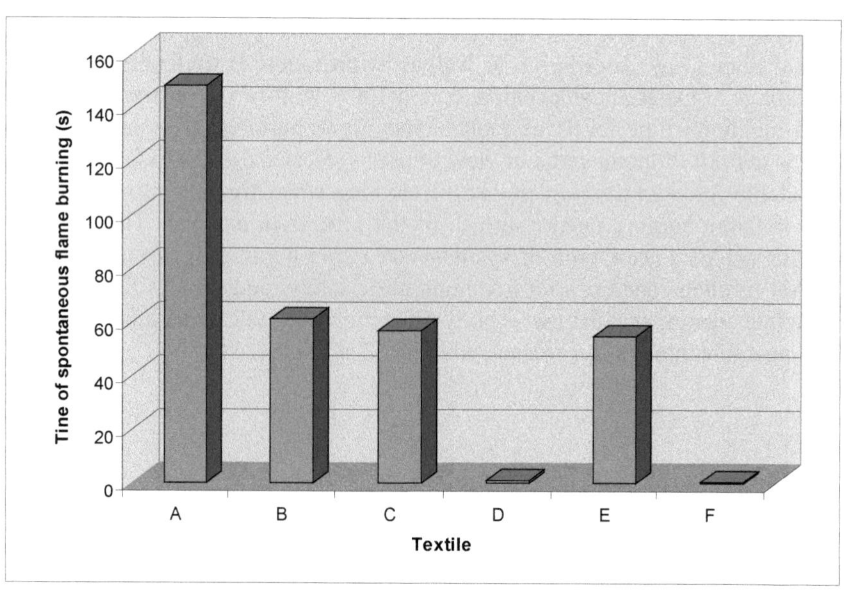

Figure 2 Chart of spontaneous flame burning of the textile fabric Procedure B

By evaluated parameter time of smoulder of individual types of textile fabric by procedure A and B the following results were discovered: Textile fabrics from material 100% cotton with surface coating PROBAN reached the parameter value time of smoulder the highest value 0.3s. Based on this finding we can evaluate the given textile fabrics as the most suitable for the production of protective clothing from the point of view of user's protection. These discovered values of parameter time of smoulder of given textile fabric fulfil conditions for minimum basic safety requirements for protective clothing that are aimed at the user's body protection by welding and other processes with comparable threat in accordance with the STN EN ISO 11611.

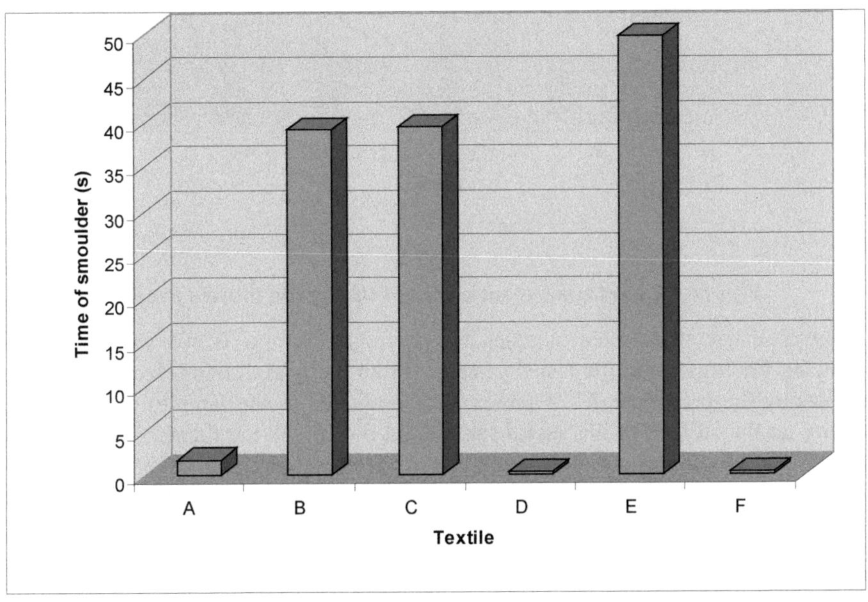

Figure 3 Chart time of smoulder of the textile fabrics Procedure A

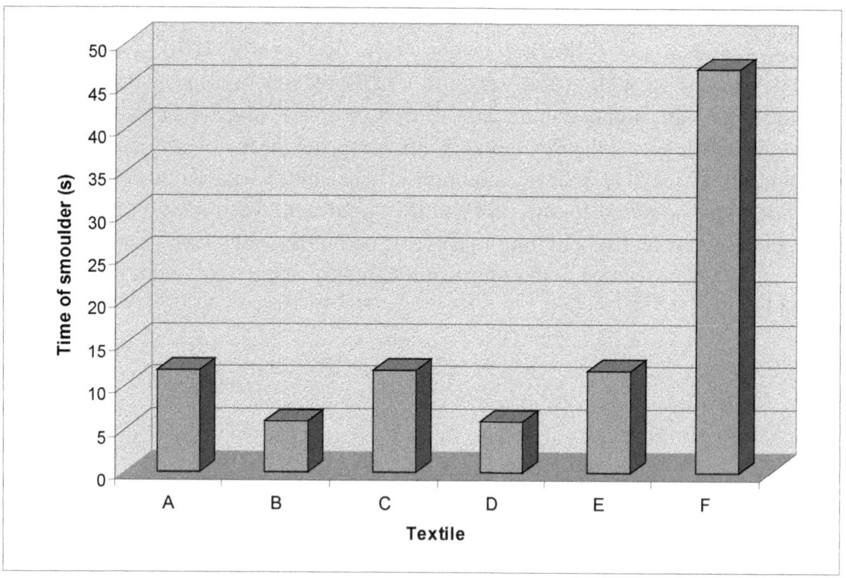

Figure 4 Chart time of smoulder of the textile fabrics Procedure B

By observing the tested samples by procedure A and B we recorded the following findings about the tested samples. From the overall number of 51 tested samples by the procedure A, 24 pieces were evaluated as the samples that by fire activity on the surface of this tested sample did not spread the flame on any of its edge. Among these samples belonged those which material composition was 100% cotton with surface coating PROBAN. The representation of these tested samples was 47.1%. The same tested samples reached similar evaluation also by the procedure B – edge ignition. The measured value for the tested samples of the material 100% cotton with the surface coating PROBAN fulfils the conditions for minimum basic safety requirements for protective clothing that are aimed at the users body protection by welding and other processes with comparable threat in accordance with the STN EN ISO 11611.

CONCLUSIONS

Based on the tests made by us we achieved the following results. The best material that we exposed to the tests for the production of protective clothing are textile fabrics of material composition 100% cotton with basic weight of the textile fabric higher than 296,49 g/m2 and with final coating PROBAN. The tested textile fabric with its qualities and also with its final coating strongly contributes to the increase of users' safety. In comparison with this textile fabric, the types of textile fabrics TERA with material composition: cotton 35% + polyester 65%, basic weight of the textile fabric 241,93 g/m2 and TINA with material composition 100%, basic weight of the textile fabric 257,03 g/m2 without final coating, it was found out that by contact with flame they created holes. It is important to show on the fact that these types of textile fabrics are from the point of view of the user's safety the least suitable for the production of protective clothing. From the aspect of complete protective clothing that underwent testing, the female boiler suit jacket of blue colour did not fulfil the requirements for protective clothing in material composition cotton 100% without final coating, basic weight of the textile fabric 240, 00 g/m^2.

ACKNOWLEDGMENTS

The authors would like to acknowledge to Jozef Holak for assistance by testing materials and evaluation the results.

REFERENCES

HIRSCHMAN, J. E.: Protective Clothing. In: Fire and Rescue. - ISSN 0964-9727 Is3 (1992), s.22-30.

JASON, N., and H.: Firefighter Thermal Exposure Workshop: Protective Clothing, Tactics, and Fire Service PPE Training Procedures. Gaithersburg: National Institute of Standards and Technology, 1997. - 41 s.

LAWSON, R., and J.: Firefighter Thermal Exposure Workshop: Protective Clothing, Tactics, and Fire Service PPE Training Procedures. Gaithersburg: National Institute of Standards and Technology, 1997. - 41 s.

Government regulation No 124/2006 Safety and occupational health and amending certain laws Internet [online] [Cit. 2011.08.08].: *http://jaspi.justice.gov.sk/jaspiw1/htm_zak/jaspiw_mini_zak_vyber_hl1.asp*

Government regulation No. 395/2006 about minimum requirements for the provision and the use of personal protective equipment Internet [online] [Cit. 2011.08.08]:*http://www.zbierka.sk/zz/predpisy/default.aspx?PredpisID=19520&FileName=06-z395&Rocnik=2006*

STN EN ISO 15025: 2003: Protective clothing method of test for limited flame spread. Slovak Institute for Standardization

CHAPTER 56

Occupational Medical Service in Practice

Andrea Lezovicova, Jan Donic, Karol Habina

BOZPO, s.r.o.
Prievidza, Slovakia
Andrea.lezovicova@boz.sk

ABSTRACT

The term "health" is placed under various polls on a regular basis and human values at the forefront, and it is striking how some employees and employers perceive Occupational Medical Service. We see there is still a misunderstanding and ignorance that are sometimes caused by too frequent changes in legislation but also personal ignorance of skilled workers but also employers.

Keywords: medical examination, doctors, workplace factors, first aid, health protection, detection of the danger and assessment of health risks

1 LEGISLATION

Legislation :

Act No. 355/2007 Coll.

- **Act on the protection, promotion and development of public health** and on amending of certain acts as amended

Act No. 124/2006 Coll.

- **Act on Occupational Health Service** and on amendment of certain acts as amended

Act No. 125/2006 Coll.

- **Act on labor inspection** and on amending Act No. 82/2005 Coll. on Illegal Work and Illegal Employment and on amendment of certain laws

Act No. 311/2001 Coll.

- **The Labor Code** as amended

Act No.377/2004 Coll.

- **Act on the protection of non-smokers** and on amending of certain acts as amended

Act No. 576/2004 Coll.

- **Act on health care, services related to health care** and on amending of certain acts as amended

Act No. 578/2004 Coll.

- **Act on health care providers, health workers and professional organizations in healthcare** and on changes and amendments to some Acts as amended

Act No. 115/2006 Coll.

- **The Government Regulation of the Slovak Republic on the minimum health and safety requirements to protect workers from risks related to exposure to noise**, as amended by the Government Regulation of the Slovak Republic No. 555/2006 Coll.

Act No. 416/2005 Coll.

- **The Government Regulation of the Slovak Republic on the minimum health and safety requirements to protect workers from risks related to exposure to vibration**, as amended by the Government Regulation of the Slovak Republic No. 629/2005 Coll.

Act No. 276/2006 Coll.

- **The Government Regulation of the Slovak Republic on the minimum safety and health requirements for work with display**

Act No. 281/2006 Coll.

- **The Government of the Slovak Republic on the minimum health and safety requirements for manual handling of loads**

Act No. 292/2008 Coll.

- **Decree of the Ministry of Health of the Slovak Republic on the details of the scope and content of the performance of occupational health services, a team composed of professionals who practice and**

requirements for their competence in the Decree of the Ministry of Health of the Slovak Republic no. 135/2010 Coll. and 124/2006 Coll.

Act No. 448/2007 Coll.

- **Decree of the Ministry of Health of the Slovak Republic on the details of the factors of work and working environment in relation to the categorization of work in terms of health risks and the particulars of the proposal for Occupational categories**

Act No. 461/2003 Coll.

- **Social Insurance Act** as amended

Act No. 470/2011 Coll. entered into force on 1 January 2012. This Act amends the Act No. 124/2006 Coll. on Occupational Health Service and on amendment of certain acts as amended and amending the Act No. 355/2007 Coll. on the protection, promotion and development of public health and on amending of certain acts as amended. This Act brought about changes in the occupational medical services compulsorily provided by employers, which has cut their funding costs. By 31 December 2011, it was the duty of employers to ensure Occupational Medical Service for all employees.

From 1 January 2012, the employer is not obliged to provide Occupational Medical Service for employees who perform work in category 1 or 2 because the amendment of Act No. 124/2006 Coll. The employer is to ensure Occupational Health Service for employees who perform hazardous jobs, which are classified in category 3 or 4.

Does not apply to an employer's obligations in the field of Occupational Health Service or his responsibility for ensuring Occupational Health Service for all employees of an employer's obligation to ensure the abolition of Occupational Medical Service for employees performing work in category 1 or 2. The employer's duties under the Act. No. 355/2007 Coll. still belongs to evaluate health risks at work, undertake an assessment of risk to categorize work in terms of health risks, submit to the Regional Public Health Authority proposals for inclusion of works in categories 3 and 4, to provide medical assessment of employees' work performance and preventive medical examinations in relation to work.

2 FACTS

The number of employees working in hazardous workplaces in 2010 in Slovakia : **103 851**

The number of employees working in hazardous workplaces in 1995 in Slovakia : **154 987**

Number of employees working in hazardous workplace :

agriculture : **3 280**
industry : **74 321**

scale of risk factors :

1. dust : **19 995**
2. vibration : **4 601**
3. disease of the limbs : **902**

The Act No. 124 – Occupational Health Service, which was issued in 2006, the term "preventive and protective services and the Occupational Medical Service" was mentioned for the first time. 122 licenses to implement of the Occupational Medical Service was issued and published on the website of the Office of Public Health to 10.04.2011.

The Public Health Authority is the supreme office for the regional public health authorities. It manages controls and coordinates the execution of state administration carried out by regional public health offices. More details on competencies and activities of the Authority in the field of protection, promotion and development of public health are stated in the Act No. 355/2007 Coll., paragraph 5.

In accordance with § 26 of Act No. 124/2006 Coll. Occupational Health Service, as amended, and Decree of Ministry of Health No. 292/2008 Coll. suggest the following contents of the Occupational Medical Service :

Occupational Health Service provided for the consumer is a specialist service related to selection, organisation and execution of specialist tasks while ensuring safety and health at work, mainly by preventing risks and protecting against them particularly in view of the influence of workplace factors.

Detection of dangers and assessment of health risks

Apprasial and measurement of the work factors and biological monitoring needed for health risks assessment. The measurement and the appraisal of workplace is ensured in collaboration with specialized laboratories of Regional Office for Public Health.

Supervision of working environment

Detection of the danger and assessment of health risks which threat the employees health, inspection of the workplace, monitoring of working activities, biological monitoring, health risks assessment, supervision of working environment factors and state of working conditions, support of employees´ work adaptation, ensuring counselling to employer and also employees.

Proffesional advisory (counselling) and help

When elaborating of employees′ health protection and support programms, for improving the working conditions and evaluation of new equipment and technologies in light of health, when elaborating the measures of working rehabilitation.

Providing of first aid trainings

Employee training of first aid is providing in the workplace. Appraisal of system, evidence and marking of first aid is providing in the workplace. Elaboration of inspections′ plan and completing the first aid kits are placed on walls.

Regular analysis of the work accidents causes

Analysis of the working incapacity, occupational diseases, diseases related working activities and health risks, participate in measures avoiding working accidents (injuries). According to § 21 paragraph 6 of the Act No. 124/2006 Coll. the amended law the employer is obliged to ensure that safety and technical services and Occupational Medical Service carried out at least once a year a joint review of its workplaces.

Preventive health program

Elaboration of the system and scope of medical examinations in relation to work conditions. Determination of rules for updating the system when there is the working conditions changing. Determination of examinations periodicity. Carrying out the compulsory annual examinations on workplaces by professionals of safety technical service. Elaboration of health risk evaluation results and measures carried out for its minimalization what is the employee obliged to submit every year within 31st December of particular year on Regional Office for Public Health (in accordance with Act No. 355/2007 Coll.)

Consulting services

Professional and qualified assessment of the compliance of health conditions in the workplace. Consulting on work hygiene observance in the workplace. Consulting on measurements providing and its evaluation on your workplaces, on employee's classification and on position changing, etc.

Our company carries out the work of Occupational Medical Service (OMS) pursuant to § 21 of the Act No. 124/2006 Coll. occupational health service laws as amended and issued on the basis of authorization No. : OPPL-6109/2007-Oj, dated 16.7.2007.

Team of health professionals:

- Preventive occupational health care – work hygiene
- Clinical occupational health care
- Common doctor for adults with experiences in the area of medical-preventive care

CHAPTER 57

Head and Facial Anthropometry of Young Chinese Male Aged 18~35 Years Old

Zheng Xiaohui[1,2], *Ding Songtao*[1], *Zhou Qianxiang*[2], *Liu Taijie*[3], *Yuan Xiaohua*[1], *Liang Guojie*[1]

1. Research Institute of Chemical Defense, Beijing, China
2. Beijing University of Aeronautics and Astronautics, Beijing, China
3. China National Institute of Standard, Beijing, China

ABSTRACT

To evaluate the head and facial form of young Chinese male, an anthropometry was carried out in China from 2010 to 2011, and 10267 young Chinese male were measured with two different methods, by hand and 3D laser-scanned graphs. All subjects were born in 31 provinces (besides Taiwan, Hongkong, and Macao) of China. Their ages ranged between 18 and 35 years. The effect of area location on head form in Chinese male was also studied and the results show that he head form becomes rounder in areas of higher latitude in China. This paper also carried out the research of Chinese men and US men based on the study of Miyo Yokota. The results show that yellow men have rounder head than white men and black men, because of larger head breadth and smaller head length. Another distinguished characterized feature is that the lip breadth is nearly 10% smaller than white and black men, and nose length is shorter than that of white and black men while nose breadth is larger than white men and smaller than black men. The instrument should be resized for different race and place to get better fitting and performance.

Keywords: anthropometry, head, face, ergonomics, young Chinese male

1. Introduction

The head and face form is essential when designing and sizing equipments for head, such as helmet, eyewear and respirator. There are many factors, such as age[1], economy[2], race[3, 4], gender[4] and so on .

Many method have been adopted to characterize the difference between people, such as marker- hand measurement[5], CT scans[6,7], laser head and facial scanning[8], Stereo-photogrammetry[9], Commercial visible-light imaging systems, and so on.

As we all known, there are great difference in appearance between different races. Meanwhile, the difference in anthropometrical data caused by race should be valued by statistics to meet the demand of design. Accordingly, many anthropologists have studied this issue[5, 10]. The effect that an increasingly admixed US Army population might have on current and future equipment sizing and design statistics was studied by Miyo Yokota[11]. Roger Ball etc. had compared the head shape difference between Chinese and Caucasian using three dimensional graphs[3].

The goal of this study was to display the head and facial form of young Chinese male and compare the difference between Chinese and white men and black men using the result of Miyo Yokota.

2. Materials and methods

A head and facial anthropometry was carried during 2010~2011 in China, and 10267 young Chinese male were measured All subjects were born in 31 provinces (besides Taiwan, Hongkong, and Macao) of China. Their ages ranged between 18 and 35 years.

Two different methods were adopted, by the traditional anthropometry measurement method and by the three dimensional laser scanning method. The traditional measurements use the traditional calipers. A laser scanner VITUS aHead is used in this measurement, which is designed to generate highly precise 3-dimensional images of the human head. This technology can be utilized for a variety of applications. VITUS aHead is based on optical triangulation, currently the most accurate method for touchless 3D imaging. The scanning volume is 1000 mm × 400 mm × 400 mm, The measurement time is approximately 10 s, and the point density is 27 Pkte./cm^2.

According to the study of Miyo Yokota, ten measurements were chosen to value the difference of head and facial form of young Chinese male and US male, and the craniofacial measurements and their definitions are showed in Table1. There are three measurements in the paper of Miyo Yokota were no included into this study, considering the result of Chinese anthropometry. All the results of ten measurements were measured by hand according to Gordon et al. (1989) and dimensions auto-extraction from 3-dimensional images.

The Statistical Package for the Social Sciences (SPSS) for Windows version 16.0 was used in the following statistical analysis. The descriptive statistics, including arithmetic means (M), standard deviations (SD) of the measurements were calculated.

Table 1 Definition of the measurements

No.	Abbreviation	Measurement	Definition
1	BIGONIAL	Bigonial breadth	Straight-line between the right and left gonion on the jaw
2	BIOCBRMH	Biocular breadth	Distance between the right and left ectoorbitale
3	BIZBDTH	Bizygomatic breadth	Maximum horizontal breadth between zygomatic arches
4	HEADBRTH	Head breadth	Maximum horizontal breadth of the head
5	HEADLGTH	Head length	Maximum distance between the glabella and back of the head
6	LIPLGTH	Lip length	Distance between the right and left cheilion on the corner of the mouth
7	NOSEBRTH	Nose breadth	Distance between the right and left alare
8	SBNSSELH	Subnasal-sellion length	Distance between the subnasal and sellion
9	SELLIONZ	Sellion	Z Distance between sellion and top of the head plane
10	SELTRAG	Facial projection	Distance in XYZ coordinates between sellion and right tragiona

3. Results and discussion

3.1 Native place distribution of samples

The native places of the measured samples are also recorded during the anthropometry. This anthropometry sampled some people of every administrative region of Chinese mainland (which include 22 provinces, 5 autonomous regions and 4 municipalities). According to GB10000-1988[12], these administrative were divided into six anthropometry areas, which are Northeast and North China, Northwest, Southeast, Central China, South China and Southwest. Comparing with the result of the 6th population census of China, the distribution of native place of samples and total people were shown in Fig 1. Every area was covered and the number of samples can meet with the need of statistical process.

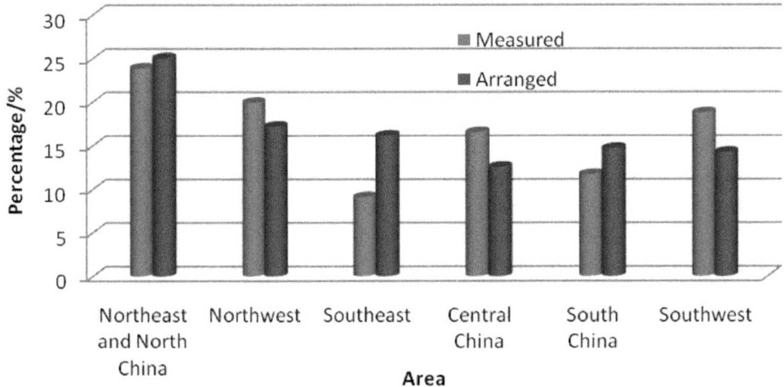

Fig. 1 The percentage of samples and total people from different areas

3.2 Head index of Chinese young male

To value the head form quantitatively, many indexes were put forward. In the present study, head type index R was adopted to value the head shape. The equation to calculate R is as follows.

$$R = (HEADBRTH/HEADLGTH) \times 100$$

The definitions of HEADBRTH and HEADLGTH are shown in Table 1. According to the value of R, the head form of Chinese young male can be divided into seven types, which are Hyperdolichocephaly, Dolichocephaly, Mesocephaly, Brachycephaly, Hyperbrachycephaly, Ultrabrachycephaly and Flat Head Type. Based on the data of Chinese head index, a now head form type named Flat Head is added, and which means that the value of head breath is close to even larger than that of head length. The value ranges of every type are shown in Table 2, and the distribution of every type is also characterized. The number of people with very long head type in samples is too little, which means that Chinese young male have little very long head type. In the following study, the very long head type is not included.

Table 2 the distribution of different head form type

Value range	Head form type	Result	
		Number	Percentage%
$71.0 \geq R$	Hyperdolichocephaly	3	0.03
$76.0 \geq R > 71.0$	Dolichocephaly	163	1.59
$81.0 \geq R > 76.0$	Mesocephaly	1508	14.72
$85.5 \geq R > 81.0$	Brachycephaly	3227	31.49
$91.0 \geq R > 85.5$	Hyperbrachycephaly	3867	37.74
$96.0 \geq R > 91.0$	Ultrabrachycephaly	1453	14.18
$R > 96.0$	Flat Head	26	0.25

There are six areas in China, and the people living in different areas differ between each other in many aspects because of the different environment, climate, economic level and living habit. As a result, the anthropometric data are also different. The population distribution of different head form type in every area was calculated and displayed in Fig. 2. The distribution in different areas varies a little, which can be ranked into two groups. Northeast and North China, Northwest, Southeast are the same group, and the other three areas are the other group. The head of the first group is rounder than the second one. In both groups, there is a slight trend that the head form becomes rounder in areas of higher latitude. The reason of this phenomenon will be studied in another paper.

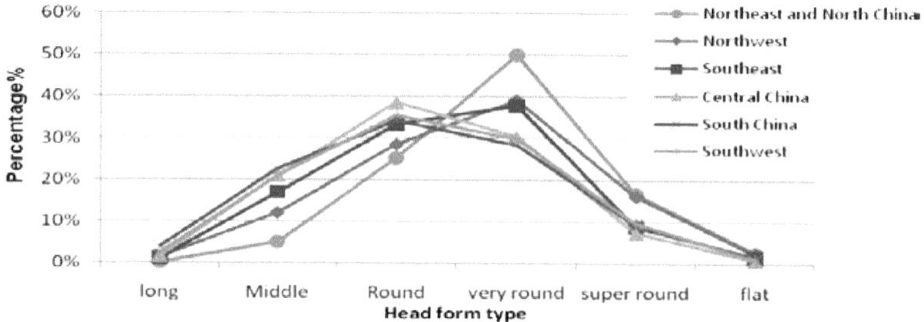

Fig. 2 Population distribution of different head form type

3.3 The head and facial form difference between different races

The means and standard deviation (SD) of the craniofacial measurements by white male, black male in US and yellow male in China were shown in Table 3. Comparing with white man and black man, head breadth is much larger and head length is smaller. Which may means that heads of yellow man are much rounder. Roger Ball et al. had also reported the result that Chinese have rounder head, comparing with Caucasian. The SD between racial groups are relatively similar, and the result indicates that most facial measurements, such as BIGONIAL, BIOCBRMH, LIPLGTH, SBNSSELH, SELTRAG , the mean measurements of those parameters for yellow men are obviously smaller than that for the White and Black groups. Meanwhile, there are three parameters, including HEADBRTH, BIZBDTH and SELLIONZ are all much larger than that of the White and Black groups. Only nose breadth falls between the White and Black groups. The mean value of nose breadth is larger than White group, smaller than Black group.

The results reveal a significant distinction between racial groups found only in the first eigenvalue ($p<0.05$). The head form of Chinese is rounder than that of white and black male. Meanwhile, most facial size is smaller, such as eye and mouth. Especially for the lip breadth, the value is nearly 10% smaller than white and black. Besides of these trends, nose is different, and the nose length is shorter than that of white and black men while larger than white men and smaller than black men.

Table 3 Descriptive summary of craniofacial measurements (mm)

No.	Measurement	Chinese male (N =10319)		White male (N =820)		Black male (N =1189)	
		Mean	SD	Mean	SD	Mean	SD
1	BIGONIAL	115.7	6.4	117.6	7.3	118.4	7.8
2	BIOCBRMH	112.2	6.7	120.3	5.2	124.3	5.2
3	BIZBDTH	144.4	5.7	139.3	5.3	140.8	5.4
4	HEADBRTH	157.48	6.1	150.9	5.1	151.1	5.4
5	HEADLGTH	184	7.2	197.3	6.7	197.6	6.4
6	LIPLGTH	49.3	3.4	54.4	3.7	59.0	3.8
7	NOSEBRTH	38.1	2.5	33.9	2.7	42.3	3.2
8	SBNSSELH	47.4	3.7	51.1	3.6	48.5	3.6
9	SELLIONZ	116.2	8.5	112.6	7.0	110.8	7.0
10	SELTRAG	118.68	4.923	124.5	4.7	122.6	4.6

4. Conclusion

Head and facial form is important for designing and sizing equipment, and many ergonomic scientists have paid attention to study on this issue. In this paper, we use the anthropometry data to value the head form of young Chinese male and the difference between Chinese male and white and black young male. The results show that the head form is rounder with the latitude ascending of area location in China. It was found that Chinese have rounder head form and smaller face than white and black man in US. Considering these difference in head form, the instrument should be resized for different race and different location to get better fitting and performance.

Acknowledgements

The authors would like to thank all the people participating in the anthropometric measurement and the Chinese Government to supporting this work.

References

1. Annis J. F. *Aging effects on anthropometric dimensions important to workplace design.* Int. J. Industrial Ergonomics 1996. **18**: 381-388.
2. Abeysekera J. D., Shahnavaz, H. *Body size variability between people in*

developed and developing countries and its impact on the use of imported goods. International Journal of Industrial Ergonomics, 1989(4): 139-149.

3. Ball R., Shu C., Xi P., et al. *A comparison between Chinese and Caucasian head shapes.* Applied Ergonomics, 2010. **41**(6): 832-839.
4. Hwang S.,Johnson P. W. *Computer input devices - Race and gender: Is there a mismatch between anthropometry and input device design.* 2011. San Francisco, CA, United states: Human Factors an Ergonomics Society Inc.
5. Hughes J. G.,Lomaev O. *An Anthropometric Survey of Australian Male Facial Sizes.* American Industrial Hygiene Association Journal, 1972. **33**(2): 71-78.
6. Whyte A., Hourihan M., Earley M., et al. *Radiologic assessment of hemifacial microsomia by three-dimensional computed tomography.* Dento-Maxillo-Facial Radiology,, 1990. **19**: 119-125.
7. Niu J. W., Li Z. Z.,Salvendy G. *Multi-resolution description of three-dimensional anthropometric data for design simplification.* Applied Ergonomics, 2009. **40**(4): 807-810.
8. Zhuang Z., Benson S.,Viscusi D. *Digital 3-D headforms with facial features representative of the current us workforce.* Ergonomics. **53**(5): 661-671.
9. Marr D.,Poggio T. A. *Cooperative Computation of Stereo Disparity.* Science, 1976. **194**(4262): 283-287.
10. Mokdad M.,Al-Ansari M. *Anthropometrics for the design of Bahraini school furniture.* International Journal of Industrial Ergonomics, 2009. **39**(5): 728-735.
11. Yokota M. *Head and facial anthropometry of mixed-race US Army male soldiers for military design and sizing: A pilot study.* Applied Ergonomics, 2005. **36**(3): 379-383.
12. GB 10000-1988. *Human Dimensions of Chinese Adult*, 1988

CHAPTER 58

The Suitable Sampling Duration for Performing Consecutive Maximum Voluntary Contraction Gripping Exertions without Fatigue

Peng-Cheng Sung, Cheng-Lung Lee

Department of Industrial Engineering and Management
Chaoyang University of Technology
Wufong District, Taichung City 41349, Taiwan
sungpc@cyut.edu.tw

ABSTRACT

Previous studies related to grip strength focused on exploring the maximum and sub-maximum grip strength or maximum endurance time for gripping exertions for bare and gloved hands. However, no literatures were found investigating suitable duration for sampling of consecutive maximum grip strength without fatigue. This study used grip dynamometer and electromyography (EMG) module to collect EMG signals for four forearm muscles when performing consecutive maximum gripping exertions for bare and gloved hands. Ten males and eight females participated in this experiment. The suitable sampling duration is calculated as the time that 90% of males and 75% of females can perform consecutive MVC gripping exertions without fatigue. Based on Joint analysis of EMG spectrum and amplitude (JASA), the subjects can follow the recovery time of 2 minutes of Caldwell's regimen between consecutive grip exertions for bare and gloved hand conditions without fatigue. Therefore, the suitable sampling duration is defined as the time that the subject can maintain 80% of the maximum grip strength during consecutive maximum gripping exertions. For each of the four different gripping conditions (bare hand/2 minutes rest, gloved hand/2 minutes rest, bare hand/4 minutes rest, gloved hand/4 minutes rest), the suitable sampling duration that 90% of males and 75% of females can perform consecutive MVC gripping exertions without fatigue were 17、8、48、38 minutes and 47、32、98、58 minutes, respectively.

Keywords: Grip strength、EMG、MVC、JASA、Suitable sampling duration

1 INTRODUCTION

Work-related musculoskeletal disorders (WRMSDs) of upper extremities are among the most prevalent lost-time injuries and illnesses in almost every industry (BLS, 2007). Forearm muscles are common sites and sources for the development of upper extremity disorders even though the normal muscles function is essential in protecting ligaments and other tissues against injury or disorders (Nicolay et al., 2007). Gripping is one of the most commonly applied hand exertion to accomplish complex tasks for industrial jobs and has been implicated as a risk factor for several WRMSDs (Bao and Silverstein, 2005). The association between gripping and occurrence of upper extremity disorders indicates the need to explore risk factors that may affect upper extremity disorders when performing gripping tasks.

Published literature on grip strength frequently assessed the effects of different variables (e.g. age, gender, glove, posture, physical activity, etc. on maximum voluntary contraction (MVC) grip strength (Nicolay and Walker, 2005). Gloves are the most used protective device for human hand in many industrial tasks to protect the hand from injury. The effect of gloves on grip strength capabilities are consistent in the sense that gloves decrease strength compared to the bare hand condition (Sung, 2006). The reduction compared to bare hand ranged from 3.7% of synthetic rubber to 50.0% of extra vehicle glove. For bare handed conditions, females produce significantly less (50-65%) grip strength than males (Nicolay and Walker, 2005; Hallbeck and Mcmullin, 2003), but, females generally have greater muscular endurance than males when worked at sub-maximal level (Hicks et al., 2001). Hallbeck and Mcmullin (2003) found that when over several glove conditions, females averaged 74% the grip strength of males.

Caldwell's regimen (1974) has been used enormously to test the isometric (static) MVC grip strength for bare and gloved hand (Hallbeck and McMullin, 2003; Muralidhar et al., 1999; Nicolay and Walker, 2005; Sung, 2006). However, no efforts have been made to discuss "what is the suitable duration for a strength sampling period without fatigue?" Bishu et al. (1995) is the only study to assess the recovery time of 2 minutes of Caldwell's regimen after a MVC grip exertion. They divided the hand into 18 regions and measured the discomfort scale at each region after 1, 2, 5, 15, 30, 60, 120, and 240 minutes of a MVC exertion. Most of the subjects reported lack of discomfort after 30 minutes. They stated that perhaps the Caldwell's regimen needs to be looked into, in order to see the validity of two minutes rest between trials. In this study, suitable duration for sampling of consecutive MVC gripping exertions without fatigue and the validity of two minutes rest between MVC exertions will be investigated.

2 METHOD

2.1 Subject

Ten male and eight female volunteers free of musculoskeletal disorders/injuries (MSDs) in the upper extremities comprised the subject pool. The subject's free of MSDs status in the upper extremities was identified through interviewing during the recruiting process. All these subjects are right handed. The mean values of age, height, hand length and maximum breadth of hand for male/female subjects are 24.2±2.1/ 23.6±0.5 years, 171.3±6.1/161.0±4.1 cm, 182.1±5.4/169.6±2.4 mm, and 104.1±4.4/89.8±4.1 mm, respectively.

2.2 Apparatus

Glove

Sung (2006) examined three commercially available glovebox gloves. He found that the 0.03" neoprene glove produced significantly less grip strength than the other gloves. Kovacs et al. (2002) reported that lack of significant effect of glove type on EMG activity combined with the significant effect of glove type on MVC grip indicated that gloves have an effect on the transfer of energy from internal muscle activity to external grip force. Therefore, the 0.03" thickness neoprene glovebox glove that caused greater reduction of the output force/muscle activity ratio will be selected for evaluation. Two hand sizes, 8.5"and 9.75" are provided and the subject will wear both gloves to pick the best fit glove.

Grip dynamometer

Grip strength with maximal efforts for bare and gloved hands will be measured for each subject using a Jamar hand dynamometer with transducer (Lafayette Instrument Company). The grip strength data will be collected and processed using the laboratory owned BIOPAC MP150 EMG Systems (described below) to be synchronized with EMG data for later analysis.

Electromyography (EMG)

The BIOPAC MP150 EMG System which includes amplifier & A/D converter (BIOPAC Systems UIM100C) and data acquisition software (Acq Knowledge 3.7.2) installed in a laptop will be used to collect and process the EMG signals. After shaving and scrubbing the recording sites with alcohol, surface EMG electrodes are positioned over the following forearm muscles of dominant arm:(1) flexor digitorum superficialis (FDS), (2) flexor carpi radialis (FCR), (3) extensor carpi radialis longus (ECR), and (4) extensor digitorum (ED) of the subjects' dominant arm involved with gripping tasks (Larivière et al., 2004). Electrodes are positioned following the instructions as described in the DHHS (NIOSH) publication no. 91-100 (NIOSH, 1991). A ground electrode is also placed at the lateral epicondile of the subjects' dominant arm. The EMG signals will be bandpass filtered (20-450 Hz) and preamplified (gain: 1000) with 1k Hz sampling rate.

Two techniques, time domain and frequency domain analysis (Klein and Ferrnandez, 1997), commonly used for analyzing the electromyography (EMG) signals were used in this experiment for analysis of the EMG signals. The root mean square (RMS) amplitude which is proportional to the energy content of the muscle activity is used in time domain analysis. Mean power frequency (MPF) value is used in frequency domain analysis. The increase in RMS value combined with decrease in median frequency will indicate possible local muscle fatigue and a relative increase in stress to the muscles (Klein and Ferrnandez, 1997) involved in gripping exertions.

The RMS amplitude for the j_{th} MVC gripping exertion (RMS_i) for i_{th} session will be normalized for each subject using equation 1. The $RMS_{max,i}$ determined from three replicates of barehanded MVC gripping and $RMS_{base,i}$ determined when the forearm is in rest will be measured 30 minutes before the starting of i_{th} session.

$$\text{Normalized RMS}_{ij} = \frac{RMS_{ij} - RMS_{base,i}}{RMS_{max,i} - RMS_{base,i}} \quad (1)$$

Linear regression will be applied to the normalized RMS (nRMS) and MPF time series to estimate their rate of change ($nRMS_{slope}$ and MPF_{slope}, slope of their respective linear regression) as muscle fatigue indices.

2.2 Procedures

In this experiment, the MVC gripping strength for bare and gloved hands was measured according to Caldwell's regimen (1974). The subjects were instructed to increase to maximum exertion (without jerk) in about one second and maintain this effort during a four second count. The strength datum is the mean score recorded during the first three seconds of the steady exertions. The rest period between consecutive MVC gripping exertions is two or four minutes. The subjects is seated with their shoulder adducted and neutrally rotated, elbow flexed at 90^0, forearm in neutral position, and wrist between 0^0 and 30^0 extended and between 0^0 and 15^0 ulnar deviations to maintain neutral posture (Mathiowetz et al., 1985).

To determine the suitable sampling duration, the subjects were asked to repeat as many MVC grip as possible to two hours maximum or until the subject voluntarily quitting the experiment. EMG signals for forearm muscles were recorded synchronously with the gripping strength data. A seven days rest period will be followed between trials to ensure full recovery of forearm muscles. A total of 4 trials, two replicates each for bare and gloved hand conditions, lasted 3~4 weeks were administered to each subject to complete this experiment. The suitable sampling duration will be the time that 90% of males and 75% of females can perform consecutive MVC gripping exertions without fatigue. At the same time, grip force reduction (the percentage change in force production between the mean of the first three and the last three repetitions), relative change of EMG activity ($1 - \frac{\text{mean of the last three normalized RMS amplitude replicates}}{\text{mean of the first three normalized RMS amplitude replicates}}$), and EMG fatigue indices (RMS_{slope} and MF_{slope}) were also measured.

2.3 Experimental Design and Statistical Analysis

The independent variables in this experiment are gender, rest time (2 and 4 minutes) and glove usage (bare hand versus gloved hand). The order of presentation of the gloved condition and rest time is randomized. The performance measures are grip force, muscle activity (nRMS), fatigue index ($nRMS_{slope}$ and MPF_{slope}). Separate analyses of variance (ANOVAs) with repeated measures were used to determine whether there are significant differences between independent variables on dependent variables. All data were analyzed for statistical significance at $p \leq 0.05$ using the SPSS 12 (SPSS Inc, Chicago, Illinois) statistical software.

3 RESULT

3.1 Effects on Grip Strength

The results of the repeated-measured ANOVA (Table 1) show that there are significant differences between glove usage ($F=11.455$, $p<0.001$) and gender ($F=41.408$, $p<0.001$) on MVC grip strength. The bare hand condition can retain significantly higher maximum grip strength (2.4 Kg) than the neoprene glove. In addition, the MVC grip strength for the males is 17.0 Kg higher than that of the females. No significant difference were found between the 2 and 4 minutes rest time although the MVC grip strength for the 4 minutes rest are 2.6±0.6 Kg higher than the MVC grip measured for the 2 minutes rest trials.

Table 1 ANOVA with repeated measures results (F and *p*-value) for effects of glove usage, gender, and rest time on MVC grip strength

Source	F	Sig.
Glove usage	11.455	0.004
Gender	41.408	0.000
Rest time	2.863	0.110

3.2 Effects on Muscle Activity

Table 2 shows the summary of significance for the factors on muscle activity (nRMS) for flexor digitorum superficialis (FDS), flexor carpi radialis (FCR), extensor carpi radialis longus (ECR), and extensor digitorum (ED) muscles. Rest time effects were observed on the ECR (F=4.511, p=0.05) and ED (F=5.508, p<0.05) muscles when performing MVC grip strength. The nRMS for the 4 minutes rest trials are 8.3% and 8.4% higher than the 2 minutes rest trials for the ECR and ED muscles, respectively.

Table 2 ANOVA with repeated measures results (F/*p*-value) for effects of glove usage, gender, and rest time on muscle activity (nRMS)

Source	FDS	FCR	ECR	ED
Glove usage	0.073/0.790	0.903/0.356	1.107/0.308	0.867/0.366
Gender	1.530/0.234	1.036/0.324	3.623/0.075	0.504/0.488
Rest time	1.408/0.253	0.582/0.457	4.511/0.050*	5.508/0.032*

*: p<0.05

3.3 Effects on Fatigue

No statistically significant effects were found on the $nRMS_{slope}$ meaning that the relative change of muscle activity were minimal during the whole gripping trials. For the effects of glove usage, gender, and rest time on the relative change of mean power frequency (MPF_{slope}), the results of the repeated-measured ANOVA (Table 3) show that there are significant differences between gender on the FCR (F=6.146, p<0.05) and ECR (F=5.392, p<0.05) muscles and between rest time on the ECR (F=5.032, p<0.05) muscle. The slopes of the MPF regression lines of the females are less steep than the males on the FCR and ECR muscles. Since all the regression lines have negative slopes, the reduction of the MPF of the males is larger than the females. In addition, the reduction of the MPF of the 2 minutes rest is larger than that of the 4 minutes rest on the ECR muscle.

Table 3 ANOVA with repeated measures results (F/*p*-value) on relative change of mean power frequency (MPF_{slope}) when performing MVC grip strength

Source	FDS	FCR	ECR	ED
Glove usage	1.220/0.304	0.044/0.836	0.107/0.747	2.621/0.125
Gender	6.146/0.025*	0.807/0.382	5.392/0.034*	1.762/0.203
Rest time	0.228/0.640	0.343/0.566	5.032/0.039*	1.391/0.255

*: p<0.05

3.4 Suitable Sampling Duration for Performing Consecutive MVP Gripping Exertions

Joint Analysis of EMG Spectrum and Amplitude (JASA) method (Lin et al., 2004) was applied to evaluate the occurrence of muscular fatigue during consecutive gripping exertions. The pair of $nRMS_{slope}$ and MPF_{slope} was presented in a two-dimensional JASA plot where the change in nRMS over time is plotted in the abscissa, and the change in MPF over time is sown in the ordinate. The results of JASA plots are classified into the following four categories (Lin et al., 2004):

1. Force increase (upper-right quadrant in the JASA plot): increase in both nRMS and MPF over the EMG recording period.
2. Recovery (upper-left quadrant): decrease in the nRMS along with increase in the MPF.
3. Force decrease (lower-left quadrant): decrease in the nRMS accompanied by a decrease in the MPF.
4. Muscular fatigue (lower-right quadrant): increase in the nRMS along with decrease in the MPF.

Figure 1 shows the JASA plot of four forearm muscles for consecutive bare-handed gripping with 2 minutes rest between trials for all 18 subjects. Similar plots were also observed for the other three experimental settings (Figures not shown here). The results indicated that the muscles of the subjects during the gripping session were categorized as "recovery" or "force decrease" when most of the pairs were plotted in the left quadrants. According to the results, all the subjects performed MVC gripping for 2 hours without fatigue. In this study, the suitable sampling duration for performing consecutive MVP gripping were decided as the time when the current MVC grip value fall below 80% of the MVC griping exertions. The rationale is that "three replicates of maximum grip strength within 10% tolerance of the mean score were recorded as the MVC grip strength" following the Caldwell's regimen (1974).

The duration for male and female subjects performing consecutive MVP gripping without fatigue (<80% MVC) ranged from 10~112 and 18~112 minutes, respectively. Table 4 shows the average and standard deviation for suitable sampling duration for 10 male and 8 female subjects. Applying Z-standardization to the original not normally distributed data, the suitable sampling duration for 90% of males and 75% of females to perform consecutive MVC gripping exertions without fatigue ranged from 8.19 to 47.88 minutes and 32.18 to 98.43 minutes, respectively.

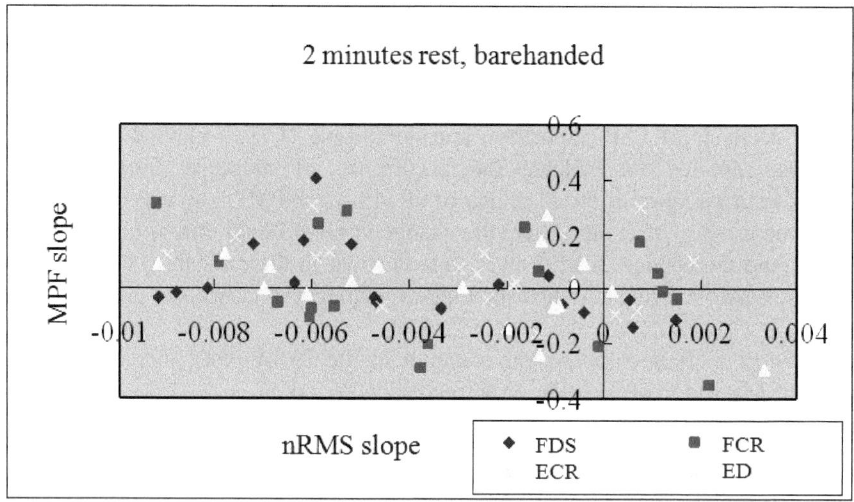

Figure 1 JASA plot of four forearm muscles for consecutive barehanded gripping with 2 minutes rest between trials for all 18 subjects

Table 4 Suitable sampling duration for performing consecutive gripping exertions w/o glove with 2 or 4 minutes rest between exertions

Gender	Rest time	Glove usage	Average	Standard Deviation	Z value	90% male/ 75% female
Male	2 minutes	Bare	53.80	28.90	1.28	16.80
		Neoprene	44.40	27.82	1.28	8.19
	4 minutes	Bare	84.00	28.22	1.28	47.88
		Neoprene	78.00	31.06	1.28	38.24
Female	2 minutes	Bare	65.00	26.23	0.68	47.16
		Neoprene	48.75	24.38	0.68	32.18
	4 minutes	Bare	107.00	12.60	0.68	98.43
		Neoprene	75.50	25.70	0.68	58.02

4 DISCUSSION and CONCLUSION

The means and standard deviations of the MVC grip strength of the bare hand measured are consistent with the normative data for male and female adults ranging from 20 to 30 years old reported by Mathiowetz et al (1985). Gloved hands impaired MVC grip strength significantly compared to that of bare hand and the observed finding are as expected and consistent based upon a review of the published literature of industrial gloves. The results also indicated that male retain higher grip strength than that of females which are similar to previous studies (Hallbeck and McMullin, 2003; Nicolay and Walker, 2005).

Based on Joint analysis of EMG spectrum and amplitude (JASA), the subjects can follow the recovery time of 2 minutes of Caldwell's regimen and 4 minutes between consecutive grip exertions without fatigue. Therefore, the suitable sampling durations for gripping exertions were not decided upon the JASA results. In this study, the suitable sampling duration is defined as the time that the subject can maintain 80% of the maximum grip strength during consecutive maximum gripping exertions. For each of the four different gripping conditions (bare hand/2 minutes rest, gloved hand/2 minutes rest, bare hand/4 minutes rest, gloved hand/4 minutes rest), the results show that the suitable sampling duration that 90% of males and 75% of females can perform consecutive MVC gripping exertions without fatigue were 17、8、48、38 minutes and 47、32、98、58 minutes, respectively.

According to the suitable sampling duration, the maximum number of gripping exertions recommended for the males and females are 8、4、12、9 times and 23、16、24、14 times for the four different gripping conditions (bare hand/2 minutes rest, gloved hand/2 minutes rest, bare hand/4 minutes rest, gloved hand/4 minutes rest) in a consecutive sampling session, respectively. For the male subjects, the 4 minutes rest protocol will be recommended which yields more exertions of gripping when consecutive sampling is required. For the female subjects, 2 minutes rest protocol advised by the Caldwell's regimen and 4 minutes rest protocol are both accepted according to the results of this study. However, the four minutes rest protocol will be recommended in this study due to the following two reasons: (1) the MVC grip strength for the 4 minutes rest protocol are 2.6±0.6 Kg higher than those of the 2 minutes rest protocol although no statistically significant differences were found, and (2) the reduction of the MPF on the ECR muscle in the 2 minutes rest protocol is larger than that of the 4 minutes rest protocol meaning the potential for ECR muscle to fatigue earlier. Adopting 4 minutes rest protocol for both genders could potentially get higher (and more accurate?) MVC gripping reading with less fatigue during consecutive sampling period although JASA results show that the subjects could performed MVC gripping for 2 hours without fatigue following both the 2 and 4 minutes rest protocols.

ACKNOWLEDGEMENTS

The authors would like to acknowledge grant NSC 98-2221-E-324-010 from the National Science Council of Taiwan for financially supporting this research.

REFERENCES

Bao, S. and Silverstein, B. (2005), "Estimation of hand force in ergonomic job evaluations." *Ergonomics*, 48, 288-301.

Bishu, R.R. and Kim, B. (1995), "Force-endurance relationship: does it matter if gloves are donned?" *Applied Ergonomics*, 26 (3), 179-185.

Bureau of Labor Statistics (2007), Nonfatal occupational injuries and illnesses requiring days away from work 2007, Table 5 and Table 6, United States Department of Labor, Washington, DC, 12-15.

Caldwell, L.S., Chaffin, D.B., Dukes-Dobos, F.N., Kroemer, K.H.E., Laubach, L.L., Snook, S.H.L. and Wasserman, D.E. (1974), "A proposed standard procedure for static muscle strength testing." *American Industrial Hygiene Association Journal*, 35, 201-206.

Hallbeck, M.S. and Mcmullin, D.L. (2003), "Maximal power grasp and three-jaw chuck pinch force as a function of wrist position, age, and glove type." *International Journal of Industrial Ergonomics*, 11, 195-206.

Hicks, A.L., Kent,-Braun, J., and Ditor, D.S. (2001), "Sex differences in human skeletal muscle fatigue", *Exercise and Sport Science Reviews*, 29, 109-112.

Klein, M.G. and Fernandez, J.E. (1997), The effect of posture, duration, and force on pinching frequency, *International Journal of Industrial Ergonomics*, 20, 267-275.

Kovacs, K, Splittstoesser, R., Maronitis, A., and Marras, W.S. (2002), "Grip force and muscle activity differences due to glove type." *American Industrial Hygiene Association Journal*, 63, 269-274.

Larivière, C, Plamondon, A., Lara, J., Tellier, C., Boutin, J., and Dagenais, A. (2004), "Biomechanical assessment of gloves. A study of the sensitivity and reliability of electromyographic parameters used to measure the activation and fatigue of different forearm muscles." *International Journal of Industrial Ergonomics*, 34, 101-116.

Lin, M.I., Liang, H.W., Lin, K.H., and Hwang, Y.H. (2004), Electromyographical assessment on muscular fatigue-an elaboration upon repetitive typing activity, *Journal of Electromyography and Kinesiology*, 14, 661-669.

Mathiowetz, V., Kashman, N., Volland, G., Weber, K., Dowe, M. (1985), "Grip and pinch strength: normative data for adults." *Arch Phys Med Rehabil*, 66, 69-74.

Muralidhar, A., Bishu, R.R., and Hallbeck, M.S. (1999), The development and evaluation of an ergonomic glove, *Applied Ergonomics*, 30, 555-563.

Nicolay, C.W., Kenney, J.L., and Lucki, N.C. (2007), "Grip strength and endurance throughout the menstrual cycle in eumenorrheic and woman using oral contraceptives." *International Journal of Industrial Ergonomics*, 37, 291-301.

Nicolay, C.W. and Walker, A.L. (2005), "Grip strength and endurance: Influences of anthropometric variation, hand dominance, and gender." *International Journal of Industrial Ergonomics*, 35, 605-618.

NIOSH (1991), Selected topics in surface electromyography for use in the Occupational setting: expert perspectives, DHHS (NIOSH) publication No. 91-100, National Institute of Occupational Safety and Health, Cincinnati, OH.

Sung, P.C. (2006), "Glovebox gloves: ergonomics guidelines for the prevention of musculoskeletal disorders." Doctoral Dissertation, Industrial Hygiene Division, Department of Environmental Health Science, School of Public Health, University of California, Los Angeles, U.S.A.

CHAPTER 59

Basic Study on Automotive Warning Presentation to Front/Rear Hazard by Vibrotactile Stimulation

Atsuo MURATA, Susumu Kemori, Takehito Hayami and Makoto Moriwaka

Graduate School of Natural Science and Technology, Okayama University
Okayama, Japan
murata@iims.sys.okayama-u.ac.jp

ABSTRACT

An automotive warning presentation system with vibrotactile device was evaluated with an empirical study. Young and older adults participated in the experiment. The participants were required to simultaneously carry out a tracking task, a switch pressing task such as a selection of light-off, and a judgment task of information which randomly appeared on the front or the rear. The vibrotactile device was located at their arms, legs, and stomach/back. The vibrotactile stimulation was presented as apparent movement and single-point vibrotactile stimulation. It was confirmed that the vibrotactile warning was effective for both young and older adults. As for young adults, the stimulation location did not affect the performance measures such as the reaction time to dangerous situations. On the other hand, concerning older adults, the stimulation to the leg led to better performance such as the reaction to dangerous situations irrespective of the method of warning presentation.

Keywords: vibrotactile, warning presentation, automobile, young and older adults

1 INTRODUCTION

As the display and control systems of automobile is becoming more and more complex, it is predicted that older drivers are distracted by these systems and cannot cope with such situations. It is regarded that the utilization of sense of touch as a medium for information representation is promising (Jones et al., 2008). They concluded that sense of touch represents a promising means for communication in human-vehicle system. In driving environment, most information is presented via a visual or auditory stimulus. If the warning signal is presented via a visual or auditory stimulus, the auditory or visual interference with other information might arise. On the other hand, if a vibtotactile warning, that is, tactile interface is used, the possibility of such interference would be sure to reduce.

Recently, the tendencies of cross-modal information processing (Spence et al., 1997, Driver et al., 1998, Ho et al., 2005, Ho et al., 2006a, Ho et al., 2006b) and design have emerged as major research topics in the design of automotive warning system. Presenting information via multiple modalities such as vision, audition, and touch has been expected to be a promising means to reduce transmission errors and enhance safety. A better understanding of cross-modal spatial and temporal links is essential to ensure a better application of this property to the automotive warning design. Ho et al., 2005 and Ho et al., 2006b showed the effectiveness of vibrotactile warning presentation in driving environment. In traffic situations, many hazards exist ubiquitously. Therefore, the effectiveness of automotive warning must be confirmed using a lot of locations where hazards potentially hide. However, few studies have examined the effectiveness of automotive warning system using more than two locations. Ho et al., 2005, Ho et al., 2006a and Ho et al., 2006b explored the effectiveness of tactile warning for front and rear locations. Murata et al., 2011 examined how the tactile warning is effective for left and right locations. Moreover, few studies investigated the effectiveness of vibrotactile warning by apparent movement. It is expected that vibrotactile warning by apparent movement can more quickly transmit the directional cues than the simultaneous stimulation of two vibrotransducers or the single-point stimulation.

Although older adults exhibit deficits in various cognitive-motor tasks (Goggin et al., 1989, Goggin et al., 1990, Imbeau et al., 1993, Smith et al., 1993, Stelmach et al., 1993, Wirewille et al., 1993, Murata et al., 2005, Murata et al., 2011), older adults' decline of tactile sense seems to be less as compared with visual or auditory sense. A few studies showed that the presentation of spatially predictive vibrotactile warning signal can facilitate driver's response to driving event seen through the windscreen or rear mirror. On the basis of the discussion above, it is expected that a vibrotactile signal would be very promising as a warning signal especially for older adults.

The aim of this study was to improve driving safety using a vibrotactile warning system. For both young and older adults, the effectiveness of vibrotactil warning system was compared between the system that made use of apparent movement using 2-point stimuli and the system of single point stimulation.

2 METHOD

2.1 Participants

Twenty participants took part in the experiment. Ten were male adults aged from 65 to 76 years. All had held a driver's license for 30 to 40 years. Ten were male undergraduate students aged from 21 to 24 years and licensed to drive from 1 to 3 years. Stature of participants ranged from 160 to 185 cm. The visual acuity of the participants in both young and older groups was matched and more than 20/20. They had no orthopedic or neurological diseases.

2.2 Apparatus

Electromechanical actuator (Audiological Engineering Corporation, Tactaid VBW32 transduces) was used for presenting tactile warning The frequency and the intensity of this transducer was adjusted using a function generator (GWInstek, SFG-2004). The switch used was the same as that used in Murata et al., 2011. For a simulated driving task, a personal computer (mouse computer, 0603Lm-i211B), a projector (EPSON, EMP-S4), and a steering wheel (Logitech, MOMO) were used. For a switch operation, a personal computer (HP, NX6120), a CRT (SONY, SDM-N50) and a digital I/O card (Interface, PIO-24W (PM)) were used.

2.3 Experimental task

2.3.1 Simulated driving (Tracking task)

The display used in the simulated driving is shown in Figure 1. The participant was required to steer, maintain the center line, and make the deviation between the center line and the actual location as small as possible.

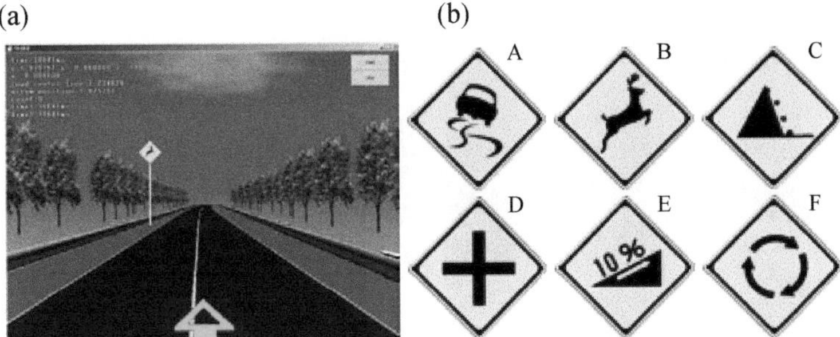

Figure 1 (a) Hazardous information displayed on front screen and (b) traffic signs used in the experiment.

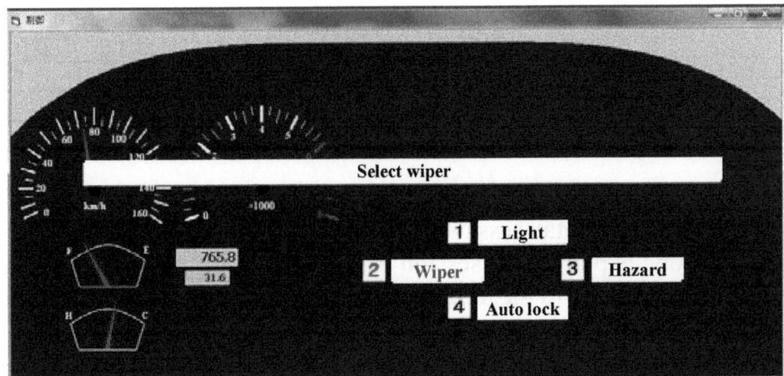

Figure 2 Display of switch pressing task.

2.3.2 Switch operation task

The display in the switch operation task is depicted in Figure 2. The participant was required to press a steering-wheel mounted switch according to the instruction on the display. When the participant was required to select "Wiper", he must press the corresponding switch using a steering-wheel mounted switch.

2.3.3 Judgment to hazardous situation

The participant was required to recognize the frontal or rear hazardous situations as fast and accurately as possible. In the judgment to frontal hazardous situations, the participant was required to react using a foot switch to hazardous traffic signs A, B, and C in Figure 1. The participant did not need to traffic signs D-F in Figure 1. In one experimental session, a total of 20 traffic signals were presented to the participant. The traffic signs A-C were randomly presented to the participant 6 times (Each of A-C was presented 2 times).

In the rear hazard judgment task, the participant was required to monitor the distance between own vehicle and the following vehicle using a back mirror. In Figure 3, the display of rear hazard judgment is depicted. In Figure 3(a) corresponding to a hazardous situation, the participant must recognize this situation and react to this using a foot switch as fast and accurately as possible. In other situations than Figure 3(a), no reaction was required to the participant. In one experimental session, such scenes as shown in Figure 3(a) and (b) were presented to the participant 20 times of which the scenes corresponding to Figure 3(a) appeared randomly 6 times.

(a) (b)

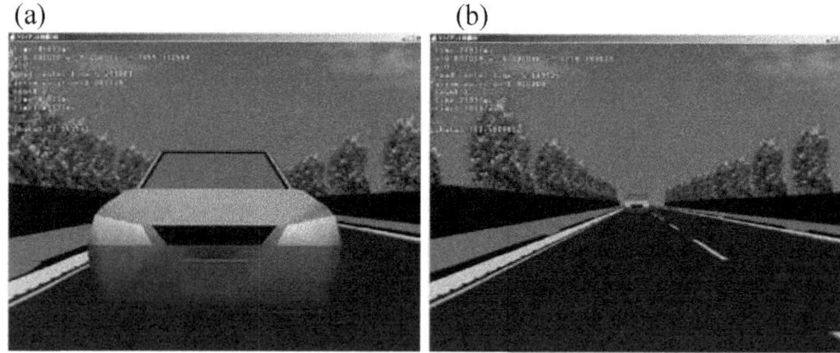

Figure 3 (a) Hazardous situation and (b) the situation not regarded as hazardous displayed on rear screen (back mirror).

Table 1 Condition for vibrotactile stimulation by apparent movement.

	d [ms]	SOA[ms]	ISI [ms]	E [V]	*f* [Hz]	*D* [mm]
Arm	125	175	50	10	281	65
Leg	75	150	75	10	251	48
Stomach and back	150	150	0	10	249	

In summary, six frontal hazards and six rear hazards were presented to the participants in one experimental session. The overlap of the switch pressing and the hazard judgment in one experimental session was, on the average, 83%. The participant was instructed to have priority to the hazard judgment when the overlap occurred.

2.4 Design

The age, the warning presentation method, and the body segment to which the warning signal was presented were experimental factors. Only the age was a between-subject factor. Other factors were within-subject factors. The warning was presented to the right lower arm, the right thigh, or the stomach/abdomen. The vibrotactile warning presentation methods (no warning, vibrotactile stimulation by apparent movement, and single-point vibrotactile stimulation) are described. In the simultaneous two-point stimulation, the warning was presented simultaneously to the body segment (right lower arm, the right thigh, and the stomach/abdomen). Two tactors (vibrotranceducers) were attached to the body segment, and the tactors were simultaneously stimulated. The condition for simultaneous two-point vibrotactile stimulation is summarized in Table 1. As for the warning presentation making use of apparent movement, the condition necessary for producing apparent movement was preliminary determined for each body segment as in Table 2. The condition included the duration of stimulation (d), the inter-stimulus interval (SOA: Stimulus Onset Asynchrony), the distance between two tactors (D), the frequency of stimulation (f), and the intensity of stimulation (E). The explanation of these parameters is shown in Figure 4.

2.5 Procedure

The participants were required to adjust the seat and the location of back mirror so that they can carry out an experimental task comfortably. For all participants, it was confirmed whether they can recognize the front-rear and the rear-front apparent movements. The experimental task was explained to the participant, and the practice session was carried out until the participant understood the task thoroughly.

The participants were required to simultaneously carry out a tracking task, a switch pressing task such as a selection of light-off, and a judgment task of information which randomly appeared on the front or the rear. In case of the frontal hazard judgment task,

Table 2 Condition for simultaneous two-point vibrotactile stimulation.

	d [ms]	E [V]	*f*[Hz]
Arm	300	10	281
Leg	225	10	251
Stomach and back	300	10	249

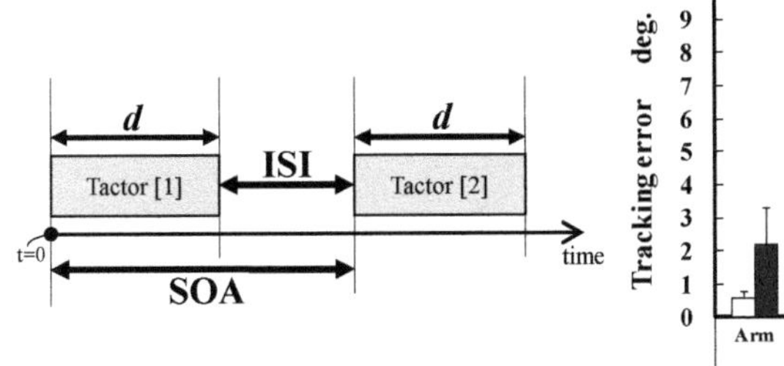

Figure 4 Explanation of stimulus duration (*d*), stimulus onset asynchrony (SOA) and inter-stimulus interval (ISI).

Figure 5 Tracking error as a function of age, type of warning and attachment location of tactors.

the participant were required to respond as quickly and accurately as possible to the pre-specified traffic sign. As for the rear hazard judgment task, the participant responded as fast and accurately as possible when the following vehicle appeared on the rear mirror as pre-specified by the experimenter.

A total of seven experimental sessions (six combinations of warning presentation and body segment (2 X 3), and no warning) were carried out by the participant. About two-minute rest was inserted between experimental sessions.

2.6 Evaluation measure

The tracking error, the correct percentage of switch pressing, and the reaction time and the percentage correct reaction to the front or rear stimulus were measured. It was examined how age, the method of warning presentation (no warning, vibrotactile stimulation by apparent movement, and stimulus two-point vibrotactile stimulation), and the attachment location of tactors (arm, leg, and stomach/back) affected the measures above.

3 RESULTS

3.1 Tracking error

In Figure 5, the tracking error is plotted as a function of age, type of warning and attachment location of tactors. A three-way (age by warning method by warning segment) ANOVA carried out on the tracking error revealed only a main effect of age (***F***(1,18)=17.929, ***p***<0.01). As a result of a two-way (age by warning method including body segment (seven levels)) conducted on the tracking error, a significant main effect of age (***F***(1,18)=19.760, ***p***<0.01) and a significant age by warning method interaction (***F***(6,18)=2.284, ***p***<0.05) was detected. Fisher's PLSD revealed a significant difference as shown in Table 3. The tracking error when older adults were presented warning via apparent movement to the stomach/back tended to be larger.

Table 3 Results of multiple comparisons for tracking error by Fisher's PLSD for each age group.

		Apparent movement			Two-point stimulation			No warning
		Arm	Leg	Stomach & back	Arm	Leg	Stomach & Leg	
Apparent movement	Arm							
	Leg	Y: n.s. O: n.s.						
	Stomach & back	Y: n.s. O:*p*<0.01	Y: n.s. O:*p*<0.05					
Two-point stimulation	Arm	Y: n.s. O: n.s.	Y: n.s. O: n.s.	Y: n.s. O:*p*<0.01				
	Leg	Y: n.s. O: n.s.	Y: n.s. O: n.s.	Y: n.s. O:*p*<0.05	Y: n.s. O: n.s.			
	Stomach & back	Y: n.s. O: n.s.	Y: n.s. O: n.s.	Y: n.s. O:*p*<0.05	Y: n.s. O: n.s.	Y: n.s. O: n.s.		
No warning		Y: n.s. O: n.s.	Y: n.s. O: n.s.	Y: n.s. O:*p*<0.01	Y: n.s. O: n.s.	Y: n.s. O: n.s.	Y: n.s. O: n.s.	

Y: Results on young adults, O: Results on older adults, n.s.: not significant

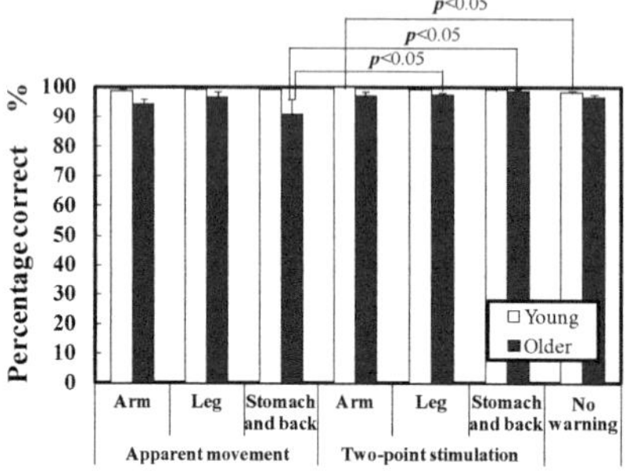

Figure 6 Percentage correct in the switch operation task as a function of age, type of warning and attachment location of tactors.

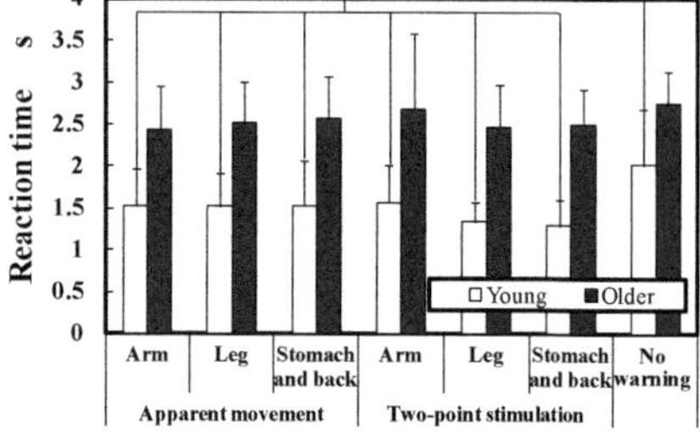

Figure 7 Reaction time to frontal hazard information as a function of age, type of warning and attachment location of tactors.

3.2 Percentage correct in switch operation

In Figure 6, the percentage correct in switch operation is compared between age groups, between types warning, and among attachment locations of tactors. A three-way (age by warning method by warning segment) ANOVA carried out on the percentage correct revealed main effects of age ($F(1,18)=10.619$, $p<0.01$) and warning method ($F(1,18)=5.088$, $p<0.05$). As a result of a two-way (age by warning method including body segment (seven levels)) conducted on the percentage correct, only a significant main effect of age ($F(1,18)=19.760$, $p<0.01$) was detected. The pairs for which significant differences were detected as a result of Fisher's PLSD are depicted in Figure 6. The percentage correct of older adults tended to be lower when the warning was presented via apparent movement to the stomach/back.

3.3 Reaction time to frontal hazardous situations

The reaction time to frontal hazardous information is plotted as a function of age, type of warning and attachment location of tactors in Figure 7. A similar three-way (age by warning method by warning segment) ANOVA carried out on the reaction time revealed only a main effects of age ($F(1,18)=50.036$, $p<0.01$). As a result of a similar two-way (age by warning method including body segment (seven levels)) conducted on the reaction time, significant main effects of age ($F(1,18)=45.788$, $p<0.01$) and

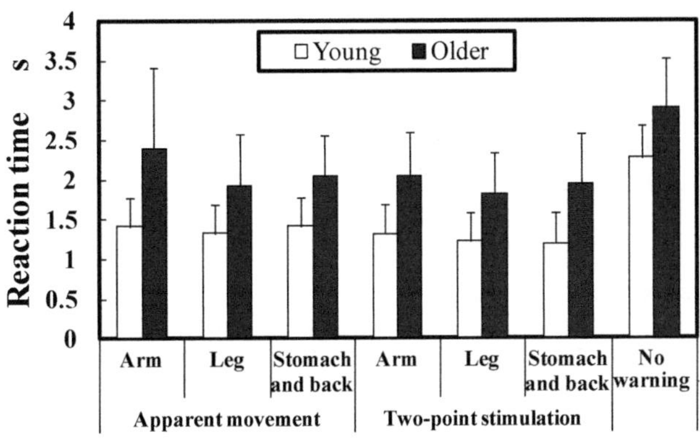

Figure 8 Reaction time to rear dangerous information as a function of age, type of warning and attachment location of tactors.

Table 4 Results of multiple comparisons for reaction time to rear dangerous information by Fisher's PLSD for each age group.

		Apparent movement			Two-point stimulation			No warning
		Arm	Leg	Stomach & back	Arm	Leg	Stomach & Leg	
Apparent movement	Arm							
	Leg	Y: n.s. O: $p<0.05$						
	Stomach & back	Y: n.s. O: n.s.	Y: n.s. O: n.s.					
Two-point stimulation	Arm	Y: n.s. O: n.s.	Y: n.s. O: n.s.	Y: n.s. O: n.s.				
	Leg	Y: n.s. O: $p<0.01$	Y: n.s. O: n.s.	Y: n.s. O: n.s.	Y: n.s. O: n.s.			
	Stomach & back	Y: $p<0.05$ O: $p<0.05$	Y: n.s. O: n.s.	Y: $p<0.05$ O: n.s.	Y: n.s. O: n.s.	Y: n.s. O: n.s.		
No warning		Y: $p<0.01$ O: $p<0.05$	Y: $p<0.01$ O: $p<0.01$	Y: $p<0.01$ O: $p<0.01$	Y: $p<0.01$ O: $p<0.01$	Y: $p<0.01$ O: $p<0.01$	Y: $p<0.01$ O: $p<0.01$	

Y: Results on young adults, O: Results on older adults, n.s.: not significant warning method ($F(6,18)=3.350$, $p<0.01$) were detected. The pairs for which significant differences were detected as a result of Fisher's PLSD are depicted in Figure 7.

3.4 Reaction time to rear hazardous situations

The reaction time to frontal hazardous information is plotted as a function of age, type of warning and attachment location of tactors in Figure 8. A similar three-way (age by warning method by warning segment) ANOVA carried out on the reaction time revealed only a main effects of age ($F(1,18)=14.902$, $p<0.01$) and warning method ($F(1,18)=13.814$, $p<0.01$). As a result of a similar two-way (age by warning method including body segment (seven levels)) conducted on the reaction time, significant main effects of age ($F(1,18)=14.563$, $p<0.01$) and warning method ($F(6,18)=19.809$, $p<0.01$) were detected. Fisher's PLSD revealed a significant difference as shown in Table 4. As for older adults, it tended that the reaction time stimulated by thigh was shorter for both warning methods.

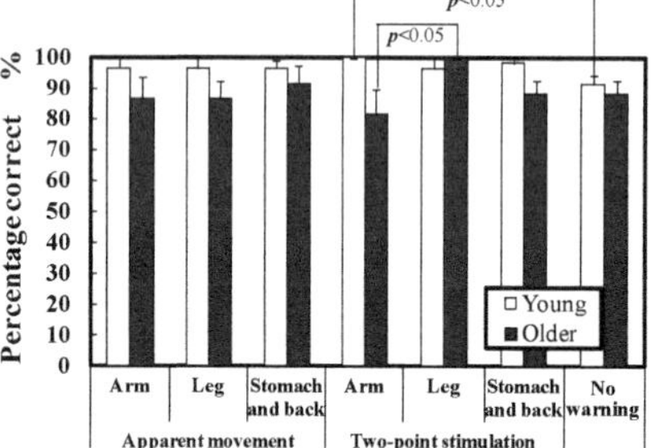

Figure 9 Percentage correct in frontal danger judgment task as a function of age, type of warning and attachment location of tactors.

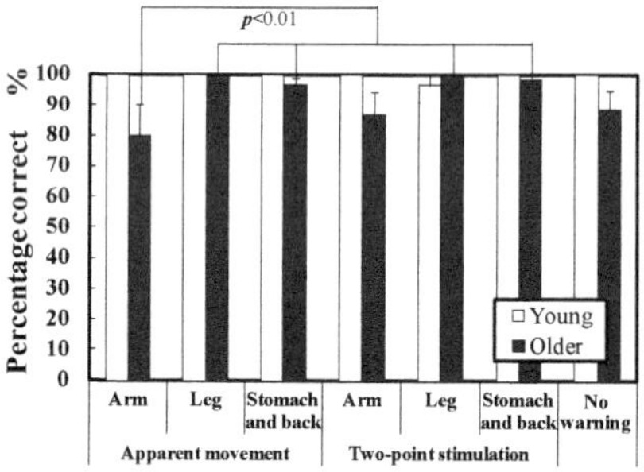

Figure 10 Percentage correct in rear danger judgment task as a function of age, type of warning and attachment location of tactors.

3.5 Percentage correct judgment to frontal hazardous situations

The percentage correct judgment to frontal hazardous situations is shown as a function of age, type of warning and attachment location of tactors in Figure 9. A similar three-way (age by warning method by warning segment) ANOVA carried out on the percentage correct revealed only a main effects of age ($F(1,18)=10.150$, $p<0.01$). As a result of a similar two-way (age by warning method including body segment (seven levels)) conducted on the reaction time, only a significant main effects of age ($F(1,18)=11.695$, $p<0.01$) was detected. The pairs for which significant differences were detected as a result of Fisher's PLSD are depicted in Figure 9.

3.6 Percentage correct judgment to rear hazardous situations

The percentage correct judgment to rear hazardous situations is shown as a function of age, type of warning and attachment location of tactors in Figure 10. A similar three-way (age by warning method by warning segment) ANOVA carried out on the percentage correct revealed no significant main effects. A age by warning method interaction was significant ($F(2,36)=3.334$, $p<0.05$). As a result of a similar two-way (age by warning method including body segment (seven levels)) conducted on the percentage correct, a significant main effects of age ($F(1,18)=4.846$, $p<0.05$) and a significant age by warning method interaction ($F(6,18)=2.538$, $p<0.05$) were detected. The pairs for which significant differences were detected as a result of Fisher's PLSD are depicted in Figure 10.

4 CONCLUSIONS

4.1 Effects of age on performance and effectiveness of warning presentation

The percentage correct in the switch pressing task, the percentage correct in the frontal hazard judgment and the percentage correct in the rear hazard judgment of older adults were 0.985 times, 0.964 times, and 0.833 times as high as those of young adults, respectively. Different from these evaluation measures, the difference of the tracking error, the frontal reaction time, and the rear reaction time between young and older adults tended to be larger. The tracking error, the frontal reaction time, and the rear reaction time were 3.434 times, 1.357 times and 1.281 times as large as those of young adults, respectively. This indicates that the degradation of perceptual, cognitive, and motor functions of older adults has been reflected in these measures.

The evaluation measures of young adults were not affected by the attachment location of tactors, while those of older adults were affected by the attachment location of tactors (See Figures 5-10). For both age groups, the warning conditions led to higher performance than the condition without warning presentation, indicating that the warning presentation is effective for safety driving. These results correspond well with Ho et al., 2005, Ho et al., 2006a, Ho et al., 2006b and Murata et al., 2011.

4.2 Effects of warning presentation method on performance

As for the frontal warning presentation, the warning presentation method (apparent movement and simultaneous two-point stimulation) did not affect the reaction time. On the other hand, concerning the rear warning presentation, the simultaneous two-point stimulation tended to lead to faster reaction than the warning presentation by apparent movement. The apparent movement did not necessarily lead to faster reaction. The reason might be that the rear hazard judgment need more complicated cognitive information processing than the frontal hazard judgment. As all participants managed to perceive the frontal-rear or the rear-frontal apparent movement, more than two vibrotranceducers were not used in this study. It might be possible that the warning presentation by apparent movement with more than two tactors leads to faster perception and reaction to the hazardous situation. This should be further investigated in future research.

It was hypothesized that apparent movement enables us to perceive the direction of hazard quickly and leads to faster reaction to hazard situation. However, the effectiveness of apparent movement as a warning presentation means has not been verified. In the range of this study, the rear warning presentation to the lower arm or the stomach/abdomen by apparent movement is not proper from the viewpoint of accuracy and speed of reaction.

Murata et al., 2011 discussed the effectiveness of warning presentation to the right or left hazard. On the basis of this result and this study, it seems that the perception and addition of directional information is easier in the right-left direction than in the rear-front direction. In future research, it must be explored how the further elaboration of rear-front warning system should be realized using a method other than apparent movement.

4.3 Effects of stimulated body parts on performance

Judging from Figure 5, Figure 8, and Figure 10, the following warning condition is not proper for older adults: apparent movement (stomach/back), apparent movement (lower arm), simultaneous two-point stimulation (lower arm), and simultaneous two-point stimulation (stomach/back). The attachment to the thigh for both apparent movement and simultaneous two-point stimulation leads to a faster reaction and should be recommended for older adults. The reason can be inferred on the basis of Jones et al. 2008 that the thigh is more sensitive to the vibrotactile stimulation than the abdomen. By contrast, the performance measures of young adults were not affected by the attachment location of tactors and the warning presentation method.

The discussion above supports a viewpoint that different design of warning presentation systems is necessary according to the age of drivers. When designing a warning presentation system from the viewpoints of universal design, the attachment location of tactor should be thigh, because the presentation of tactile warning to the thigh led to a faster reaction of older adults.

It was confirmed that the tactile warning were effective for both young and older adults. The vibrotactile stimulation on leg led to higher hit rate, in particular, for older adults irrespective of the method of warning presentation. The warning presentation by vibrotactile apparent movement led to the slower reaction to the rear danger than the single-point vibrotactile stimulation. In the range of our experiment, the single-point stimulation by vibrotactile sense led to faster response for both age groups. As for young adults, the stimulation location did not affect the performance measures such as the reaction time to dangerous situations. This means that any stimulation location is possible for young adults. On the other hand, concerning older adults, the stimulation to the leg led to better performance such as the reaction to dangerous situations irrespective of the method of warning presentation.

REFERENCES

Driver,J. and Spence,C. 1998. Attention and the cross-modal construction of space, *Trends in Cognitive Science*, 2, 254-262.

Goggin,N.L., Stelmach,G.E. and Amrhein, P.C. 1989. Effects of age on motor preparation and restructuring, *Bulletin of the Psychonomic Society*, 27, 199-202.

Goggin,N.L., and Stelmach,G.E. 1990. Age-related differences in kinematic analysis of perceptual movements, *Canadian Journal on Aging*, 9, 371-385.

Ho,C., Reed,N. and Spence,C. 2006a. Assessing the effectiveness of "intuitive" vibrotactile warning signals in preventing front-to-rear-end collisions in a driving simulator, *Accident Analysis and Prevention*, 38(5), 988-996.

Ho,C., Tan,H.Z. and Spence,C. 2005. Using spatial vibrotactile cues to direct visual attention in driving scenes, *Transportation Research*, Part F, 8, 397-412.

Ho,C., Tan,H.Z. and Spence,C. 2006b. The differential effect of vibrotactile and auditory cues on visual spatial attention, *Ergonomics*, 7(10), 724-738, 2006.

Imbeau,D., Wierwille,W.W. and Beauchamp,Y. 1993. Age, display design and driving performance, in *Automotive Ergonomics*, eds. Peacock,B. and Karwowski,W., Taylor & Francis, London, 339-355.

Jones,L.A. and Sarter,N.B. 2008. Tactile displays: Guidance for their design and application, *Human Factors*, 50(1), 90-111.

Murata,A. and Moriwaka,M. 2005. Ergonomics of Steering Wheel Mounted Switch -How Number of Arrangement of Steering Wheel Mounted Switches Interactively Affects Performance-, *International Journal of Industrial Ergonomics*, 35, 1011-1020.

Murata,A. and Moriwaka,M. 2008. Evaluation of automotive control-display system by means of mental workload, *Proc. of IWICA2008*, 52-57.

Murata,A., Tanaka,K., and Moriwaka,M. 2011. Basic study on effectiveness of tactile interface for warning presentation in driving environment, *International Journal of Knowledge Engineering and Software Data Paradigm*, 3, 112-120.

Smith,D.B.D., Meshkait,N., and Robertson,M.M. 1993. The older driver and passenger, in *Automotive Ergonomics*, eds. Peacock,B. and Karwowski,W., Taylor & Francis, London,453-467.

Spence,C. and Driver,J. 1997. Cross-modal links in attention between audition, vision, and touch: Implications for interface design, *International Journal of Cognitive Ergonomics*, 1, 351-373.

Stelmach,G.E. and Nahom,A. 1993. The effects of age on driving skill cognitive-motor capabilities, in *Automotive Ergonomics*, eds. Peacock,B. and Karwowski,W., Taylor & Francis, London, 219-233.

Wierwille,W.W. 1993. Visual and manual demands of in-car controls and displays, in *Automotive Ergonomics*, eds. Peacock,B. and Karwowski,W., Taylor & Francis, London, 299-320.

CHAPTER 60

Characterization of the Hand Bike Sport Gesture: A Quantitative Kinematic Analysis and an Ergonomic Assessment of Different Vehicle Adjustments

Marco Mazzola, Fiammetta Costa, Giuseppe Andreoni.

Dipartimento di Industrial Design, Arts and Communication (INDACO), Politecnico di Milano, Milano, Italy

ABSTRACT

The aim of the presented paper is to propose an ergonomic assessment through a biomechanics description of the hand bike gesture of a Paralympics Athletes that will challenge in London 2012 Paralympics competition.

The research investigates the joint kinematics of the athlete to produce a significant postural ergonomic scores in dynamic condition, and to compare different vehicle adjustments to determinate the most preferable from a biomechanical point of view. The 3D total-body kinematics of one top Italian Paralympics athlete was recorded by an optoelectronic system. Joint angles patterns and segmental mass distribution were used to estimate the comfort/discomfort scores. The ergonomic assessment has been provided with the application of the Method for Movement and Gesture Assessment (MMGA), an innovative index that allows the quantitative determination of discomfort perception in dynamic condition. Three vehicle adjustments have been evaluated: the backseat inclination, the handlebar height and the handle grip length.

The main relevance of the results is that a new and more complete set of biomechanical data is provided. The ergonomic analysis presents results both coherent with the observation about the influence of the backseat inclination, and both surprising about the strong variation measured for the handgrip length.

INTRODUCTION

The HandBike (HB) is officially classified as a Paralympics sport discipline since 2004. It is a profile in the Sports class classification according to the para-cycling specific classification system [1] that assesses the athlete's ability based on the level of the impairment relevant to their specific injury. Although the hand bike as a sport discipline is growing for popularity also in leisure and rehabilitation purposes, there is still a lack of quantitative data to characterize the specific gesture. It is possible to affirm that in the last years the scientific community has started to produce literature for the investigation of the physiological data for human performance and sport medicine. It is possible to classify three main areas of interests. The first one is related to the evaluation of the energetic costs of the athletes during the performance, that is characterized by biosignals or parameters derived from heart frequency, breath frequency and metabolic consumption[2-5]. The second area is related to the comprehension of the effects that vehicle adjustments produce on the posture and the performance of the athlete (optimization and ergonomic of the HandBike-human system [6]. The third area consists of the kinematics analysis of the athletes in order to quantify possible muscle-skeletal disorder induced by the effort and it is based on the traditional rehabilitation approach to biomechanics [7-9]. These works are probably the most relevant studies on the characterization of the kinematics of the HandBike disciplines, but are not exhaustive and demonstrates that there is still a need of quantitative information for the evaluation of the high level athletes performance. The aim of this study is to study the kinematics of the HandBike propulsion of an Italian Paralympics Athletes that will challenge during the 2012 London Paralympics competition in the HB category H2 [9] . Objective of the investigation is the kinematic of the upper limb extremity of the athlete while performing on its personal vehicle with its personal configuration, and the comparison of the results with the kinematics observed after three specific vehicle regulations. The quantification of the effects produced by the regulations has computed through a specific Ergonomic Index based on the joints angle kinematics[10]. This work is focused on a single athletes in response of the need to produce specific information about the biomechanics of HB propulsion to improve the vehicle assessment for the competition.

METHODS

Experimental set-up

Upper-body kinematics was recorded on an Italian Paralympics athlete through a six-cameras (TVCs) optoelectronic system (Vicon M460, Vicon Motion System

Ltd, Oxford Metrics, Oxford, UK) working at 120 Hz. TVCs were placed so that a volume of about 3 x 2 x 2 m was covered. Calibration procedures were carried out before each experimental session and a maximum mean error of 0.6 mm concerning markers placed on a rigid wand was obtained.

The determination of the joints angles pattern is based on the upper limb biomechanical model proposed by Schmidt [11], with the addition of specific markers for the evaluation of shoulder and trunk joint kinematics. 32 retro-reflective spherical markers were used for the kinematic computation (Figure 1). Among these markers, 14 are glued onto the subject's anatomical landmarks and 18 are technical markers fixed with elastic bands on the arms and the forearms of the subjects.

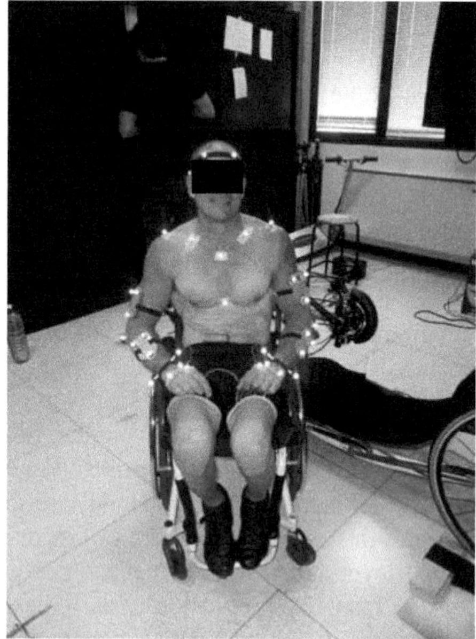

Figure 1: marker placement for the quantitative motion capture analysis

The following variables were considered for this study: wrist and elbow flex-extension; shoulder flex-extension, intra-extra rotation and abd-adduction; trunk flex-extension, rotation and lateral bending

Experimental protocol

The experimental protocol consists of 32 experimental conditions. Each condition is related to a specific vehicle regulation:

- The athlete performs the gesture on his own HandBike to produce the reference value for the comparison (1 cond. – Figure 2a)
- The athlete performs the gesture on a different vehicle of the same but capable to be regulated. This measure produce the background offset between the two vehicles. (1 cond. - Figure 2b)

- 5 different regulations of the backseat inclination (5 cond. - Figure 3)
- 4 different height of the handlebar (3 conditions – Figure 4)
- 2 different length of the hand grips, 175 mm and 170 mm (2 conditions)

Each one of the different regulation has been tested as a single variable with a total number of 32 experimental conditions. In each trial, the athlete has been asked to reach a speed of 15 km/h.

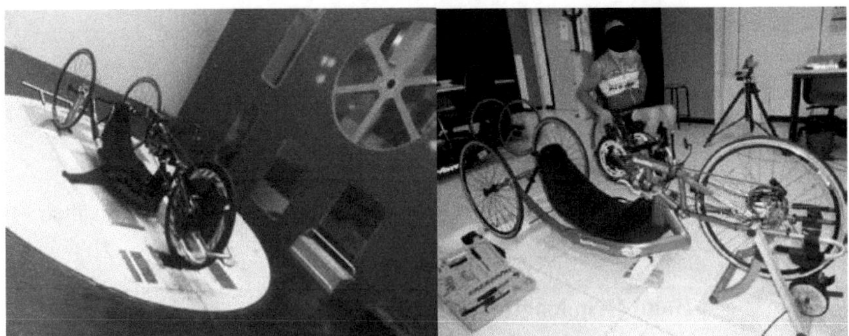

Figure 2. Left: The athlete's vehicle, for the reference acquisition. Right: same model of the athlete's vehicle with the adjunct of the different possibility of regulations.

Figure 3. Three different backseat inclination.

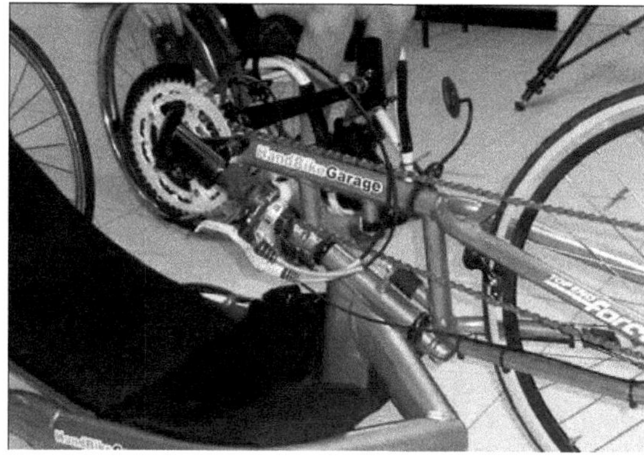

Figure 4. The HandBike handlebar.

Table 1 describes the different regulations tested in this study and their label codes

Hand Grip length = 170mm	**L1**
Hand Grip length = 175 mm	L2
Handlebar heigh: 1 higher to 4 lower	M1
	M2
	M3
	M4
Reference backseat inclination. It is computed fixing the distance between seat and handlebar at 165 cm	S3
Backseat Inclination lower intermediate	S1
Backseat Inclination higher intermediate	S2
Backseat Inclination higher	S4
Backseat Inclination lower	S5

Table 1: List of regulations for the experimental conditions and label codes.

Data analysis

The kinematic analysis is the computation of the joints centers' trajectories of the trunk, left and right shoulders and left and right elbows and the flexion-extension angles of the elbows and shoulders joints This is due to the fact that this is the most relevant movement for the analyzed task. Abd-Adduction and Intra-extra rotation are not evaluated in this specific study as stated as not relevant by the athlete himself. The joints trajectories are reported as the projections on the sagittal planes of the different cycles acquired during a single acquisition. The joints angles are reported as a single cycle obtained by the average of the different cycles performed in the experimental trial of each condition.

The ergonomic analysis is based on the Method of Movement and Gesture Assessment proposed by the authors (10). The method comes from the composition of three factors:

a) the joints kinematics
b) a joint coefficient of discomfort defined as Discomfort score
c) a body normalization coefficient estimating the "weight" of the ergonomic c contribution of each joint to the movement.

The MMGA method is capable to compare different vehicle adjustments to determine the best ergonomic advantage during the hand bike competition and has been computed for each experimental condition

RESULTS

Results of the kinematic analysis produce a quite clear characterization of the hand bike movement, with a significant movement of the shoulder's joints trajectory on the sagittal and transversal plane and a circle movement of the elbow joint center.

Figure 5 and 6 represent the shoulder and elbow 3D-trajectories in the reference configuration (the athlete's vehicle) and their projection on the sagittal plane (xz), the frontal plane (yz) and the transversal plane (xy). Due to the symmetry of the analysed movement, results from the right part of the body are reported.

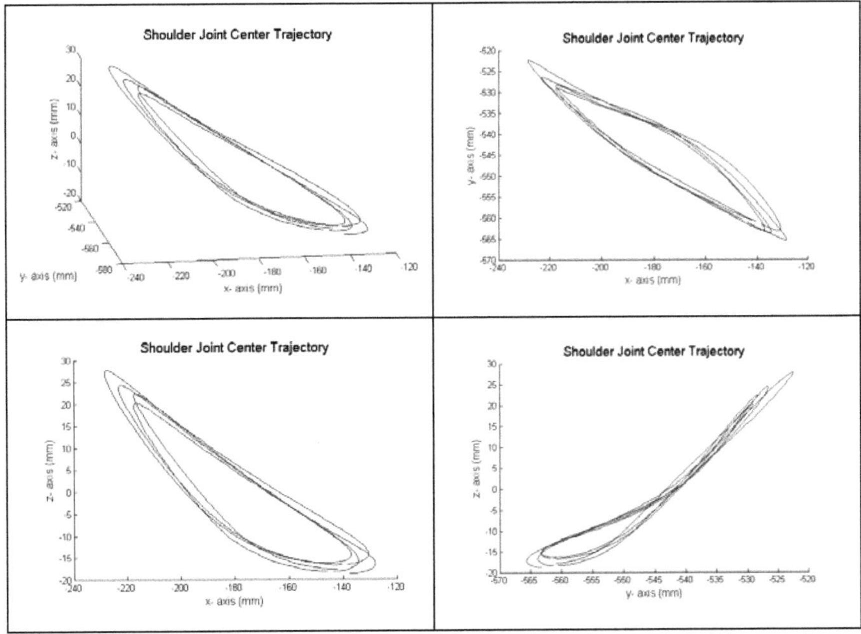

Figure 5: Shoulder Joint trajectory. Starting from the upper left image, in a clockwise sense: 3D, transversal, frontal and sagittal planes projection.

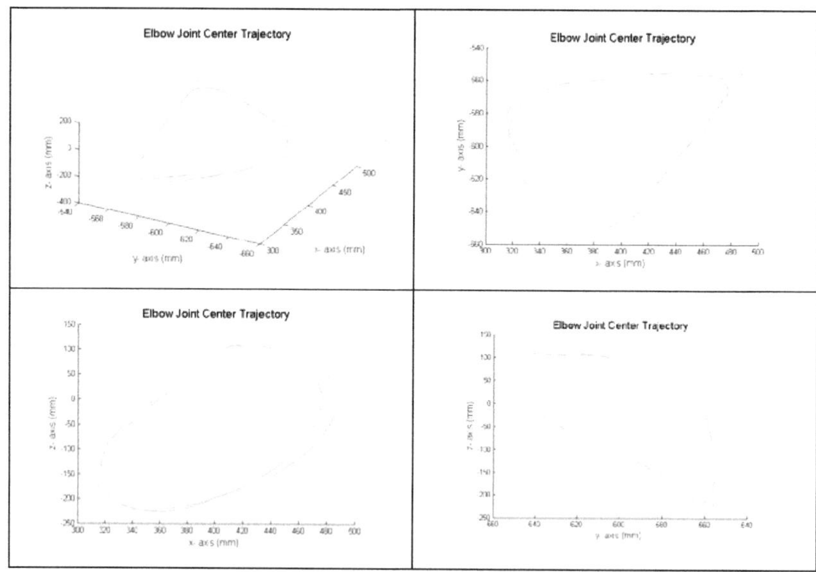

Figure 6: Elbow Joint trajectory. Starting from the upper left image, in a clockwise sense: 3D, transversal, frontal and sagittal planes projection.

The shoulder joint trajectory results stretched on the x and y direction (transversal and sagittal planes) while is reduced on the frontal plane.

The elbow joint trajectory presents a round path in all the three planes. This morphology is maintained for all the other experimental conditions with no relevant exception.

The joints angles analysis presents three different patterns for the shoulder joint kinematics, and one curve for elbow and trunk.

The first shoulder flex-extension angle results bell-shaped and asymmetric (Figure 7a), with a slow increase for the first half of the cycle followed by a rapid increase with a peak corresponding at the 70% of motion.

The second pattern is symmetric and bell-shaped, centered on the 50% of the motion (Figure 7b)

The third pattern (Figure 7c) presents a rapid increase of the angle values in the first 15% of motion, followed a plateau (50%) and another rapid increase with a peak at 70%.

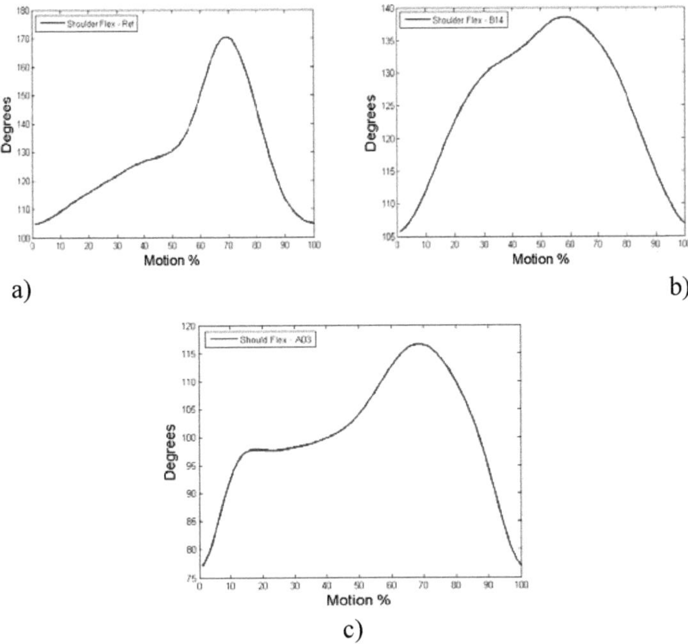

Figure 7. the three different Shoulder angles pattern. On the top-left (a), the reference movement; on the top right (b) is presented the condition L2-m1-s2 (long handle grip, higher handlebar height and higher intermediate backseat inclination); on the bottom line (c) is presented the angles for the configuration L1,m1,s2 (short handle grip, higher handlebar height and higher intermediate backseat inclination.

Figure 8 presents the elbow and the trunk joint angles observed. The elbow present a quasi-sine wave pattern with an irregular ripple at the 20% of the motion. The Trunk movement is sine waved with a maximum excursion of 10 degrees.

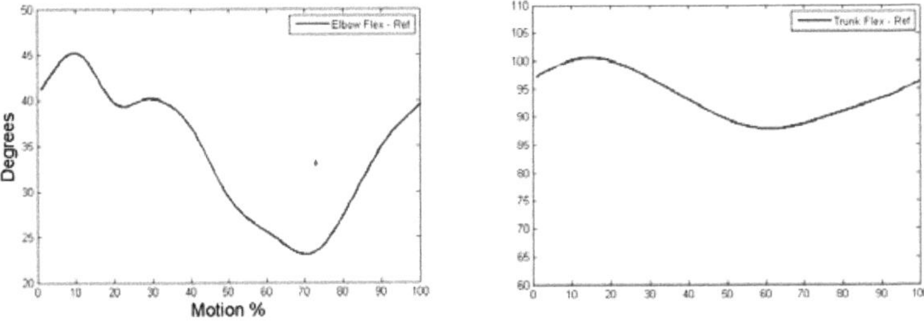

Figure 8: The reference joints angles for the elbow (left) and the trunk (right)

The ergonomic analysis demonstrates three main evidences:

1 – The handlebar height does not influence the ergonomics of the hand bike gesture

2 – The backseat inclination has a direct relation with the ergonomic perception measured by the score of the Ergonomic Index; the more the seat is vertical, the more the MMGA index present lower values of discomfort. This result is confirmed by the subjective evaluation of the athlete.

3 – The handgrip length influences the ergonomic perception with an unexpected evidence.

In Figure 9 these results are summarized with the evidence of the relation between backseat inclination and discomfort evaluation.

EI	HG	HB	BS
31,06	L1	m1	s4
33,9	L1	m1	s3
33,19	L1	m1	s1
37,8	L1	m1	s2
43,21	L1	m1	s5
52,8			Ref

HB	HG	BS	EI
m1	L1	s3	**33,9**
m2	L1	s3	**34,8**
m3	L1	s3	**36,3**
m4	L1	s3	**37,8**
m4	L2	s3	**47,3**
m3	L2	s3	**49,1**
m1	L2	s3	**51,1**
m2	L2	s3	**51,2**

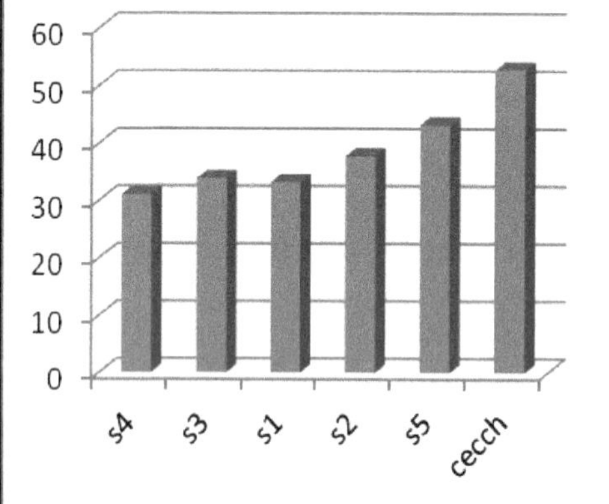

HB= Handle Bar Height
HG= Handle Grip Length
BS= Bakseat Inclination
EI= Ergonomic Index

Figure 9: Results of the Ergonomic Analysis. On the left: effect of the Backseat inclination on the final score. On the right: effect of the Handle Grip length on the final score.

DISCUSSION AND CONCLUSION

In this paper a characterization of the HandBike sport gesture has been presented. The kinematic analysis of the joints centers trajectories demonstrates that the sagittal and transversal plane are the most important planes of motion, while the movement in the frontal plane is not relevant.

The joints angles analysis demonstrates that there are three different patterns for the shoulder movement, depending on the experimental condition, while for the trunk and the elbow there are no significant relevancies.

The Ergonomic evaluation demonstrates how the length of the Handle Grip influences the biomechanical efficiency. This is an unexpected result, considering the low difference in measurement (5 mm).

The inclination of the backseat seems to affect the athlete's ergonomic. The higher is the value of the inclination, the higher is the MMGA score.

In conclusion, in this paper quantitative data on the biomechanics of HandBike sport gesture has presented, concerning a high level athlete that will challenge in London 2012 Paralympics competition. In addition, it is possible to affirm that, from an ergonomical point of view, the athlete is suggested to use the short hand grip levers and to assume a more vertical backseat position to improve his comfort during the challenge. Further analysis on the effective benefits for energy expenditure in this new assessment are required to validate these data definitely.

REFERENCES

[1]http://www.paralympic.org/Classification/Sports

[2] Dallmeijer, A.J., Zentgraaff, I.D., Zijp, N.I., Van der Woude, L.H., 2004. Submaximal physical strain and peak performance in handcycling versus handrim wheelchair propulsion. Spinal Cord 42 (2), 91– 98.

[3] Verellen J, Meyer C, Janssens L, Vanlandewijck Y. *Peak and submaximal steady-state metabolic and cardiorespiratory responses during arm-powered and arm-trunk-powered handbike ergometry in able-bodied participants.* Eur J Appl Physiol. 2011 Jun 30.

[4] Meyer C, Weissland T, Watelain E, Ribadeau Dumas S, Baudinet MC, Faupin A. *Physiological responses in handcycling. Preliminary study.* Ann Phys Rehabil Med. 2009 May;52(4):311-8. Epub 2009 May 7. English, French.

[5] Knechtle B, Müller G, Knecht H. *Optimal exercise intensities for fat metabolism in handbike cycling and cycling.* Spinal Cord. 2004 Oct;42(10):564-72.

[6] Krämer C, Schneider G, Böhm H, Klöpfer-Krämer I, Senner V. *Effect of different handgrip angles on work distribution during hand cycling at submaximal power levels.* Ergonomics. 2009 Oct;52(10):1276-86.

[7] Arnet U, van Drongelen S, van der Woude LH, Veeger DH. *Shoulder load during handcycling at different incline and speed conditions.* Clin Biomech (Bristol, Avon). 2011 Aug 8.

[8] Maki KC, Langbein WE, Reid-Lokos C. *Energy cost and locomotive economy of handbike and rowcycle propulsion by persons with spinal cord injury.* J Rehabil Res Dev. 1995 May;32(2):170-8.

[9] http://www.uci.ch – UCI para-cycling classification guide

[10] Andreoni G., Mazzola M., Ciani O., Zambetti M., Romero M., Costa F., Preatoni E. *Method for Movement and Gesture Assessment (MMGA) in ergonomics.* International Journal of Human Factors Modeling and Simulation (IJHFMS) – 2011

[11] Schmidt R, Disselhorst-Klug C, Silny J, Rau G. *A marker-based measurement procedure for unconstrained wrist and elbow motions.* J Biomech. 1999 Jun;32(6):615-21.

CHAPTER 61

Parameters in Phases of Causal Dependency in the Occurrence of Negative Phenomena

Juraj Sinay, Anna Nagyova, Slavomira Vargová, Frantisek Kalafut

Technical University of Košice
Košice, Slovakia, EU
juraj.sinay@tuke.sk

ABSTRACT

Negative phenomena, such as accidents, failures or occupational diseases do not occur randomly. There are certain inherent laws that form their causal dependency. The aim of all safety-enhancing or risk minimization measures is identifying the causal dependency of negative phenomena occurrence and consequently its discontinuance. The performed analyses must be based on the identification of the properties of individual phases of causal dependency for each particular negative phenomenon. The identification involves defining their properties and, if possible, parameters characteristic of particular phases. Precise identification of the phases requires knowing the actual operational condition of particular machinery and corresponding manufacturing processes. The importance of this procedure is increasing with the emergence of new technologies and newly developed machinery, such as mechatronic systems, nanotechnologies, biotechnologies, new renewable energy sources, which bring about new risks and potential hazards.

Keywords: causal dependency, danger, hazard, initiation, parameter, failure

1 INTRODUCTION

The key task of all preventative measures within an efficient system of risk management and occupational health and safety is to analyze all phases of causal dependency of the occurrence of a negative phenomenon during the service life of machines and systems – Figure 1 (Sinay and Nagyova, 2004). The results of analyses will enable us to develop risk management procedures that will allow for disconnecting the causal dependency in its early stages and eliminate negative phenomena such as failures, accidents or injuries.

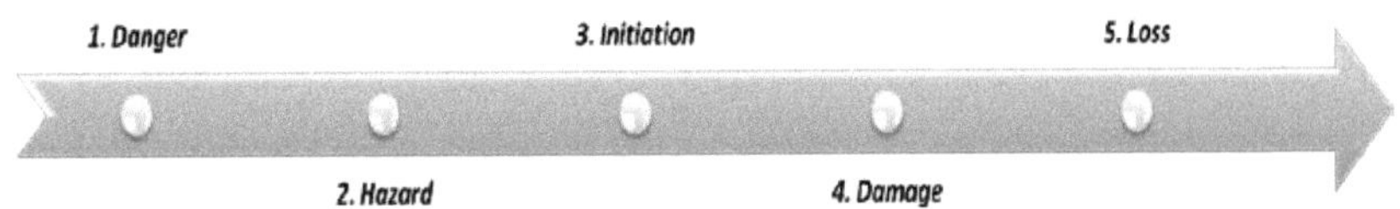

Figure 1 Causal dependency of the occurrence of a negative phenomenon (Sinay and Nagyova, 2004)

2 CHARACTERISTICS OF PHASES OF CAUSAL DEPENDENCY IN THE OCCURRENCE OF NEGATIVE PHENOMENA

Designing effective procedures for interrupting the causal dependency requires deep knowledge of parameters of individual phases. Efficient and economical risk minimization measures have to be implemented in the initial stages of causal dependency leading to failures or accidents resulting in injuries, i.e. in danger, threat, or initiation phases. Our analyses will therefore focus on the initial phases. The factors and parameters must be analyzed for a particular subject, e.g. for a particular machine, technology, process or social phenomenon. It is difficult to draw general conclusions applicable to the entire spectrum of subjects with the current knowledge in risk management. This is caused by the subjects' wide spectrum of parameters and properties existing in the functional dependency of multiparametric systems.

2.1 Danger

The first phase of causal dependency, i.e. danger, is characterized by factors or parameters, as defined in the object-subject-environment system, similarly to the well-known machine-man-environment system (Sinay, 2011). The environment is the system in which the subject interacts with the object.

The term **object** (machine) denotes all tangible elements that are parts of the manufacturing process (machines, appliances, material, consumables, chemicals, etc.). The **subject** represents the human factor, the person in operation and the **macro environment** denotes the space where the elements (man and machine) are located (either a closed workplace or an outdoor environment with all corresponding influences).

Each element in the mentioned system, typical of the danger phase of the causal dependency, is influenced by factors such as (see Figure 2):

- ***properties*** of the given element (properties of a human, macro environment machines and appliances),
- ***microenvironment*** – the immediate environment in which the element is situated (the surroundings of the machine and worker),
- ***other elements*** that may affect the danger (other machines and appliances, other workers).

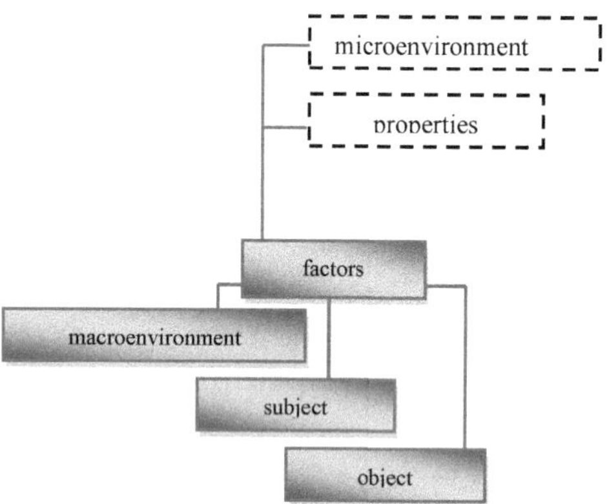

Figure 1 General diagram of factors having impact in the danger phase

In terms of the risk minimization implementation as part of prevention, effective measures focus on the mentioned elements; e.g. the layout of machines and appliances in a plant should prevent negative impact caused by domino effect.

2.2 Hazard

Hazard is the phase that follows the danger phase in the causal dependency. It is the part in which the factors and parameters typical of the danger phase are activated. There are three possible sources of hazard in the hazard phase – object, subject and environment.

This phase can also be influenced by several parameters and factors, such as:

- ***time,***
- ***activation of properties*** (of the object or subject),
- ***activation of macro environment,***

in consequence of which there occurs

- ***change in properties*** (of the object, subject or the environment),

The application of effective preventative measures requires creating conditions for direct manipulation with parameters (time), or factors (activation of properties and macro environment) in the analyzed system. This involves development of a tool that will enable the technology user or machine operator to discontinue the causal dependency. The aim is to create conditions to minimize the duration of the danger to the shortest possible period, i.e. the minimum time value "T".

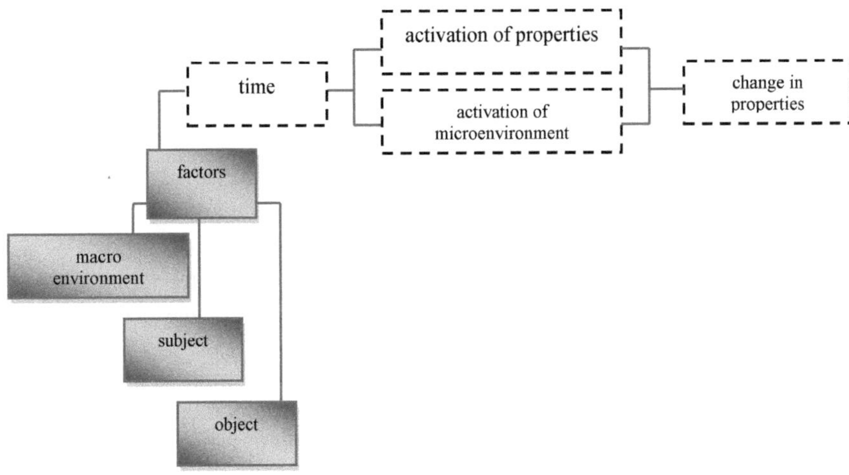

Figure 3 General diagram of factors having impact in the hazard phase

2.3 Initiation

The initiation phase is considered to be the crucial part of causal dependency form the risk management perspective. It represents the active part of the causal dependency that "triggers" a negative phenomenon.

The parameters and factors characteristic of the initiation phase that provide opportunities for discontinuance of causal dependency and implementation of effective preventative measures include:

- ***time,***
- ***microenvironment,***
- ***other elements*** (objects, subjects).

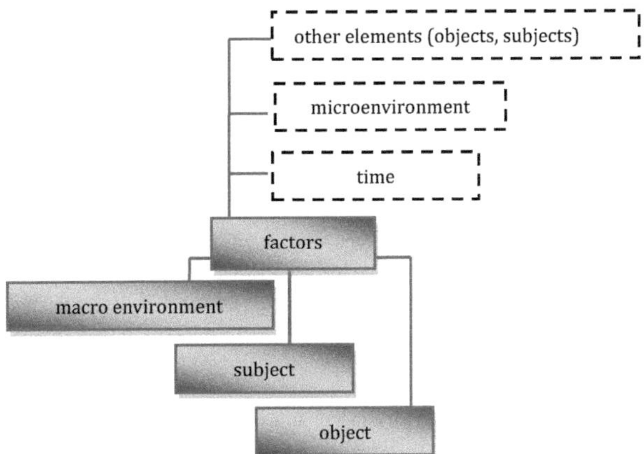

Figure 4 General diagram of factors having impact in the initiation phase

3 DAMAGE IN THE MACHINE STRUCTURE RESULTING FROM FATIGUE FRACTURE – PARAMETERS AND FACTORS OF CAUSAL DEPENDENCY PHASES

The strength of machine structure is one of key properties that facilitate reliable operation and manufacture of quality end products that meet the customers' requirements.

Sturdy and functional structure (Sinay, Maier and Hoeborn, 2008) is an important part of modern machines in mechatronic systems. Its failure causes interruption of their operation but also provides conditions for the occurrence of accidents and injuries with fatal consequences.

Modern machinery and systems frequently perform non-stationary operation movements with significant dynamic effects during their operation load. Dynamic load applied to bearing members of the structure can lead to formation and growth of fatigue fractures in its critical parts and thus damage the structure.

The possibility of serious consequences resulting from the damage is the argument for using modern methods of discontinuing the causal dependency of the occurrence of failures/injuries/accidents. These methods must be based on the causal dependency phases and on the detailed knowledge of their factors or parameters. Qualification and quantification of typical parameters of individual causal dependency phases is a condition for the selection of effective methods able to stop the causal dependency. Preventative measures will consequently be implemented and safe operation of machines and systems facilitated.

3.1 Fatigue fracture formation theory and safe condition of the structural node in operation load

Damage to the strength of a structural node of the functional structure (supporting steel structure) is a multiparametric system based on two characteristic properties of the operation conditions. The materials used in structural nodes and their formation into functional shapes have to resist operation load throughout the entire service life of machinery. The interaction of the two parameters facilitates safe operation only if their internal balance requirements are met. The procedure defining safe operation conditions for structural nodes, as parts of functional structure of a machine or complex machine systems in new industrial technologies with high hazard (risk) potential, is shown in the diagram (Figure 5). The hazards are related to manipulation with dangerous substances, application of nanotechnologies, the production of nanomaterials, biotechnologies, and generation of electricity using nuclear power technologies.

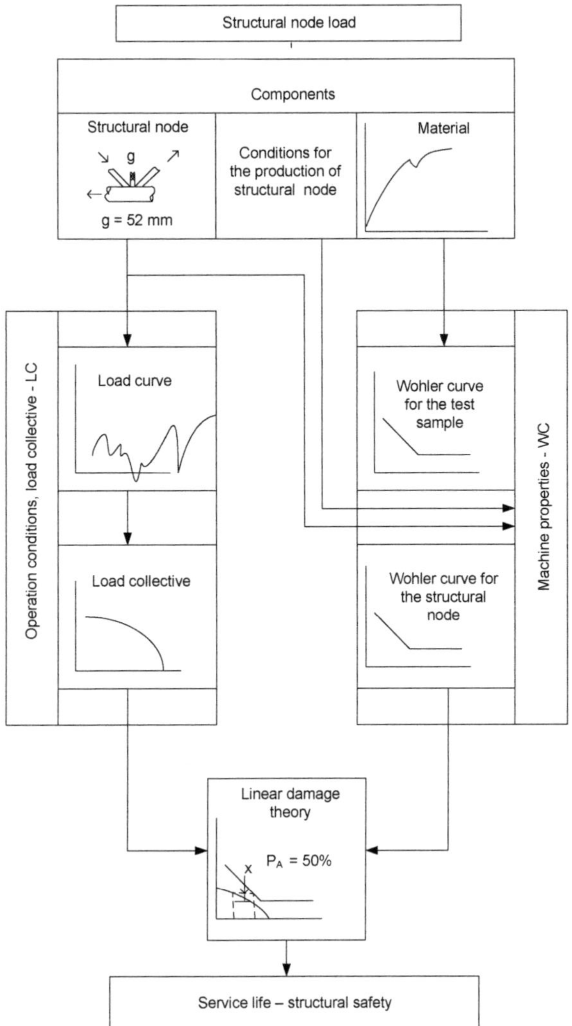

Figure 5 Model of interdependency between material factors and external load for the operational strength assessment of the structural node

It is obvious from Figure 5 that in order to ensure operational strength of the machine structure, it is desirable to know the interaction of two parameters – operation conditions, which are defined as load collective (LC) and properties of the machine or its structural nodes that are determined by the Wöhler curve (WC) for a particular type of material, approximated for the actual shape of structure. The interrelation of these parameters is the subject of research for respective causal dependency phases. This enables risk management professionals to cooperate with design engineers on a selection of effective methods of the causal dependency discontinuance in its early stages.

3.2 Danger – the first phase of causal dependency in the failure (damage) of the bearing structure

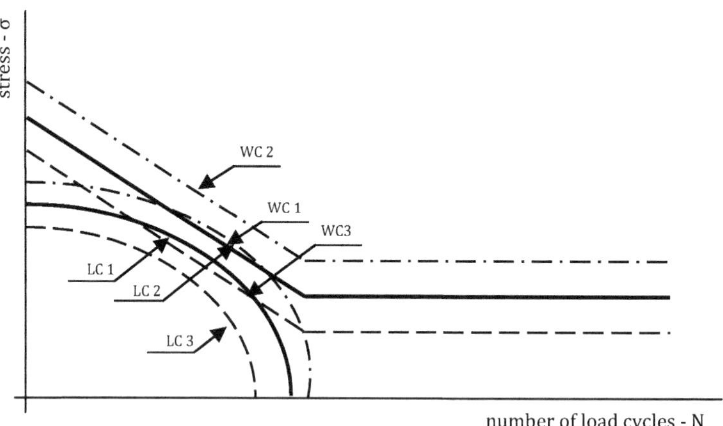

Figure 6 Parameters and/or factors of the danger phase

The group of machine structure parameters typical of the danger phase includes:
Machine properties - Wöhler curve of an actual node – WC, which depends on

1. material,
2. construction shape,
3. manufacturing technology – e.g. welds,
4. quality of assembly.

Operation conditions (load) – LC

5. calculation method including assessment of operation load.

The danger phase parameters of the operation load – LC include factors with a potential positive impact on this phase:

- formulation of technical conditions,
- machine or system operating instruction.

It is necessary to take into consideration the consequences of market globalization and the resulting "multilingualism" (Sinay, 2011).

The mentioned parameters and factors depend on activities and capacities of the human factor – subject, and do not represent a direct function of time.

The implementation of risk minimization measures and preventing the continuance of causal dependency of the occurrence of a negative phenomenon – breakdown, accident and/or injury is possible by using the following measures for the parameter *Machine properties - Wöhler curve of an actual node – WC*

a. inspection of the materials before they are used in manufacture,
b. verifying the calculation method by experiments or simulation of operation conditions,
c. ensuring the appropriate implementation of manufacturing technology,

and for the parameter *Operation conditions (load) – LC*

d. load monitoring,

e. review of the machine operating instructions.

3.3 The hazard phase in relation to the damage of a structural node

The activation of danger leads to the hazard phase. In terms of bearing structure damage analysis, this means setting the machine or system into operation. The operation conditions are related to the function of time – f (T) and the nature of operation, especially the actual operation load.

In case the machine is set into operation with incorrectly selected structural node parameter properties, the position of Wöhler curve changes from WC 1 into WC 2 or WC 3 – see Figure 7.

Wrong assessment of operation conditions (load) – LC affects the parameter (see Figure 7), which causes imbalance in the system and thus creates conditions for the impact of the hazard on the structural element of the machine or system.

The group of parameters related to potential impact of the hazard on the machine structure includes (in accordance with Figure 6):

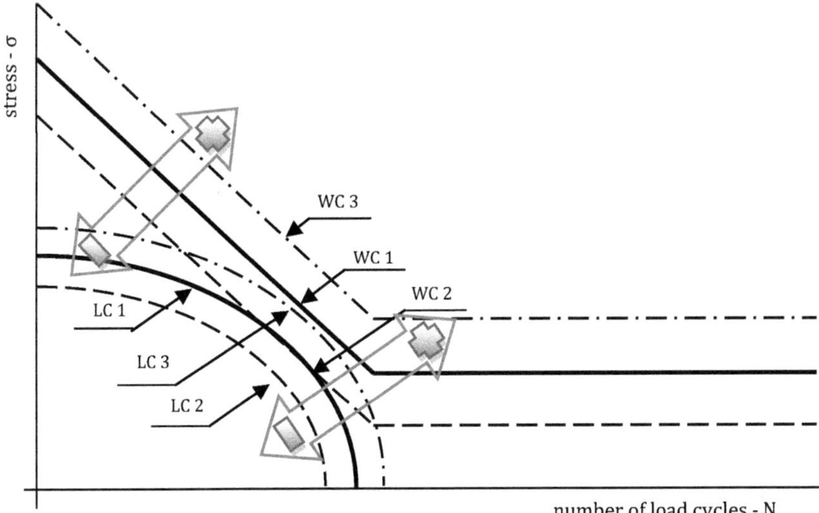

Figure 7 Parameters and/or factors of the hazard phase

Machine properties - Wöhler curve of the actual node – WC

a. driving mechanisms – their properties as defined by the manufacturer,

b. surface finish of the machine structural part,

c. elements of safe functional structure (Sinay, 2011),

d. mechanism of the formation and growth of microfracture – as the function of the operation time, and for the parameter *Operation conditions (load) – LC*

a. load collectives as the function of time – f(T) – Figure 7,
b. working environment – macro environment (snow, wind, etc.)
c. human factor at the machine operation – subject,
d. machine or system operation resulting from incorrect formulation of instructions in the manual,

The implementation of risk-minimization measures and thus preventing continuance of the causal dependency of the occurrence of a negative phenomenon, such as failure, accident and/or injury, is possible by utilizing the following measures for the parameter *Machine properties - Wöhler curve of an actual node – W*

a. inspection of critical points – using technical diagnostic methods

and for the parameter *Operation conditions (load) – LC*

b. monitoring the actual operation load of a machine as f(T)
c. machine operator training.

3.4 Initiation as a part of causal dependency in the structural node damage

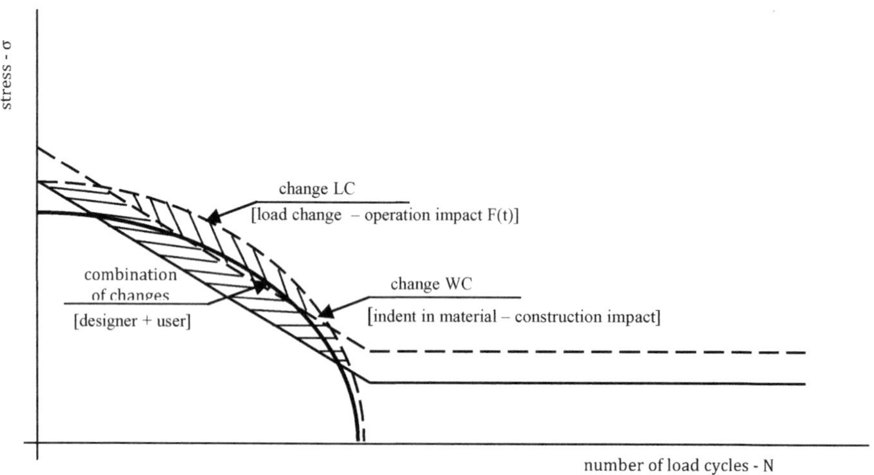

Figure 8 Parameters and/or factors of the initiation phase

From the standpoint of risk management, the initiation phase of the occurrence of a failure, accident and/or injury, is the last chance to discontinue the causal dependency without potential damage to the machine or system.

In case of failure caused by fatigue fracture in the machine functional structure, we can define minimum number of factors that can be manipulated in the initiation phase in order to discontinue the causal dependency.

The size of microfracture growing into an actual crack is a critical factor of the 'machine properties' parameter. Should the size of the bearing area of the steel structure be reduced, its resistance to operation conditions decreases.

The abovementioned facts determine the possibility of discontinuing the causal dependency of failure occurrence. The discontinuance requires identification of the properties of the actual fracture and assessment of its size, or creating a model of its growth in the structural node, e.g. by methods of simulation or technical diagnostics, as well as regular inspection of particular structural nodes.

4 SUMMARY

Knowing how failures, accidents and/or injuries arise, i.e. knowing the particular causal dependencies is essential for effective prevention. Appropriate and effective discontinuance of the causal dependency before its final phase and preventing the occurrence of an injury requires knowledge of relevant parameters or factors in respective phases. In the phases of danger, hazard and initiation, it is possible to take measures that prevent the causal dependency from proceeding to its final phases. The procedures and methods presented in the paper are designed to be applicable in occupational health and safety in workplaces with different types of technologies, machines and systems. The procedures can be extended to include also civil security.

ACKNOWLEDGMENTS

This paper is the result of the project implementation "Center for Research into control of technical, environmental and human risks for sustainable development of mechanical engineering production and products" (ITMS 26220120060), supported by the Research & Development Operational Program funded by the European Regional Development Fund.

REFERENCES

Sinay, J. 2011. *Bezpečná technika, bezpečné pracoviská – atribúty prosperujúcej spoločnosti.* Edícia SjF TUKE, 2011, ISBN 978-80-553-0750-3

Sinay,J., ,I.Majer and G.Hoeborn. 2008. Risk in mechatronical systems XVIII. World Congress on Safety and Health at Work, June 29 – July 2, 2008, Seoul Korea - Section 26. Sicherheit von High-Tech Kontrollsystemen übernehmen die Führung bei der Sicherheit am Arbeitsplatz.

Sinay, J. and A. Nagyová. 2004. Causal relation of negative event occurrence – injury and/or failure. In: *Advances Factors, Ergonomics, and Safety in Manufacturing and Service Industries*, AHFE Conference 2010. pp 818 - 827, CRC Press Florida 2010, ISBN 978-1-4398 -3499 -2

CHAPTER 62

Determination of Muscles Fatigue for Production Packers and Sawing Machine Operators in Furniture Enterprise Using Myotonometry

Valdis Kalkis, Zenija Roja, Henrijs Kalkis

University of Latvia, Ergonomics Research Centre,
Latvian Ergonomics Society
Valdis.Kalkis@lu.lv

ABSTRACT

Wood-processing is one of the biggest branches in Latvia where workers suffer from ergonomics risks and unhealthy environment. This research work is dedicated to occupational health problems caused by ergonomic risks of production packers and sawing machine operators working in furniture enterprise. In relation to these occupations problems mostly are associated with chronic pain in neck, shoulders, and legs (NSL) region. The aim of this study was to investigate the NSL muscles fatigue using myotonometry and to estimate the workload hardness categories using methods possible to identify overloads when lifting and moving heavy loads or performing other dynamic operations. The investigation was done in the one year period and with focus on the packers and sawing machine operators who suffered from chronic and repetitive pains of NSL.

Keywords: furniture, packers, sawyers, muscle fatigue, myotonometry

1 INTRODUCTION

Injuries and muscle pain in NSL region are common problems for employees in the furniture enterprises. It is associated with repetitive movements working in the same position during a longer period of time. The respective muscles group tension is considered to be the main cause of pain. Persons with chronic pain are less active and report greater disability and interference with daily activities. It is serious social, psychological and economical consequences. To a certain extent it also influences workers' work ability and life quality in general. Therefore, the investigation of the NSL muscles fatigue is necessary to estimate the workload hardness and to choose the convenient preventive solutions.

The several medium sized furniture enterprises were chosen for the research. The main problems of previously described ergonomics risks were founded in the process of production packaging and sawing. In the objective myotonometric measurements participated finished production packers (n=30) and sawing machine operators (n=20) with chronic pain (for four month and more) in the neck, shoulders and legs.

Literature data show that furniture packers are exposed to heavy workload (Ferreira, 2001). The heavy load is applied also to the sawing machine operators. The process of communication between workers and sawing machine is of great importance because frequently the control panel is not located ergonomically correct and this causes the difficulty of vision and leads to inadequate positions of the sawing machine operators, increasing the risks to health and also may increase the cognitive overload (see Figure 1). Operators needed to physically assist in relocation, moving and lifting actions of board's. It could cause much absenteeism among workers due to musculoskeletal illnesses that resulted in slowdown sawing and packing processes as well as creating a waiting section of other auxiliary processes.

Figure 1 Communication between the workers and the sawing machine

2 METHODS

Study population

The inclusion criteria of study population were: age and length of service; presence of chronic NSL pain (medical examination data); full consent to participate in the research. The exclusion criteria were: acute pain in the NSL regions; inflammatory rheumatic disease and disorders caused by trauma, having not visited doctor for mandatory medical examinations. Participants, all males, were divided into two groups accordingly to the length of service in occupation: group I, 1–10 years, and group II, more than 10 years.

Participants completed questionnaires providing us with the information regarding body parts suffering from pain, duration of pain, length of service in occupation, age, education, as well as their physical and other activities. Background factors of the reference groups are shown in Table 1.

Table 1 Background factors of the reference groups, length of service, number of participants (n), mean age and range, and standard deviation (SD)

Occupation (length of service)	n	Mean age ±SD	Range
Production packers	30	37.3 ± 9.5	18-65
(1-10 years)	20	25.2 ± 6.3	18-35
(> 10 years)	10	53.6 ± 7.7	46-65
Saw operators	20	35.9 ± 9.2	22-65
(1-10 years)	12	32.8 ± 6.1	22-40
(> 10 years)	8	56.9 ± 7.6	42-65

Data Collection – Nordic Musculoskeletal Questionnaire (NMQ-E)

In the present research the extended version of Standardised Nordic Musculoskeletal Questionnaire (NMQ-E) was used to assess musculoskeletal problems of sawyers and packers. NMQ has been widely used to assess the nature and severity of self-rated musculoskeletal symptoms (Kuorinka et al, 1987) and includes items, inquiring about the experience of problems in nine body areas. In our research the extended version comprises some additionally questions regarding body postures, job demands and social support.

The Key Indicator method

The Key Indicator Method for assessment of the manual handling of heavy loads developed by the German Federal Institution for Industrial Safety and Occupational Medicine was used to assess workers ergonomics risks (Steinberg, 2006). By means of this method possible overloads lifting or moving heavy loads or performing other dynamic operations are identified.

Key indicators (criteria) to be taken into account are: object mass rating points (*M*); the employee's posture rating points (*P*); working conditions rating points (*C*); working time/intensity value points (*I*). Risk assessment is carried out by physical workload risk score (*RS*) using the following formula: $RS = (M + P + C) \times I$. According to this method work hardness categories (or risk range) are: I – light work or low load situation (*RS* < 10); II – moderate work or increased load situation (*RS* = 10…25); III – hard work or highly increased load situation (*RS* = 25…50); IV – very hard work or physical overload (*RS* > 50). If the risk range is II, physical overload is possible for persons older than 40 or younger than 21 years, newcomers in the job or people suffering from illness. In all other cases (risk range III or IV) redesign of the workplace is recommended or it is necessary. Design requirements can be determinate by reducing the weight, improving the execution conditions or shortening the strain time, also elevated stress can be avoided.

Myotonometry (MYO)

Assessment of the functional state of skeletal muscles and muscle fatigue was carried out using myotonometric measurements with the MYOTON-3 device created in Estonia, the University of Tartu (Vain, 1995). The complete theoretical concepts of myotonometry (MYO) are described in reference (Vain and Kums, 2002).

The principles of the MYO lies in using acceleration probe to record the reaction of the peripheral skeletal muscle or its part to the mechanical impact and the following analysis of the resulting signal with the aid of the personal computer. Myoton exerts a local impact on the biological tissue by means of a brief impulse which is shortly followed by a quick release. The force of the impact is chosen such that it does not create changes in the biological tissue or precipitate the neurological reactions. The testing end (mass 20 grams) of the computerized myotonometry (CMYO) device was in contact with muscle belly area (see Figure 2), and the effective weight was employed on the surface of the measuring tissue. As a result, the tissues were in a compressed state. The CMYO device was fired in response to a fixed posture at the testing sensor end. For our study, the duration of the impact on the muscle belly of studied muscles was 15 milliseconds.

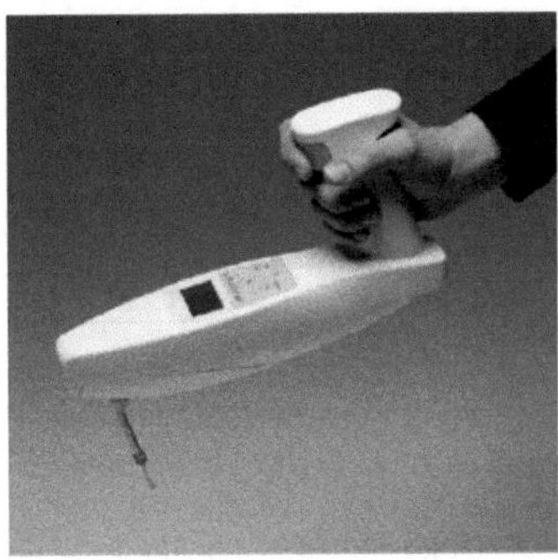

Figure 2 Myotonometrical testing

Measurements for determination of muscles tone during one week work cycle were carried out with relaxed muscles before the beginning of the work cycle. Myotonometry testing of the following muscles was performed in relaxed state: *m. gastrocnemius* and *m. trapezius* (upper part). The procedure of muscles testing was performed in a sitting position; the muscles length was middle; for all measurements the subject took the same position.

Statistical analysis

The acquired results were processed, using statistical data processing software SPSS.16 (SPSS Inc., Chicago, IL) according to popular descriptive statistical methods. Reliability interval (interrater agreement) was also calculated determining Cohen's Kappa (κ) coefficient (Landis and Koch, 1977). This coefficient identifies connectivity of the experimental data, the number of participants and the proportion or correlation of the participants' acceptance of the experimental data: $\kappa = (P_O - P_{C)} / (1 - P_C)$, where: P_O – correspondence proportion of objective experimental data with respondents' responses (yes or no), P_C – correspondence proportion of data with number of participants ($P_C = \Sigma p_i^2$, where p_i is acceptance of each participant expressed in percent or as fractional number).

3 RESULTS

Figure 3 presents the main interrelationships of musculoskeletal problems in different body regions, as well as distribution of postures, job demands and social support for packers' and saw operators' for both reference groups.

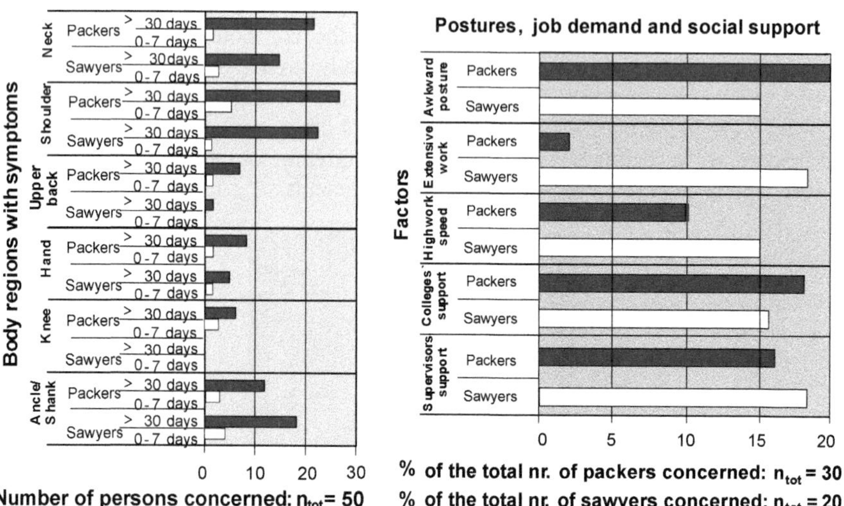

Figure 3 Distribution of musculoskeletal problems, postures, job demand and social support

Table 2 show the results of possible risk range and rating points of workload key indicators for packers and sawing machine operators.

Table 2 Risk range performing lifting and other dynamic operations

Occupation	Key indicators					Risk range
	M	P	C	I	RS*	
	Rating points					
Packers	3	4	1	5	56	IV
Saw operators	2	2	1	6	30	III

*Mean values

Although the work heaviness was identified, this study doesn't show the objective muscles fatigue. Therefore, a further research work was carried out using MYO measurements which allow to determine the values of the tone (frequency of the muscle oscillation, Hz), and stiffness (N/m) of the muscle.

According to regression analysis of MYO data several muscle tone levels were identified allowing subdivision of packers and saw operators into different conditional categories basing on muscle tone: – I a state of equilibrium, when muscles are able to adapt to the work load and are partly able to relax (no significant changes, muscles frequency and stiffness not exceeds the norm); and – II muscle fatigue and increased tone (frequency and stiffness exceeds the norm). Example of MYO measurement data *for m. trapezius* is reflected in Figure 4.

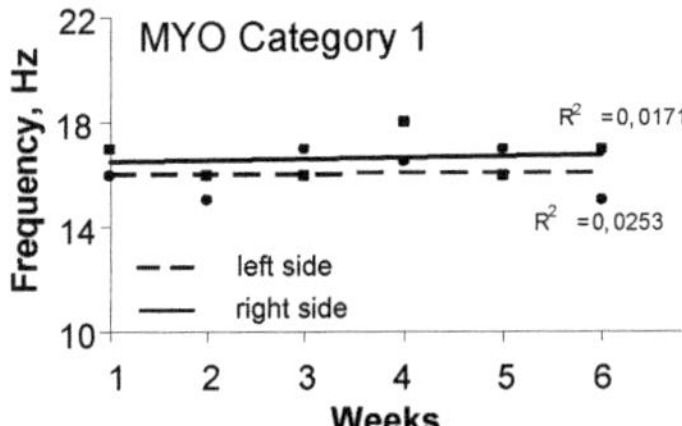

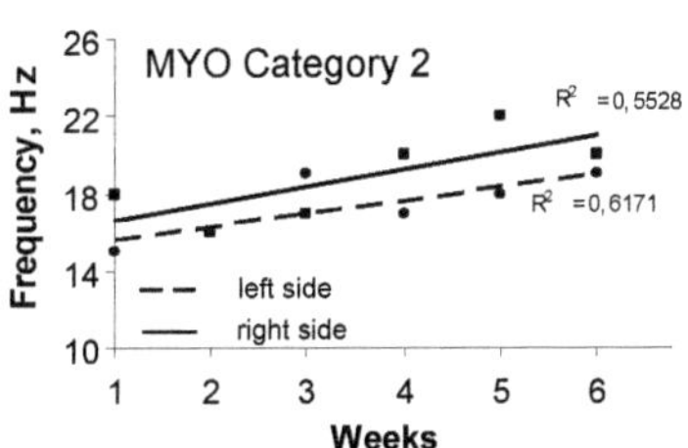

Figure 4 Results of the regression analysis of *m. trapezius* frequency during consecutive 6 work weeks

It was also found out that increase of muscle tone and fatigue mainly depends on workers physical preparedness and length of service (in this study – in proportion with age), see Table 3.

Table 3 Percent of packers and sawyers with differences in their muscle tone depending on the length of service (MYO Categories I-II), Cohen's Kappa (*κ*)

Length of service in the occupation, years			
1-10		> 10	
MYO Category	*κ*	MYO Category	*κ*
Packers (n=20):		Packers (n=10):	
I – 70 %	0.80	I – 50 %	0.85
II – 30 %	0.75	II – 50 %	0.70
Sawyers (n=12):		Sawyers (n=8):	
I – 75 %	0.80	I – 60 %	0.80
II – 25 %	0.68	II – 40 %	0.75

4 DISCUSSIONS

It is well known that physical risk factors in the workplace, or "ergonomic stressors," along with personal characteristics and social factors, are thought to contribute to the development of musculoskeletal disorders (Cohen, 1997). This is a topical problem also in the furniture industry because sometimes the furniture raw materials, finished products and pallets are on the floor, making the operators bend the spine causing health problems.

Nordic Musculoskeletal Questionnaire inquiry data showed that production packers and sawing machine operators (65% from n=50) most frequently complain on discomfort after the work, particularly, fatigue or muscle pain in the neck, shoulders, and legs. Inquiry data showed that packers (90%) and sawyers (75%) work in awkward postures. Both study groups admitted low support at the work from colleagues and supervisors.

It was stated that packers and saw operators in the process of lifting and moving the load are exposed to most severe loads, what corresponds with the risk range III and IV. Therefore, such workload is an endangerment to the workers' health. For this reason special attention has to be paid to necessary preventive measures in order to allow fatigued muscles to relax (relaxing exercises, increasing length and frequency of rest breaks) and further, more detailed investigation of physical load.

Analysis of the MYO data shows that for the production packers the greatest load was put on neck and shoulder muscles and less on legs muscles. Muscles tone at the end of the investigation period (also at the end of each working week) increased in the *m. trapezius* upper part, both sides. Sawing machine operator's muscles tone in the same time increased in the legs: *m. gastrocnemius*. It is reflected in Figure 5.

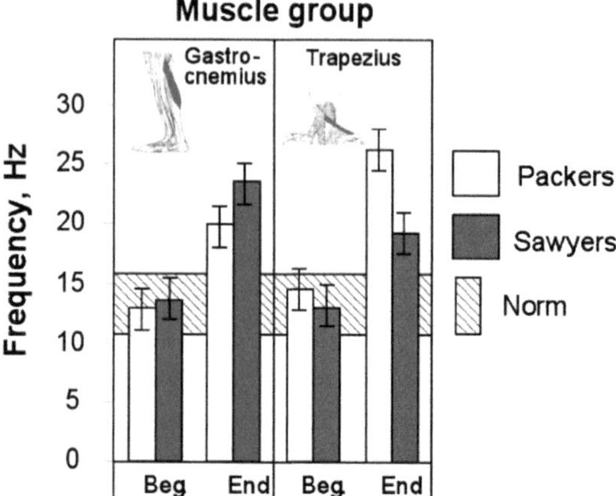

Figure 5 Illustration of frequency changing in separate muscles groups while performing the sawing and production packing at the beginning and at the end of the work week – for workers who are not able to adapt with the workload and whose muscles frequency exceeds the norm after the investigation period

It is noted that the natural oscillation frequency of muscles in their functional state of relaxation is usually 11–16 Hz (contracted 18–40 Hz), depending on the muscles wise. The stiffness values depend significantly on the muscles to be examined; their usual range is 150–300 N/m. For contracted muscles the stiffness value may be higher than 1000 N/m.

The muscles frequency exceeds the norm and muscles fatigue (MYO Category II) can be referred to 30% finished production packers with length of service 1-10 years and to 50% packers with the length of service more than 10 years. The muscle fatigue is referred also to 25% sawing machine operators with length of service 1-10 years and to 40% with the length of service more than 10 years. Hence for both reference groups muscles relax partly or are not able to relax. For reference group's length of service in occupation was used as a measure of exposure, including time spent in the relevant furniture factory and time of work in other wood-processing enterprises.

MYO measurements show that the cause of chronic pain is not always related to great workload, muscles fatigue and increase of muscles tone. During the research (investigation period 6 weeks) it was stated that, at the end of each week work cycle, muscle tone for 63% packers (from n=30) and 55% saw operators (from n=20) met MYO category I. Consequently, their workload has adapted to the work speed and working circumstances. Frequency of muscles contractions didn't exceed norm and muscles fatigue couldn't be considered to be the reason of chronic pain. Thus there emerged a question regarding the real cause of chronic pain. During our research low muscle frequencies were not examined, because the device doesn't allow measuring the deep muscles groups.

In our research we established that persons with chronic pain involving in the investigation were exposing by negative life habits (tobacco smoke, alcohol), low supervisors and colleagues support. Possibly these are psychosocial pains associated with psychological and social factors when mental or emotional problems can cause, increase, or prolong pain (Tyrer, 2006). Psychosocial pain may increase the overtime. The excess of overtime can cause postural overload, injury (back pain a.o.), fatigue, increased risk of accidents and a decrease in productivity (Roja, Kalkis et al., 2006).

The necessary preventive activities to reduce the heavy workload can be achieved by improving work organization, as well as workers` health (Wærsted and Westgaard, 1991). Our recommendations are: production packers and sawing machine operators should be allowed to select most appropriate working methods individually; appropriate time for rest breaks for individuals according to determined work heaviness category. The early multidisciplinary rehabilitation (relief exercises, behavior therapy a.o.) could also significantly improve workers health who suffers from chronic and subacute muscular skeletal pain caused by heavy workload (Tan, Fukui, Jensen et al., 2010).

5 CONCLUSIONS

Myotonometric measurements are suitable for objective determination of the fatigue of various muscles groups. According to the character of muscles tone determined by muscles contraction frequency production packers and sawing machine operators can be subdivided into two myotonometric categories basing on the dynamics of tone changes during one work week and investigating irregularity and different degrees of fatigue. Myotonometric measurements suggested that the causes of chronic pain are not always related to great workload and muscles fatigue.

REFERENCES

Cohen, A. L., C. C. Gjessing, L. J. Fine, J. D. McGlothin and B. P. Bernard. 1997. Elements of Ergonomics Programs: A Primer Based on Workplace Evaluations of Musculoskeletal Disorders. *DHHS (NIOSH) Publication No. 97–117*. Cincinnati: NIOSH, U.S. Department of Health & Human Service.

Korhonen, R. K., Vain, A., Vanninen, E., Viir, R., Jurvelin, J.S. 2005. Can mechanical myotonometry or electromyography be used for the prediction of intramuscular pressure? *Physiological Measurement*, 26: 951–963.

Ferreira, M. C., Freire, O. N. 2001. A empresa treina, mas na prática é outra coisa: carga de trabalho e rotatividade na função de frentista. *ACT*, 5(2): 175–200.

Landis, J. R., Koch, G. G. 1977. The Measurement of Observer Agreement for Categorical Data. *Biometrics*, 33: 59–174.

Roja, Z., Kalkis, V., Eglite, M., Vain, A., Kalkis, H. 2006. Assessment of skeletal muscles fatigue of road maintenance workers based on heart rate monitoring and myotonometry, *Journal of Occupational medicine and Toxicology*, 1: 20–28.

Steinberg, U.; Caffier, G.; Liebers, F. 2006. Assessment of Manual Material Handling based on Key Indicators – German Guidelines. In: *Handbook of Standards in Ergonomics and Human Factors*, eds. W. Karwowski. Lawrenz Erlbaum Associates. Mahwah, New Jersey, London. S. 319–338.

Thompson, M. L., Myers, J. E., Kriebel, D. 1998. Prevalence odds ratio or prevalence ratio in the analysis of cross sectional data: what is to be done? *Occup Environ Med*. 55:272–7.

Tyrer, St. 2006. Psychosomatic pain. *The British J. of Psychiatry*, 188: 91–93.

Vain, A. 1995. Estimation of the Functional State of Skeletal Muscle. In: *Control of ambulation using Functional Neuromuscular Stimulation*, Veltink, P.H. (ed), Boom HBK, Enschede, University of Twente Press, pp. 51–55.

Tan, G., Fukui, T., Jensen, M. P., Thornby, J., Waldman, K. L. 2010. Hypnosis treatment for chronic low back pain. *Int. Journal of Clinical and Experimental Hypnosis*, 53-64.

Vain, A., Kums, T. 2002. Criteria for Preventing Overtraining of the Musculoskeletal System of Gymnasts. *Biology of Sport*, 19 (4): 329–345.

Wærsted, M., Westgaard, R. H. 1991. Working hours as a risk factor in the development of musculoskeletal complaints. *Ergonomics*, 34: 265–76.

CHAPTER 63

Heart Rate Monitoring – Physical Load Objective Evaluation Method for Nurses and Assistants of Nurses

Zenija Roja[1], *Valdis Kalkis*[1], *Inara Roja*[2], *Henrijs Kalkis*[3]

[1]University of Latvia, Ergonomics Research Centre
[2]Riga First Hospital, Outpatient Department, Latvia
[3]Riga Stradins University, Faculty of European Studies, Latvia
Zenija.Roja@lu.lv

ABSTRACT

This investigation focuses on physical workload assessment of health care stuff. The aim of this study was to evaluate the work load during shift work of ward nurses and assistants of nurses using the heart rate monitoring. The study group involved 30 ward nurses and 30 assistants of nurses, who complained about hard working conditions. The work heaviness degree depending on workers physical activity (intensity) was estimated by objective method heart rate monitoring (HRM). HRM was performed using Heart Rate Monitor device. The rating of perceived exertion (RPE) for ward nurses and assistants of nurses was also assessed using Borg rating scale. The results show that the total energy expenditure of assistants of nurses in day and night shifts was significantly higher than ward nurses. Therefore, the heart rate monitoring is effective objective method for evaluating physical load for health care stuff.

Keywords: heart rate monitoring, physical load, nurses, assistants of nurses

1 INTRODUCTION

Nowadays work is getting more intensive and leads to the increase of total workload. In Latvia health problems related to physical load for health care staff, especially for nurses and assistants of nurses are very up to date. In many cases nurses and assistants of nurses are working in forced/constrained postures and have overload during the work shift. The main reasons are the patient lifting and moving, shift work, long working hours. It is known that stress at work can also reinforce physical load and that leads to the risk of work related health complaints. In such cases the reduction of the work load and improvement of working conditions for nurses and assistants of nurses is necessary. It should be noted that the health care stuff are not familiar with proper patient lifting and moving techniques, do not use technical aids. They themselves indicate that the work load is very hard. However, studies regarding the work load and energy expenditure during shift work of nurses and assistants of nurses engaged in hospitals are not enough.

The heart rate is one of the most effective physical work load indicators. Evaluation of heart rate measures is used to estimate physical intensity levels of a task (Johansson and Borg, 1993; Louhevaara, 1995). There are many factors to consider when determine the physical workload experienced by people while performing a task, each of which influences the energy output of the individual in some way. These factors include the nature of the work, training, motivation, and environmental factors (Astrand and Rodahl, 1986). The reaction of the circulatory system to physical workload has led to the consideration of heart rate as indicator of workload (Meshkati, 1988). It should also be noted that heart rate may be less reliable than oxygen consumption as an indicator of workload, because it can be influenced by other factors, including emotional stress, nervousness, apprehension, caffeine, or working in a hot environment (Spurr, Prentice, Murgatroyd et al., 1988).

The aim of this study was to evaluate the work load during shift work of ward nurses and assistants of nurses using the heart rate monitoring.

This study was carried out at the one of the largest hospitals in Latvia. The study group involved 30 ward nurses and 30 assistants of nurses, all females, who complained about hard working conditions. The inclusion criteria were: age and length of service; having discomfort and pain in the different body parts (medical examination data); full consent to participate in the study.

Table 1: Background factors of the subjects, mean, standard deviation (SD) and range

Variable	Ward nurses (n = 30)		Assistants of nurses (n = 30)	
	Mean ± SD	Range	Mean ± SD	Range
Age (years)	27 ± 4	23-33	29 ± 5	22-50
Height (cm)	168 ± 4	164-173	164 ± 5	165-176
Weight (kg)	66 ± 5	58-79	68 ± 4	67-88
Length of service (years)	5 ± 2	2-7	6 ± 3	1-10
Body mass index (kg/m^2)	25 ± 4	16–29	26 ± 3	17–30

Table 1 describes the baseline characteristics of the study group. The mean age of the ward nurses was 27 years (SD ± 4), ranging from 23-33 years, assistant of nurses – 29 (SD ± 5), ranging from 22-50 years. The mean BMI of the nurses were 25 (SD ± 4); 35.4% were overweight, but assistant of nurses this ratios was 26 (SD ± 3); 24.7% were overweight and 6.6% were obese. The mean length of service for ward nurses were 5 years (SD ± 2), ranging 2-7 years, but assistant of nurses – 6 years (SD ± 3) ranging 1-10 years.

2 METHODS

Questionnaire about work-related factors consisted of questions on physical activities (the regular presence in the current job of manual handling, awkward back postures) and psychosocial indicators (high work demands, lack of support) at work shift. Study group recorded the time and durations of their breaks during the shifts.

The work heaviness degree depending on physical intensity was estimated by HRM using device POLAR S810i^{TM} and data processing software Polar Precision Performance, which transforms HRM data into energy expenditure (kcal/min) (Jackson, Blair, et al., 1990). The device sums up the acquired heart rate (HR) data and transforms them into metabolic energy consumption (kcal/min). The relative range of the HR (%HRR) was calculated using a following equation: $100\times\{(HR_{work}-HR_{rest})/(HR_{max}-HR_{rest})\}$. Maximal heart rate was calculated as the most common formula HR_{max}=220-age, although there exist most accurate formulas (Karvonen, M. et al., 1957). HRM data correlates with oxygen consumption and allows quantifying the objective energy expenditure for each work phase including rest periods (Karvonen, et al., 1957). The Polar^{TM} S810 Heart Rate Monitor was placed on the participant's torso against the skin and the receiver (watch) on the wrist for full day and night shift work.

The HRM was done during one month period. Work heaviness in terms of kcal/min was classified according to ISO 28996 and NIOSH standard (Mantoe, Kemper, et al., 1996). Corresponding scale is shown in Table 2.

Table 2 Work heaviness classification in terms of energy expenditure

Work heaviness categories (WHC)		Energy expenditure*	
		Male, kcal/min	Female, kcal/min
Light work	I	2.0 – 4.9	1.5 – 3.4
Moderate work	II	5.0 – 7.4	3.5 – 5.4
Hard work	III	7.5 – 9.9	5.5 – 7.4
Very hard work	IV	10.0 – 12.4	7.5 – 9.4
Ultimate work	V	more 12.5	more 9.5

* Energy expenditure can transform using coherence: 1 W = 1 J/s = 0.0143 kcal/min

The rating of perceived exertion (RPE) for ward nurses and assistants of nurses was also assessed using Borg rating scale, ranging from 6 to 20 (Borg, 1970; Borg, 1998). Data were gathered with questionnaires and interviews, considering the age, physical conditions, and subjective view of increased heart rate. The RPE scale is a category scale that relates the intensity of an individual's perception to a perceptive range (Borg, 1985). The RPE scale was constructed in relation to an objective physical measure, heart rate. The scale contains verbal anchors that allow individuals to choose a number that represents their feelings of exertion.

The reliability of the statistical processing of HRM was determined using popular descriptive correlating analysis (Pearson's correlation coefficient *r*). Reliability interval was also calculated determining Cohen's Kappa coefficient (κ), which identifies connectivity of the experimental data, the number of participants and the participants' acceptance proportion or correlation of the experimental data (Landis & Koch, 1977). The results acquired were processed by applying statistical data processing program SPSS.16.

3 RESULTS AND DISCUSSION

Accordingly to the survey data, the nurses and assistants of nurses complain frequently about feelings of discomfort after the work (ward nurses: 73%, assistants of nurses: 83%), subjectively marked hard work (accordingly 80% and 86%), and all of them noted that they felt very tired after the work shift and felt discomfort in low back area. Work activities for ward nurses and assistants of nurses are similar and included patient's lifting and moving, moving on wheelchair, lifting and holding in sitting position on the bed, injections, a.o. Results from interviews about psychosocial working conditions showed that main risk factors were time pressure, high work demands and sometimes poor social support in the team.

HRM for ward nurses and assistants of nurses was done in day shift work (8:00 AM – 4:00 PM) and night shift work (4:00 PM – 8:00 AM). Research results of HRM and subjective RPE by Borg scale are summarized in Table 3. Results of HRM are shown by taking into account average heart rate and energy expenditure of each person, standard deviation (SD), Pearson's correlation (r), and Cohen's Kappa coefficient (κ).

Table 3 Heart rate (HR), objective energy expenditure (E), perceived exertion (RPE), work heaviness category (WHC), Pearson's (*r*), and Cohen's Kappa (*κ*)

Occupation	Heart rate monitoring data				Objective E±SD, kcal/min	WHC	Mean RPE±SD (range)
	Mean HR±SD, beats/min	Range HR, beats/min	*r*	*κ*			
			day shift				
Ward nurses (n=30)	98±8	81...131	0.95	0.65	4.1±1.3	II	13±2 (10–16)
As. nurses (n=30)	101±14	84...135	0.95	0.61	6.5±0.7	III	16±3 (12–19)
			night shift				
Ward nurses (n=30)	9 ± 2	79...124	0.95	0.80	2.0±1.2	I	11±2 (8–14)
As. nurses (n=30)	103± 3	82...126	0.95	0.68	4.8±1.5	II	14±2 (11–16)

Objective measurements, using the HRM, showed that work hardness categories accordingly to NIOSH standards are category II (moderate work) for ward nurses and category III (hard work) for assistants of nurses in the day shift, but during the night shift – category I (light work) for ward nurses and category II (moderate work) for assistant of nurses. It can be explained with decrease of work amount and intensity during the night shift. Worth to mention that heart rate of assistants of nurses (102 ± 13 beats/min) was higher than ward nurses (94 ± 9 beats/min) during the whole working time. The total energy expenditure of assistants of nurses in day (6.5 ± 0.7 kcal/min) and night shifts (4.8 ± 1.5 kcal/min) also was significantly higher than ward nurses (day shift: 4.1 ± 1.3 kcal/min; night shift: 2.0 ± 1.2 kcal/min). Our opinion is that study groups have been exposed not only to physical load, but also to psychoemotional stress at work, respectively ward nurses are subjected to higher level of psychoemotional strain (high responsibility for performance, time limits a.o.), but assistants of nurses – to physical load (care, moving and lifting of patients, cleaning of wards a.o.). The further research is necessary to analyze the interaction of psychoemotional and physical load on ward nurses and assistants of nurses during the day and night shift work.

4 CONCLUSIONS

Subjective work heaviness assessment method should be approved with objective evaluation methods. Heart rate monitoring is effective objective method for evaluating physical load for health care stuff: nurses and assistants of nurses. It was stated that assistants of nurses are exposed to higher workload, especially during the day shift compared with ward nurses.

REFERENCES

Astrand, P. and Rodahl, K. 1986. *Textbook of work physiology*. (3rd ed.). New York: McGraw-Hill.

Borg, G. 1970. Perceived exertion as an indicator of somatic stress. *Scandinavian Journal of Rehabilitation Medicine*, 2: 92–98.

Borg, G. 1998. *Borg's Perceived Exertion and Pain Scales*. Human Kinetics Publishers, Champaign, Illinois, USA.

Borg, G., Ljunggren, G., and Ceci, R. 1985. The increase of perceived exertion aches and pain in the legs, heart rate and blood lactate during exercise on a bicycle ergo meter. *European Journal of Applied Physiology*, 54: 343–349.

Borg, G., Hassmen, P., and Lagerstrom, M. (1987). Perceived exertion related to heart rate and blood lactate during arm and leg exercise. European Journal of Applied Physiology, 65, 679-685.

Jackson, A. S., Blair, S. N, Mahar, M. T, Wier, L. T, Ross, R. M., Stuteville, J. E.,1990. Prediction of functional aerobic capacity without exercise testing. *Medicine and Science in Sports and exercise*, 22 (6): 863–870.

Johansson, S.-E. and Borg, G. 1993. Perception of heavy work operations by tank truck drivers. *Applied Ergonomics*, 24(6): 421–426.

Karvonen, M., Kentala, E. and Mustala, O. 1957. The effects of training on heart rate: A longitudinal study. *Annales Medicinae Experimentalis et Biologiae Fenniac*, 35: 307–315.

Landis, J. R., Koch, G. G. 1977. The Measurement of Observer Agreement for Categorical Data. *Biometrics*, 33: 159–174.

Louhevaara, V. 1995. Assessment of physical load at work sites: a Finnish-German concept. International Journal of Occupational Safety and Ergonomics, 1(2): 144–152.

Mantoe, H. I., Kemper, W. M., Saris, M. & Wasshburn, R. A. 1996. *Measuring Physical Activity and Energy Expenditure*, Human Kinetics Publishers, Champaign, Illinois, USA.

Meshkati, N. 1988. Heart rate variability and mental workload assessment. In: *Human Mental Workload*, eds. Hancock P. A. and Meshkati N., pp. 101–115. Amsterdam: Elsevier Science Publishers.

Noble, B. J., Borg, G. A. V., Jacobs, I., Ceci, R., and Kaiser, P. 1983. A category-ratio perceived exertion scale: relationship to blood and muscle lactates and heart rate. *Medicine and Science in Sports and Exercise*, 15(6): 523–528.

CHAPTER 64

Ergonomic Analysis of Rice Field Plowing

Manida Swangnetr [1], *Ploypailin Namkorn* [2], *Chatchai Phimphasak* [2], *Krittaya Saenlee* [2], *David Kaber* [3], *Orawan Buranruk* [2], & *Rungthip Puntumetakul* [2]

[1] Back, Neck and Other Joint Pain Research Group, Department of Production Technology, Faculty of Technology, Khon Kaen University, Khon Kaen, 40002 Thailand
manida@kku.ac.th

[2] Back, Neck and Other Joint Pain Research Group, School of Physical Therapy, Faculty of Associated Medical Sciences, Khon Kaen University, Khon Kaen, 40002 Thailand

[3] Edwards P. Fitts Department of Industrial and Systems Engineering, North Carolina State University, Raleigh, NC, 27695-7906, USA

ABSTRACT

Thailand is the largest exporter in the world. Rice farmer health and safety have become increasingly important concerns in ensuring a sufficient work force to meet high production demands. Rice cultivation involves processes from plowing to harvesting. Some work is performed with heavy machinery, such as power tillers, which has been found to pose high energy demands. In addition, most tasks are performed with bare hands and feet and involve awkward postures. These work factors have previously been found to lead to musculoskeletal disorders (MSDs). A task analysis was conducted on plowing using a power tiller machine for initial field preparation at a large farm in the Khon Kaen area. Results of the analysis were used to structure a job screening for ergonomics-related risk factors. Multiple expert analysts evaluated plowing subtasks in a lab setting. Subjective ratings of risk of injury were made for each body part for motion, force and posture. The screening revealed farmer exposure to repetitive motion, high forces and extreme postures at the shoulders, hands and legs when performing straight plowing and clearing debris from the machine. The results indicated all farmer body parts were exposed to high risk conditions. The plowing job was found to have a high overall potential for causing MSDs. Specific ergonomic interventions are recommended in order to reduce the risk of occupational injury of rice farm workers. Other rice operations, beyond field plowing, need to be investigated for potential ergonomics-related risks, given high future production demands in Thailand.

Keywords: rice cultivation, musculoskeletal disorders, task analysis, job screening

1 INTRODUCTION

Second to China and India in rice production, Thailand is the largest rice exporter in the world. Since the turn of the last century, there has been a significant increase in Thailand's rice production from 25.8 million tons in 2000 to 32 million tons in 2009. This has been attributed to an increase in farming activity in the northeastern region of the country (Office of Agricultural Economics, 2010). This area primarily produces Jasmine rice, which has unique flavor and texture compared with other products. There have been recent increases in demand for Jasmine rice in Europe and the U.S. (Chataigner, 1992; USDA, 2001; Suwansri et al., 2002) as part of increased rice consumption. From a rice farm perspective, Jasmine rice can be sold at higher prices and with fewer competitors (Suwannaporn and Linnemann, 2008). The annual export value of Jasmine rice has represented up to 38% of total rice export value for Thailand (Office of Agricultural Economics, 2010). Therefore, Thailand is likely to export more Jasmine rice in the future. Unfortunately, recent major flooding has caused severe damage to crops estimated to amount of 3.5 to 4 million tons in the central region and some parts of the northeast (Kasikorn Research Center, 2011). Due to expected increases in rice prices in the global market, it is also forecasted that Thailand will dramatically increase rice production in 2012 in order to compensate for current losses (Kasikorn Research Center, 2011). Consequently, rice farmer health and safety have become increasingly important issues to ensure a sufficient work force to meet high production demands.

Rice cultivation involves many processes including: plowing, soil digestion and harrowing, seeding, planting, nursing and fertilization, and harvesting. Most tasks are performed with bare hands and feet and involve awkward postures as well as highly repetitive movements. These work factors have previously been found to lead to musculoskeletal disorders (MSDs) (NIOSH, 2001). Such disorders remain the most common occupational non-fatal injuries and illnesses for farm workers, in general (Kirkhorn et al., 2010; Fathallah, 2010). Some common MSDs in rice cultivation include low back pain and hand and wrist disorders (Kirkhorn et al., 2010; Fathallah, 2010; Kar and Dhara, 2007).

The rice field plowing process is performed in multiple stages with heavy machinery, such as power tillers. The first stage is an initial field preparation with the use of a single-blade harrow for trenching. The second stage involves the use of a disc harrow to till the soil with less depth. A third stage uses on tine harrow for creating fine planting rows. These plowing activities have been found to pose high energy-consumption demands on farmers (Mamansari and Salokhe, 1995). Each stage also requires several plowing laps around a field over an extended period of time. Furthermore, the use of the vibrating machinery with bare hands and feet can lead to an ischemic effect in body parts causing fatigue. This type of body loading and vibration exposure had been found to represent a major occupational injury risk for workers (Walker-Bone and Palmer, 2002).

There is a need to conduct structured and systematic screening of rice operations for ergonomics-related risk factor exposures. Furthermore, there is a need to quantify risks as a basis for recommending interventions. This study focused on the rice plowing activity in an attempt to identify critical areas of body exposure to hazards and to identify methods that would reduce farmer MSDs.

2 METHODOLOGY

A field investigation of rice plot preparation was performed at a large farm in Khon Kaen province, Thailand. Our team visited the farm and made observations on: 1) farmer personal protective equipment, 2) farm implements/machinery, 3) farmer work methods in plowing, and 4) motion patterns at joints (using video). Interviews were also conducted with the farm manager and workers regarding: 1) locations of body discomfort during plowing, 2) intensity of discomfort, including identification of the worst areas of pain, and 3) methods used by the farmers in an attempt to reduce discomfort in work.

Subsequently, a task analysis was generated based on the field observations of the plowing task when using a power tiller. The analysis involved identifying: 1) the overarching goal of the plowing activity; 2) all subtasks to the goal; 3) the plan or strategy used by the farmer to complete the tasks; and 4) environmental conditions or cues triggering farmer performance of specific tasks or use of certain methods. As a basis for the task analysis, three views of motion were recorded including front, back and side for each subtask (see Figure 1 for straight plowing).

Figure 1 Illustration of side and back view during straight plow.

Based on the video analysis, seven subtasks were identified as part of plowing process, including: 1) start motor of plowing machine; 2) start plow (release the break and shift gear); 3) adjust plow level; 4) straight plowing; 5) curve plowing (turn left at the corner of field); 6) clear debris from the plow; and 7) stop the machine (shift gear and lift up the handles). Partial results of the task analysis are presented in Figure 2 and were used to structure and guide the ergonomics job screening.

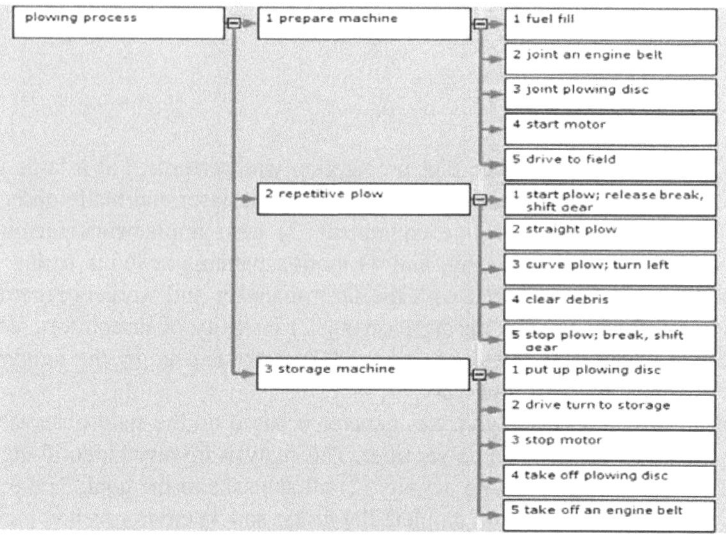

Figure 2 Hiearchical task analysis diagram for the rice plowing activity.

The job screening method was based on an "Industrial Ergonomics Screening Tool" developed by the Ergonomics Center of North Carolina. This tool has been used in several prior studies for identifying ergonomics-related risk factors in, for example, maintenance operations and veterinary tasks (Gangakhedkar, Kaber, and Mosaly, 2011; Rogers, Gangakhedkar, and Kaber, 2011). The tool was developed based on the Hand Activity Level (HAL; Armstrong, 2006) measure and Rapid Upper Limb Assessment (RULA; McAtamney and Cortlett, 1993) methodology. Both the HAL and RULA have been previously used to evaluate workplaces in which workers are exposed to risk factors resulting in upper limb disorders.

The screening tool was applied in a lab setting by at least two expert analysts (ergonomists and physiotherapists) for each of the subtasks appearing in the rice field plowing task analysis. The tool involved:

1) Rating the risk of potential injury for various body parts, based on exposure to extreme postures, force and repetition in subtask performance - Body parts covered by the method included the neck, trunk, shoulders, arms/elbows, hands/wrists and legs. Unfortunately, the tool did not

specifically elicit ratings for the feet and ankles; however, analysts were instructed to consider these body parts when rating the legs. A rating scale ranging from 0 to 10 is used to evaluate each part with 0 to 3 representing low risk priority, 4 to 6 representing moderate risk priority, and 7 to 10 representing high risk priority.

2) Using ratings to determine a total risk priority for each body area and to identify the worst subtasks for exposures.
3) Determining an overall task risk score as "low", "moderate" or "high", based on the average of analyst risk priorities across body parts for each subtask.

3 RESULT

In general, the screening results revealed the plowing activity to pose an overall high risk potential for MSDs (see Table 1). Specifically, the straight plowing subtask was rated to be the worst subtask for body parts, including the left shoulder, both hands and wrists and the right leg. The table also reveals that repetitive motion and extreme posture were considered the main underlying factors in the injury risk to these parts. Straight plowing was followed by clearing debris from the plowing machine in terms of severity of risk for potential injury, particularly for both arms and elbows. The main underlying risk factor for this subtask was awkward posture. The left leg appeared to be at greatest risk during curve plowing due to awkward posture and repetitive movement, and the right shoulder was considered at risk in cranking the machine upon start due to awkward posture and high forces. With respect to the risk priority level for each body part, we found all farmer body parts, besides the neck, to be exposed to high risk conditions.

Table 1 Worst task identification for each body part based on risk of injury

Body part	Motion	Force	Posture	Total priority	Worst task
Neck	L	L	H	L	Start motor
Back	H	M	H	H	Start motor
Rt. Shoulder	M	H	H	H	Start plow
Lt. Shoulder	H	H	M	H	Straight plow
Rt. Arm & elbow	M	M	H	H	Clear debris
Lt. Arm & elbow	M	M	H	H	Clear debris
Rt. Hand & wrist	H	M	H	H	Straight plow
Lt. Hand & wrist	H	M	H	H	Straight plow
Rt. Leg	M	H	H	H	Straight plow
Lt. Leg	M	H	H	H	Curve plow

Key: Rt. = right side, Lt. = left side, L = low, M= moderate and H = high risk

When considering the aggregate risk levels for each body part across risk factors and within a subtask (see Table 2), again the straight plowing and clearing debris subtasks emerged as having the greatest priority. For straight plowing, only the neck and back were not considered to be a high risk of injury. For clearing debris, only the neck and right shoulder were considered to be at lower risk by the analysts. From the risk ratings, the overall scores for the various subtasks included: 1) Start plowing =20; 2) Start machine = 17; 3) Adjust plow level = 14; 4) Straight plowing = 31; 5) Curve plowing = 23; 6) Clear debris = 33 and; 7) Stop machine = 19. Any score exceeding 25 points was considered to represent a high risk job. It is clear from Table 2 that the hands and wrists as well as the back were the most at risk body parts across the subtasks. It was logical that the hands emerged as critical areas due to the lack of protection and use of machinery. These subtasks and body areas were considered to represent focal points for ergonomic interventions.

Table 2 The risk priority levels for subtasks across body parts

Body part	Start motor	Start plow	Adjust plow level	Straight plow	Cure plow	Clear debris	Stop plow
Neck	L	L	L	L	L	L	L
Back	H	L	M	M	H	H	L
Rt. Shoulder	H	H	L	H	L	M	H
Lt. Shoulder	L	L	L	H	M	M	L
Rt. Arm & elbow	M	M	L	M	L	H	H
Lt. Arm & elbow	L	M	L	M	H	H	L
Rt. Hand & wrist	H	M	H	H	M	H	H
Lt. Hand & wrist	L	M	L	H	H	H	L
Rt. Leg	L	L	L	H	M	H	L
Lt. Leg	L	L	L	H	H	H	L

Key: Rt. = right side, Lt. = left side, L = low, M= moderate and H = high risk

4 DISCUSSION

The job screening analysis revealed a high risk of injury across body parts for the rice plowing activity due to various ergonomic issues. The straight plowing and clearing debris subtasks appeared to have the highest risk priorities. In both of these subtasks, analysts considered the shoulders, hands and legs to be at high risks for MSDs. The debris clearing task also led to a high risk of back injury. Previous studies on MSDs in agricultural workers in the U.S. (e.g., Rosecrance et al., 2006) found prevalence to be highest for the low back (37.5%) followed by the shoulders (25.9%), knees (23.6%) and wrists/hands (12.0%). In Thailand, Puntumetakul et al. (2011) demonstrated similar results for 12 month period prevalence of body part pain among rice farmers, including the highest for the low back (73.31%) followed by the shoulders (36.0%), knees (35.4%) and wrists/hands (12.5%). The findings from current study on the rice plowing activity are in agreement with the trends of these prevalence rates. High risks of MSDs were observed for farmers in the same areas of the body as reported by the previous studies.

With respect to the origin of the stresses placed on the farmers body in the critical subtasks of the plowing activity, the field observations on straight plowing revealed asymmetrical whole-body postures. This was primarily due to the farmer walking with one foot in a furrow behind a tiller. Such posture created an imbalance in force loading among the hands, feet and limbs (Hoffman, 2007), and was considered in the job screening. Screening results revealed awkward upper-extremity postures, specifically shoulder abduction and extreme ulnar deviation at wrist. These postures were due, in part, to the horizontal design of handles on the tiller. The wrist posture position created the potential for high compression in the carpal tunnel due to the associated hand gripping force (NIOSH, 2001). Awkward wrist postures can also lead to a decrease in maximum grip force (Kattel et al., 1996; Donheny et al., 2008) and safe control of machine. The risk of back injury for the debris clearing task was primarily attributed to farmer extreme forward flexion of the spine for an extended duration. In addition, high forces were required for the farmer to tear vines and plants from the plow by hand.

Other related safety issues not captured by the job screening analysis included the fact that the debris clearing task occurred while the tiller was running and a farmer's head, arms and hands were positioned next to an unguarded drive belt during removal of unwanted material. from the single-blade harrow.

5 CONCLUSION

The findings of this study revealed the rice plowing process, especially straight plowing and clearing debris from a mechanical plow, to lead to a high potential for pain in the upper and lower extremities and low back. There is a need to conduct detailed ergonomic analyses on these tasks, including quantification of posture positions at the hand and wrists, shoulders and back, as well as forces applied to task objects at the hands, and cycle times for repetitive motions. Such data should be compared with established ergonomics criteria to objectively identify the level of risk for MSDs due to the plowing activity.

Ergonomics interventions using engineering, administrative, and/or personal protection controls should be applied to the rice field plowing operation in order to reduce risk of occupational injury for rice farm workers. Based on the findings of the present study, such interventions should include:

1) Redesign of plowing machinery - In the use of power tillers during plowing, farmer height appears to be an important factor in wrist posture position. The height of a power tiller is not adjustable (only the cut-depth of the plow can be adjusted) and, therefore, shorter farmers experience more extreme ulnar deviation at the wrists as compared to taller framers. Most of the farmers that we observed in this study were relatively short and the unnatural wrist posture was readily observable and supported the need for a handle redesign at the plowing machine.

2) Use of machine guarding technology – In subtasks of the plowing activity, all components of power tillers should be guarded to protect farmers from inadvertent exposure to fly wheels, drive belts, etc. During the debris clearing task, farmers do not keep the plow motor in view and there is a high potential for a body part to get caught in the machine leading to an acute injury.
3) Instruction on appropriate work methods to prevent awkward whole-body posture – The level of farmer expertise also appeared to be a factor in behavior during the plowing activity. More experienced farmers have specific strategies for controlling the plowing machinery that appears to reduce muscle exertion and for and walking inside or outside the furrow to prevent awkward body postures. The risk of injuries to body parts is likely far less for more experienced farmers due to these behaviors. Therefore, experienced farmer motion partners should be further investigated in order to develop guidance for novice farmers on physical behaviors towards minimizing risk of injury.
4) Use of farmer personal protective equipment at the hands – Farmer ability to grip the handles of a power tiller during plowing appeared to be comprised by the awkward wrist posture position. Hand protection, such as lightweight gloves with grip pads should be provided to increase stability in machine control and reduce hand vibration exposure potentially leading to an ischemic effect.

Other rice operations, beyond rough and fine field plowing, need to be investigated for potential ergonomics-related risks, given the anticipated high future production demands in Thailand. Our research team also conducted task analyses on: 1) soil digestion and harrowing; 2) seeding; 3) planting; 4) insect prevention and crop fertilization; and 5) harvesting. These analyses are to be used as a basis for additional ergonomics job screenings and to identify risks of MSDs for specific body parts. For example, the planting operation requires farmers to maintain a stooping posture with extreme forward flexion of the spine for extended periods. This posture and the duration of exposure is expected to pose a very high risk of low-back pain and disability. The additional job screenings will also be used to make recommendations of ergonomic interventions for the entire rice cultivation process in order to further promote rice worker safety and health.

ACKNOWLEDGMENTS

This study was supported by grants from the Back Neck and Other Joint Pain Research Group and Khon Kaen University. David Kaber's work on the study was supported by a grant from the U.S. National Institute for Occupational Safety & Health (NIOSH) (No. 2 T42 OH008673-06). The opinions expressed in this paper are those of the authors and do not necessarily reflect the views of NIOSH.

REFERENCES

Armstrong T. 2006. The ACGIH TLV for hand activity level. In: Marras WS, Karwowski W, editors. Fundamentals and assessment tools for occupational ergonomics. Boca Raton (FL): CRC Press. 41:1-14.

Chataigner, J. 1992. Simultaneous growth and diversification of rice consumption in Europe and the USA. *Proceedings of the Prospects for Rice Consumption in Europe Symposium,* October 24, Verona, Italy.

Doheny, E. P., Lowery, M. M., FitzPatrick, D. P., and O'Malley, M. J. 2008. Effect of elbow joint angle on force–EMG relationships in human elbow flexor and extensor muscles. *Journal of Electromyography and Kinesiolog* 18: 760–770.

Fathallah, F. A. 2010. Musculoskeletal disorders in labor-intensive agriculture. *Applied Ergonomics* 41: 738-743.

Gangakhekar, S., Kaber, D. B., Mosaly, P. & Diering, M. 2011. Effects of scaffolding equipment interventions on muscle activation and task performance. *Proc. of the 55th Annual Meeting of the Human Factors and Ergonomics Society* (CD-ROM). Santa Monica, CA: Human Factors and Ergonomics Society.

Hoffman, S.G., Reed, M.P., and Chaffin, D.B. 2007. The relationship between hand force direction and posture during two-handed pushing tasks. *Proceedings of the Human Factors and Ergonomics Society 51st annual meeting 2007*: 928-32.

Kar, S. K. and Dhara, P. C. 2007. An evaluation of musculoskeletal disorder and socioeconomic status of farmers in West Bangal, India. *Nepal Medical College Journal* 9(4): 245-249.

Kasikorn Research Center. 2011. "K-Econ Analysis," Accessed November, 2011, http://www.kasikornresearch.com/EN/K-EconAnalysis/Pages/Search.aspx?cid=4

Kattel, B. P., Fredericks, T. K., Fernandez, J. E., and Lee, D. C. 1996. The effect of upper-extremity posture on maximum grip strength. *International Journal of Industrial Ergonomics* 18: 423-429.

Kirkhorn, S. R., Earle-Richardson, G., and Banks, R. J. 2010. Ergonomic risks and musculoskeletal disorders in production agriculture: recommendations for effective research to practice. *Journal of Agromedicine* 15: 281–299.

Mamansari, D. U. and Salokhe, V. M. 1995. The need for ergonomics considerations for the design and development of agricultural machinery in Thailand. *Journal of Human Ergology (Tokyo)* 24 (1): 61-72.

McAtamney, L. and Corlett, E.N. 1993. RULA -: A survey method for investigation of work-related upper limb disorders. *Applied Ergonomics* 24(2): 91-99

National Institute for Occupational Safety and Health (NIOSH). 2001. *Simple solutions: ergonomics for farm workers.* Report No. 2001-111: 1–53.

Office of Agricultural Economics. 2010. "Agricultural Statics," Accessed November, 2011, http://www.oae.go.th/main.php?filename=index

Puntumetakul, R., Siritaratiwat, W., Boonprakob, Y. and Puntuatakul, M. 2011. Prevaleance of Musculoskeletal Disorder in Farmer: Case study in Sila, Muang Khon Kaen, Khon Kaen Province. *J Med Tech Phy Ther.* 23: 298-303.

Rogers, M., Gangakhedkar, S. & Kaber, D. B. 2011. Ergonomic evaluation of emergency veterinary clinic operations. *Proceedings of the 2011 Applied Ergonomics Conference* (CD-ROM). Orlando, FL (March 21-23): IIE.

Rosecrance, J., Rodgers, G., and Merlino, L. 2006. Low back pain and musculoskeletal symptoms among Kansas farmers. *Am J Ind Med.* 49:547-56.

Suwannaporn, P. and Linnemann, A. 2008. Rice-eating quality among consumers in different rice grain preference countries. *Journal of Sensory Studies* 23(1): 1-13.

Suwansri, S., Meullenet, J. F., Hankins, J. A., and Griffin, K. 2002. Preference mapping of domestic/imported Jasmine rice for U.S.-Asian consumers. *Journal of Food Science* 67: 2420–2431.

USDA, 2001. *Rice Situation and Outlook Year Book.* Washington, DC, November 2001.

Walker-Bone, K. and Palmer, K. T. 2002. Musculoskeletal disorders in farmers and farm workers. *Occupational Medicine* 52 (8): 441-450.

CHAPTER 65

Automatic Landmark Identification from Three Dimensional (3D) Human Body Data using Geometric Characteristics

J.W. Niu[1], *Y.M. Wu*[1], *F. Hou*[1], *A.L. Feng*[1], *X. Chen*[2]

[1] School of Mechanical Engineering, University of Science and Technology Beijing, China

[2] Quartermaster Research Institute, General Logistics Department, CPLA, Beijing, China

niujw@ustb.edu.cn

ABSTRACT

The development of three dimensional (3D) scan technologies makes the anthropometric data collection more efficient. While taking measurements from 3D scanning data, markers are usually placed on human body surface to facilitate landmarking. But the procedure of placing markers is very tedious and error-prone. Human body feature automatic identification from 3D models provides a fast and easy approach to collect anthropometric measurements. Geometric characteristics can be used to develop the algorithms for automatic identifying the landmarks. The purpose of this research is to develop an automatic extraction method by using human body geometrical characteristics such as curvature extreme value. The method was evaluated on 30 head models as a case study. The automatically identified landmarks are compared with those manually palpated by anthropometry experts. The mean deviation in identification of landmarks was found acceptable for sportswear industry, etc. This showed good accuracy of the proposed method for some typical landmarks on head. This method can be extended to landmark automatic identification of other human body segments, such as torso, thigh, legs, arms, etc. What this paper has done is expected to benefit future fitting design based on 3D anthropometric data.

Keywords: Human body data, automatic landmark identification, three dimensional (3D), geometric characteristics

1. INTRODUCTION

The application of the 3D anthropometric data was very widespread. It was an important fundamental of the product design in the industry of garment, building, and animation, etc. 3D anthropometry aroused the researchers' interest and researchers have devoted themselves to the 3D data processing research. Some international representative 3D anthropometric surveys include Civilian American and European Surface Anthropometry Resource (CAESAR) (Robinette et al., 2002), SizeUK, Japan Ergonomics Institute of 3D anthropometric surveys project, SizeGermany (Seidl, A. 2009), etc.

Although 3D anthropometric data can be obtained more easily than several years ago, to process 3D data is still quite challenging and landmark identification is one of the most challenges. In general, markers or stickers were placed on the surface of human body to highlight the positions of the landmarks, since landmarks are mostly bony protrusions. In this way, the landmarks can be easily identified on the scanned data by using naked eyes. However, the procedure of placing markers on body surface is a tedious process and may involve human errors. It greatly impacts the efficiency of future data analysis. It is suggested why not use the color information obtained from the CCD cameras in the scanned data. If so, the landmarks can be identified by analyzing the RGB information in the scanning image. However there exist some difficulties in reality. First, the color information may expose the real identity of the scanned subjects, especially when the face was scanned. This can cause the criticism from the human rights activists. Second, though it's not a hard task for contemporary 3D scanning device to obtain the color information of the human body surface, there are lots of legacy 3D scanned data without color information. Such kind of data were collected by using some old fashioned scanning devices, where the color information may have been ignored during the scanning. So there is still no widely recognized method of 3D anthropometric landmark automatic identification, and even few related researches have ever been reported. Cristina et al (2006) adopted spin image together with SVM classifier to identify the relevant landmarks of nose and eye. Zouhour et al (2006) used spin image and Hidden Markov Model to identify human landmarks by using the CAESAR anthropometric data. In contrast, landmark automatic identification methods focusing on 2D data, usually originated from the area of image processing, have gradually matured. The combination of 2D landmark automatic identification with 3D data opens a new way and deserves thorough investigation.

Various approaches have been proposed for the automatic identification algorithm of landmarks. They could be classified into three groups as follows.

The first group is based on prior knowledge. This method summarized some experiences and rules according to the general characteristics of human feature. According to these rules the users selected candidate points or areas from the human models. For example, Levy-Mandel et al (1986) tried to use a knowledge-based line tracker method to extract the model edge. Burnsides et al

(2001) tried to use the feature location rule such as "difference in the z direction of left acromion and right acromion does not exceed a certain threshold", but the rule was too simple and largely depends on the data samples. Consequently, the adaptability of this method was in question.

The second group usually adopts the concept of template matching. This method, such as features face, human color model, probabilistic reasoning model, could set up a model for each landmark, then search the degree of corresponding between the model and the target. Cardillo and Sid-Ahmed (1994) used style matching algorithm which was based on mathematical morphology to distinguish head landmarks. The human color model used the statistical methods to establish the color model of human feature. When positioning traversal candidate area, according to the point color and the matching degree of the model, the user could select the candidate landmarks (Hsu, et al., 2002). This method was sensitive to the illumination condition and the characteristics of the image acquisition devices, it was vulnerable to environmental factors, and it was difficult for stabilizing of the precision. Grayscale value model was one kind of human color model. The method was mainly used to identify facial landmark. It aims to narrow the scope of candidates (Yow and Cipolla, 1996) on the basis of local information of relative position between face features. Probabilistic reasoning model was a hotspot in recent years, such as neural network, Markov model and so on. They were usually based on a certain statistics to study related information of landmarks and information between landmarks. This method adopted the probabilistic reasoning model (probabilistic reasoning) to obtain the relative degrees, thus realized the target landmark identification.

The third group utilizes geometry information. For example, pronasale was located in the most prominent nose position, so by using the minimum circumference value method, this landmark can be easily obtained (Wang, et al., 2007). Curvature was another frequently used important geometric feature. Lu et al (Lu, et al., 2004) mapped the 3D model of human face into a cylindrical coordinates, recollected samples to calculate the features of Shape Index, but uniformity along with the net on the model surface did not conform the true feature distribution of 3D models. Hose et al (2007) defined the geometry information of the tip of nose, and identified some landmarks comparative effectively.

This paper is organized as follows: the method will be introduced in chapter 2. The results of our research will be introduced in chapter 3. In chapter 4, we will summarize our paper and give some discussions on this paper. We will conclude our work in chapter 5.

2. METHOD

The geometric characteristic of human body such as the silhouette is a logical approach for landmark automatic identification. Figure 1 presents the schematic flow of the proposed method, from point cloud projection to landmark identification. To validate the system, 30 heads were used to compare the identification performance between this proposed method and manual landmarking. The system was achieved with the aid of Unigraphics CAD software and Visual C++ 6.0.

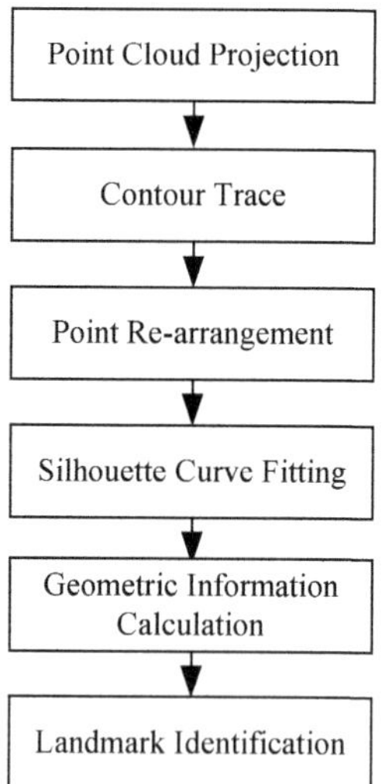

Figure 1 Schematic flow of the method

2.1 Point Cloud Projection

To extract the geometric characteristics of head data can be realized by silhouette analysis. Silhouette analysis is enabled by projecting the 3D head scanned data onto a 2D plane, such as coronal plane, sagittal plane and horizontal plane. That is to say, we can get the front (back) view, side view and top view (bottom) view of the head data through point cloud projection. Since it is easier to compute the geometric characteristics of the silhouette when the head data was projected to the coronal plane, we adopt the side view projection to conduct our scheme. By analyzing the variation in curvature of the silhouette, landmarks can be located.

2.2 Contour Trace

There are three sub-steps for contour trace, i.e., point cloud binarization, preliminary silhouette gridding extraction and contour extraction.

In point cloud binarization, first perform traversal through all point cloud and find out the four extreme coordinate values, x_{max}, x_{min}, y_{max}, and y_{min}. Second calculate the side length of each grid as follows,

$$side\_length=\sqrt{(x_{max}-x_{min})(y_{max}-y_{min})/n} \tag{1}$$

Third, we can get the number of grids in x and y direction, i.e., x_num and y_num, respectively.

$$x\_num=\left\lfloor \frac{x_{max}-x_{min}}{side\_length} \right\rfloor+1 \tag{2}$$

$$y\_num = \left\lfloor \frac{y_{\max} - y_{\min}}{side\_length} \right\rfloor + 1 \tag{3}$$

Fourth, we map each point into the gridding. For example, if there is no point falling into one grid, then the binary value of such a grid equals zero. Otherwise its value equals one. This procedure is called point cloud binarization.

The 2nd sub-step of contour trace, preliminary silhouette gridding extraction, aims to help the user identify the silhouette gridding. This resultant silhouette is not the final contour, it stays a coarse level. Some simple rules can be followed to help accomplish this step. If the binary value of the current grid equals zero, this grid must not be part of the silhouette gridding. If the binary value of the current grid equals one, and among its eight neighbor grids (right, right down, down, left down, left, left up, up, right up), there is at least one grid whose binary value equal to zero, then such grid can attribute to the silhouette gridding.

The 3rd sub-step, contour extraction, plays the core role of contour trace. Contour extraction is based on the neighbor points' distribution uniformity of the undetermined point. It's common sense that if the K-nearest points of one point have a bias distribution, then the point can be judged as one contour point; or else, if the K-nearest points of one point have a uniform distribution, then the point can be judged as one inner point. This judgment procedure can be done as follows,

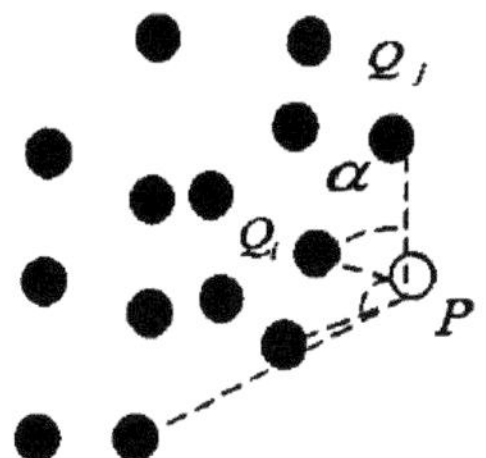

Figure 2 Principle of contour extraction

Choose the point P as the undetermined one. First, draw a directed line $P\overline{Q}_i$, where Q_i is the most nearest point of point P. Second, calculate the angle between $P\overline{Q}_i$ and $P\overline{Q}_j$ (j=1, …, n), where Q_j stands for any other point among the K-nearest points of point P, and P has n+1 K-nearest points. Arrange the angle values ascending, we can get a new angle sequence, $S' = (\alpha_1', \ldots, \alpha_n')$. We define another sequence L, where

$$S' = \begin{cases} \alpha_{i+1}' - \alpha_1' (1 \leq i < n) \\ \alpha_i' - \alpha_1' (i = n) \end{cases} \tag{4}$$

and calculate the standard deviation of sequence L,

$$E = \sqrt{\frac{\sum_{i=1}^{n}(L_i - \bar{L})^2}{n}} \tag{5}$$

where

$$\bar{L} = \frac{1}{n}\sum_{i=1}^{n} L_i \tag{6}$$

Given the threshold of E, i.e., the standard deviation of sequence L, we can determine whether point P belongs to the contour or not. This procedure can be optimized literally.

2.3 Point Re-arrangement

Assuming the contour point set is $\{P_1, ..., P_n\}$. First, select one point randomly as the first reference point, P_1'. Find out the reference point's most nearest neighbor, P_2', in the contour point set except the reference point itself. Then take P_2' as the new reference point, continue to find the new reference point's most nearest neighbor, P_3', in the contour point set except the points which have been found out. This procedure literally goes on till all points have been studied. There may appear two extreme situations. One is some 3D points may coincide each other once they are projected on a 2D plane. This may affect the future curvature calculation badly. So we can choose a threshold for distance between different projected points. If the distance between them is small enough (smaller then the threshold), then the reduplicated projected points can be regarded as one point. Another situation is retracing may take place during the arrangement. This can be avoided based on the judgment of angle between the points.

2.4 Silhouette Curve Fitting

Curve fitting is the calculation of a curve that most closely approaches a number of points in a plane. It is a procedure to pass a curve through a set of points, in such a way that the curve shows as well as possible the relationship between the quantities plotted (wikipedia, 2011). This work was done under the help of implementation of some APIs of UG CAD software.

2.5 Geometric Information Calculation

Intuitively, curvature is the amount by which a geometric object deviates from being flat, or straight in the case of a line. The curvature of a smooth curve is defined as the curvature of its osculating circle at each point (wikipedia, 2011). UG CAD software provides some useful APIs to help calculate the geometric characteristics such as curvature of a smooth curve once the user specifies the parameter of the position on the curve.

2.6 Landmark Identification

Totally eight landmarks were selected for this study. They are aresellion, pronasale, subnasale, labrale superius, stomion, labrale inferius, under tip and gnathion. Variation of the curvature of such landmarks can be observed easily via naked eyes. Some anthropologist experts and/or engineers were asked to manually identify the eight facial landmarks in a soft package developed by our team. Also the eight landmarks of each head were also automatically extracted by using the method proposed in this paper.

3. RESULTS

We applied our method to 30 head scans, in which eight anthropometric landmarks were identified. Figure 3 illustrates the point cloud of a head sample. The left image corresponds to the isometric view, while the right image shows the projection effect to a 2D plane.

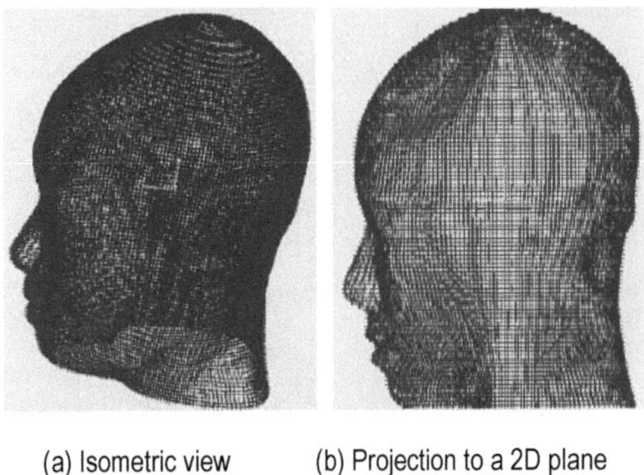

(a) Isometric view (b) Projection to a 2D plane

Figure 3 Point cloud of a head sample

As shown in Figure 4, the left image displays the points lying on the silhouette after contour tracing, and the right image demonstrates the curve fitting result of the contour points. Cubic B-spline method was adopted for curve fitting in this paper.

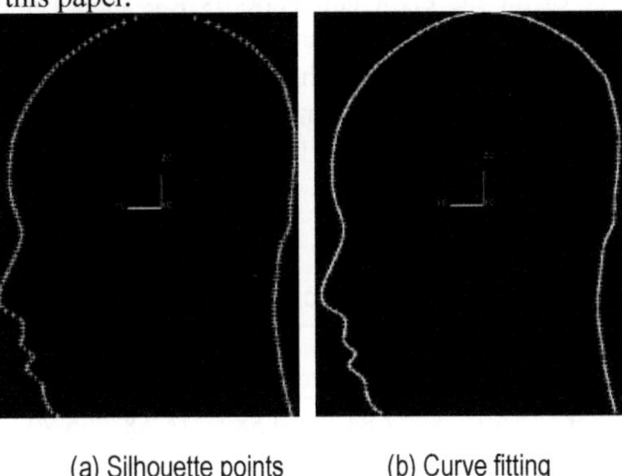

(a) Silhouette points (b) Curve fitting

Figure 4 Contour tracing and curve fitting

Results of landmark identification on a head sample were shown in Figure 5. The left correspond to landmarks identified manually and the right ones correspond to the landmarks identified using our approach. Each row corresponds to a different view of the head sample. Except the pronasale, the positions of most of the landmarks are identified very similar between the two methods.

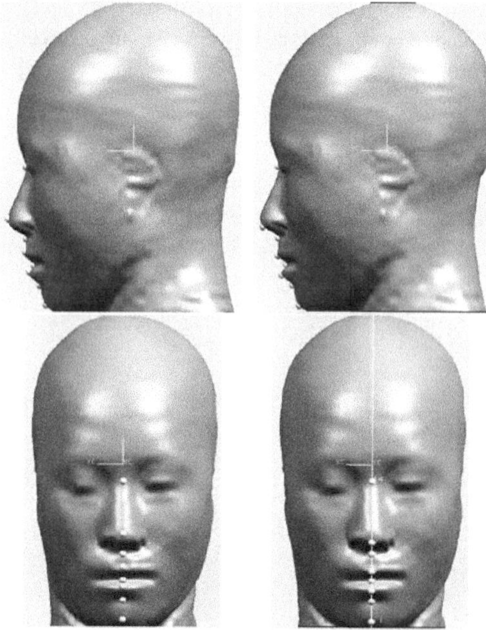

(a) Manual landmarking (b) Automatic landmarking

Figure 5 Landmark identification on a head sample.

We calculated the Euclidean distances between the identified landmark positions and their corresponding positions manually placed. Table 1 shows the average and standard deviation of the identification errors for each landmark. It can be seen the error of all the eight landmarks is less than 10mm. This is acceptable for some application such as sportswear industry. But the standard derivations seem too high, compared with the average values. The variation of results may mainly result from manual landmarking. Despite they had been told to obey the rule that a landmark usually has an extreme curvature locally, different measurer varies in interpreting the meaning of the extreme curvature when they locate the landmarks.

Table 1. Error of facial landmark automatic identification over 30 heads.

No.	Landmark	M. (mm)	Min. (mm)	Max. (mm)	Std. (mm)
1	sellion	7.84	0.58	16.58	5.20
2	pronasale	5.19	0.02	17.43	4.84
3	subnasale	2.54	0.31	5.60	1.30

4	labrale superius	2.11	0.05	4.55	1.43
5	stomion	2.57	0.02	5.13	1.40
6	labrale inferius	2.37	0.06	6.46	1.64
7	under tip	3.79	0.04	8.12	1.81
8	gnathion	4.81	0.07	15.97	4.28

4. DISCUSSIONS

In this paper, instead of identifying the landmarks directly from 3D human body data, we proposed a method in which 3D data were projected onto a 2D plane first. Some researchers may argue why not use 2D scan device to get the 2D pictures or images of the human body. Actually there are some studies adopting such kind of 2D scan methods to conduct landmark identification, for instance, to extract the landmarks of the hands. But our former studies show 2D scan method was sensitive to the illumination condition and the characteristics of the image acquisition devices. Also it was vulnerable to environmental factors. The last but not the least, it's usually too difficult to utilize 2D scan for some human body segments, such as the torso, the head, etc.

It must be pointed out that there are still several shortcoming related with this method. First, there exist numerous geometric characteristics which can be utilized to extract the landmarks, not only the curvature information. For example, it's quite easy to distinguish the nose tip when using another geometric characteristic, extreme coordinate value. Consequently, other geometric characteristics should be also taken into account combined with curvature. Second, some landmarks can not be identified through silhouette analysis, such as ectocanthion and zygion. To identify such kind of landmarks, other information such as color, contour plot, texture or even some expertise should be taken into consideration. Third, contour tracing is not an easy task, especially when noise or holes exist in the scanned data. This may call for great effort in pre-processing. Fourth, the computation efficiency of contour tracing may be a big problem and should be improved.

5. CONCLUSION

This paper presents a method of 3D human landmark automatic identification based on geometric characteristics. This work can help the researchers get rid of unnecessary pre-marking prior to scanning. Under the aid of UG CAD software, eight landmarks were identified automatically. To evaluate the validity and reliability of the proposed, 30 heads were tested, and a comparison between the proposed method and manual landmarking was conducted. Results suggest that the method was effective and robust.

This method bridges the gap between 2D image processing and 3D landmark automatic identification. However, how to extend this method to other human segments is worthy further research. Automatic landmark identification from 3D data directly using geometric characteristics, which means without projection to a 2D plane, may be a promising direction.

ACKNOWLEDGMENT

The study is supported by the National Natural Science Foundation of China (No.51005016).

REFERENCE

Burnsides, D.B., Boehmer, M., and Robinette, K.M. 2001. 3-D Landmark Detection and Identification in the CAESAR Project. In: Proceedings of the 3rd International Conference on 3-D Digital Imaging and Modeling Conference, Quebec City, Canada: 393-398.

Cardillo, J. and Sid-Ahmed, M.A. 1994. An image processing system for locating craniofacial landmarks. IEEE Trans. Med. Imag. 13, 275-289.

Cristina, C., Licesio, J.R., Enrique, C. 2006. Automatic 3D Face Feature Points Extraction with spin images. International Conference on Image Analysis and Recognition, vol. 4142: 317-328.

Hose, J. D', Colineau, J., Bichon, C., etc. 2007. Precise Localization of Landmarks on 3D faces using Gabor Wavelets. 2007 First IEEE International Conference on Biometrics Theory Applications and Systems, September 27-29, Washington DC.

Hsu, R.L., Mottaleb, M.A., Jain, A.K. 2002. Face detection in color images. IEEE Transactions on Pattern Analysis and Machine Intelligence. 24(5): 696-706.

http://en.wikipedia.org, [Accessed 2011.12.1]

Levy-Mandel, A., Venetsanopoulos, A., etc. 1986. Knowledge-based landmarking of cephalograms. Compute, Biomed, 19: 282-309.

Lu, X., Colbry, D., Jain, A. 2004. Three dimensional model based face recognition. In Proceedings of International Conference on Pattern Recognition, Cambridge, UK, 362-366.

Robinette,K. M., Blackwell, S., Daanen, H., et at. 2002. CAESAR, Final Report, Volume I: Summary, AFRL HE WP TR 2002 0169. United States Air Force Research Lab., Human Effectiveness Directorate, Crew System Interface Division, Dayton, Ohio.

Seidl, A. 2009. SizeGERMANY-the new German Anthropometric Survey Conceptual Design, Implementation and Results. In: Proceedings of 17th World Congress on Ergonomics, Beijing, China.

Wang, M.J.J., Wu, W.Y., Lin, K.C., etc. 2007. Automated anthropometric data collection from three-dimensional digital human models. International Journal of Advanced Manufacturing Technology, 32: 109-115.

Yow, K.C., Cipolla, R. 1996. A probabilistic framework for perceptual grouping of features for human face detection. Proceedings of the 2nd International Conference on Automatic Face and Gesture Recognition. Washington, USA, 16- 21.

Zouhour, B.A., Chang, S., Anja, M. 2006. Automatic Locating of Anthropometric Landmarks on 3D Human Models. 3th International Symposium on 3D Data Processing, Visualization and Transmission (3DPVT 2006). Chapel Hill, North Carolina, USA, June 13-16, 2006.

CHAPTER 66

Evaluation of Office Chairs by Measuring the Seat Pressure Variations in Daily Office Work

Jae Hee Park, Seung Hee Kim, Min Uk Kim, Hanbum Jung, Young Soo Shim**

Hankyong National University
Anseong, Korea
maro@hknu.ac.kr

* Sidiz Inc.
Anseong, Korea

ABSTRACT

Office chair companies and consumers want to know what chairs are ergonomically good. Furthermore, they both want to quantify the evaluation process. The body interface pressure data has been dominantly applied to quantitative chair evaluation. In this study, we also tried to develop chair evaluation method based on body pressure data. For this objective, we tested six office chairs with a body pressure measurement system and subjective questionnaires on discomfort and comfort. The experiment results showed that there were correlations between body pressure data and subjective evaluation scores. Especially average back pressure data explained well the subjective discomfort and comfort level. Also, most variables on pressure data and subjective evaluation scores could significantly discriminate office chairs. Therefore, The interface pressure data combined with subjective evaluation can be used office chair evaluation process.

Keywords: body pressure, office chair, comfort, discomfort

1. INTRODUCTION

Work-related musculoskeletal disorders(MSDs) of office workers as well as physical workers increase due to the VDT work(Lewis et al, 2002). To prevent the MSDs of VDT workers, first above all, ergonomically good chairs should be provided. Ergonomic chairs should give sitters proper seat cushion on seat pan and good lumber support on backrest.

However, it is not easy to discriminate which chair is ergonomically good. Especially, chair companies want to quantify the chair evaluation. Until now, body interface pressure measurement method mainly has been used in car seat evaluations (Na et al., 2003; Kyung and Nussbaum, 2008). Relatively, office chair evaluation cases are rare.

In this study, we aimed to develop office chair evaluation method which combined quantitative body pressure data and qualitative subjective discomfort and comfort scores. For this objective, we tried to test six office chairs with a variety of measures.

2. METHOD

Eight subjects, seven males and one female, participated in the experiment. Their average age was 23 years. Their average height was 174 cm and average weight was 70.5 kg.

As test material, six office chairs were prepared(Figure 1). They all could be adjusted in seat height. Some chairs have soft cushion while the others have hard cushion. For measuring the interface pressure, a body pressure measurement system, X-sensor PX100 with 36×36 cells, was used. It gathered pressure data in 10 Hz.

Parallel with the body pressure measurement, we also evaluated subjunctives' subjective comfort and discomfort level by using questionnaires(Helander and Zhang, 1997). Each questionnaire includes seven terms on nine point Likert scale(Table 1). The body regional discomfort level was rated for seven segments of human body(Corlett and Bishop, 1976).

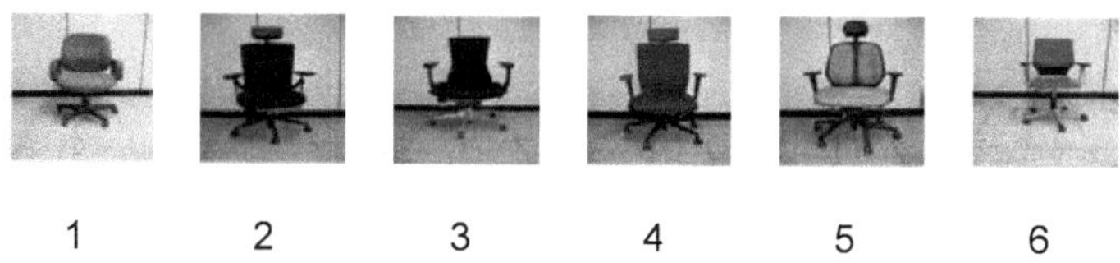

1 2 3 4 5 6

Figure 1 Six chairs tested

Table 1 Evaluation terms of comfort and discomfort questionnaires

Discomfort rating terms	Comfort rating terms
I have sore muscles	I feel relaxed
I have heavy legs	I feel fit
I feel uneven pressure	The chair feels soft
I feel stiff	The chair is spacious
I feel restless	The seat looks nice
I feel tired	I like the chair
I feel uncomfortable	I feel comfortable

In the experiment a subject was asked to sit on a chair and to adjust a desk and the chair height fitting to his/her body. After the subject posed ergonomically right posture, he/she was asked to stabilize the posture in five minutes. During the time, we measured interface pressure and used it as static pressure of seat. After gathering the static pressure data, the subject performed fifty minutes Internet surfing task. During the time, a subject's posture had been changed to perform Internet surfing task. The pressure data also was collected. We used the data as dynamic pressure data in evaluation. After finishing the task the subject were asked to evaluate his/her comfort and discomfort of a chair(Table 1). The average score of seven terms in discomfort questionnaire was used as subjective discomfort level. Similarly, the average score of seven terms in comfort questionnaire was used as subjective comfort level.

3. RESULTS

For the convenience, the data was sorted by ascending order of discomfort level and then the chairs were compared for each measure(Figure 2,3,4).

For the static seat pressure, there was significant difference among chairs ($P=0.000$). The seats could be grouped into three subgroups; (5,2,3), (3,1), and (1,6,4). Dynamic seat pressure of chairs shows very similar pattern with static pressure(Figure 3). The correlation coefficient between static and dynamic seat pressure is very high ($r=0.770$). In terms of dynamic seat pressure, the chairs could be grouped as (2,5,3,1), (1,4), and (4,6).

For the back pressure, the static and back pressure data showed no significant difference based on the pair-wised t-test ($p=0.836$)(Figure 4). Also the correlation coefficient was very high($r=0.736$). Based on the static back pressure, the chair could be grouped as (3,2,5), (2,5,1,4), and (6).

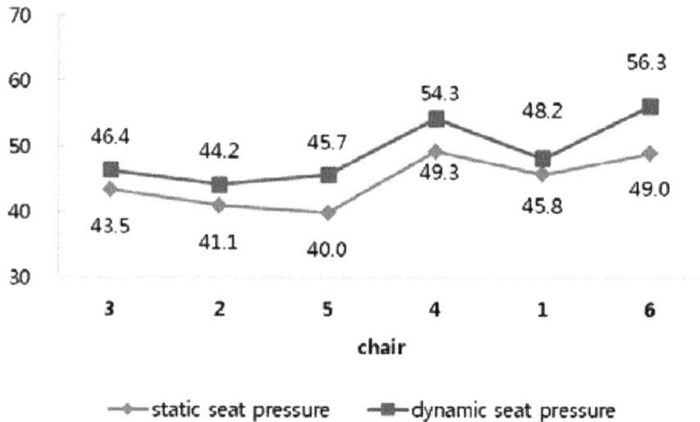

Figure 3 Static and dynamic seat pressure of chairs

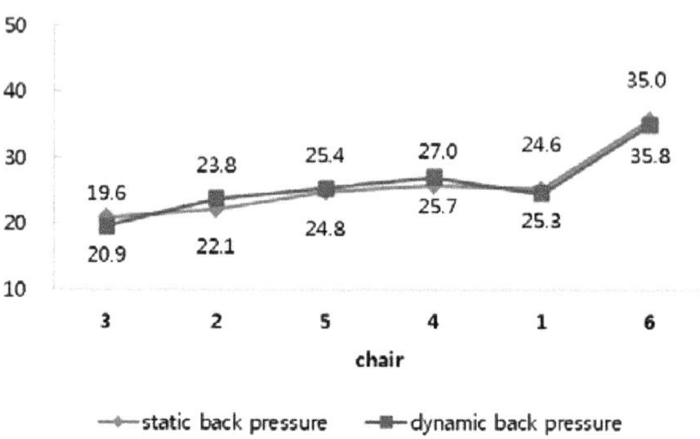

Figure 4 Static and dynamic back pressure of chairs

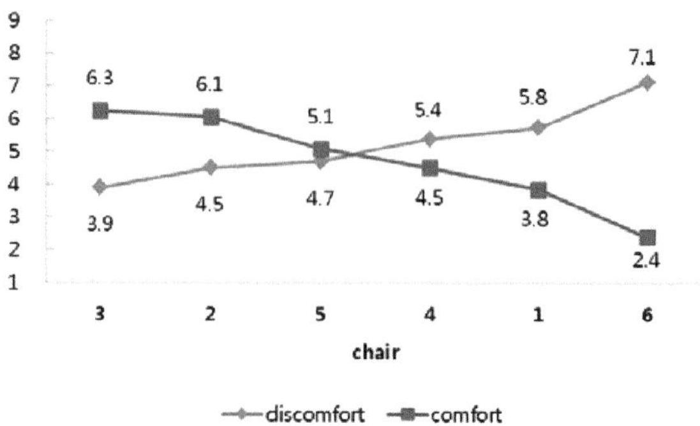

Figure 5 Discomfort and comfort levels of chairs

In addition to quantitative pressure data, subjective discomfort and comfort scores were analyzed(Figure 5). For the discomfort and comfort scores there were also significant differences among chairs (p=0.000). The chairs could be grouped into (3,2,5), (2,5,4,1), and (6) in terms of discomfort. For comfort score, there was negative correlation to discomfort score (r=-0.814). Based on the comfort score, the order of chairs was same, but the subgroups were (3,2,5), (5,4),(4,1) and (6).

The correlation coefficients between six measures are shown in Table 2. Based on this result, the seat discomfort(r=0.412) and comfort(-0.495) may be affected by static back pressure. Therefore seat comfort and discomfort can be predicted by seat back pressure somehow (discomfort= 0.112×static back pressure +2.351, comfort =-0.141×static back pressure +8.313) . However, contrary to our expectation, seat pressure data could not explain the variation of comfort or discomfort.

Table 2 Correlation coefficients between measurements

correlation coefficient	dis	com	st_seat	dy_seat	st_back	dy_back
discomfort	1	-.814**	.173	.116	.412**	.330*
comfort	-.814**	1	-.179	-.106	-.495**	-.335*
static seat pressure	.173	-.179	1	.770**	.391**	.386**
dynamic seat pressure	.116	-.106	.770**	1	.426**	.507**
static back pressure	.412**	-.495**	.391**	.426**	1	.736**
dynamic back pressure	.330*	-.335*	.386**	.507**	.736**	1

4. CONCLUSIONS

In this study, we wanted to find good interface pressure measures for evaluating office chairs. For this object, we tried to evaluate six chairs with four pressure variables and two subjective evaluation scores. The result of the test showed that some correlation existed between subjective evaluation scores and pressure data. Therefore interface pressure measurement data can be used in office chair evaluation. Especially the back pressure data was significantly correlated with subjective evaluation results. However, seat pressure data did not show significant correlation. For

ACKNOWLEDGEMENTS

This work was funded by grants from Gyeonggi Science & Technology Promotion.

REFERENCES

Corlett, E.N., Bishop, R.P., 1976. A technique for assessing postural discomfort. Ergonomics,19 (2), 175-182.

Ellegast, R. at al, 2012, Comparison of four specific dynamic office chairs with a conventional office chair: Impact upon muscle activation, physical activity and posture, Applied Ergonomics, 43(2), 296-307.

Helander, M,G. and Zhang, L., 1997,Field studies of comfort and discomfort in sitting, Ergonomics, 40; 895-915.

Kyung G., and Nussbaum, M.A., 2008, Driver sitting comfort and discomfort (part II): Relationships with and prediction from interface pressure, Industrial Ergonomics, 38, 526-538.

Lewis , J. et al., 2002, Musculoskeletal disorder worker compensation costs and injuries before and after an office ergonomics program, International Journal of Industrial Ergonomics, 29(2), Pages 95-99.

Na S.H., Lim S.H. and Chung M.K., 2003, Quantitative Evaluation of Driver`s Postural Change and Lumbar Support Using Dynamic Body Pressure Distribution Quantitative Evaluation of Driver`s Postural Change and Lumbar Support Using Dynamic Body Pressure Distribution, J. of Korean society of ergonomics, 22(3), 57-73.

Author Index

For Product Safety Concerns and Information please contact our EU representative GPSR@taylorandfrancis.com Taylor & Francis Verlag GmbH, Kaufingerstraße 24, 80331 München, Germany

T - #0168 - 160425 - C0 - 234/156/28 [30] - CB - 9781439870389 - Gloss Lamination